Lecture Notes in Civil Engineering

Volume 245

Lecture Notes in Civil Engineering (LNCE) publishes the latest developments in Civil Engineering - quickly, informally and in top quality. Though original research reported in proceedings and post-proceedings represents the core of LNCE, edited volumes of exceptionally high quality and interest may also be considered for publication. Volumes published in LNCE embrace all aspects and subfields of, as well as new challenges in, Civil Engineering. Topics in the series include:

- Construction and Structural Mechanics
- Building Materials
- Concrete, Steel and Timber Structures
- Geotechnical Engineering
- Earthquake Engineering
- Coastal Engineering
- Ocean and Offshore Engineering; Ships and Floating Structures
- Hydraulics, Hydrology and Water Resources Engineering
- Environmental Engineering and Sustainability
- Structural Health and Monitoring
- Surveying and Geographical Information Systems
- Indoor Environments
- Transportation and Traffic
- Risk Analysis
- Safety and Security

To submit a proposal or request further information, please contact the appropriate Springer Editor:

- Pierpaolo Riva at pierpaolo.riva@springer.com (Europe and Americas);
- Swati Meherishi at swati.meherishi@springer.com (Asia - except China, and Australia, New Zealand);
- Wayne Hu at wayne.hu@springer.com (China).

All books in the series now indexed by Scopus and EI Compendex database!

More information about this series at https://link.springer.com/bookseries/15087

Theo C. Haupt · Mariam Akinlolu ·
Fredrick Simpeh · Christopher Amoah ·
Zakheeya Armoed
Editors

Construction in 5D: Deconstruction, Digitalization, Disruption, Disaster, Development

Proceedings of the 15th Built Environment Conference

Springer

Editors
Theo C. Haupt
Mangosuthu University of Technology
Durban, South Africa

Mariam Akinlolu
Mangosuthu University of Technology
Durban, South Africa

Fredrick Simpeh
Akenten Appiah-Menka University of Skills
Training and Entrepreneurial Development
Kumasi, Ghana

Christopher Amoah
University of the Free State
Bloemfontein, South Africa

Zakheeya Armoed
Durban University of Technology
Durban, South Africa

ISSN 2366-2557 ISSN 2366-2565 (electronic)
Lecture Notes in Civil Engineering
ISBN 978-3-030-97747-4 ISBN 978-3-030-97748-1 (eBook)
https://doi.org/10.1007/978-3-030-97748-1

This Springer imprint is published by the registered company Springer Nature Switzerland AG
The registered company address is: Gewerbestrasse 11, 6330 Cham, Switzerland

Preface

This internationally peer-reviewed and edited proceedings of the 15th Built Environment Conference held virtually from Durban, South Africa, is aimed at contributing significantly to the body of knowledge relative to the science and practice of construction not only in South Africa but everywhere where the products of construction are produced even in these new challenging times of fear and uncertainty. The theme is particularly relevant in these times of rapid exponential knowledge and information growth accompanied by high levels of uncertainty, fear, and the unknown. There is little doubt that the construction industry has experienced exponential change and development in recent years. The 15th Built Environment Conference examined these cutting-edge concepts to determine their state of the art in the construction sector in both practice and academic research. This proceedings therefore suggests responses to questions related to current conversations, debates, and empirical research on:

Deconstruction—the dismantling or 'unbuilding' of buildings to maximize reusing and preserving the demolished fragments and involve taking a building apart piece by piece, essentially reversing the order of its construction.

Digitalization—the conversion and transformation of construction business processes to use digital technologies and embrace the ability of digital technology to collect data, establish trends, and make better business decisions.

Disruption—displacement of well-established construction technologies, techniques, or products to disruptively affect the normal operation or function of the construction industry while potentially creating a new industry or market. Artificial intelligence, virtual/ augmented reality, Internet of things, blockchain technology, and e-commerce are some of the disruptive technologies that are significantly influencing the future of the construction industry.

Disaster—an occurrence that disrupts the normal conditions of existence and operation causing a level of suffering and challenge that exceeds the capacity of adjustment of the affected community and the construction industry.

Development—in the context of construction refers to an industry that possesses the vision, leadership, and capacity to bring about a positive transformation of itself within a condensed period of time.

The proceedings is a must reference work for anyone serious about staying abreast of cutting-edge and revolutionary developments in the construction industry present and future. They provide the opportunity for researchers and practitioners from developed and developing nations to network with chapter contributors and deliberate pressing and burning issues that impact the built environment and will potentially change the way it responds to rapid growth in technological advancement.

The proceedings includes papers that addressed inter alia,

- Current trends and developments
- Policies
- Legislation and regulations
- Practices
- Case studies.

The proceedings can be read as a whole or individual chapters selected depending on the interest of the reader, and all interests in the five areas are accommodated in this comprehensive collection of peer-reviewed research.

Durban, South Africa

Prof. Theo C. Haupt

Contents

Disruption

Disaster

Development

Contributors

Oluseyi Julius Adebowale Department of Building Sciences, Tshwane University of Technology, Pretoria, South Africa

Ghada Adel Architectural Engineering Department, Faculty of Engineering, The British University in Egypt (BUE), Cairo, Egypt

Adeleye Ayoade Adeniran Department of Construction Management, Nelson Mandela University, Gqeberha, South Africa

Justus Ngala Agumba Department of Building Sciences, Tshwane University of Technology, Pretoria, South Africa

Mariam Akinlolu Faculty of Engineering, Mangosuthu University of Technology, Durban, South Africa

Brink Botha Faculty of Engineering, The Built Environment and Information Technology, Nelson Mandela University, Port Elizabeth, South Africa

Nick Brown Sustainability Oriented + Cyber Research Unit for the Built Environment, Department of Construction Economics and Management, University of Cape Town, Cape Town, South Africa

Progress S. Chigangacha School of Engineering, University of KwaZulu-Natal, Durban, South Africa

Fabio Companie Department of Construction Economics and Management, University of Cape Town, Cape Town, South Africa

F. Dowelani Department of Construction Economics, University of Pretoria, Gauteng, South Africa

Wayne Draai Faculty of Engineering, the Built Environment and Information Technology, Nelson Mandela University, Port Elizabeth, South Africa

Obas John Ebohon Sustainability and Environmental Law, London South Bank University, London, England;

School of the Built Environment and Architecture, London South Bank University, London, UK

David J. Edwards Department of the Built Environment, Birmingham City University, Birmingham, UK

Charles Egbu Leeds Trinity University, Leeds, England

Cyril Chinonso Ejidike Department of Construction Management and Quantity Surveying, Durban University of Technology, Durban, South Africa

Daniel Fong School of the Built Environment and Architecture, London South Bank University, London, UK

Viloshin Govender School of Built Environment and Development Studies, University of KwaZulu-Natal, Durban, South Africa

Nishani Harinarain The School of Engineering, Construction Studies Discipline, The University of KwaZulu-Natal, Howard College, Durban, South Africa

Theo C. Haupt Department of Engineering, Mangosuthu University of Technology, Umlazi, Durban, South Africa

Danie J. Hoffman Department of Construction Economics, University of Pretoria, Gauteng, South Africa

Olubisi Ige School of Built Environment and Architecture, London South Bank University, London, UK

Comfort Olubukola Iyiola Department of Construction Management and Quantity Surveying, Durban University of Technology, Durban, South Africa

Dawood Joosab University of KwaZulu-Natal, Durban, South Africa

Onele Kala Department of Construction Economics and Management, Faculty of Engineering and the Built Environment, University of Cape Town, Cape Town, South Africa

A. Kelani J Browne Construction Co. Ltd, London, England, UK

Geraldine J. Kikwasi Department of Building Economics, Ardhi University, Dar es Salaam, Tanzania

Asianoah Rexford Kofi Koforidua Technical University, Koforidua, Ghana

Jake. Le Roux Urban Real Estate Research Unit, Department of Construction Economics and Management, University of Cape Town, Cape Town, South Africa

Claudia Loggia School of Built Environment and Development Studies, University of KwaZulu-Natal, Durban, South Africa

Lawrence Maboreke Department of Construction Economics and Management, Faculty of Engineering and the Built Environment, University of Cape Town, Cape Town, South Africa

Upeksha Madanayake Course Director - BSc (Hons) Quantity Surveying, Department of Construction, Property and Surveying, London South Bank University, London, UK

Sebata Mafereka Faculty of Engineering, The Built Environment and Information Technology, Nelson Mandela University, Port Elizabeth, South Africa

Rashen Maharaj Department of Quantity Surveying and Construction Management, University of Johannesburg, Johannesburg, South Africa

Nomusa Makhathini Center of Applied Research and Innovation in the Built Environment (CARINBE), Faculty of Engineering and the Built Environment, University of Johannesburg, Johannesburg, South Africa

Ayesha Mall Department of Construction Management and Quantity Surveying, Durban University of Technology, Durban, South Africa

Makoye B. Mapani School of Construction, Economics and Management, University of the Witwatersrand 2000, Johannesburg, South Africa

Lungie Maseko School of Construction, Economics and Management, University of the Witwatersrand 2000, Johannesburg, South Africa

Lindelani Matshidze School of Construction Economics and Management, University of the Witwatersrand, Johannesburg, South Africa

Modupe Cecilia Mewomo Department of Construction Management and Quantity Surveying, Durban University of Technology, Durban, South Africa

Jacob Mhlanga Department of Landscape Architecture and Quantity Surveying, BA ISAGO University, Gaborone, Botswana

Kathy Michell Sustainability Oriented + Cyber Research Unit for the Built Environment, Department of Construction Economics and Management, University of Cape Town, Cape Town, South Africa

Tandokazi Mkhwe Faculty of Engineering, The Built Environment and Information Technology, Nelson Mandela University, Port Elizabeth, South Africa

Ghada N. Mossad Department of Architectural Engineering, Faculty of Engineering, The British University in Egypt, Cairo, Egypt

Abdul Razak Musah Faculty of Engineering, the Built Environment and Information Technology, Nelson Mandela University, Port Elizabeth, South Africa

Innocent Musonda Center of Applied Research and Innovation in the Built Environment (CARINBE), Faculty of Engineering and the Built Environment, University of Johannesburg, Johannesburg, South Africa

W. M. S. Muthimba School of Construction, Economics and Management, University of the Witwatersrand 2000, Johannesburg, South Africa

Tebogo H. Ngele Department of Construction Economics, University of Pretoria, Gauteng, South Africa

T. S. Ngobeni Department of Construction Economics, University of Pretoria, Gauteng, South Africa

Mbangi Nozipho Zipporah Mandela University, Gqeberha, South Africa

Kgashane Stephen Nyakala Department of Operations Management, Tshwane University of Technology, Pretoria, South Africa

Chioma Okoro Department of Finance and Investment Management, College of Business and Economics, University of Johannesburg, Johannesburg, South Africa

Adetayo Onososen Center of Applied Research and Innovation in the Built Environment, Faculty of Engineering and the Built Environment, University of Johannesburg, Johannesburg, South Africa

Uzor Onyia Department of Civil Engineering, University of Nigeria, Nsukka, Nigeria

Ayman Ahmed Ezzat Othman Architectural Engineering Department, Faculty of Engineering, The British University in Egypt (BUE), Cairo, Egypt

John J. Posillico Department of the Built Environment, Birmingham City University, Birmingham, UK

Jan-Harm Pretorius School of Engineering Management, Faculty of Engineering and the Built Environment, University of Johannesburg, Johannesburg, South Africa

Mohlomi Terah Raliile Department of Quantity Surveying and Construction Management, University of the Free State, Bloemfontein, Free State, South Africa

Herbert Robinson Knowledge & Learning Department, The African Capacity Building Foundation, Harare, Zimbabwe

Rafiu Deji Seidu School of the Built Environment and Architecture, London South Bank University, London, UK

Patricia Sekalo Department of Construction Economics, University of Pretoria, Gauteng, South Africa

Winston Shakantu Department of Construction Management, Nelson Mandela University, Gqeberha, South Africa

Y. D. Sikidi School of Construction, Economics and Management, University of the Witwatersrand 2000, Johannesburg, South Africa

Neil Sirbadhoo University of KwaZulu-Natal, Durban, South Africa

Geraldine Smith Department of the Built Environment, Birmingham City University, Birmingham, UK

Nyamagere Gladys Sospeter Department of Building Economics, Ardhi University, Dar es Salaam, Tanzania;
SBL, University of South Africa, Pretoria, South Africa

Jason Terblanche Sustainability Oriented + Cyber Research Unit for the Built Environment, Department of Construction Economics and Management, University of Cape Town, Cape Town, South Africa

Motheo Meta Tjebane Center of Applied Research and Innovation in the Built Environment, Faculty of Engineering and the Built Environment, University of Johannesburg, Johannesburg, South Africa

James Olaonipekun Toyin Department of Construction Management and Quantity Surveying, Durban University of Technology, Durban, South Africa

James Tucker Sustainability Oriented + Cyber Research Unit for the Built Environment, Department of Construction Economics and Management, University of Cape Town, Cape Town, South Africa

Hamza Vawda University of KwaZulu-Natal, Durban, South Africa

Andre Vermeulen School of Engineering Management, Faculty of Engineering and the Built Environment, University of Johannesburg, Johannesburg, South Africa

Derick Vosloo Department of Building and Human Settlement Development, Nelson Mandela University, Port Elizabeth, South Africa

Abimbola Windapo Department of Construction Economics and Management, Faculty of Engineering and the Built Environment, University of Cape Town, Cape Town, South Africa

Msindisi Yende Department of Construction Economics and Management, Faculty of Engineering and the Built Environment, University of Cape Town, Cape Town, South Africa

Bert Ediale Young School of the Built Environment and Architecture, London South Bank University, London, UK

Deconstruction

Circular Economy: A Sustainable Approach to Waste Management in the City of Johannesburg

Make B. Mapani, Won-Hyang M. S. Muthimba, Yulandi D. Sikidi, and Lungie Maseko

Abstract Circular Economy (CE) is a fresh approach to a systematic waste management strategy. It establishes principles which align with a net-zero waste agenda, promoting a sustainable environment, society, and economy. Furthermore, the concept aims to reduce carbon emissions by using waste as a resource. Johannesburg (JHB) is facing social and economic problems, which need effective waste management techniques. Current waste management methods implemented by the public sector are not sufficient in combating waste accumulated daily. As part of a comprehensive research into how systematic waste management strategies are implemented in other countries, a qualitative study was conducted. Pikitup was used as a case study, and empirical data from the interviews of experienced practitioners were used. An inductive approach provided explanations of the continuous interplay between theories, business proposals and real-life examples. To bridge the current research gap, benchmarking currently implemented methods in the JHB city to those of leading countries in CE allows for enhanced developments in terms of efficiency and sustainable living. The results further suggested that waste production is a serious problem that not only the country is dealing with but the entire world. Therefore, many barriers should be overcome, concerning the introduction of environmental policies, effective investments, and social inclusion. JHB city should ensure that CE is adaptable and flexible to the standards set out by government and private sector waste management service provider; guaranteeing the technical capacities to implement CE. The model should accommodate activities like the informal sector but boost the principles of sustainable development.

Keywords Circular economy · Pikitup · Sustainability · Technology-know-how · Waste management

---

M. B. Mapani · W.-H. M. S. Muthimba · Y. D. Sikidi · L. Maseko (✉)
School of Construction, Economics and Management, University of the Witwatersrand 2000, Johannesburg, South Africa
e-mail: lungie.maseko@wits.ac.za

T. C. Haupt et al. (eds.), *Construction in 5D: Deconstruction, Digitalization, Disruption, Disaster, Development*, Lecture Notes in Civil Engineering 245,
https://doi.org/10.1007/978-3-030-97748-1_1

1 Introduction

There are 58.78 million people in the country of South Africa of which about 5.05 million reside in Johannesburg. Pikitup is the main Integrated Waste Management (IWM) service provider collecting around 1.4 million tons of waste a year of which 94,355 tons was diverted from landfills [21]. Only 6.7% of the waste collected is recycled which is a portrayal of some underlying issues within the municipal waste management industry. One resident in Johannesburg produces about 2.5 kg of waste per day.

The city of Johannesburg (CoJ) is facing a crisis with solid waste disposal and efficient management of waste. The current and traditional linear extract-produce-usedump waste model is unsustainable [17]. Johannesburg has 12 waste management depots, of which only four are landfills and have limited options on how to manage it.

South Africa generates millions of municipal solid waste annually of which 93% is unable to be diverted. This is primarily due to the lack of landfill space because of population growth as a result of urbanization for better job opportunities (Rasmeni and Madyira 2019). The diversion statistic alone is overly concerning not only for the city and the environment but, for its inhabitants as well. Non-recycled waste produces a wide range of toxins such as carbon gas emissions, heavy or rusty metals, pathogens, or radiation. Metropolitan cities such as Johannesburg are unable to easily license new landfill sites causing the remainder of the 4 landfill sites to pile up to maximum capacity. Waste that is not diverted, sorted, or recycled contributes to ongoing waste management practices creating a cause-and-effect chain reaction. Landfill sites that have reached maximum capacity have no other choice but to shut down their facilities. This presents room to introduce a new waste management strategy [23].

The research aims to implement and adopt the Circular Economy to help combat and mitigate waste collection on landfill sites. MacArthur Foundation [10] defines the Circular Economy (CE) as a design that uses the industrial system to restore and regenerate waste by promoting its useful life for the purpose of producing sustainable energy, the elimination of harmful chemicals by reusing or repurposing waste for a useful second life product. These factors give context and origin to modern construction, highlighting waste management strategies as enablers for greater adoption within the construction sector. The key aspects in applying the circular economy across a building's life cycle is the preservation of a building material's added value, by recirculating them to close their loops and manufacture new products. How waste can be implemented within the construction industry's adaptability through eco-bricking and eco-asphalt initiatives researched within this study to promote circularity.

2 Literature Review

Sustainable development (WCED) [25] was originally defined as development that meets the needs of the present without compromising the ability of future generations to meet their own needs. But the issue that hinders this initiative is the traditional, linear output flow of materials and energy between nature and the human economy [17]. This output flow is running down the system in which it operates, from an ideal model. A system that does not recommend advancements of new forms of safe employment, encouraging the activities that are still in action (i.e. informal sector) and boosting the principles of sustainable development.

By critically considering the concept of CE from the perspective of sustainability, CE can be understood as an economy constructed using cyclical materials flows, renewable energy sources and recycling energy flows to provide innovative waste management solutions [18]. Successful CE contributes to economic, environmental, and social dimensions of sustainable development. And, according to CE, landfill disposal should be the last option.

2.1 Current Municipal Solid Waste Practices in the City of Johannesburg

According to the information available from the City of Johannesburg, the city produces approximately 1.7 million tons of waste per year which includes approximately 291,751 kg in the form of illegal dumping, 82,839 kg collected from the streets and only 568,893 kg of waste is diverted from the 4 landfill sites [21]. Pikitup's strategic focus is to reduce the amount of waste generated by providing sustainable and innovative waste management solutions, as well as keeping the city clean through the following initiatives:

- Waste collection services and separation at source—This strategy was implemented in 2009 to combat the strain on the city's landfill sites. Although the strategy initially envisaged the diversion of 120,000 tons of dry recyclables by 2018/19 (subject to budget availability), the current diversion targets have been revised to 40,000 tons in 2017/18, 50,000 tons in 2018/19 and 55,000 tons in 2019/20 in line with the current available resource.
- A Re Sebetseng campaign—Like the Uganda Campaign in Rwanda, the A Re Sebetseng campaign thrives to encourage the city's residents to take care of the environment, and seeks to promote a culture of reducing, reusing, and recycling waste [6]. The campaign is facilitated by Pikitup, but it is the responsibility of all councilors, city departments, and regional offices to drive the programme.
- Waste diversion from landfills strategies—Several strategies have been implemented in the city to avoid landfilling of waste, increasing available landfill airspace, reducing greenhouse gas emissions, prevention of ground contamination

and other landfill associated problems. Solutions such as waste buyback programs have been implemented.

Pikitup has been struggling to keep up with the volumes of waste produced within Johannesburg. With only four landfills and limited options on how to manage it, there is clearly room to introduce a new waste management strategy into the picture. The authors would like to introduce a new method of handling this waste while they strive for a greener future. There is a deficiency within the waste management system within Johannesburg, with a lack of technology, strategies, and facilities [14]. These deficiencies are detrimental to the environment through pollution. The authors may not be able to solve the problem but may be able to present an option to assist in mitigating this issue.

2.2 *The Circular Economy (CE)*

The circular economy is a concept that has slowly emerged as a greening tool and economic approach to waste management in many European countries attracting scholars, practitioners, and businesses alike.

From this, the authors can deduce that the circular economy is an economic and environmental concept that restores, regenerates, or revitalizes the use of a waste product's 'end-of-life', which is a shift from the conventional take-make-dispose model, by reducing the harmful toxins in waste through superior design concepts.

The CE restores primary materials into secondary reusable materials or energy through their visionary mission of innovative elimination of plastics through redesigning, reusing, recyclable or even compostable plastics [11]. Reusing waste reduces pollution and landfill waste, making it safer for the environment and a safer approach to sustainable development. Rather, the CE model embraces innovative concepts such as designing out waste and pursuing eco-effectiveness instead of eco-efficiency [15].

2.3 *Comparison to European Countries*

The implementation of principles concerning CE is intensifying in developed countries [16] while developing countries are still suffering inappropriate waste management due to the lack of economic funds, social/public awareness, and political will. The most common practice is the take-make-dispose approach which leads to unsuitable landfill conditions. The CE is a common waste management practice in developing countries.

The EU called for waste management to be transformed into sustainable material management which embeds the principles of the CE, enhancing the diffusion of renewable energy, increasing energy efficiency, reducing the dependence of the Union on imported resources, to provide economic opportunities and long-term competitiveness [13]. As part of its Action Plan, the waste services are distributed between direct and indirect municipal companies, with some corporations owned by the municipalities [24]. Whereas, in most developing countries, municipal solid waste is handled by the municipalities proving unsuccessful and detrimental to the environment [20].

Given a clear focus from legislature, most EU countries plan on achieving full CE by 2030 by means of the following initiatives: waste to energy initiatives, active participation by residents, recycling, district heating, waste prevention, separation of waste and specific recycling circuits, Waste and Resource Action Programme (also known as WRAP) which aims at showing how businesses, organizations, and consumers can be part of a resource revolution. Waste to Energy is a well proven and preferred waste management practice in the EU and worldwide since it results in both the minimization of environmental impacts in landfilling and renewable energy generation [19]. Developing countries like China, Serbia and India have started to implement CE, because of low-middle income, they are introducing projects or management plans with no effective changes [12]. Thus, the reason of comparing JHB City with first-world countries. To see what improvements can be adopted (Table 1).

Table 1 Comparison of the city of Johannesburg to European countries

City	Waste generated (G) collected © and landfilled (L) in tons	IWM service provider (s)	Amount of waste diverted to landfills via CE
Johannesburg	G = 59 Million tons C = 69% of G L = 90% of G	CoJ municipality via Pikitup	Aim of study
Rotterdam	G = 500 kg per person per year C = 100% of G L = 3% of G	The municipality of Rotterdam	By 2030 the city wants circularity to become common practice, aiming to reduce primary resource use by 50%
Oslo	G = 321 kg per person per year C = 100% of G L = 3% of G	Local municipality and private entities	The city of Oslo has set an overall target to reduce its CO_2 emissions by 50% by 2030 and to become carbon neutral by 2050
London	G = 7 million tons C = 100% of G L = 20% of G	Local municipality	Plans for achieving net zero waste by 2026

3 Methodology

In order to execute this research, a case study approach was adopted, to obtain an in-depth appreciation of an issue, in its natural real-life context [22]. The case studied was Pikitup, JHB city's official waste management service provider. The primary data was collected using semi-structured interviews containing a mixture of open- and closed-ended questions, where participants were asked about the legal framework of South Africa allowing for the introduction of a new waste management strategy approach towards net zero waste. Whether JHB has the technological know-how of repurposing waste readily available and, can something new be produced from the waste in the loop of the CE? The aim was to provide a detailed holistic description that illuminates the participants' understanding of the CE initiative.

The inductive style used, aimed to generate meanings from the data set collected to identify patterns and relationships to build a theory by using the participants' views to build broader themes and generate a theoretical interconnection of themes [26]. The participants were selected based on their role in the waste management process and their familiarity with CE efforts. Using a non-probability, snowballing sampling technique [22], interviews were conducted with 3 waste minimization team's at Pikitup, 5 Pikitup regional managers across JHB, and 2 other recycling companies (*see* Fig. 1, *in* Sect. 4.1). This population was chosen based on the need to eliminate waste on landfill sites. The questions related to their experiences on how they controlled unexpected barriers of new waste management strategies currently implemented.

The CE is a relatively innovative concept in South Africa, and, in this study, it was important to obtain a detailed and comprehensive view on how the City of Johannesburg can implement the CE to improve existing waste management plans in the approach to net zero waste.

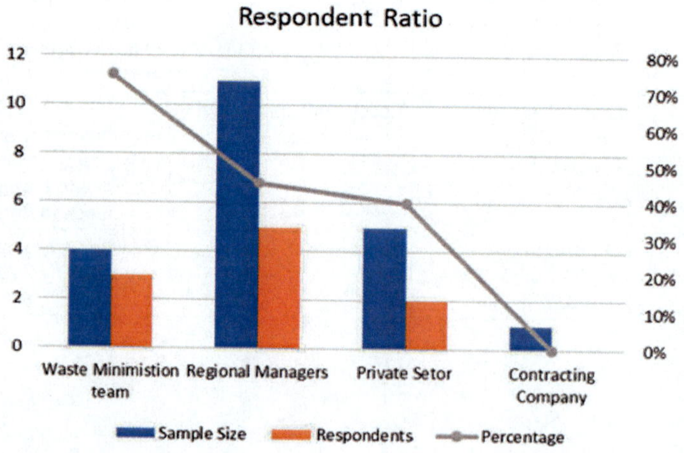

Fig. 1 A low respondent ratio due to Covid-19

4 Results and Discussion

4.1 *Interview Analysis*

Landfill disposal is the most preferred method for waste management. Respondents saw a need to have more options in either recycling or re-use, rather than straight disposal. From the anticipated sample size, a rather low respondent rate was achieved. During this research, most participants were not available, even through Zoom or on mobile calls. Based on the current respondent rate, the following information was established; that few people looked at the CE strategy as a business venture into having another form of recycling. They viewed CE as a process whereby waste is transformed into a new waste product often altering the products chemical characteristics, whilst others saw a waste economy solution that can benefit to solve common challenges—only collaboratively. This collaboration should be formed by the government, local municipalities, product retailers and other third parties. Figure 1 shows the poor response rate of the interviews, compared to the anticipated sample size. And the unwillingness of contracting companies to invest in recycling facilities.

4.2 *Legal Framework*

Currently, SA has a fragmented approach on waste management responsibilities. Participants from Pikitup had thought provoking responses. They pointed out that Pikitup sets out goals or corporate strategies that need to be met at the end of each year, however, these strategies have not been able to eradicate waste efficiently. Because many barriers still exist for developing CE strategies, there are no proper follow-up strategies on the progress. The SA government is not at the forefront in driving the legislation as it is in the EU. In the CoJ, Pikitup has only four active landfill sites. These land-fills sites cannot accommodate the daily waste intake that the city's inhabitants produce. The landfill sites have piled up to the point of no return, forcing a few landfills to shut down completely. The implementation of CE in European countries provides better waste management techniques that can move from over-packed landfill sites to not utilizing them at all. In SA, this solution would not only slowly and effectively reduce waste on landfill sites but would reduce the need for landfill sites. But, until collaboration between the government, municipalities, and the private sector to enhance the transition towards CE, improvements or plans such as converting waste-to-energy as fuel, addressing the issue of solid waste management will only be in the pipelines for solving over-packed landfill sites. The main tool is the introduction of reliable regulations/legislations that can support the development of the CE through sufficient economic support using public–private partnership, technical assistance, and international collaboration.

4.3 Technical Capacity

The choices of technology primarily depend on a country's income level. Technology is constantly evolving, and most countries do not rely on a single waste-to-energy technology but use multiple technological methods. When asked whether South Africa as a developing country has the technological-know-how to implement or adopt a new waste management benchmark, some respondents answered that it would be possible if knowledge sharing on recirculation and treatment solutions on technical facilities is made known, with awareness on improved social behaviors. But others felt that South Africa would not be able to accomplish this due to high electrical rates, grid interconnection costs and high capital start-up costs. However, technology industries, come in forms of choosing the most efficient and environmentally friendly type of technology.

From these responses, the researchers were able to deduce that although the CE is an international benchmark, the technique, knowledge, and strategy need to be adapted or adjusted according to the capabilities of the country. A new waste management benchmark "is not as challenging as one would think" said one respondent. 82% of the respondents had said that a new waste management strategy is possible, while 18% were skeptical. If the benchmark can benefit the country, economy, and environment, then it can be implemented with the necessary research and development said another respondent. Technology is ever evolving, and the implementation of the benchmark may take a while to implement however, people adapt easily to new ideas or concepts particularly now during the twenty-first century.

4.4 Role of the Private Sector

The private sector is a key sector in addressing the challenges of vulnerable communities, and it has much to contribute to the planning, development, and implementation of climate adaptation strategies, including sector-specific expertise, technology, efficiency, financing, and entrepreneurship [2]. A participant from Pikitup agreed with this statement stating that the private sector has more resources, technology and especially funds as municipal budgets are spread thin across other departments. Integrated waste management with all stakeholders is key to the successful implementation of the circular economy. Interviewees noted that the current strategy is effective, however this did not deliberate on the full scope of the challenges within the industry.

The issues relating to lack of integration between the government and the private sector delays the appropriate policy formulation and authority allocation. The obstacle of lack of communication, has caused slow progress in ensuring environmental safety and facility efficiency. Though these challenges are present, other respondents took the neutral route and stated that effectiveness should encompass measures formulated with a multi-disciplinary approach. And there should be guidelines on comprehensiveness, consistency, and flexibility of the policies. Central to the CE in SA, is the informal waste pickers and dealing with illegal dumping within communities. CE should deal with poverty reduction and social inclusion. Informal waste pickers are responsible for a large part of the recycling activities, without a contribution to public financing. A participant stated that, the experience and knowledge of waste pickers can create more effectiveness in the CE. A successful implementation of the CE should adapt to the countries' needs.

From the results obtained above, the collaboration of the private and public sectors is vital to obtaining optimum results in waste management efficiency. Not only is it vital for their collaboration to succeed but it is just as important for this collaboration to take effect in ensuring the possible implementation of the CE which has been proven effective in countries abroad reducing waste all round.

4.5 Second Phase: Eco-Bricking and Eco Asphalt as a Circularity Solution in the City of Johannesburg

Factors affecting the growth of the CoJ in terms of waste management can be mitigated with the CE; the repurposing of waste, the reusing of waste or the recycling of waste by providing old waste with a new secondary life consisting of various forms of waste management (technologies, systems or methods promoting healthy waste management). Waste collected on landfill sites remains a hazard to the environment and the citizens of Johannesburg. Waste uncollected or unused accumulates requiring more space, more energy, and more resources. CE may be costly in the beginning but in the long-term, is highly effective and a worthy investment to make. Complicated techniques are not necessary for the CE to become a success in a country. This strategy is adapted to each country's capabilities, standards, and resources like that of the city of Oslo who only landfills 3% of its waste and has aimed for zero waste on landfill sites by 2030.

Originating from Guatemala, an eco-brick is a building block made up of a PET (Polyethylene Terephthalate) plastic bottle that is packed tightly with plastic waste. Filler material that can be used to make an eco-brick include paper, cardboard, tetra pack, metal, etc. It is a low-cost construction material and an effective method of plastic waste management [1]. Eco-bricks are made from 500 mL plastic bottles and have the potential to save about 200 g of waste and as such helps divert waste from landfills. A waste management method such as this is in-line with Pikitup's waste prevention strategy–reusing, recycling and recovery activities and net-zero waste plan by 2030.

In the year 2019, the first trial of the eco asphalt road was successfully trialed and implemented in Kwa-Zulu Natal. The trial included a paving mix that utilised a binder called HDPE (High-Density Polyethylene) to incorporate a portion of locally sourced recycled plastic waste. It was proven that the use of the binding agent in the asphalt mixed increased the properties of the plastic in the asphalt mix, better enhancing the capabilities of the paving mixture (Shisalanga Construction 2019). Such initiatives, requires radical innovations and disruptive changes.

The circular economy is an innovative approach to resolving waste problems in the city by reducing natural resource exploitation, greenhouse gas emissions through optimum agricultural protection. For the circular economy to be as successful as EU countries, a waste hierarchy needs to be implemented in South Africa's current legislation laws for waste management to reduce waste production effectively. A country that wishes to adopt the CE successfully needs to [15]:

1. Develop a communal culture to environmental issues: such as a hierarchy on waste management as well as stating the importance of sustainable waste management and promoting public health,
2. Establish financial support or funding from the private and public sectors,
3. Governmental cooperation,
4. Ensure that adequate information on the CE is readily available,
5. Establish a demand and supply: this will help in developing the CE, acquiring of services to assist in the production of the CE and finding a need for secondary reused materials produced from the CE concept,
6. Protect the environment in terms or agriculture or non-renewable resources; and
7. Improve the quality of the developing products (Govindan and Hasanagic 2015).

5 Conclusion

The CE principles include the 3Rs (reduce, reuse, and recycle), but have been extended to include the 6Rs (reuse, recycle, redesign, remanufacture, reduce, and recover). Collaborative governance at national, municipal, and public sector levels on legal frameworks to the introduction of a new waste management strategy in RSA must be adopted to deal with financial, technical, and social obstacles and be steered to direct the cities' significant resources of physical, human, natural and intellectual capital towards the objective of a CE. This concept views waste as a sustainable resource. It has proven beneficial in European countries; hence the proposal to benchmark CoJ's waste solution with these different EU countries. Incorporating insight from a wide range of disciplines and accounting for externalities that can influence CE is a synergistic approach.

The drive to create a net-zero waste environment is heightened due to the lack of landfill space which the city no longer has. JHB has the opportunity to make advances in exploring potentials for and benefit from new CE initiatives. Adopting analyzed solutions for the CE transition entails good interrelations, however, education and training to raise awareness about the CE and to promote CE solutions will enable the city's approach to facilitate this transition and help overcome existing challenges.

The built environment is always changing. There is always room for innovative, positive, and sustainable methods for waste management to evolve or develop. Within the CE, a construction material such as eco-bricking was made, by re-using waste.

Eco-bricking will aid in addressing housing challenges in metrsopolitan cities such as Johannesburg where there is an influx of residents migrating for better living and working conditions. This construction material is durable, strong and complies with building standards. Hence, the CE can potentially prove beneficial for the built environment as a whole and not limited to only building management.

References

1. Antico F, Araya-Letelier G, Retamal R, Wiener M (2017) Eco-bricks: A sustainable substitute for construction materials. https://doi.org/10.7764/RDLC.16.3.518
2. Ansah RH, Sorooshian S (2019) Green economy: Private sectors' response to climate change. Environ Qual Manage 28(3):63–69
3. Brown L (2006) Eco-economy update, 2006–11 from earth policy institute. www.earthpolicy.org/Books/PB2/Contents.htm
4. Chakraborty A, Mehta S (2015) Utilization of waste plastic in construction of pavement. Int J Advan Eng Res Develop 3(02):354–358
5. City of Johannesburg (CoJ), 2018/19, end of term report
6. City of Johannesburg (CoJ) (2019) End of Term Report
7. City of Johannesburg (CoJ), 2018/19, Integrated annual report
8. City of Johannesburg (2020). What Is A Re Sebetseng. www.joburg.org.za Available at: [Accessed 29 June 2020]
9. Cresswell J (2014) Research design: qualitative, quantitative, and mixed methods approaches, 4th edn. SAGE, London
10. Ellen MacArthur Foundation (2016) Circular Economy–UK, Europe, Asia, South America & USA. http://www.ellenmacarthurfoundation.org/circulareconomy
11. Ellen McArthur Foundation (2017) Plastics and the circular economy. www.ellenmacarthurfoundation.org [Accessed 10 May 2021]
12. Esbensen KH, Velis C (2016) Transition to circular economy requires reliable statistical quantification and control of uncertainty and variability in waste, Waste Management & Research 2016, Vol. 34(12):1197–1200
13. European Parliament Amendments (2017) Adopted on 14 March 2017 on the proposal for a directive of the European parliament and of the council amending directive 2008/98/EC on waste (COM (2015)0595-C8-0382/2015-2015/0275(COD)). Strasbourg
14. Godfrey L, Oelofse S (2017) Historical review of waste management and recycling in South Africa. Resources 6(57):1–11
15. Heyes G, Sharmina M, Mendoza J, Gallego-Schmid A, Azapagic A (2018) Developing and implementing circular economy business models in service-oriented technology companies. J Clean Prod 177:621–632
16. Kirkman R, Voulvoulis N (2016) The role of public communication in decision making for waste management infrastructure. J Environ Manage, 203(2):640–647
17. Korhonen J, Honkasalo A, Seppälä J (Jan 2018) Circular economy: the concept and its limitations. Ecol Econ 143:37–46
18. Korhonen J, Snäkin J-P (2015) Quantifying the relationship of resilience and eco-efficiency in complex adaptive energy systems. Ecol Econ 120:83–92
19. Malinauskaite J, Jouhara H, Czajczyńska D, Stanchev P, Katsou E, Rostkowski P, Thorne R, Colón J, Ponsá S, Al-Mansour F, Anguilano L, Krzyżyńska R, López I, Vlasopoulos A, Spencer N (2017) Municipal solid waste management and waste-to-energy in the context of a circular economy and energy recycling in Europe. Energy 141:2013–2044
20. McAllister J (2015) Factors influencing solid-waste management in the developing world. All Graduate Plan B Other Rep 528:10–14

21. Pikitup (2020). City of Johannesburg Pikitup (SOC) Ltd Business Plan 2020–21.
22. Saunders M, Lewis P, Thornhill A (2016) Research Methods for Business Students. 7th Edition, Pearson, Harlow. www.shisalanga.co.za
23. Taiwo OE, Venter C, Oso KA (2016) Treating waste as resource in the cities of Lagos and Johannesburg. In: 23rd Wastecon conference. [Online] institute of waste management of Southern Africa, pp 466–474
24. Weghmann V (2017) Waste management in Europe. Good jobs in the circular economy? European circular economy stakeholder platform. www.circulareconomy.europa.eu
25. World Commission on Environment and Development (1987) Our common future. Oxford University Press, New York
26. Zainal Z (2007) Case study as a research method, Jurnal Kemanusiaan bil.9

Digitalisation

Digitalization of the South African Construction Industry: A Potential for Further Future Research

Adeleye Ayoade Adeniran and Winston Shakantu

Abstract This paper reviewed recent researches on digitalization of the South African construction industry in comparison with countries like United Kingdom and Korea among others with a view to identify the potential of subsequent research that will further enhance the digitalization level of the South African construction industry. It was pursued via a review of existing and recent literature retrieved from various search engines such as GoogleScholar® within the years 2017–2020 which is assumed to be the years of active research activities in digitalization in South African Construction industry based on Deloitte South Africa-White Paper Report. The research accessed the various researches conducted and the result proposed five broad research directions among which are: broad trends in digitalization of the construction industry; perceptions and methods used by the scholars; government guidelines; digital technologies; and apprehension around security. It is on the premises of this direction that this paper suggests that subsequent research should be challenged in this direction as it was observed as an obvious study/research gap.

Keywords Digitalization · Construction · Construction industries · South Africa

1 Introduction

One of the key discourses in the contemporary research is the development of digital technologies and their introduction into every sector of the economy inclusive of the construction industry [1–4] and South Africa is no exception. Issues of digitalization of the country's construction industry are addressed by scholars and industry commentators at all levels [5]. Although there are no specific government policies on building a digital economy in the country as well as ensuring digitalization of the construction industry, it is worthy to note that the most substantial work of research on the South African construction industry, including issues of its digitalization, is published in journals by academics who seem to be in the fore of the subject.

A. A. Adeniran (✉) · W. Shakantu
Department of Construction Management, Nelson Mandela University, Gqeberha, South Africa
e-mail: adeniranayoa@mandela.ac.za

T. C. Haupt et al. (eds.), *Construction in 5D: Deconstruction, Digitalization, Disruption, Disaster, Development*, Lecture Notes in Civil Engineering 245,
https://doi.org/10.1007/978-3-030-97748-1_2

However, in order to understand and evaluate the notion of the digitalization of an important sector of the economy, one should have an understanding of the current vital trends in the South African scholarship, which will give expert evaluations and may further dovetail into policy documentation to be adopted by the government.

Auriga [6] indicated that the notion of digital products, services, and mediums were already well-understood as early as the nineteenth century. However, between 2000 and 2015, the advancement of technology in the form of smart devices and social media platforms resulted in a significant shift in the expectation and medium of communication of consumers with businesses in terms of reaction times and multi-channel availability [7]. Businesses began to notice that they could now engage with their clients digitally on an individual basis, and often in real time and faster at that [7].

As it is with all other industries globally, the construction industry is not left out in the digitization process as the opportunities are infinite spanning the entire value chain of the construction process up until with management of built environment assets [8]. Although they belong to the same value chain, construction site automation, digital design and utilization of Big Data each have their own distinctive characteristics hence standards and best industry practices make their environment challenging for digital tools and processes to pierce through to become industry-wide accepted practices [9]. The governments of the United Kingdom and South Korea support the use of digital technology in the delivery of building projects and have further supported research in this area hence resulting in its recognition [10].

Based on the foci of the papers, our preliminary analysis enables us to allot them in the following research themes viz: broad trends in digitalization of the construction industry; perceptions and methods used by the scholars; government guidelines; digital technologies; and apprehension around security. Hence, the paper gives a narrative of the South African scholarship on the subject under review and defines the key research areas covered.

This was done by first reviewing the South African scholarship on digitalization of the construction industry according to the sections and with focus on each of five research directions outlined above. Based on the review, gaps and existing drawbacks were identified, conclusions drawn and used in proposing further future research areas. It is anticipated that the findings will further inform policy decision as well as stimulate the construction industry towards funding digitization research.

2 Research Methodology

A conventional literature search was undertaken using terms 'digitization", "construction industry" and "South Africa" from Google® Scholar® because of time limitation. The papers were included if they were published since 2017 (see reason in limitation section) and had either qualitative or quantitative research focus. This approach is called 'berry picking' [11] and to select the published papers on digitalization of the South African construction industry, we searched for specific words such as "digitalization of the construction industry" 4IR, IoT, among others in reviewed articles published between 2017 to 2020 in journals, conference proceedings and theses. Some of the journal/proceeding titles are Building Information Systems in the Construction Industry; Nordic Conference on Construction Economics and Organization; Engineering, Construction and Architectural Management; Construction Innovation; Journal of Public Affairs; Buildings; Built Environment Project and Asset Management; Proceedings of the Creative Construction Conference and International Journal of Construction Management among others. In total, eighteen papers are covered in our review.

3 Limitation

This study is limited to the South African literature on subject related to digitization of the construction industry from 2017 to 2020. The 2017 date was chosen because as of [12] reported that impact of industry 4.0 on the African continent was comparatively low and South Africa as an emerging economy could be an early adopter.

4 Presentation and Discussion of Findings

This section of the paper reviews the works of scholars on digitalization of the South African construction industry by focusing on broad trends in digitalization of the construction industry; concepts used by the scholars; government blueprints; digital technologies; and apprehension around security. After reviewing the literature, we discuss potential opportunities for future research on construction industry digitalization.

4.1 Broad Trends in the Digitalization of the South African Construction Industry

This section contains general trends in digitalization of the South African construction industry with focus on what literature reveals and with the significant effect of digital technologies on almost all the areas of human life. The construction sector plays a vital role in the South African economy as it results in infrastructure stock/capital accumulation which stimulates increased socio-economic activities by contributing to employment, household income and economic growth. StatSA [13] reports that the gross domestic product (GDP) from construction in South Africa increased to ZAR80129.25 Million in the third quarter of 2020 from ZAR 70,059.94

Million in the second quarter of 2020. Hence, given the primary role of the construction sector in the South African economy, the importance of digitalizing it cannot be overemphasized.

Reference [14] indicated that the efficiency presented by the implementation of information technology is yet to be harnessed by the South African construction industry as it has concentrated more on the use of information and communication technology software. Meanwhile, [15] highlighted that the digitalization of the construction industry involves sophisticated assembling of smart and intelligent data by use of new gadgets which could facilitate easy data analysis aimed at making quick resolutions that enable the creation of a smarter, responsive and efficient built environment. Ref. [16] in their quantitative assessment of Gauteng Province construction professionals' perspective on digitalization within the South African construction industry revealed that the major phases of construction where digitalization is evident are the design and the feasibility stage. They further highlighted that the major construction procedures where it is evident are at processes of preliminary cost estimate, planning and control as well as building system analysis.

Reference [17] indicated that, characterized by poor project delivery and technological backwardness, the construction organizations in developing countries and particularly South Africa are lagging in digital technology usage. Their study, using a Delphi approach, concluded that construction organizations are lagging in digital technologies adoption due to several challenges while this partnership is a necessity. They recommended that top management in construction organizations must make an informed decision of partnerships with the main aim of improving their digital capability and competitive advantage.

Also, [18, 19] affirmed that in order to harness the benefit of digitalization, developing effective 4IR leadership framework for construction organizations is a significant for South Africa's future. Hence, with the South African construction industry contributing a large chunk to its GDP [20], the use of digital technologies will create new opportunities as well as economic benefits in all sectors attached to it and can according to [21] further stimulate financial investments in technologies such as robotics, the Internet of Things (IoT) and artificial intelligence among others.

Several papers discuss prospects of South Africa's construction industry digitalization, focusing on practical applications of digital technologies in almost all spheres. Ref. [22] argues that it can provide a remedy for the challenge of sustainability of the industrial and manufacturing process. In addition, [16] highlighted that it could stimulate timesaving in projects delivery, increased productivity, response time and work speed as well as enhanced document quality using simpler working approaches.

4.2 Concepts Used by Scholars

With the purpose to describe and explain processes of digitalization of the South African construction industry, scholars rely on a variety of concepts and approaches.

Hence this section will review the concepts used by authors such as Industry 4.0, ICT, 4IR, BIM, "3D", virtual reality, Internet of things and smart systems.

Ref. [23] observed that the modern-day era of information and communication technology, ICT, also referred to severally as the fourth industrial era was originally invented in Germany around 2011 and is grounded on concepts and technologies such as cyber-physical systems, the Internet of Things (IoT) and the Internet of Service among others. IoT is a notion denoting major deviations which causes basic economic and societal change within a short time [24, 25]. As such, the fourth industrial revolution has been referred to as Industry 4.0, digital sustainability, Smart Factory, and the Internet of Things (IoT) among other names [24, 26, 27].

According to [2, 24, 28], with the attention garnered over time by BIM (Building Information Modelling), the use of digital technologies has only lately become a common concept among construction industry participants.

As emphasized by [15], the use of Industry 4.0 concepts in the construction industry is a new idea by means of the Internet of Things (IoT) for the incorporation of data from diverse platforms which adopts innovative devices such as drones, laser scanning, enhanced 3D printing which helps to monitor and deliver sustainable and smart building construction at all its various phases.

In another study, [29] established that Industry 4.0 uses internet of things and smart technologies to facilitate the integration of different technologies like; cloud data, robotics, and virtualization for manufacturing that enables the connectivity of people with machines. Ref. [24, 29] further applied this to the construction industry and christened it Construction4.0 with the aim of digitalizing the construction industry by adopting Industry 4.0 powered technologies such as 3D printing and RFID among others.

Ref. [30] acknowledges Construction 4.0 as the administration of modern technology to inspire the digitalization of the construction industry as well as the supply chain for efficient and effective performance of the sector. This initiates the prospect of the intelligent control of construction tools grounded on man-to-machine-to-machine (M2M2M) interface. Thus, authors conclude that digitalization is an operational apparatus for managing a complex system such as the construction industry inclusive of its material(s) distribution chain(s) [24, 31].

As documented by [14, 32] who reported that distinctive physiognomies of the construction industry make it challenging for the application of the industry 4.0. Some of the physiognomies include the site-based nature of construction projects and the existence of a large number of small and medium-sized firms with little or low potentials for investments in new technologies [14]; high cost of education and training of construction workers [32].

4.3 Government Blueprints

The governments of the United Kingdom and South Korea support the use of digital technology in the construction project delivery and have further funded researches towards this, hence its recognition [33–35]. However, there is no paper devoted in the South African scholarship to analyzing government policies aimed at digitalization of the construction industry, a few papers cover this topic superficially.

Reference [36] stated that the post-apartheid government in a bid to evolving the construction industry designed new policies and established relevant bodies such as the construction industry development board (CIDB) to ensure the standardization of the sector. The government is also adjudged the biggest client of the construction industry in the development of RDP houses and other infrastructures [37] and it has further experienced significant growth since the 2010 World Cup soccer fiesta projects but has slowed down with the crash of the Rand [38].

Despite governments patronage of this sector, [33] established that there is no visible support from government towards the digitalization of the industry. The study of [32] was stimulated by the call of the South Africa Government for the use of industry 4.0 technologies in the construction industry and they recommended the subsidization of the costs of technologies that support construction 4.0 technologies.

Ref. [39] posited that while policy design and the governance framework are critical, the effective implementation and enforcement of any industrial policy will depend on enhanced government capacity and more effective cooperation with the private sector but that it will also raise potential trade-offs and new conflicts in the economy.

4.4 Digital Technologies

A number of papers are devoted to various digital technologies that can be used in the South African construction industry, including ICT tools such as video conferencing, convergence networking, data mining, and project management and they are used to design, construct, operate, and manage construction, as well as implementing ICT in the completed projects.

Ref. [23] identified e-procurement technologies, Project management tools, Big data tools, Communication software, Simulation tools/3D modelling/BIM tools, Estimation and quantity take-off tools, Robotics/Augmented reality, Convergence network and Cloud computing as digital technologies adapted for use in the construction industry.

The use of Radio Frequency Identification (RFID) as alluded by [14] could aid the construction industry evolve from reactive maintenance as it will afford an opportunity for equipment and machines to automatically send performance data to engineers.

Practicing construction professionals in Johannesburg identified automation, product lifecycle management, robotics, additive manufacturing, internet of things, internet of services and human–computer interaction as important for Construction 4.0 for delivery of sustainable housing [29].

Ref. [40] also identified that building information modelling BIM, promises better project delivery; the Internet of Things can resolve the issue of poor communication;

big data analytics will aid future construction project delivery predictions; 3D printing and autonomous robots will reduce labor and material costs, injuries and fatalities while improving productivity and creation of more job opportunities.

They include Industry 4.0 elements, simulation and modelling tools (BIM, augmented reality, virtual reality, simulation tools, and mixed reality), digitization and virtualization tools (e.g. cloud computing, big data), and smart factory-inclined ICT tools (for instance, the product-lifecycle management [PLM] tool, robotics, Internet of Things (IoT), cyber-physical systems) [23].

4.5 Cyber-Security Concerns

The accelerating processes of digitalization of the global construction industry is stimulating the attention to security risks that are associated with the use of digital technologies. While not much has been done on this concern by South African Scholars save for [41] who reviewed the emerging risks in the digitization of the South African construction industry and the issue of security issues came out strong.

However, many authors elsewhere have also raised concerns around cyber-attacks. Ref. [42] argued that the dimension of the complexity which may arise from unintended and accidental disruptions or deliberate threats may leave the digitized construction industry to contend with organized crime, labor union, government, and hacktivists. They posited that the resultant effect of this complexity cannot be neglected because it can be the enabling tool for terrorist subversion and terrorism.

Reference [43] also observed that the security and legal risks associated with digitalization are critical since information storage and access is made simpler via technology. This simplicity however makes data vulnerability to viruses and hackers and this hence calls for more attention as against when things are done manually.

In giving recommendation, [44] posited that to protect data, other intelligent techniques can be used with very careful considering given to cyber security issues and data protection.

5 Potential for Further Future Research

According to Wyman [45], digitalization offers numerous opportunities to increase operational efficiency (direct costs, delays, security, and environmental impact) and beyond operations, digitalization in the construction industry triggers opportunities to raise sales and enhance customer satisfaction. As presented in Fig. 1, digital opportunities in operations are numerous: for industrial players they mainly fall under the "Industry 4.0" framework, whereas for distributors they are related to "omnichannel operations" while for contractors there are three specific "smart operations" lever categories. The interactive work processes of digitization allow for

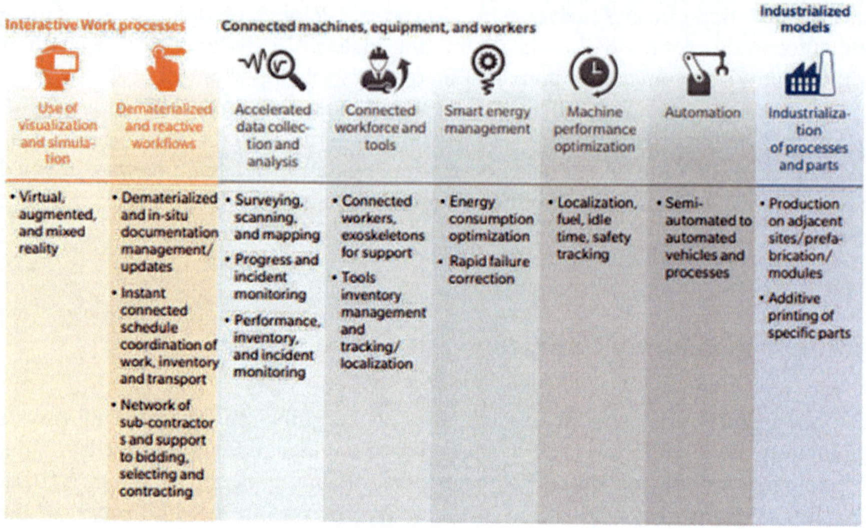

Fig. 1 Digital operations efficiency levers

virtual, augmented and mixed realities as well as dematerialization, in-situ documentation, connected schedules and immediate planning adjustments, vision of historical changes. Process efficiency has the potential to drastically increase through more efficient, transparent and rapid collaboration.

The adoption of digitization is expected to lead to better project delivery [46], improved communication between stakeholders, informed decision-making [47], reduced on-site injuries and fatalities [48] and improved productivity [49]. However, there is still little research that explores the potential applications of UAVs and immersive technologies, either individually or integrated with other technologies in South Africa. Our study however reveals that the South African scholars have begun to study the array of issues related to digitalization of the construction industry in terms of trends in construction industry digitalization; its perceptions and methods; government guidelines; digital technologies; and apprehension around security although not much has currently been done leaving room for gaps in the body of knowledge.

Given the significance of global digitalization of the construction sector, there is no paper which has examined government policies intended for digitalization, hence this is an item for research. Furthermore, the increase in digital platforms that offer resources for digitalization of the construction industry is of great academic interest and practical importance but there are suboptimal studies on them. Also, a wide assessment of the process of digitization and digital technologies in the overall assembly the construction industries in South Africa is germane as this will further stimulate the development and implementation of macro-policies.

6 Conclusion

The research conducted allowed us to review South African scholar literature on the general trends in digitalization of the construction industry; concepts used by the scholars; government blueprint; digital technologies; and cyber-security concerns on the digitalization of construction industry. The study, based on this, proposed further future research to: review current and prospective construction industry digitization policies, focus on the platforms that offer digitalization, and assess processes of digital and digitization technologies in the overall structure of construction industry in South Africa.

References

1. Yoo Y, Boland RJ Jr, Lyytinen K, Majchrzak A (2012) Organizing for innovation in the digitized world. Organ Sci 23(5):1398–1408. https://doi.org/10.1287/orsc.1120.0771
2. Ibem EO, Laryea S (2014) Survey of digital technologies in procurement of construction projects. Autom Constr 46(2014):11–21. https://doi.org/10.1016/j.autcon.2014.07.003
3. Fitzgerald M, Kruschwitz N, Bonnet D, Welch M (2014) Embracing digital technology: a new strategic imperative. MIT Sloan Manag Rev 55(2):1
4. Elia G, Margherita A, Passiante G (2020) Digital entrepreneurship ecosystem: how digital technologies and collective intelligence are reshaping the entrepreneurial process. Technol Forecast Soc Change 150:119791
5. Crampton N (2017) The digitalisation of the construction industry available online at https://bizconnect.standardbank.co.za/sector-news/construction/the-digitalisation-of-the-construction-industry.aspx. Retrieved 23 Dec 2020
6. Auriga (2016) Digital transformation: history, present, and future trends available online at https://auriga.com/blog/digital-transformation-history-present-and-future-trends/
7. Dwivedi YK, Ismagilova E, Hughes DL, Carlson J, Filieri R, Jacobson J, Jain V, Karjaluoto H, Kefi H, Krishen AS, Kumar V (2021) Setting the future of digital and social media marketing research: Perspectives and research propositions. Int J Inf Manag 59:102168
8. Leminen S, Rajahonka M, Wendelin R, Westerlund M (2020) Industrial internet of things business models in the machine-to-machine context. Ind Mark Manag 84:298–311
9. Sivarajah U, Kamal MM, Irani Z, Weerakkody V (2017) Critical analysis of big data challenges and analytical methods. J Bus Res 70:263–286
10. Matarneh R, Hamed S (2017) Barriers to the adoption of building information modeling in the Jordanian building industry. Open J Civil Eng 7(3):325–335
11. Sutton A, Clowes M, Preston L, Booth A (2019) Meeting the review family: exploring review types and associated information retrieval requirements. Health Info Libr J 36(3):202–222
12. Pillay K, Ori A, Merkofer P (2017) Industry 4.0: is Africa ready for digital transformation. A case study of Deloitte South Africa-white paper report
13. StatSA (2020) South Africa GDP from construction available online at https://tradingeconomics.com/south-africa/gdp-from-construction
14. Osunsanmi T, Aigbavboa C, Oke A (2018) Construction 4.0: the future of the construction industry in South Africa. World Acad Sci, Eng Technol, Int J Civil Environ Eng 12(3):206–212
15. Montgomery D (2017) The future for construction product manufacturing. Digitalisation, industry and the circular economy, available online at http://www.thefis.org/wpcontent/uploads/2016/10/cpa-future-of-construction-productmanufacturing-report-2016.pdf. Retrieved 11 Nov 2020

16. Aghimien D, Aigbavboa C, Oke A, Koloko N (2018) Digitalisation in construction industry: construction professionals' perspective. In: Shiau J, Vimonsatit V, Yazdani S, Singh A (eds) Streamlining information transfer between construction and structural engineering. ISEC
17. Aghimien DO, Aigbavboa CO, Oke AE (2020) Critical success factors for digital partnering of construction organisations–A Delphi study. Eng, Constr Architectural Manag
18. Alade K, Windapo AO (2020) Developing effective 4IR leadership framework for construction organisations. Eng, Constr Architectural Manag
19. Adepoju OO, Aigbavboa CO (2020) Assessing knowledge and skills gap for construction 4.0 in a developing economy. J Public Affairs e2264
20. Mosenogi JM (2014) An impact analysis of construction sector on economic growth and household income in South Africa (Doctoral dissertation)
21. Allam Z, Dhunny ZA (2019) On big data, artificial intelligence and smart cities. Cities 89:80–91
22. Amusan L, Aigbavboa C, Ojo SIF, Owolabi J (2020) Lean thinking and industrial 4.0 approach to achieving construction 4.0 for industrialization and technological development. Buildings 10(12):221
23. Odubiyi TB, Aigbavboa CO, Thwala WD (2019) A concise review of the evolution of information and communication technologies for engineering innovations. University of Johannesburg
24. Mohammad-Akhriza T, Ma Y, Li J (2017) Revealing the gap between skills of students and the evolving skills required by the industry of information and communication technology. Int J Software Eng Knowl Eng 27(05):675–698
25. Olsson NO, Shafqat A, Arica E, Økland A (May 2019) 3D-printing technology in construction: results from a survey. In: 10th Nordic conference on construction economics and organization. Emerald Publishing Limited
26. Lekan A, Aigbavboa C, Babatunde O, Olabosipo F, Christiana A (2020) Disruptive technological innovations in construction field and fourth industrial revolution intervention in the achievement of the sustainable development goal 9. Int J Constr Manag 1–12
27. Lekan A, Aigbavboa C, Fayomi OSI, James O (2020) Lean thinking and industrial 4.0 approach to achieving construction 4.0 for industrialization and technological development. Buildings 10(12):221
28. Govender K, Nyagwachi J, Smallwood J, Allen C (2018) The awareness of integrated project delivery and building information modelling-facilitating construction projects. Build Inf Syst Constr Ind 13(1):121–129
29. Osunsanmi T, Aigbavboa C, Oke AE, Ohiomah I (2018) Construction 4.0: its impact towards delivering quality and sustainable houses in South Africa. In: Creative construction conference 2018. Budapest University of Technology and Economics, pp 147–156
30. Pereira AC, Romero F (2017) A review of the meanings and the implications of the industry 4.0 concept. Procedia Manufact 13:1206–1214
31. Osunsanmi TO, Aigbavboa CO, Oke EA, Liphadzi M (2020) Appraisal of stakeholders' willingness to adopt construction 4.0 technologies for construction projects. Built Environ Project Asset Manag 10(4):547–565. https://doi.org/10.1108/BEPAM-12-2018-0159
32. Ozumba AOU, Shakantu W (2018) Exploring challenges to ICT utilisation in construction site management. Constr Innov
33. Howard R, Restrepo L, Chang CY (2017) Addressing individual perceptions: an application of the unified theory of acceptance and use of technology to building information modelling. Int J Project Manage 35(2):107–120
34. Jung W, Lee G (2015) The status of BIM adoption on six continents. Int J Civil Environ Eng 9(5):512–516
35. Lee S, Yu J (2016) Comparative study of BIM acceptance between Korea and the United States. J Constr Eng Manag 142(3):1–9
36. Wood L (2019) South Africa construction industry report 2019—available at https://www.businesswire.com/news/home/20190802005286/en/South-Africa-Construction-Industry-Report-2019---ResearchAndMarkets.com. Retrieved 31 January 2021

37. Gnade H, Blaauw PF, Greyling T (2017) The impact of basic and social infrastructure investment on South African economic growth and development. Dev South Afr 34(3):347–364
38. Takawira M (2019) Addressing skills shortages in the South African residential construction industry through automation: a case study of Gauteng region (Doctoral dissertation)
39. The department of trade industry and competition, DTIC (2019) Towards a digital industrial policy for South Africa: a review of the issues industrial development think tank available at http://www.thedtic.gov.za/wp-content/uploads/DPIP.pdf. Retrieved 31 Jan 2021
40. Aghimien D, Aigbavboa C, Oke A, Thwala W, Moripe P (2020) Digitalization of construction organisations—a case for digital partnering. Int J Constr Manag 1–10
41. Meno T (2020) An assessment of risk associated with digitalisation in the South African construction industry. University of Johannesburg (South Africa)
42. de Soto BG, Georgescu A, Mantha B, Turk Ž, Maciel A (2020) Construction cybersecurity and critical infrastructure protection: significance, overlaps, and proposed action plan
43. Sengupta J, Ruj S, Bit SD (2020) A comprehensive survey on attacks, security issues and blockchain solutions for IoT and IIoT. J Netw Comput Appl 149:102481
44. Hossain A, Nadeem A (2019) Towards digitizing the construction industry: state of the art of construction 4.0. Struct Eng Constr Manag 1–6
45. Wyman O (2018) Digitalization of the construction industry: the revolution is underway https://www.mmc.com/content/dam/mmc-web/Files/OliverWyman_Digitalization_in_the_construction_industry_web_final.pdf. Retrieved 10 Aug 2021
46. Ammar M, Russello G, Crispo B (2018) Internet of things: A survey on the security of IoT frameworks. J Inf Sec Appl 38:8–27
47. Elghaish F, Abrishami S, Hosseini MR, Abu-Samra S, Gaterell M (2019) Integrated project delivery with BIM: an automated EVM-based approach. Autom Constr 106:102907
48. Aghimien DO, Aigbavboa CO, Oke AE, Thwala WD (2019) Mapping out research focus for robotics and automation research in construction-related studies: a bibliometric approach. J Eng, Des Technol 18. https://doi.org/10.1108/JEDT-09-2019-0237
49. Leviäkangas P, Paik SM, Moon S (2017) Keeping up with the pace of digitization: the case of the Australian construction industry. Technol Soc 50:33–43

A Review of Barriers to the Adoption of Smart Building Concepts (SBCs) in Developing Countries

Cyril Chinonso Ejidike and Modupe Cecilia Mewomo

Abstract The smart building concept is becoming more popular in construction in recent times. This is because the world is embracing digitization in response to the advent of new technologies. The call for high building performance and sustainable development has brought about the adoption of smart building concepts in the construction industry. The Smart Building Concept adoption has not been without hindrances occasioned by several constraints undermining the efforts that aid its adoption. This paper intends to review systematically extant literature on barriers undermining the adoption of SBCs, considering papers published in peer-reviewed journals and conferences. The study reviewed 30 relevant articles through topical analysis of the article then followed by abstract and the finding of the paper selected concerning barriers to the adoption of SBCs, and barrier frequency was employed to select the most reported ones. The study revealed the most reported barriers undermining the adoption of SBCs as the high cost of initial construction, vague objectives, lack of guidelines to manage SBC, lack of government incentive and policy, lack of knowledge on the smart building by contractors and professionals, and resistance to change from the use of traditional technologies among others. For SBCs to thrive in developing countries, there is a need to mitigate the identified barriers to their adoption.

Keywords Smart building concepts · Sustainable construction · Smart building barriers and Smart building technologies

1 Introduction

The smart building concept is gaining more attention as a means of sustainability in the building sector of the construction industry due to growing public concern about the environmental impact [1]. The smart building concept has gained popularity

C. C. Ejidike · M. C. Mewomo (✉)
Department of Construction Management and Quantity Surveying, Durban University of Technology, Durban 4001, South Africa
e-mail: Modupem@dut.ac.za

T. C. Haupt et al. (eds.), *Construction in 5D: Deconstruction, Digitalization, Disruption, Disaster, Development*, Lecture Notes in Civil Engineering 245,
https://doi.org/10.1007/978-3-030-97748-1_3

among construction professionals and academics who are working to preserve the environment and advance the construction industry's building sector [2]. Sustainable construction is an essential strategy to minimize a building's environmental impact, and it entails the integration of new methods based on inter-disciplinary knowledge. The smart building concept has been recognized by professionals, the government, and the global community as a blueprint for improving the sustainability of the building construction industry. The smart building concept is considered the practice of using smart technology in creating structure, energy reduction, improvement of occupant comfort, and satisfaction that are environmentally sustainable [3]. According to [4] smart building concept forms part of sustainability a key driver in the economic efficiency, ecology balance system in protecting and restoring as well as improving human life.

1.1 Smart Building Concept

A smart building is a building that provides dynamic real-time control in managing the internal and external activities of the building [5]. According to [6] a smart building has an integrated platform for monitoring energy use, intelligent facilities management, and the use of surveillance video for security systems. Smart buildings contain intelligence with building automation as an important aspect of their management [4].

Presently, the increasing number of smart devices, smart sensors, smart materials, and smart cities have been adopted and gradually been implemented in developing countries such as South Africa [7]. Research has indicated that there are difficulties in defining in clear terms the concept and target of smart buildings [8]. Notwithstanding, its concept could be established concerning the time, culture and needs within a particular country or continent. While some continents concentrate on the adoption of SBC for the improvement of environmental quality, some other countries concentrate on its economic benefits in achieving low-carbon economic goals [8]. In addition, some countries embrace the smart building concept due to the increased environmental performance, operation and safety as well as reliability of various technologies in automation, control and communication associated with its adoption. The benefit of sustainability over the years has brought about the promotion of the smart building concept adoption which has been of great importance to many countries. However, the smart building concept is not free from barriers hindering the adoption in the building sector of the construction industry. According to [9] the barriers must be first recognized and understood. However, few studies have been conducted on the barriers to the adoption of smart building concepts in developing countries [4]. In a similar study by [10], they called for the incorporation of intelligent buildings in the architectural curriculum in tertiary institutions in Nigeria. There is a gap in the literature in terms of smart building adoption barriers in developing countries [11, 12].

2 Literature Review

The smart building concept allows a transition from traditional construction methods to industrial production by increasing the use of standardised components and production offsite [13]. Adopting the smart building concept offers a significant sustainable benefit that is greater than that of conventional traditional building technique practice in developing countries such as Nigeria [10]. Ref. [14] revealed the smart building learns from the experience to make the most effective decisions in real-time to maximize comfort and productivity while using the least amount of energy possible. By implementing SBC many benefits such as improved technologies in communication within a building as well as increased technologies in automation and control can be achieved.

2.1 Barriers to Smart Building Concept Adoption

Despite the increased perception of SBC benefits, there are still barriers to their adoption [4]. The barriers to the smart building concept in the building construction industry have been investigated in the developed countries but not so much in developing countries. According to [15] lack of financial incentives and a regulatory environment favorable to adopting the smart building concept. Similarly, [16, 17] identified the following as barriers to smart building, the high initial cost of smart building construction compared to conventional building, lacks of financing schemes, lacks of government incentives, unavailability of smart building suppliers, lack of institutes for research development of smart building technologies, a high degree of distrust about Smart Building concept among professionals and client, resistance to change from the use of traditional technology, lack of interest from client and market demand [1]. Identified a higher cost of a smart building, lack of technical expertise/skill, interoperability, difficulty to change consumer behaviour as barriers to SBC adoption. Ref. [10] reported that low quality of maintenance culture, lack of awareness, insufficient power supply in the country are the major barrier to the adoption and implementation of smart buildings. In the same vein, [8] discovered a lack of guidelines and framework, lack of client involvement in the design process, high cost of smart building materials, and lack of smart building definition as a major barrier facing the Egyptian building construction industry. Ref. [9, 16, 18] identify lack of smart building promotion by the government, long payback period, lack of financing schemes, and risks and uncertainties involved in adopting new technologies. Ref. [4, 19] discovered that extension of project schedules, resistance to change from traditional practices, unavailable smart building material, and equipment, planning of different construction techniques, lack of government policy, level of risk the client is willing to take our part of the challenges in smart building technologies. In Indian, [20] discovered that lack of ability to meet electric power demand, unavailability of solar radiation data, and lack of political commitment are the major barriers to smart building.

3 Research Methodology

The essential basis of progressing in the knowledge of a subject is a relevant review of literature on the subject [21]. This essentially eases theory development that is beneficial to industrial and academia at large. This study is based on the review of literature on barriers to Smart Building concept adoption. The review focused on the issues surrounding smart building adoption that have been published in academic publications and conference papers. A literature search was undertaken with the use of the Scopus search engine. This method has always been employed to find relevant papers for studies because of its recovering precision and accuracy in performance [4, 22]. The search keywords used are "smart building barriers", "smart building concept", and "smart building technologies". A total of 150 Articles were found during the initial search that was conducted systematically [searched on April 2021]. However, not all of the publications found are related to the research of smart building adoption barriers. It was necessary to filter out unrelated papers because the goal of the study was to analyze literature studies on the barriers to smart building concept adoption in the construction industry's building sector. Following the screening process, 30 articles were identified as being relevant for further investigation. To review the 30 relevant articles, the study used a systematic review technique through topical analysis of the article then followed by abstract and the finding of the paper selected to discover common barriers to SBC adoption. This study focuses on reviewing and making conclusions from relevant publications that were found, rather than a comprehensive evaluation of the entire population of literature on the subject (Fig. 1).

4 Discussion

About 60 barriers to smart building concepts were discovered after a systematic review of 30 articles. The identified barriers were crystallized into four major categories namely: economy-related barriers, government-policy-related barriers, technical related barriers, and Social- human-related barriers [1, 9]. many barriers to adoption of smart building concepts in developing countries as seen but the most commonly reported barriers are the high cost of initial construction from adopting smart building concepts, higher cost of smart building technologies, lack of government incentive and policy, and lack of knowledge on smart building technologies.

The categorized barriers are presented in Table 1

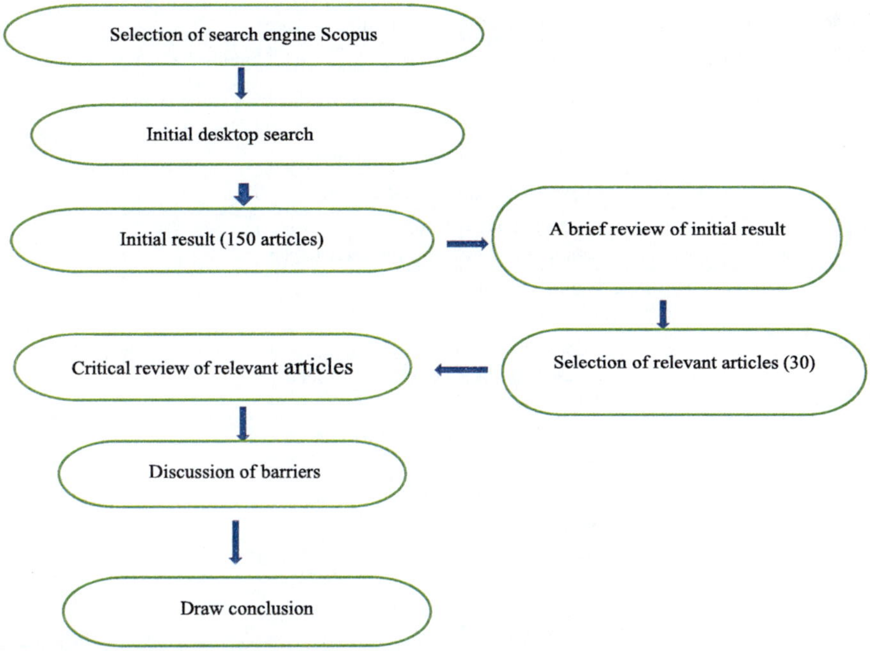

Fig. 1 Systematic research flow chat. *Source* Adapted source: Ghansah et al. [4]

5 Conclusion and Recommendation

This study was conducted as a systematic review of literature on the barriers to smart building concept adoption in developing countries, with the help of data-bases that were used to collect relevant academic journals and conference pa-pers. From the comprehensive literature review, many barriers are preventing the adoption of smart building concepts in developing countries as seen but the most commonly reported barriers are the high cost of initial construction from adopting smart building concepts, higher cost of smart building technologies, lack of government incentive and policy, and lack of knowledge on smart building technologies. To overcome these barriers, this study recommends that government should establish a common platform for collaboration of all stakeholders such as professionals in the construction industry by way of policymaking and provision of funds for research and development towards implementation of these smart technologies. This will go a long way for employment creation and better improve the economy of the countries. Also, government support may be a very good instrument in promoting smart building concepts. Government should consider, for example, waiver on tariff cost of importation of smart building materials, provision of the financial scheme to encourage the usage of the smart building. According to [4] government play a vital role in the promotion and adoption of smart building in the building sector of the construction industry. As a result, the

Table 1 Common barriers to the adoption of smart building concept

N/S	Categories	Barriers	Number of times barrier is reported
Economy-Related			
1		Delays of procurement requirements	4
2		high cost of initial construction	10
3		Long payback periods from adopting the smart building concept	12
4		Higher cost of smart building technologies	12
5		Adoption of a smart building is time-consuming and causes project delays	3
6		Lack of interest from clients and market demand	5
7		High cost of smart building equipment	1
8		Lack of financing schemes	11
Government-Related			
1		Lack of framework	6
2		Lack of Government incentives and policy	12
3		Lack of local institutes and facilities for research and development of smart building technologies	5
4		Lack of promotion of smart building practice	6
5		Risk and uncertainties involved in implementing new technologies	4
5		Limited procurement method of a smart building	2
Technical-Related			
1		Failure of electric power supply	3
2		Lack of knowledge on the smart building by contractors and professionals,	11
3		Lack of awareness of professionals and clients	8
4		Lack of technical and clear understanding of smart building by professionals, clients, and sub-contractors	2
5		Extension of project schedules	3

(continued)

Table 1 (continued)

N/S	Categories	Barriers	Number of times barrier is reported
6		Unfamiliarity with smart building technology	6
7		Lack of smart building databases and information	8
8		Risk and uncertainties involved in implementing new technologies	9
Social-Human-Related			
1		Lack of expressed interest from the public	3
2		Lack of awareness of clients and professionals	3
3		Lack of well-trained labour	6
4		lack of integrated work environment among professionals	5
5		Resistance to change from the use of traditional technologies	10
6		Limited procurement method of a smart building	8
7		Aesthetically less pleasing	1
8		Lack of importance attached to the smart building by senior management	3
9		Risks and uncertainties involved in adopting new technologies	10
10		Lack of poor maintenance culture	2

government is in charge of developing an effective smart building concept policy framework. As a result, this study contributes to the understanding of barriers to smart building concept adoption by identifying the most commonly reported barriers in the literature, which can be linked to a global perspective in finding, creating better insight and understanding for professionals and government on what is preventing the adoption of smart building concepts in the building sector of the construction industry.

References

1. Junior AJ, Nunes de Souza M (2017) Barriers and challenges to smart buildings' concepts. Int J Sustainable Real Estate Constr Econ 1(1):31–50
2. Lim BT, Liu AZ, Oo LB (2019) Awareness and practices of sustainable construction in Australia: consultant quantity surveyors' perception. Cite as: AIP conference proceedings 2124, pp 1–10. https://doi.org/10.1063/1.5117120

3. Attoue N, Shahrour I (2018) Smart building: use of the artificial neural network approach for indoor temperature forecasting. Energies https://doi.org/10.3390/en11020395
4. Ghansah FA, Owusu-Manu D-G, Ayarkwa J (2021) Project management processes in the adoption of smart building technologies: a systematic review of constraints. Smart Sustain Built Environ 10(2): pp 208–226. https://doi.org/10.1108/SASBE-12-2019-0161
5. Smart cities council (2015) Retrieved 25 July 2021, from https://india.smartcitiescouncil.com/article/smart-buildings-key-smart-cities
6. Vattano S (2014) Smart buildings for a sustainable development. Econ Worl 2(2328–7144):310–324. Retrieved 12 June 2021, from https://www.academia.edu/9734998/Smart_Buildings_for_A_Sustainable_Development
7. Oke EA, Aghimien DO, Akinradewo O, Aigbavboa C (2020) Improvng resilience of cities through smart city drivers. Constr-Econ Build 20:45–64. https://doi.org/10.5130/AJCEB.v20i2.6647
8. El-Motasem S, Khodeir LM, Eid FA (2021) Analysis of challenges facing smart buildings projects in Egypt. Ain Shams Eng J 1–13. https://doi.org/10.1016/j.asej.2020.09.028
9. Chan A, Darko A, Ameyaw E, Owusu-Manu D-G (2016) Barriers affecting the adoption of green building technologies. J Manag Eng https://doi.org/10.1061/(ASCE)ME.1943-5479.0000507
10. Iwuagwu BU, Iwuagwu MC (2014) Adopting intelligent buildings in Nigeria: the hopes and fears. 2nd International conference on emerging trends in engineering and technology (ICETET'2014). London (UK), pp 160–163. https://doi.org/10.15242/IIE.E0514560
11. Ogunde OA, Emmanuel I, Nduka D (2018) Assessment of integration of building automation systems in residential buildings in developing countries: professionals' perspectives. Int J Civil Eng Technol (IJCIET) 9(0976–6316):577–585
12. Oyewole M, Araloyin M, Oyewole P (2019) TResidents' awareness and aspiration for smart building features: the case of Okota, Lagos, Nigeria. Nigerian J Environ Sci Technol (NIJEST) 30–40
13. Gbadamosi A-Q (2019) The role of internet of things in delivering smart construction. CIB world building congress 2019. Hong Kong SAR, China, pp 1–10
14. Apanaviciene R, Vanagas A, Fokaides P (2020) Smart building integration into a smart city (SBISC): development of a new evaluation framework. Energies 1–19
15. Gobbo Jr, JA, Nunes de Souza MG (2016) Barriers and challenges to smart buildings concepts and technologies in Brazilian social housing projects. Int J Sustain Real Estate Constr Econ https://doi.org/10.1504/IJSRECE.2017.10005278
16. Ahn YH, Annie RP, Yuhong W, George W (2013) Drivers and barriers of sustainable design and construction: the perception of green building experience. Int J Sustain Build Technol Urban Dev 4(1):35–45. https://doi.org/10.1080/2093761X.2012.759887
17. Chan AP, Darko A, Olanipekun AO (2017) Critical barriers to green building technologies adoption in developing countries: the case of Ghana. J Cleaner Prod 1067–1079
18. Aiello S (2010) Addressing financial objections to sustainable design and construction. J Green Build 67–77
19. Darko A, Chan AP, Ameyaw EE, He B-J (2017) Examining issues influencing green building technologies adoption: the United States green building experts' perspectives. Energy Build 320–332 https://doi.org/10.1016/j.enbuild.2017.03.060
20. Luthra S, Kumar S, Garg D, Haleem A (2015) Barriers to renewable/sustainable energy technologies adoption: Indian perspective. Renew Sustain Energy Rev 762–776. https://doi.org/10.1016/j.rser.2014.08.077
21. Webster J, Watson R (2002) Analyzing the past to prepare for the future: writing a literature review. MIS Quarterly, 13–23. Retrieved 12 July 2021, from https://web.njit.edu/~egan/Writing_A_Literature_Review.pdf
22. Osei-Kyei R, Chan AP (2015) Review of studies on the critical success factors for public–private partnership (PPP) projects from 1990 to 2013. Int J Proj Manag 33(6):1334–1346. https://doi.org/10.1016/j.ijproman.2015.02.008

23. Frank OL (2007) Intelligent building concept: the challenges for building practitioners in the 21st century. J Assoc Architectural Educators Nigeria. (AARCHES J) 107–113
24. Gou Z, Siu-Yu Lau S, Prasad D (2013) Market readiness and policy implications for green buildings: case study from Hong Kong. J Green Build 8(3):162–173
25. Hakkinen T, Belloni K (2011) Barriers and drivers for sustainable building. Build Res Inf 239–255
26. Lam P, Chan E, Poon C, Chun K (2009) Integrating green specifications in construction and overcoming barriers in their use. J Manag Eng (10523928):142–152. https://doi.org/10.1061/(ASCE)1052-3928(2009)135:4(142)
27. Omopariola ED, Albert I, Windapo A (2019) Appropriate drivers for sustainable construction practices on construction sites in Nigeria. In: Laryea S (ed), West Africa built environment research (WABER) conference. Accra, Ghana, pp 103–115. https://doi.org/10.33796/waberconference2019.07
28. Owajionyi (2007) Intelligent building concept: the challenges for building practitioners in the 21st century. J Assoc Architectural Educators Nigeria. (AARCHES J) 107–113
29. Oyedele LO, Ajayi SO, Kadiri KO (2014) Use of recycled products in UK construction industry: an empirical investigation into critical impediments and strategies for improvement. Resour Conserv Recycl 93:0921–3449. https://doi.org/10.1016/j.resconrec.2014.09.011
30. Shen L, Zhang Z, Long Z (2017) Significant barriers to green procurement in real estate development. Resour, Conserv Recycl 160–168

Adaptive Co-Design for Self-Help Housing in Durban

Viloshin Govender and Claudia Loggia

Abstract In Africa, informal settlements are home to more than half the urban population. Because of their spontaneous and unplanned nature, such neigh bourhoods represent one of the most complex challenges worldwide as they are characterised by a lack of basic services and infrastructure, poorly performing building materials, without any building plans approved and often on illegally accessed and hazardous land. As a result, informal settlements are more vulner able to natural disasters (such as storms, flash floods). Nevertheless, they show a unique resilience (social resilience) and natural capacity to cope with these catastrophic events. It can be argued that vulnerability and resilience can co-exist in those settlements. This research has two key objectives: mapping and characterizing vulnerability to natural hazards in spontaneous settlements and co-designing adaptive solutions for self-help housing upgrading in such neigh bourhoods. By combining drone imagery with collaborative mapping, the authors suggest a new approach to enhance community resilience and co-produce practical strategies for disaster management in South African informal settlements. Participatory action research methods are followed to co-produce knowledge with local residents, in a case study in the Durban Metropolitan area. The ultimate goal of this study is to build capacity in local communities seeking to improve their quality of life and assist local authorities in enhancing their intervention towards more resilient futures. Findings show the importance of a meaningful participation of the local dwellers to co-design solutions tailored to the local context and community's priorities, which lead to adaptive interventions.

Keywords Informal settlements · Self-help housing · Resilience · Collaborative mapping · Co-design

V. Govender (✉) · C. Loggia
School of Built Environment and Development Studies, University of KwaZulu-Natal, Durban, South Africa
e-mail: govenderv3@ukzn.ac.za

T. C. Haupt et al. (eds.), *Construction in 5D: Deconstruction, Digitalization, Disruption, Disaster, Development*, Lecture Notes in Civil Engineering 245,
https://doi.org/10.1007/978-3-030-97748-1_4

1 Introduction

Current estimates show that informal settlements are home to over half of the urban poor in developing nations, with a figure of 61, 7% in the African cities [1]. Despite the global experience of the informal settlement phenomenon, the UNHabitat [2] acknowledged that the phenomenon is not adequately addressed. As a consequence of rapid urbanisation and population growth, informal settlements form a major part of the urban landscape globally and therefore constitute a major challenge [1, 3]. At the same time, more than half of the global population already lives already in urban areas with a significant increase in projections by 2050,this is likely to increase the impact of issues related to, which is inevitably linked to issues of poverty, inadequate infrastructure, housing and poor living conditions [4, 5].

As part of a transdisciplinary research project on informal settlement upgrading,[1] this study proposes a hybrid methodology to map and characterise vulnerability to natural hazards in spontaneous settlements and to co-produce adaptive responses with the local residents. The overarching aim of this research is to build capacity in local communities seeking to improve their quality of life and assist local authorities in enhancing their intervention towards resilient futures. The authors applied community action research tools to co-produce knowledge with the residents in a case study in the Durban Metropolitan area, in South Africa. Findings show the importance of a meaningful participation of the local dwellers to co-design solutions which can adapt to the local context and to community's priorities. Assessing existing skills, informal networks and procurement systems is key to support informal dwellers in the process of upgrading their settlement.

2 Informality and Self-Help Housing

Informal settlements have been defined and interpreted very differently across the Global South. The meaning and connotations behind this concept vary considerably with the context and other specific parameters. According to Srinivas [6] informal settlements may be defined as residential areas which developed without formal legal standing or claim to land. They have some common features such as inadequate services and infrastructure [7]. Informal settlements are characterised by self-help efforts, often illegal, and considered 'informal' as they do not align with prevailing regulations. In the self-help efforts residents make use of the limited resources available to them for the purposes of erecting a shelter on interstitial or marginal land [8] often close to economic, social or survival benefits. Dovey and King [9] recorded that informal settlements often form small pockets of irregular, unconventional, substandard, unregulated human settlements, with few instances where informal settlements

[1] ISULabaNtu (www.isulabantu.org) is a multidisciplinary and collaborative research project aimed at developing an environmental and construction management toolkit for upgrading informal settlements in South Africa.

take up larger geographical area, as in the case of Dharavi in Mumbai, Rocinha in Rio de Janeiro, Kibera in Nairobi, and Ezbet el-Haggana in Cairo. They are an integral part of the city and its economy, yet they are invisible to most in the city.

Unplanned settlements may be defined by certain characteristics: physical, social and legal. Physical characteristics refer to poor services of networks (water, sanitation, electricity, and roads) and inadequate physical infrastructure constructed from a variety of materials including corrugated iron, plastic, clay, timber and metal sheeting [10]. Some literature attempts to distinguish between 'slums' and 'informal settlements'. Slums often refer to physical aspects i.e., buildings, facilities and services (particularly sanitation and waste management) in inner cities that gradually deteriorate into slummed conditions, due to overcrowding. Informal settlements are related mostly to the legal standing of the scheme,namely, settlements that mushroom on vacant land, within and around places of opportunities, without proper planning, building regulations or standard construction methods [11]. In South Africa, however, a clear departure from Apartheid terminology included the term 'slum' being replaced by 'informal settlements' [12]. Most of the definitions emphasize the dwelling type as a dominant feature of informal settlements (e.g., shacks built of temporary and poor materials) [13]. Moreover, several definitions refer to ownership of the land, the nature of land tenure and formal demarcation. The SA Department of Human Settlements [14] defines informal settlements on the basis of the following characteristics:

- illegality and informality;
- inappropriate locations;
- restricted public and private sector investment;
- poverty and vulnerability; and
- social stress.

3 Vulnerability and Resilience in Informal Settlements

According to UN-Habitat [1], most of the urban population in Africa live in slums—striving for better job opportunities and decent living conditions. This figure is predicted to rise significantly by 2030 [2], with negative impacts on the vulnerability of urban areas in terms of natural disasters, warfare, and environmental and health issues related to the poor living conditions of residents. Often built-in high-risk areas, informal settlements are mainly composed of makeshift dwellings that do not comply with building standards. Because of the unplanned nature of the settlement, informal neighbourhoods are characterised by different levels of vulnerability to natural hazard, which range from physical, economic, environmental and social vulnerabilities [15].

It is evident that rapid urbanisation and climate change increase the vulnerability of urban poor [16]. Spontaneous and unplanned settlements are exposed to risk to multiple environmental hazards ranging from flooding, landslides, earthquakes to fire, as well as other risks such as legal and social challenges stemming from lack of access to services, land tenure issues and violence. Despite this evident vulnerability, it can be envisaged an inbuilt resilience of some communities (in terms of social structures/networks, strong bonds and relationships) in responding to natural disasters. It can be argued that vulnerability and resilience can actually co-exist, since the strength of social relationships—social bond- can help lessening community vulnerability [17].

Resilience refers to the capability of individuals, social groups or socio-ecological systems, including towns and cities, to not only survive changes, disturbances, adversities or disasters but also to adapt, innovate and transform into new desirable configurations [18]. Interestingly enough, the concept of resilience incorporates also the ability to self-organize and learn from past experiences and changing or adapting to new needs [19].

The literature on urban resilience is new in the field of the built environment. In fact, the concept of resilience in the urban context was borrowed from Ecology studies referring to the capacity of ecological systems to cope with stresses and disturbances caused by external factors [20]. Most cities around the world are adapting and challenging their urban frameworks to create resilient cities. According to the Rockefeller Foundation's 100 Resilient Cities project,[2] resilience can be defined as '*The capacity of individuals, communities, institutions, businesses and systems within a city to survive, adapt and grow no matter what kinds of chronic stresses and acute shocks they experience*'. The concept of resilient city focuses primarily on disaster management and vulnerability but more recently it has evolved to accommodate a multitude of factors that affect cities today, such as informality.

In fact, upgrading of informal settlements provides a set of challenges for Municipalities, NGOs, communities and built environment Professionals. Issues related to resource constraints, high densities, poor waste and sanitation management, contestation of land and power challenge the everyday life in unplanned settlements. Resilience concepts have been applied to slum upgrading as a strategy to understand risk, disaster management, ecological and organizational management.

Factors such as informality, ecological systems, spatial urban form and spatial planning need to be integrated and studied through a transdisciplinary approach to co-design and co-produce knowledge with the local marginalized communities. Urban resilience will not be built and achieved by Government and professionals alone but will need the active partnership of local communities in defining the vulnerabilities they face and developing strategies to reduce exposure to hazards, reduce sensitivity to impacts and strengthen capacity to cope and adapt [21]. Urban resilience is often confused with sustainability concepts, both influencing policy making decisions, however, resilience focus on more than climate and ecological factors. Resilience encompasses drivers that affect change such as governance and planning, innovation and competiveness and lifestyle and consumption. The main challenge of resilience is to broaden views and go beyond resilience in relation to climate change and natural disasters, but to go further and understand the social, cultural, economic and spatial factors that make up a city [22].

[2] The City Resilience Framework, source: http://100resilientcities.org/resources/.

The concept of resilience is useful in seeking to understand communities and the risk to which they are exposed in a holistic manner, revealing how their economic, social and environmental stresses are interconnected, in the slum context these dynamics are important factors to understand how communities live to create a bottom-up approach towards urban resilience [23].

It is interesting to note the progressive approach proposed by the eThekwini Municipality (2017) for the Durban's resilience strategy as part of the 100 Resilient Cities Programme (100RC). The Durban's strategy recognised the need to rethink new perspectives on informality and accepting it as part of the city. It envisages the collaborative action for informal settlement upgrading as the key Building Resilience Option.

Spontaneous settlements are such dynamic spaces that change continuously and require appropriate planning strategies that involve (in a meaningful way) the residents. Currently, the housing targets (performance goals) are reducing the approach to informal settlements to a mere set of numbers (delivery target). Thus, the key strategy to address this challenge is represented by collaboration and partnership between the local Government and all the other stakeholders (including NGOs, CBOs and academia). There is a lack of understanding about the dynamics of informal settlements and there is need for coordinating all the interventions.

4 A Hybrid Methodology to Map Vulnerability and Resilience

UN-Habitat [24] stated there is urgent need for a radical re-imagination of African approaches to urbanism, to better understand and redirect the ongoing urban transitions and to improve the living and working conditions of inhabitants. In addition, Parnell et al. ([25], p. 233) claim the need to formulate new approaches to planning and development that are '... grounded in the realities of urban settlement in South Africa'. Given the dynamic nature of informal settlements, under continuous change, the maps usually produced to represent a settlement are mostly outdated and quite hard to understand for non-professionals (e.g. GIS maps). Moreover, it would be too expensive to get live satellite images on a monthly (or even weekly basis) for all the existing unplanned settlements.

Monitoring informal settlement expansion is fundamental for mitigating the risks related to population growth and economic disparities. Given the extreme complexity of these spontaneous spaces—interdisciplinary and mixed method approaches that combine qualitative and quantitative data have proved to be more effective. According to Hachmann et al. [26], a spatial tool for informal settlement upgrading should empower the residents through meaningful participation in data gathering and interpretation. Informal settlement mapping can be an effective tool for communities to document their structures, making 'the invisible visible', and in doing so can protect them from evictions [27]. As pointed out by Dobson et al. [21], low-income informal dwellers can make a meaningful contribution to urban resilience and adaptive capacity through several participatory initiatives—such as savings groups, self-enumeration and profiling of the settlements.

In response to this, the authors suggest a hybrid methodology (Loggia & Govender, [28] which combines the use of Unmanned Aerial Vehicle (UAV)—drone technology—with community mapping exercises. The first part of the proposed mapping process involves *sensing the place*; this is a transect walk to trigger individual perceptions and affective responses. This fundamental phase assists in understanding what underlying social, historical and personal connotations are associated to spaces that define or can help defining space. The transect walk also enables first-hand experiences of available materials and resources that exist within the site that might considered for recycling and re-use. Then, detailed maps, 3D topographical models and Birdeye views of the settlement are generated by means of the drone technology, with a 15-min drone-survey. In particular, for this study a quadcopter drone, called the DJI mavic pro, which utilizes a standard 12 MP/4 k camera has been used. The drone software used to chart the mission around the sites are Drone deploy and 3D_survey. The reconstructed 3D-survey map shows immense detail and real-time imagery since the resolution of the picture is big enough to reprint and analyse, which can assist municipality and communities in understanding factors such as topography, climatic condition, spatial elevation heights, available materials and structures that can be analysed qualitatively from the bird's eye view, which can be key in understanding the actual layout of informal settlements and the climatic and environmental conditions that affect it. The drone generated maps and images can assist the community as well as the municipality in understanding how heavy rain or fires can affect the site area.

The drone-generated maps, printed at the scale of the settlement, have proved to be effective and readable by non-professional participants since the quality of the images (both 2- and 3-dimensional) allows for an easy depiction of each shack. The most crucial phase of this methodology is the collaborative mapping exercise, whereby the drone-generated maps are combined with the insights of participants in a 'lived 4Dmap'. This process allows to fully understand and analyse the projected needs of the community and consider potential adaption to new requirements, through a resilience perspective. These maps provide not only a real-time representation of the settlement, but also reveal new, critical attributes (e.g. socio-economic, cultural, political dynamics, etc.), which refer to the invisible fourth 'lived dimension' (hence called '4Dmap') [28].

Community action research was applied to this study as a methodology that involves the community as integral part of the data gathering and analysis. This approach builds relationships with relevant stakeholders in the community to explore and develop solutions to local problems [29]. The community action research also reveals the spatial, social and political dynamics that exist within settlements.

In this study, one of the first methodologies applied through action research was community mapping. Community mapping is relatively new and focuses on how the local inhabitants identify and record the elements in their neighbourhood that they wish to focus on. Community mapping allows the community to produce a visual depiction of an area that reflects their social, environmental and cultural resources and is an important tool for community learning and planning [30]. The maps created through community mapping are a visual tool that includes marginalized communities in the decision-making processes, by sharing their lived experiences, raising awareness and co-producing knowledge [30]. Some of the advantages of the community mapping is understanding how the community relates to environment, availability of resources, skills and level of resilience.

As shown by Williams et al. [16] the co-production process leading to the creation and design of the collective causal map generates a sense of ownership amongst the participants, enhancing the credibility of the results.

5 A Case Study in Durban

The above-mentioned hybrid methodology was applied in an informal settlement called Havelock, in Durban. Located in the northern suburb of Durban called Greenwood Park, Havelock is a very dense settlement of 400 residents living in shacks, surrounded by formal housing and located on a steep hillside comprising a mixture of municipality and privately-owned land. It has several environmental hazards, including illegal connection to electricity, flooding, naked electrical wires around the settlement, and a polluted stream with a sewage pipe at the lower end of the settlement. Nevertheless, the strength of this settlement is on its mobilised community, which is actively campaigning for an improvement to their living conditions.

A team of trained community researchers (CRs), selected amongst the local residents in Havelock settlement, participated in the process of collaborative mapping and co-design of the resilient dwelling, highlighting the community's priorities, main challenges but also the strengths in terms of knowledge and skills. Four participatory workshops have been conducted by the authors and the CRs, between October 2018 and May 2019, as part of Phase 5 of the ISULABANTU project.

The CRs clearly indicated as their favourite priority the re-blocking of the settlement and they discussed the key challenges related to that, namely the issue of land tenure—part of that is privately owned—the environmental challenges—steep and dense site prone to natural hazards—floods, fires and mudslides. They said that they would like to start with a small (pilot) project just to show how things can be done, to mobilize the rest of the community. For example, one of the respondents said: "*[…] I think we can start by cleaning the stream because it not expensive and it is affecting many dwellings.*"

Unfortunately, in such spontaneous settlements there is still a rooted dependency syndrome whereby people are expecting that the municipality does things *for* them (delivering houses and services) instead of being proactive and doing the development themselves. That is the reason why the CRs indicated the need for a pilot project to use as a showcase for the rest of the community (Figs. 1 and 2).

Fig. 1 Aerial view of the Havelock settlement and the river affecting dwellings

Fig. 2 Aerial view of the Havelock settlement and the river affecting dwellings

An initial drone flight was done to start preliminary drawing and to get a live aerial view of the settlement. A transect walk was also done with members of the community so that the researchers understood the features of the settlement as well as familiarized themselves with the aerial map. The transect walk revealed first-hand experiences of the community, revealing their issues as well as their resources available. This data was then visually represented onto the drone map, making it the '*lived*' experience of the community. This map then served as the base map of all attributes of the community for future discussions.

By using the drone map of the settlement as a base map, CRs mapped the key attributes of the settlement and identified the developable areas around Havelock.

Through heavy rains and flooding, a near stream was found to be blocked, this caused many shacks to be affected. A second flight was conducted immediately after the event to assess the damage to structures as well as act as a visual representation, mapping the hazards affecting the community.

The overlay of the new data onto the existing map, showed how social, recreational and economic spaces of the settlement were affected by the flooding. The map now served as a tool to mediate between municipality and the community for disaster relief and dialogue. For the community, the drone-based map was in fact a proof that their structures were affected by the flood. The drone flight also captured a video to show the blockages of the stream and the extent of damage.

The extent of damage was mapped as a visual line on the map, showing the flood plain of the river, this information affects the spatial layout of the settlement. The community can now plan their layout with this line in mind and also other hazards that were mapped. The maps also served the basis for the reblocking exercise.

CRs showed a good understanding of the process and the related issues and seemed very determined in doing something that can be used as a pilot for the rest of the community. They said that it is crucial to show the Havelock residents that reblocking can be effective. For them the drone-based maps and the various pictures and videos taken with the drone are useful instruments to negotiate with the local authorities.

Community researchers completed a site analysis by mapping vulnerabilities and attributes of the site. Then a reblocking exercise was initiated to respond to these vulnerabilities and attributes, adapting to suit the community's needs. The community prioritized their needs and what issues should be tackled first. Environmental hazards such as the stream, spatial organization and reblocking were some of the challenges noted. Interestingly, one of the community researchers highlighted the importance of public space and walking paths '*to make the settlement conducive*'. Land use management was then discussed further, contestation of land as well as describing the surrounding plots. This allowed the researchers as well as the community to understand what is needed for the reblocking process. One noted problem with reblocking is that it has to be a community effort and requires a buy in from all. Most times, the reblocking is not accepted by the community because of political influence or a lack of understanding. However, one of the community researchers argued that:

> [...] if we can agree with some community members and do reblocking just to show other people, that's where the whole community can agree with reblocking, because they want to see it rather than being told about it.

The researchers understood that the reblocking exercise should be tested through a visual methodology tool such as model building first, so that the community could engage, make suggestions and visualize the changes proposed in their settlement.

There was consensus amongst the community that the co-production process that they went through has empowered them to make decisions on their spaces and homes. At the same time, this process revealed the in-built resilience of the dwellers to respond to disasters and challenges, as a result of their self-organization and adaptive capacity.

6 Co-Designing Adaptive Solutions for Self-Help Housing

The '*4D*' lived map co-produced with the community researchers acted as a design generator or integrated toolkit of available resources, experiences, skills, materials and climatic conditions that inform the co-design of a climate responsive unit. The co-design process is based on community collaboration and inputs aimed at creating an architecture that responds to the socio-economic needs of the community, as well as being resilient to climatic conditions. Thus, the ultimate aim of this methodology is to explore how the culture of *informal* communities can inform the dwelling design as a means of creating better and more responsive upgrading interventions in spontaneous and unplanned settlements.

The co-design process entailed a workshop and two brainstorming sessions where the *lived 4D* map produced, forms the basis of the design. Through first hand experiences and accurate representations of the site, a 'climate proof dwelling' was co-designed with local residents based on the availability of resources (materials and skills) and in response to the environmental context. This ensured community capacity building as well as a socially acceptable and inclusive design that promotes urban resilience (Figs. 3 and 4).

The dwelling's design chosen is simple, easy to build and cheap. As emerged during the participatory workshops, preferably the dwelling should use materials that are easily available in such contexts and, more in particular, the potential of reusing

Fig. 3 Two co-designed typologies for a climate proof dwelling in Havelock informal settlement

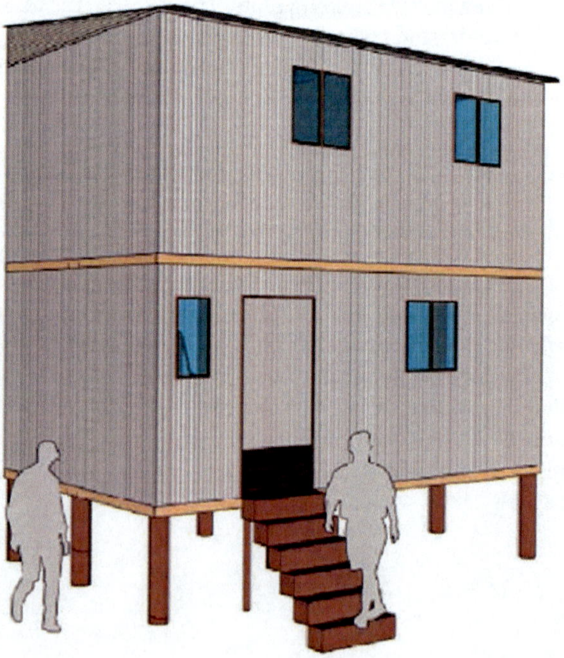

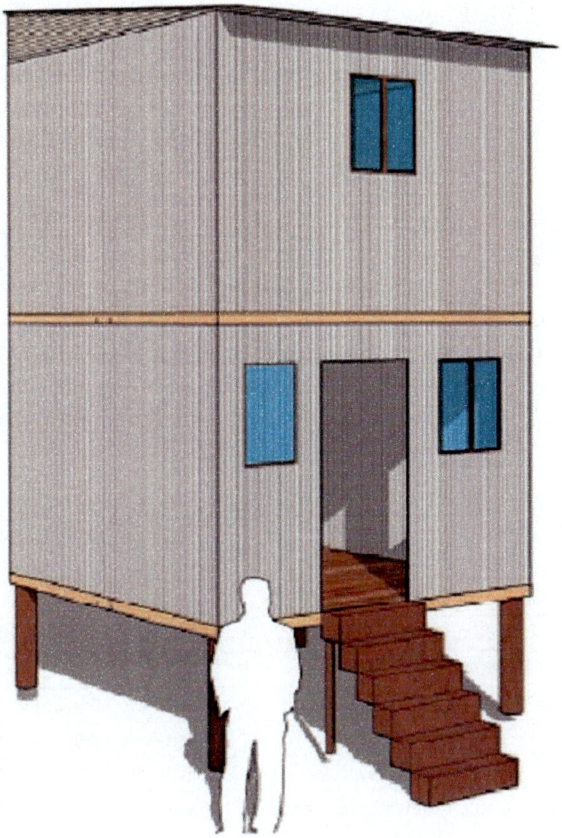

Fig. 4 Two co-designed typologies for a climate proof dwelling in Havelock informal settlement

urban waste should be explored. For example, it is important to select materials that are durable, easily available and with good performances in terms of fire resistance and thermal comfort. Many examples of settlement prototypes were looked at, based on how they can respond to climate, topography, resources available and community self-building capacity. A stilt 'climate proof dwelling' was proposed to respond to the steep topography of the settlement. By raising it off the ground, the structure answers to the heavy rains and flooding that affect the settlement. It was also observed that a lack of collective space is a challenge in high density settlements. This space can be used by local dwellers or social, economic and recreational activities. By picking the proposed unit off the ground, the underneath space can be used by the community to generate an income or for family purposes. The materials used on the dwelling also were chosen for their availability and performance; since one of the main problems in these dense settlements is fire risk, the materials chosen were suitably rated for fire hazards.

Two typologies were looked at, one to accommodate larger families (40 sqm) and one for smaller households (20 sqm).

The structures were designed with material performance and cost in mind. This list was then shown to the community and changed according to their availability, cost and re-use properties. Some of the material proposed are (Table 1):

The structures also show potential for family expansion. Often families have no space to expand, the space beneath could be used as an additional room for the family. The structure is also designed to be flexible. The frame can be added onto to expand the configuration to create and additional room.

The overall dimensions of the structure were designed to fit exactly the rolled sheets of corrugated iron, eliminating the cost of cutting and joining for the community. A co-design workshop was held with the community to engage on the advantages and disadvantages of the proposed structure and select the best available materials to build it. This helped to realize the limitations of the design as well as adapting

Table 1 Materials proposed for the structure

200 Mmm Ø Gum Pole	Easily available and easy to cut to suit to the site conditions
Corrugated sheeting and flashing	IBR sheeting being the best material to use
150 × 38 mm timber sections	Easily available at local hardware store
114X38mm timber sections	Easily available at local hardware store
76X76mm timber sections	Easily available at local hardware store
76 × 38 mm timber sections	Easily available at local hardware store
38 × 38 mm timber sections	Easily available at local hardware store
150 × 50 mm timber sections	Easily available at local hardware store
250 × 50 mm timber sections	Easily available at local hardware store
Shutter board (Floor)	Can be substituted with Masonite or plywood sheets
Shutter board or equal (Wall)	Can be substituted with Masonite boards
38 mm thick timber board	Easily available at local hardware store
Mild steel angle 75 × 75 mm	Easily available at local hardware store
Mild steel angle 75X50mm	Easily available at local hardware store
30 mmØ mild steel tube	Expensive material, both sourcing and cost
10 mmØ mild steel rod	Expensive material, both sourcing and cost
Windows	Can be substituted with thrown away window frames
Door	Can be substituted with thrown away wood panels or constructed by community
Aluminum Louvre	Expensive material, both sourcing and cost
Concrete and reinforcing steel	Expensive material, both sourcing and cost
Aluminum gutter and down pipe	Can be substituted or omitted

this to the community needs and key priority issues. The workshop also consisted of constructing prototype scale models to show the community the necessary steps taken to construct the dwelling previously designed. The process of construction was simulated through the scaled models and discussions about material procurement and skills enhancement have been undertook.

The simulation of building a model, allowed the community researchers to realize the steps necessary to build a climate proof dwelling, as well as understanding the correct building practices to use when constructing the structure. The model also served as a visual communication tool which allowed the community to substitute materials as well as construction processes to suit their needs (Figs. 5 and 6).

Fig. 5 Model building workshop to simulate the construction process of the 'climate proof dwelling'

Fig. 6 Model building workshop to simulate the construction process of the 'climate proof dwelling'

7 Conclusion and Recommendations

Spontaneous and unplanned settlements are prone to a wide range of natural hazards (e.g. fire, floods, landslides, etc.).

This study showed how to map and characterize vulnerability to natural hazards in an informal settlement in Durban and co-design adaptive solutions with the local residents. By combining drone imagery with collaborative mapping, the authors suggest a new hybrid approach to enhance community resilience and co-produce practical strategies for disaster management in South African informal settlements. A key part of the mapping methodology is the co-production process, whereby different stakeholders—with diverse levels of knowledge- work together towards improving the living conditions of the informal dwellers. The co-produced maps proved to be useful tools to negotiate the level of intervention with the local government.

A climate proof dwelling was co-designed with trained community researchers to respond to actual needs and priorities of the local (informal) community. The simple design proposed offers a flexible response to the environmental and construction constraints faced by the informal dwellers.

Findings show the importance of a meaningful participation of the marginalised community to better understand the contextual priorities and shape more sustainable and resilient interventions. The co-production process between academia, local community, NGO and local government revealed the crucial role of the university in supporting local residents while negotiating with the municipality. In fact, this process has led to an upcoming pilot project with the Municipality in response to one of the biggest challenges identified by the community in the participatory workshops, namely the river rehabilitation. This, in turn, demonstrates the efficacy of the applied action research methods.

Complex dynamics within the community and settlement politics hinder the communication between informal dwellers and local government and in turn delay the response of the latter to the real needs of the residents.

It is therefore fundamental to build capacity and invest in further training in both communities and local authorities by understanding the minimum preconditions that unlock community participation in an upgrading project.

Acknowledgements The study is part of two ongoing research-funded projects: a Royal Society Newton Advanced Fellowship (ref.NA150082) and an Economic and Social Research Council (ESRC) award as part of the Newton Fund, and in collaboration with the National Research Foundation (NRF) (ref.UTSA150520118179/n.101581) in South Africa (www.isulabantu.org). The authors would like to thank Dr Judith Ojo-Aromokudu for her precious assistance and advice, the community researchers in Havelock for being our co-investigators in this study and the officers of uTshani Fund (SASDI alliance) for their important assistance and support during the fieldwork activities.

References

1. UN-Habitat (2015) Habitat III issue papers. 22—informal settlements. New York, 31 May 2015. Available at: https://unhabitat.org/wp-content/uploads/2015/04/Habitat-III-Issue-Paper-22_Informal-Settlements.pdf [Accessed 14 March 2019]
2. UN-Habitat (2013) State of the world's cities 2012/2013 prosperity of cities. New York: Routledge. Available at: https://sustainabledevelopment.un.org/content/documents/745habitat.pdf [Accessed 14 March 2019]
3. Marutlulle NK (2017) Causes of informal settlements in Ekurhuleni metropolitan municipality: an exploration. Africa's Public Serv Deliv Perform Rev 5(1):a131. https://doi.org/10.4102/apsdpr.v5i1.131
4. Majale M (2008) Employment creation through participatory urban planning and slum upgrade: the case of Kitale, Kenya. Habitat Int 32(2):270–282. https://doi.org/10.1016/j.habitatint.2007.08.005
5. El Menshawy A, Aly SS, Salman AM (2011) Sustainable upgrading of informal settlements in the developing world, case study: Ezzbet Abd El Meniem Riyadh, Alexandria, Egypt. Procedia Eng 21:168–177
6. Srinivas H (1991) Viability of informal credit to finance low-income housing: case study of three squatter settlements in Bangalore, India, Unpublished Masters Thesis Report. Bangkok: Division of Human Settlements Development, Asian Institute of Technology
7. Klug N, Vawda S (2009) Upgrading of informal settlements: an assessment with reference to the application of 'Breaking New Ground' in Cape Town and Johannesburg, town and regional planning, 54. Available: http://journals.ufs.ac.za/index.php/trp/article/view/603/601. [Accessed 23 April 2019]
8. Ojo-Aromokudu JT, Loggia C (2017) Self-help consolidation challenges in low-income housing in South Africa. J Constr Proj Manage Innov (JCPMI) 7(1):1843–1858. University of Johannesburg
9. Dovey K, King R (2011) Forms of informality: morphology and visibility of informal settlements. Built Environ 37(1):11–29. Retrieved from http://www.jstor.org/stable/23289768
10. Marx C, Charlton S (2003) Urban Slums report: the case of Durban, South Africa (online). London: Global Report. Available: https://www.ucl.ac.uk/dpuprojects/Global_Report/pdfs/Durban.pdf [Accessed 14 March 2019]
11. Khalifa MA (2015) Evolution of informal settlements upgrading strategies in Egypt: from negligence to participatory development. Ain Shams Eng J 6:1151–1159. https://doi.org/10.1016/j.asej.2015.04.008
12. Huchzermeyer M (2011) Cities with "Slums": from informal settlement eradication to a right to the city in Africa. UCT Press, Claremont
13. Housing Development Agency (HDA) (2013) KwaZulu-Natal: informal settlements status. Research series published by the housing development agency. http://thehda.co.za/pdf/uploads/multimedia/HDA_KwaZuluNatal_Report_lr.pdf. [Accessed 23 April 2019]
14. Department of Human Settlements (DoHS) National Housing Code (2009) Available at: http://sacitiesnetwork.co.za/wp-content/uploads/2014/05/Part-3-NHC-2009-UpgradingInformal-Settlements.pdf. [Accessed 23 April 2019]
15. Risks (2018) Procedia engineering 212:238–245. https://doi.org/10.1016/j.proeng.2018.01.031
16. Williams D, Manez Costa M, Sutherland C, Celliers L, Scheffran J (2019) Vulnerability of informal settlements in the context of rapid urbanization and climate change. Environ Urbanization. 31(1):157–176. https://doi.org/10.1177/0956247818819694
17. Usamah M, Handmer J, Mitchell D, Ahmed I (2014) Can the vulnerable be resilient? Coexistence of vulnerability and disaster resilience: informal settlements in the Philippines. Int J Disaster Risk Reduction 10:178–189. https://doi.org/10.1016/j.ijdrr.2014.08.007
18. City-region Observatory (2014). Available at http://www.gcro.ac.za/media/reports/URreport_1201MR_9BDHWFr_cUW0tn0.pdf [Accessed 14 March 2019]
19. Shaw K (2012) The rise of the resilient local authority? Local Gov Stud 38(3):281–300. https://doi.org/10.1080/03003930.2011.642869

20. Chmutina K, Lizarralde G, Dainty A, Bosher L (2016) Unpacking resilience policy discourse. Cities 58:70–79. https://doi.org/10.1016/j.cities.2016.05.017
21. Dobson S, Nyamweru H, Dodman D (2015) Local and participatory approaches to building resilience in informal settlements in Uganda. Environ Urban 27(2):605–620. https://doi.org/10.1177/0956247815598520
22. Wikström A (2013) The challenge of change: planning for social urban resilience. An analysis of contemporary planning aims and practice. Unpublished master's dissertation in urban and regional planning, Department of Human Geography, Stockholm University. Available at: https://www.diva-portal.org/smash/get/diva2:636112/FULLTEXT01.pdf [Accessed 23 April 2018]
23. Woolf S, Twigg J, Parikh P, Karaoglou A, Cheaib T (2016) Towards measurable resilience: a novel framework tool for the assessment of resilience levels in slums. Int J Disaster Risk Reduction 19:280–302. https://doi.org/10.1016/j.ijdrr.2016.08.003
24. UN-Habitat (2014) The state of African cities 2014. Re-imagining sustainable urban transitions. United Nations human settlements programme, UN-Habitat, Nairobi, Kenya. Available at: https://unhabitat.org/books/state-of-african-cities-2014-reimagining-sustainable-urban-transitions/ [Accessed 23 April 2019]
25. Parnell S, Pieterse E, Watson V (2009) Planning for cities in the global South: a research agenda for sustainable human settlements. Prog Plan 72:232–240. https://doi.org/10.1016/j.progress.2009.09.001
26. Hachmann S, Arsanjani JJ, Vaz E (2018) Spatial data for slum upgrading: volunteered geographic Information and the role of citizen science. Habitat Int 72:18–26. https://doi.org/10.1016/j.habitatint.2017.04.011
27. Beukes A (2015) Making the invisible visible: generating data on 'Slums' at local, city and global scales. London, UK: International Institute for Environment and Development. https://knowyourcity.info/wp-content/uploads/2016/02/10757IIED-1.pdf [Accessed 23 April 2019]
28. Loggia C, Govender V (2020) A hybrid methodology to map informal settlements in Durban, South Africa. Proceedings of the institution of civil engineers—engineering sustainability 173(5):257–268. https://doi.org/10.1680/jensu.19.00005
29. Ozanne JL, Anderson L (2010) Community action research. J Public Policy Mark 29(1):123–137. https://doi.org/10.1509/jppm.29.1.123
30. Fahy F, Cinneida MO (2009) Re-constructing the urban landscape through community mapping: an attractive prospect for sustainability? Area 41(1):167–175. https://doi.org/10.1111/j.1475-4762.2008.00860.x

Risk Assessment in Construction Businesses

Mbangi Nozipho Zipporah, Brink Botha, and Asianoah Rexford Kofi

Abstract The South African construction industry is growing, where new contracting businesses are rapidly emerging. However, statistics show that some of these emerging businesses seem to fail on a more rapid rate. Studies have shown that adequate risk management has a huge impact on sustaining and increasing both productivity and profitability in construction businesses. The rise of the 4th industrial revolution has brought about automated technology and, it is moving the industry forward, simplifying tasks and changing the construction industry as a way of assessing and managing risk. This study aimed to determine the effects of risk and the effective ways of conducting risk assessment in construction businesses. This exercise may help to establish efficiency in construction business functional areas in order to maximise productivity. The study further highlights on the importance of using BIM (AI) as a tool to assess risks in improving construction business efficiency. Quantitative methodology is used to conduct the study where questionnaires were administered to receive field data from respondents. The study is analysed based on the perspective of the various construction contractors in small and medium sized construction businesses in Port Elizabeth, Republic of South Africa. The purpose is to create an understanding on why most of the construction businesses fail in their operations and establish whether the solution could be the effective implementation of BIM as risk assessment tool.

Keywords Business · Construction · Management and Risk

M. N. Zipporah (✉) · B. Botha
Faculty of Engineering, The Built Environment and Information Technology, Nelson Mandela University, Port Elizabeth, South Africa
e-mail: S214274209@mandela.ac.za

A. R. Kofi
Koforidua Technical University, Koforidua, Ghana

T. C. Haupt et al. (eds.), *Construction in 5D: Deconstruction, Digitalization, Disruption, Disaster, Development*, Lecture Notes in Civil Engineering 245,
https://doi.org/10.1007/978-3-030-97748-1_5

1 Introduction

Construction business is highly risky due to the project environment issues and its complexity in nature. It is due to the involvement of many parties such as owners, contractors, subcontractors and suppliers in the construction business. In addition, there are other risk factors including political, social and economic that affect construction businesses. In practice, construction project managers usually adopt reactive, semi-permanent, casual and unstructured risk assessment systems and this attribute to the lack of capacity and competence in managing risks appropriately. The main barriers that are found for the implementation of ineffective risk management systems are the lack of formality of the system and the lack of integrative mechanisms of risk management among the parties involved in the project [1]. Construction businesses in developing countries approach to risk assessment and risk management over the years are known to be inadequate [1]. This has resulted into limited success in several construction projects. Hence there is a need to adopt a credible and adequate risk assessment system that can be used to assess and manage risk in construction businesses. Conducting effective risk assessment among construction businesses will provide contribution to the body of knowledge in construction sector.

Building Information Modeling (BIM) as an emerging digital technology has been increasingly adopted in many first world countries to support construction project life cycle. It plays a role in facilitating risk assessment, design management, construction, and maintenance phase of projects [2]. Different models for risk assessment processes have been proposed in literature by different researchers. The intended output of this study framework is to determine the importance of conducting risk assessment and to establish whether the use of BIM can help in controlling risks to an acceptable degree in construction businesses. The research further narrows the discussion into the challenges encountered by risk management personnel when implementing a risk management system [3].

2 Review of the Literature

Achieving success in construction business is a key factor and that is why project managers recognize that risk management is essential to carry out a good project management [4]. One can assert that old data cannot give a comprehensive and accurate view of tomorrow's project. However potential solution to the critical success factors affects construction projects during both project planning and execution phase [5]. Risk assessment and risk management are usually based on information collected from traditional source or analogy to well-known cases, common sense perspective, results of experiments or tests and reviewing of inadvertent exposure. Once the database exists, it can be processed and mined some useful information to help managers to analyse risks and make decisions [6]. The application of BIM has significant benefits in the construction processes by ensuring effective cost and resource saving, shorter project life cycle, improving project communication, ensuring quality and the like. This may imply that the adoption of BIM could help to assess and manage risks in construction process.

The combination of machine learning with simulation can help risk managers to improve the evaluation and simulation of risk [7]. Artificial Intelligence (AI) improves the quantitative risk assessment through understanding the risk emergence in a specific project context [8]. Every industry needs to understand where it is in terms of technological maturity in order to choose appropriate techniques that can help to improve it [9]. Project success on the other hand can be defined as the act of meeting goals and objectives as prescribed in the project plan. By identifying possible delay factors, there is a better chance to manage and control possible causes of risks throughout the project life cycle [4].

The existence of project delay problem is related to the interdependent factors that affect construction projects and construction activities. Providing an efficient tool for assessing delay factors is on the other hand the key for estimating an accurate duration in construction projects. Several statistical models and sample tests were developed to predict the cost and duration of construction business projects [5]. An important force for strengthening the overall national strength by promoting political, economic, social development and ensuring the improvement of people's livelihood, businesses to satisfy common interests of the society through construction projects is critical [10]. Therefore, risk management activities must be set out and to become habitual routine work within construction businesses. Also, risk management should be expanded to other construction projects within every organisation. It is vital to note that human error, inadequate communication and lack of adequate risk assessment among others in the construction sector are the key challenges project managers face regularly. In order to address these challenges, one can argue that a tailored risk assessment programme should be designed and run for construction project managers so that well technical reviews can be conducted at each project phase [13].

Multi-disciplinary knowledge in project management, experience, and mathematical analysis play a key role in risk assessment [14]. Project risk assessor's ability to demonstrate such skills give him/her advantage to employ the use of BIM and to conduct PESTEL analysis in determining and calculating risks that affect construction businesses. This promotes health and functionality in the construction business both in the medium and long term.

3 Research Methodology

3.1 Research Method and Sample Stratum Analysis

This study follows a quantitative data collection approach and analysis procedures were conducted sequentially. A structured questionnaire instrument consisting close and open-ended questions was designed for data collection. A pilot test was conducted and the questionnaire was revised for the main data collection of the study in order to ensure validity and reliability. With the data collection procedures and instrument distribution, 125 questionnaires were distributed to the respondents (construction contractors) and 95 of them were received as credible representing 76% out of the total. The researcher adopted emailed approach in administering and receiving questionnaires from the respondents. This approach is viewed as best to prevent COVID-19 spreading that might have affected the researcher and the study participants. Questionnaires were received on a stipulated date, being approximately 72 h after the date of drop-off. Descriptive statistics was employed. The findings of this paper include the viewpoints of almost all the respondents with the use of the following decision rules of the Mean Scores (MSs).

$<1.00 \leq 1.44$ Strongly disagree/Never.
$<1.45 \leq 2.44$ Disagree/Rarely.
$<2.45 \leq 3.44$ Neutral/Undecided/Sometimes.
$<3.45 \leq 4.44$ Agree/Often.
$<4.45 \leq 5.00$ Strongly agree/Always.

Approximately 65% of the respondents have tertiary built environment qualification whereas 35% took short courses in their related fields as far as built environment is concerned. The breakdown of the tertiary qualification percentage is made up of 23.5%-National Diploma holders, 32.5%-BSc Degree, and 9%-Master's degree. Seventy-Six percent (76.0%) of the respondents confirmed that they were registered members of the built environment professionals such as SAPOA, SACPVP, EAAB, SACQSP and RICS whiles 24.0% were registered with the NHBRC. This illustrates that the population sample of this study has a good representation of people who are professionally registered.

4 Limitations

The main sources of data input for this study includes but not limited to data gathered from the literature and survey of questionnaires administered to a group of qualified contractors in the construction industry as respondents of the study. Theses respondents were chosen in Port Elizabeth city, South Africa. They include main contractors and sub-contractors who undertake construction projects. Furthermore, the study focuses on the small to medium construction businesses operators. One Hundred and Twenty-Five (125) respondents were sample sized through the uses of systematic sampling technique due to the nature of the study. The registered construction contractors lists obtained from the various Municipalities in Port Elizabeth were used as sampling frame. Such number of respondents chosen helped in obtaining a true view from the respondents as far as this study is concerned.

5 Discussion of Findings

This section discusses the findings of the study in using BIM as risk assessment tool (from Tables 1, 2, 3, 4 and 5) against PESTEL analysis framework (Table 6). Based on the respondents qualification results analysis above, it is pre-supposes that majority of the respondents may know the importance of BIM and the concept of PESTEL analysis.

Table 1 illustrates the level of discord between an effective risk assessment system and high occurrence of insolvency among construction businesses on a scale of 1 (very rare) to 5 (very often) and a MS ranging between 1.00 and 5.00. The MS >2.00 < 4.00 where 65% claim to have an 'effective risk assessment' however 53% deal

Table 1 Health and functionality of business

Parameter	Unsure		Very Rare		Very Often		MS	Rank
			Frequency (%)					
Effective risk assessment	0.0	0.0	11.8	11.8	58.8	5.9	**3.67**	**1**
Experience scope creeps	0.0	0.0	23.5	23.5	41.2	11.8	**3.41**	**2**
Delayed by third Party	0.0	11.8	23.5	17.7	35.3	11.8	**3.12**	**3**
Experience cost Overruns	0.0	5.9	35.3	11.8	52.9	0.0	**3.06**	**4**
Measurement throughout all Department	11.8	0.0	23.5	29.4	29.4	5.9	**2.82**	**5**
Experience schedule Delays	0.0	23.5	29.4	11.8	23.5	11.8	**2.71**	**6**
Almost went insolvent	0.0	23.5	41.2	17.7	11.8	5.9	**2.35**	**7**

Table 2 Effectiveness of risk calculation methods

Parameter	Unsure	Not good		Very good			MS	Rank
		Frequency (%)						
Anticipating and managing emerging risks	0.0	0.0	0.0	17.7	58.8	23.5	**4.06**	**1**
Clearly articulating risk appetite and tolerances	17.7	0.0	0.0	29.4	47.1	17.7	**3.88**	**2**
Taking action on identified important and relevant risks	0.0	0.0	0.0	17.7	70.6	5.9	**3.87**	**3**
Determine each project cost with high accuracy	0.0	0.0	11.8	5.9	70.6	11.8	**3.82**	**4**
Accurately estimate the required time to complete the project	0.0	0.0	11.8	5.9	70.6	11.8	**3.82**	**5**
Linking risk management with corporate strategy and planning	23.5	0.0	5.9	17.7	41.2	11.8	**2.88**	**6**
Instilling awareness of risk as a decision-making discipline	17.7	0.0	11.8	41.2	23.5	5.9	**2.71**	**7**

Table 3 Effectiveness of project knowledge in business

Parameter	Unsure		Minor	Major			MS	Rank
			Frequency (%)					
Project quality management	0.0	0.0	0.0	23.5	47.1	29.4	**4.05**	**1**
Project cost management	11.8	0.0	0.0	5.9	35.3	47.1	**3.94**	**2**
Project time management	0.0	0.0	0.0	23.5	58.1	17.7	**3.94**	**3**
Project risk management	0.0	0.0	0.0	23.5	64.7	11.8	**3.88**	**4**
Project communication management	23.5	0.0	5.9	17.7	35.3	17.7	**2.95**	**5**
Project scope management	29.4	0.0	0.0	11.8	47.1	11.8	**2.82**	**6**
Project integration management	35.3	0.0	0.0	23.5	29.4	11.8	**2.47**	**7**
Project human resource management	29.4	0.0	11.8	23.5	23.5	11.8	**2.47**	**8**

Table 4 Risk assessment measurement

Parameter	Unsure	Very Rare		Very Often			MS	Rank
		Frequency (%)						
Use a risk assessment manual/template	5.9	0.0	23.5	0.0	58.9	29.4	**3.65**	**1**
Mis-measurement of known risks	17.7	17.7	11.8	47.1	5.9	0.0	**2.06**	**2**

with constant scope creeps. The ranking of 5 and MS of 2.82 illustrates relatively high uncertainty with regards to how other functions that are not directly linked to the physical construction of the business. Although the results show a low ranking of 7 and a MS of 2.35 for insolvency, the respondents seemed to be experienced in construction business and those with a regular occurrence of scope creeps, schedule delays and third-party delays.

Table 2 illustrates that what affects the risk assessor the most is his/her inability to instil awareness as a decision-making discipline and linking risk management with the corporate strategy and planning on a scale of 1 (not good) to 5 (very good) and with a MS ranging between 2.45 and 4.44. Although the ranking (4.06) seems high for the business ability to manage risk with appropriate risk measures, this seems to be the case with well established businesses operating on experience. Majority of 70.6% claim accuracy with cost determination and project completion. The lowest ranking with a mean of 2.71 have difficulty instilling risk awareness as a decision-making discipline. Explanation above indicates that risk assessment in construction process is important.

Table 3 illustrates how effective the project knowledge areas are perceived to be on a scale of 1 (minor) to 5 (major) with a MS ranging between 2.45 and 4.44. From the results the researcher deduced that 16.2% did not have in depth knowledge of the ‘knowledge area processes. Not a single respondent answered 1 (minor). The lowest ranking with an average mean of 2.47 is project integration and human

Table 5 Effect of risk management on business

Statement	Response %							
	Unsure	Strongly Disagree	Disagree	Neutral	Agree	Strongly Agree	MS	Rank
Mitigating risk	5.9	0.0	0.0	17.7	58.8	17.7	**3.76**	**4**
Compliance with regulatory and legal requirements	0.0	0.0	0.0	0.0	58.8	41.2	**4.41**	**1**
Accurately estimate the required time to complete the project	0.0	0.0	5.9	23.5	52.9	17.7	**3.83**	**2**
Increasing risk awareness	0.0	0.0	0.0	29.4	58.8	11.8	**3.82**	**3**
Mitigating risk	5.9	0.0	0.0	17.7	58.8	17.7	**3.76**	**4**
Repeat previous reports on similar projects	0.0	0.0	17.4	23.5	52.9	17.7	**3.64**	**5**
Increasing certainty in meeting strategic and operational	3.5	0.0	0.0	29.4	23.5	11.8	**2.41**	**6**

Table 6 Impact of PESTEL analysis on business

Parameter	Unsure	Minor		Major			MS	Rank
		Frequency (%)						
Population demographic	0.0	0.0	0.0	17.7	64.7	17.7	**4.00**	**1**
Technology for day-to-day operations	5.9	0.0	0.0	5.9	76.5	11.8	**3.82**	**2**
Third party sub-contracting	0.0	18.0	0.0	17.7	47.1	23.5	**3.70**	**3**
Current economy fluctuations	5.9	0.0	17.7	11.8	47.1	23.5	**3.56**	**4**
Trading policies	11.8	0.0	11.8	17.7	41.2	17.7	**3.29**	**5**
Project susceptibility to instances of instability	17.7	0.0	11.8	5.9	47.1	17.7	**3.17**	**6**
Globalisation	5.9	0.0	11.8	35.3	29.4	17.7	**3.00**	**7**
Environmental, social and cultural factors	23.5	0.0	11.8	11.8	41.2	11.8	**2.82**	**8**

resource management. It is important to point out that this was made up of mainly 35.3% respondents that don't have a tertiary level qualification with in the built environment and therefore are not familiar with the project lifecycle stages and possibly management processes. Therefore having in-depth knowledge in the project management areas may help project managers to assess risk that may emerge in such areas during operations.

Table 4 illustrates the use of checklists as well as individual experience and experience from previous projects. A combined number of 88% stated their risk assessment is done by means of a checklist 58.9% often and 29.4% very often, and reports are repeated for every project, hardly taking into account the changes resulting by the changing scope of work for each project. A MS score of 2.06 stated a rare occurrence of known mis-measurement of risk.

Table 5 illustrates the effect of risk assessment has on their business on a scale of 1 (strongly agree) to 5 (strongly disagree) and with a MS ranging between 2.00 and 4.44. A 100% of the respondents felt their risk assessment method played a major part in complying with regulations. Accurate project time had a MS score of 3.83 with 70.9% agreeing and strongly agreeing. The results clearly show that effective ways to manage risk helps to ensure project success.

Table 6 illustrates the impact the PESTEL analysis has on their business and from a scale of 1 (minor) to 5 (major) and with a MS ranging between 2.00 and 4.44. An average of 8.8% respondents indicated unsure. The results therefore indicate the respondents fair understanding on PESTEL analysis.

6 Conclusions and Recommendations

Although a construction business belongs to the owners or partners registered for it, however the entire business is linked to the national level and is rated as a whole. An increase in turnover of a business produces benefits to the entire industry and increases the GDP. It is therefore important to improve the risk assessment of businesses with proper strategies and systems in place so as to ensure the growth and sustainability of businesses. The importance of realising exactly which factors to focus and optimise will determine the results and success rate for a construction project and ultimately benefit the entire construction business. A whole different conversation needs to arise as to how BIM could be optimised to suit construction businesses and not only focusing on the design element.

A recommendation for this study includes the integration of BIM in the built environment professional activities. Making management support BIM implementation by devoting the time and resources needed to improve transparency regarding risks is of prime important. Embed BIM at corporate level of construction business to uncover and manage risks with cross-functional levels among professional teams will allow the teams to establish a common language for risk. Defining risks consistently and continuously could help construction business owners to establish common focus on risk with their employees. BIM is uncharted waters and a lot of improvements can be made to maximise the output it yields as far as risk assessment and risk management are concerned. Further research can be conducted on how to develop an effective framework model for improving risk management in construction sector in the Republic of South Africa.

References

1. Choudhry RM, Iqbal K (2013) Identification of risk management system in construction industry in Pakistan. J Manag Eng 29:42–49
2. Goh CS, Abdul-Rahaman H, Abdul Samad Z (2013) Applying risk management workshop for public construction projects: case study. J Constr Eng Manag 139:572–580
3. Rohaninejad M, Bagherpour M (2013) Application of risk analysis within value management: a case study in DAM engineering. J Civ Eng Manag 19:364–374
4. Yang L-R, Chen J-H, Huang C-F (2012) Requirements definition and management practice to improve project outcomes. J Civ Eng Manag 18(1):114–124
5. Avdoshin S, Pesotskaya E (2011) Software risk management. 1–6. https://doi.org/10.1109/CEE-SECR.2011.6188471
6. Assaf SA, Al-Hejji S (2016) Causes of delay in large construction projects. Int J Proj Manag 24:349–357. [CrossRef]
7. Johnsonbabu A (2017) Reinventing the role of project manager in the artificial intelligence era. Project management national conference, India
8. Afshari A, Khosravi S, Ghorbanali A, Borzabadi M, Valipour M (2011) Identification of causes of non-excusable delays of construction projects. Proceedings of the international conference on e-business, management and economics
9. Hammad AAA, Ali SMA, Sweis GJ, Bashir A (2008) Prediction model for construction cost and duration in Jordan. Jordan J Civ Eng 2:250–266
10. Seo J, Lee S, Kim J (2013) Analysis of correlation between construction business and insolvency of construction company. KJCEM 14(3):003–011

Critical Review of the Impacts of Successful BIM Technology Application on Construction Projects

James Olaonipekun Toyin and Modupe Cecilia Mewomo

Abstract Building Information Modelling technology (BIM-t) continues to gain more attention. Its adoption creates a platform that allows the built environment professionals to have a common database for project information sharing. While there is an increased perception/impression that the implementation of BIM-t on construction projects positively influences the construction project delivery, the critical analysis of such impacts is still missing. This paper, therefore, conducts a critical review of the impacts of BIM-t application on construction projects delivery and provides reports on research gaps and possible future research directions. This paper employed a systematic examination of related literature on the subject of BIM-t between the years 2008 to 2021. The search includes published journal articles, thesis, books, documents, and conference proceedings. Different databases including; ResearchGate, Taylor and Francis, ScienceDirect, Springer, and Google scholar were explored. The findings indicate seventh (17) positive impacts gathered from 41 reviewed publications. The listed positive impacts were grouped under the different construction phases. The implications of the findings were discussed and future research directions were suggested.

Keywords BIM-t application · BIM impact · Built environment · Construction project phases

1 Introduction

The increased adoption of Building Information Modelling Technology (BIM-t), by client and construction stakeholders, indicates a significant advancement from the traditional method of design, build, and maintenance of construction projects,

J. O. Toyin (✉) · M. C. Mewomo
Department of Construction Management and Quantity Surveying, Durban University of Technology, Durban, South Africa
e-mail: 22175967@dut4life.ac.za

M. C. Mewomo
e-mail: ModupeM@dut.ac.za

T. C. Haupt et al. (eds.), *Construction in 5D: Deconstruction, Digitalization, Disruption, Disaster, Development*, Lecture Notes in Civil Engineering 245,
https://doi.org/10.1007/978-3-030-97748-1_6

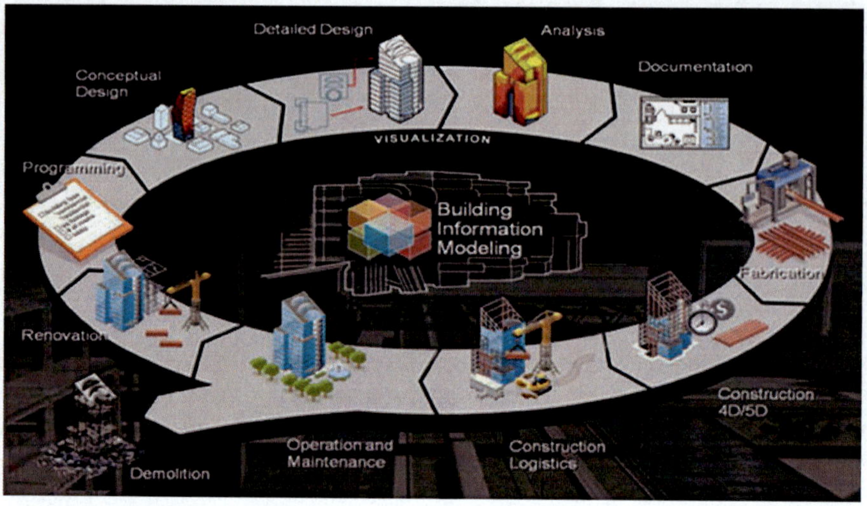

Fig. 1 BIM lifespan sequence [3]

to the current trend of BIM-t use. BIM-t generally grants experts reliable building data and accurate information about the project at any given time [1]. In essence, this enhances the efficiency, buildability, ease of material schedule, and conformity to the proposed 3D virtual design of the construction project. The data collected to model the building requires adequate attention and accuracy to transform the building [2]. Fernandes [3] in his research affirmed that the result of the data and information used during the project construction phases will produce a physical representation of the 3D virtual design as seen in Fig. 1. Therefore, BIM-t has the potentials to facilitate the production of construction projects.

BIM-t accelerates teamwork among the built environment professionals throughout the life cycle of a construction project. When BIM-t is successfully implemented, it enhances interaction and proper management of information, owing to the well-organized procedure involved [1, 4–10]. Construction professionals need adequate collaboration to effectively utilize BIM-t-based tools [11–15]. In addition, in order to successfully use BIM-t, collaborative effort is required from all project personnel from the inception of the project [1, 2, 16, 17]. Therefore, collaboration among various team members of the project is grounded on ideologies that comprise; genuinely-based decision creation, trust, effective interaction, adequate transparency, uniform reward, amenable information distribution, risk-taking, use of BIM-based tools, and supportive documents [18].

Previous research has shown that BIM-t can be viewed from different angles such as; BIM-t as a process, digital means, information means, policy-making, technology, and methodology for information conduct in a construction project, throughout the whole project life cycle. BIM-t as a digital and information means will grant a significant impact on construction projects throughout their lifecycle [19–21].

The construction industry is notably tagged as unpredictable in its product lifecycle. The industry is also noted for issues regarding lack of timely delivery, reworks, lack of collaborative understanding among its professionals [11, 12, 22]. Crowther and Ajayi [22] (as cited in Egan, [23]) affirmed that inefficiency, increase in cost and delay are major issues during the design and planning stage of construction projects. Currently, a remarkable percentage of projects still fail to meet up with completion on their deadline or handover dates, this has left clients with unfinished/unusable building construction projects, contractors are also faced with liquidated and ascertained damages to reimburse. Crowther and Ajayi, [22] (as cited In National Audit Office, [24], p. 3) reported that in the U.K, 70% of government construction projects were still lately delivered'. Also, an earlier report indicated that around one-third of the U.K construction projects defaulted to catch up with their deadline finish dates. However, the expert experience of successful BIM-t applications in different countries show that BIM-t related approaches provide series of solutions that reduced project completion time, improve the overall quality of delivered projects and cooperation in the construction industry [4, 8, 19, 25]. Notwithstanding, the increased adoption and application of BIM-t by the built professionals should have a huge positive impact on the productivity of the construction lifecycle. According to [26] BIM-t application is accustomed to more advantages than the conventional method of construction as it helps to reduce expected time to resolve an incident occurrence and also saves substantial cost in all activities of the building throughout the construction phases of the project. Li et al., [13] also affirmed that when BIM-t is applied, it permits the real-time monitoring of facility status and construction progress. In addition, the study of [4, 16, 26–30] confirmed that BIM-t facilitates communication and cooperation among building construction professionals. However, [7] opined that in order to judiciously acquire the positive impact of BIM-t during construction and post-construction phases, effective collaboration of professionals and the successful implementation of BIM-t is highly needed. In view of this, this study is designed to investigate the impact of successful BIM-t application on building construction projects in other to fill the research gap in its adoption.

2 Research Methodology

One of the essential steps to build up decent scientific research is through the literature review [31]. Therefore, this paper adopted a systematic review of existing literature on the subject of BIM-t focusing on the impact of successful BIM-t application on residential and commercial building projects. To achieve the aim of this research, the methodology review comprises of the search and collection of published articles between 2008 and 2021 on the impact of successful BIM-t application on building residential/commercial construction projects. Consequently, published articles on BIM-t impact for a period of 15 years were examined (i.e., 2008–2021). Evidence from published articles suggests that the full adoption of the BIM-t in the developed countries started around 2008.

In the search phase, search conditions used to locate relevant academic journals, are; "Building information modelling Impact', 'BIM application Impact', 'BIM impact on construction life cycle' 'BIM benefit', 'Building information modelling concept', 'BIM current situation' and 'BIM impact on construction project'. These search conditions were reiterated in selected academic databases such as Research Gate, Google Scholar, Elsevier (science direct), Springer, Francis and Taylor, etc. the retrieved papers consisted of journals, conferences proceeding, books, and thesis, excluding editorials and forums. Consequently, the initial 50 papers that were retrieved were filtered down to 41. The adoption of the exclusion criteria in the filtration of the retrieved articles yielded a final population of 41 which were considered in this paper. After retrieval of the selected articles which were critically reviewed and analyzed, the result and the impacts were addressed. The final phase involved discussions, conclusions, and recommendations based on the findings from the literature.

3 Research Discussion

3.1 Building Information Modelling Technology (BIM-t) Versus Conventional Mode of Contracting and Construction

For a well understanding of BIM-t, it is important to compare BIM-t with the conventional means of contracting and constructing projects. According to Parvan [32], the BIM package is classified into three vital fundamental technologies (Fig. 2): which are; 3D CAD technology, object-oriented technology, and parametric modeling technology. The combination of the three technologies creates an excellent platform thereby, generating good data management, improved modification management and improved interoperability for the BIM technology tools managers. Basically, the 3D CAD technology forms a collaborative virtual working atmosphere, grounded on the organized 3D orderly system. Thereby, the virtual model component of the 3D model are the drawing objects of the CAD technology. Nevertheless, the drawing objects have been scraped, transformed to object-oriented technology. The objects are captured as pipes, doors, windows, walls, beams, roof, floor, etc., into the architectural design objects and engineering design objects, The objects are the fundamental added modelling objects in the virtual object-oriented technology.

3.2 Impacts of BIM Application

According to the reviewed literature, the continuous adoption and application of BIM technology have justified positive and productive benefits to the built environment

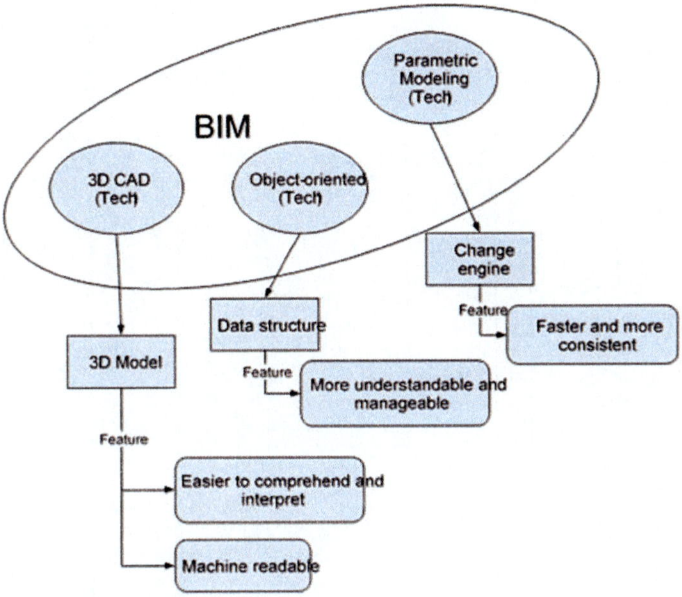

Fig. 2 BIM technologies functional decomposition diagram [32]

professionals and clients. The table below shows the impact of BIM-t on building construction projects as reviewed in the literature.

3.3 Classification of Impact of BIM-Technology Application on Construction Project

The identified impact derived from the reviewed literature in Table 1, were further classified according to phases involved in building construction project lifecycle as shown in Fig. 3. The stages according to Hoang et al. [26] include; Conception phase, Design phase, Pre-construction phase, procurement phase, post-construction phase, maintenance, and demolition phase.

3.3.1 Conception Phase

The conception phase of the building construction project deals with the planning and development of the proposed project. This is the genesis of the construction process. The client comes up with a brief of the project or facility intended to build. Some of the steps involved in this phase include; finding a portion of land for the project, choosing an architect/required built consultant professional, and initial concept of

Table 1 BIM-t impact on construction project phases

S/N	Impact of BIM technology	Reference	Frequency
1	Improved ROI. (PP1)1	[1, 6, 7, 9, 12, 15, 17, 26, 27, 30, 33–37, 39, 40]	17
2	Improved quality design and productivity. (DP1)1	[1–9, 12, 13, 15–18, 20, 22, 25–28, 30, 33, 35–44]	33
3	Enhance project speed and save time (lessen the time for routine data gathering and recording) (CP2)4	[1–9, 11, 12, 14–18, 20, 22, 26–28, 30, 32–35, 37, 38, 40, 41, 43, 44]	32
4	Overall project cost reduction (PP1)2	[4–6, 8–13, 15–17, 22, 26, 27, 30, 33–35, 37–39, 43]	23
5	Enhance scheduling of work (Facilitates the planning of resource and its allocation) (PCP1)1	[1, 2, 4–10, 13, 17, 26, 27, 29]	14
6	Improved day-to-day construction work progress and ease of tracking construction activities (Enhance communication and collaboration). (CP2)3	[2, 4–6, 9, 12, 15, 16, 26–30, 39, 44]	15
7	Eradicate design clashes amid professionals and contractors. (PCP1)2	[1, 2, 5, 6, 9, 13, 15, 16, 22, 26–30, 37, 40, 43]	17
8	Improved level of readiness for emergency occurrence (Enhance safety) (CP2)6	[2, 3, 6, 10, 14–16, 26, 27, 45]	10
9	Enhance management of construction credentials in respect to early collaboration. (CP2)7	[2, 6, 8–15, 17, 25, 27, 29, 30, 34]	16

(continued)

Table 1 (continued)

S/N	Impact of BIM technology	Reference	Frequency
10	Enhance design self-confidence (DP1)2	[1, 2, 5, 6, 8–10, 12, 15, 16, 26, 27, 30, 37, 41, 43, 44]	17
11	Ease of design discrepancy and omission detection (Support construction and project management) (PCP1)3	[1–3, 5–8, 10, 11, 15, 17, 26, 30]	13
12	Eliminate/Reduce design errors (DP1)3	[2, 6–9, 11, 15–18, 26, 30, 32, 37, 41, 44]	16
13	Enhance proper proactive maintenance scheduling and improve the management and maintenance of construction project infrastructures (PCP2)1	[6, 7, 14, 26, 32, 36, 43]	7
14	Improve construction scheduling/Planning (CP2)1	[2, 3, 5, 6, 11, 13, 14, 26, 27]	9
15	Decreased risk and cost incurred by contractor and subcontractors in a project (CP2)5	[3, 5–7, 13–15, 17, 26]	9
16	Improve efficiency and safety of works. (CP2)2	[2, 6, 9, 11, 15, 16, 18, 25–27, 32, 34–36, 39, 43]	16
17	Improve client brief drafting (CP1)1	[2, 3, 5, 6, 8–10, 12, 13, 15, 16, 18–20, 22, 26, 27, 30, 32, 35, 37, 40, 43, 44]	24

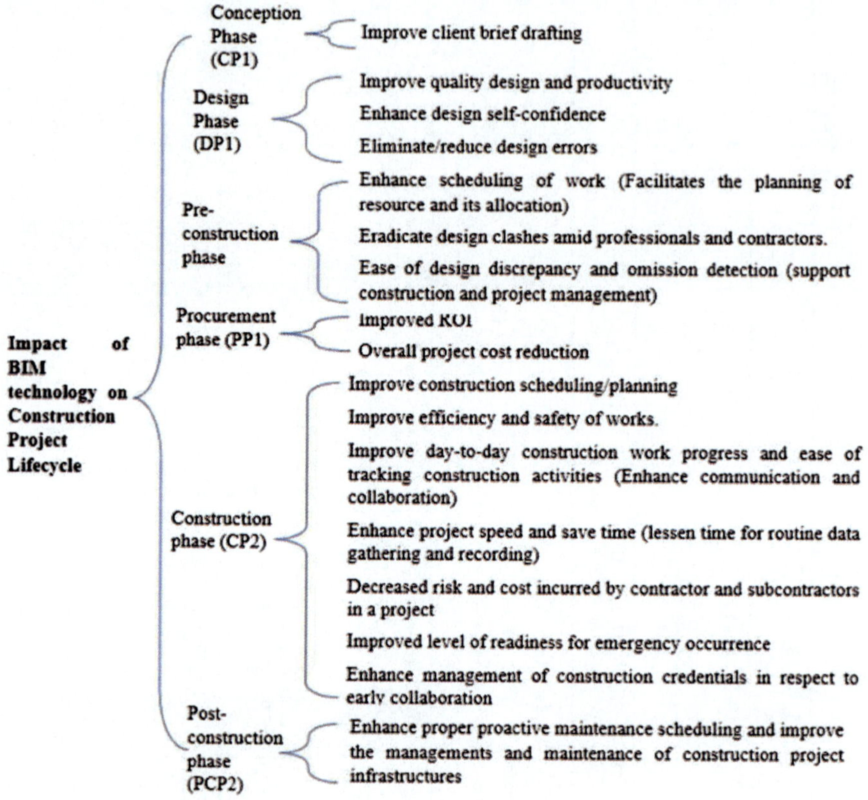

Fig. 3 Impact of BIM technology on construction project lifecycle

pre-designs [5, 9, 26, 35]. This stage is very crucial, it will determine the success of the project about client satisfaction. Therefore, adequate care must be taken to ensure a successful outcome as work progress. Researchers were able to affirm that BIM technology has a productive impact in this phase [2, 3, 8, 27]. This helps to speed up the pre-design of the client briefing.

3.3.2 Design Phase

The design phase, being the nest stage is the stage where the clients' dreams meet what is doable. Thereby, the client's hopes and dreams should not be neglected while putting up the various design or plans. Also, the 3D design of the project when it is finished will be seen and a design model can be provided base on request. BIM technology successfully applied in a project has the potential benefit to improve the quality design and productivity; enhance design self-confidence and eliminate/reduce design errors [7, 12, 16, 37, 46].

3.3.3 Pre-Construction Phase

The pre-construction phase is the stage that deals primarily with getting ready to commence the project, finalizing the design, creation of strategic plans to carry out the project, organizing the project management team. BIM application in this phase helps in proper scheduling of works, eradication of design clashes amid built professionals and contractors, also helps in design discrepancy and omission detection (Support construction and project management) [1, 28, 29, 40].

3.3.4 Procurement Phase

This stage deals with the buying/renting of materials, tools, equipment, machines and all kinds of stuff required for the construction of the project. The application of BIM-t makes it easier to schedule materials and labour for each construction stage thereby reduces overall project cost and improve return on investment for the client and contractor [6, 27, 33, 34].

3.3.5 Construction Phase

The construction phase, this stage deals with the transformation of paper/various CAD designs into the physical reality world. The stage has the most concentrated task, those tasks are assigned with completion deadlines/milestone date for each building grouped element(s). BIM has a great impact in this phase of construction; by improving construction scheduling/planning; improve efficiency and safety of works; improve day-to-day construction work progress and ease of tracking construction activities; enhance project speed and save time (lessen the time for routine data gathering and recording); decreased risk and cost incurred by contractor and subcontractors in a project; improve the level of readiness for emergency occurrence; enhance management of construction credentials in respect to early collaboration [8, 11–14, 22, 25–27, 30, 32, 35].

3.3.6 Post-Construction Phase

In the post-construction phase the project has been completely built. This is where the final checklist of all works done correctly will be checked and signed by the required built professional(s). The check is to confirm its conformity to all the design documents. Therefore, the professional will issue a certificate of substantial completion. After which the client can then move in. at this stage BIM application will enhance proper proactive maintenance scheduling which will be presented to the client [6, 7, 14, 36, 43].

However, the table also indicates that there is no consideration to check the impact of BIM-t in the 41-reviewed articles under the maintenance phase and demolition phase of construction works. This may be due to the lack of client interest in BIM-t after acquiring the completed project.

Table 2 below shows the frequency at which scholars identify the Impact of BIM-t on each phase of the construction project lifecycle. This was based on the 41 reviewed literature in reference to the year of publication of the consulted journals. The statistics shown in Table 2 have the following implications; The most frequently identified impact of BIM-t was based on the lifecycle phases of the construction project. From most frequent to least frequent, are Construction phase (37/41 = 91.24%), Design phase (35/41 = 85.36%), Procurement phase (26/41 = 64.41%), Conception phase (24/41 = 58.54%), Pre-construction phase (23/41 = 56.09%) and post-construction phase (7/41 = 17.03%).

The statistics imply two phenomena: First, it indicates that the published reviewed journal from the year 2008–2021 all identify one or two positive impacts of BIM-t relating to the design phase and construction phase. The result indicated that these two phases are the major stages where BIM-t has been judiciously used by building professionals, which has a significant impact on construction productivity. Furthermore, the construction phase ranked the highest which indicates that when BIM-t is successfully applied during the construction phase of a building project, it will be more advantageous to all parties involved.

From the result, the Procurement phase and the Conception phase ranked 64.41% and 58.54% respectively. "Pre-construction phase' which is the fifth with 56.09% is next to the procurement phase. This shows that BIM-t application is still at the average level. Lastly, Post construction phase being the least frequently identified impact of BIM-t, the result made it clear that BIM-t has not really been in use at this phase or scholars do not show consign to check the impact of BIM-t at the phase as identified from the 41 reviewed publications.

Among the 41 reviewed articles, 15 articles were reviewed between 2008 and 2012 with the following breakdown: 2008 (2 articles), 2009 (2 articles), 2010 (3 articles), 2011 (3 articles), 2012 (5 articles), 11 articles were reviewed between 2013 and 2017; 2013 (2 articles), 2014 (3 articles), 2015 (1 article), 2016 (2 articles), 2017 (3 articles), 15 articles were reviewed between 2018 and 2021 (2018–3 articles), 2019 (3 articles), 2020 (8 articles) and 2021 (1 article). The result indicates that scholars try to find out the impact of BIM-t in 2020 since it has the highest frequency. The

Table 2 Impact of BIM-t application on construction project statistics

Year of publication (Impact)	CPI	DPI	PCPI	PPI	CP2	PCP2
2008–2012	4	12	7	9	12	2
2013–2017	8	10	5	6	10	2
2018–2021	9	13	11	11	15	3
Frequency sum	**24**	**35**	**23**	**26**	**37**	**7**

* CPI (Conception phase); DPI (Design phase); PCPI (Pre-construction phase); PPI (Procurement phase) CP2 (Construction phase and PCP2 (Post-construction phase)

concentration of the scholars in 2020 is to justify the relevance of BIM-t in the built industry.

4 Conclusion and Recommendation

Building information modelling technology has proven its importance in the built environment with the positive advancement in the construction industry. Also, it was observed from the reviewed literature that various construction professionals in the built environment agree to the adoption of BIM-t for construction projects. The outcome of this paper has the following implication; the literature review has shown that BIM-t is fully loaded with numerous positive impacts on construction projects, owing to proper application and understanding of the BIM-t. Furthermore, the study had shown that implementing BIM-t is more beneficial during the design and construction phases, however, BIM-t does not have a significant impact at the post-construction phase according to the literature reviewed. In addition, it was observed from the review that professionals have not shown significant interest in the application of BIM-t at the maintenance and demolition phases of construction projects. It is therefore recommended that the benefits and impacts of BIM-t should be widely disseminated among professionals and clients, this will not only reduce design errors but increase quality management and productivity in the building construction industry.

References

1. Akintoye A, Goulding J, Zawdie G, (eds) (2012) Construction innovation and process improvement. Wiley
2. Chen Y, John D, Cox RF (2018) Qualitatively exploring the impact of BIM on construction performance. In: ICCREM 2018: innovative technology and intelligent construction, pp 60–71
3. Fernandes RPL (2013) Advantages and disadvantages of BIM platforms on construction site
4. Nawaz A, Su X, Nasir IM (2021) BIM adoption and its impact on planning and scheduling influencing mega plan projects- (CPEC-) quantitative approach, complexity, vol 2021. https://doi.org/10.1155/2021/8818296
5. Al-Ashmori YY, Othman I, Rahmawati Y, Amran YM, Sabah SA, Rafindadi AD, Mikic M (2020) BIM benefits and its influence on the BIM implementation in Malaysia. Ain Shams Eng J 11(4):1013–1019
6. Olawumi TO, Chan DW (2018) Identifying and prioritizing the benefits of integrating BIM and sustainability practices in construction projects: a delphi survey of international experts. Sustain Cities Soc 40:16–27
7. Doumbouya L, Gao G, Guan C (2016) Adoption of the building information modeling (BIM) for construction project effectiveness: the review of BIM benefits. Am J Civil Eng Archit 4(3):74–79
8. Muñoz V, Arayici Y (2015) Using free tools to support the BIM coordination process into SMEs. Build Inf Modell (BIM) Des, Constr Oper 149:33–41

9. Clevenger CM, Ozbek ME, Mahmoud H, Fanning B (2014) Impacts and benefits of implementing building information modeling on bridge infrastructure projects (No. MPC 14–272)
10. Suermann PC, Issa RR (2009) Evaluating industry perceptions of building information modelling (BIM) impact on construction. J Inf Technol Constr 14(37):574–594
11. Linderoth H, Peansupap V, Johnny W (7–8 Dec 2020) Determinants for students perceived potential of BIM use. In: ICCEPM 2020: the 8th international conference on construction engineering and project management. Hong Kong SAR, pp 291–300
12. Smits W, Buiten MV, Hartmann T (2016) Yield-to-BIM: impacts of BIM maturity on project performance. Build Res Inf 336–346
13. Smits W, van Buiten M, Hartmann T (2017) Yield-to-BIM: impacts of BIM maturity on project performance. Build Res Inf 45(3):336–346
14. Li J, Wang Y, Wang X, Luo H, Kang SC, Wang J, Guo J, Jiao Y (2014) Benefits of building information modelling in the project lifecycle: construction projects in Asia. Int J Adv Robot Syst 11(8):124
15. Hungu C (2013) Utilization of BIM from early design. Stage to facilitate efficient FM operations (Master's thesis)
16. McAuley B, Hore AV, West R (2012) Implementing building information modeling in public works projects in Ireland. In: Proceedings of the 9th European conference on product and process modelling
17. Zhang L, Chu Z, Song H (2020) Understanding the relation between BIM application behavior and sustainable construction: a case study in China. Sustainability 12(1):306
18. Mahamadu AM, Mahdjoubi L, Booth C, Manu P, Manu E (2019) Building information modelling (BIM) capability and delivery success on construction projects. Constr Innov
19. Sarvari H, Chan DW, Rakhshanifar M, Banaitiene N, Banaitis A (2020) Evaluating the impact of building information modeling (BIM) on mass house building projects. Buildings 10(2):35
20. Meng Y, Wu N, Li Y, Zhu C, Kong L (2020) BIM impact assessment of landscape architecture design. J Phys: Conf Ser 1533(4):042093. IOP Publishing
21. Matějka P, Tomek A (2017) Ontology of BIM in a construction project life cycle. Procedia Eng 196:1080–1087
22. Vanlande R, Nicolle C, Cruz C (2008) IFC and building lifecycle management. Autom Constr 18(1):70–78
23. Crowther J, Ajayi SO (2019) Impacts of 4D BIM on construction project performance. Int J Constr Manag 1–14
24. Egan J (1998) Rethinking construction, construction task force report for department of the environment, transport and the regions. HMSO, London
25. UK National Audit Office (UKNAO) (2003) PFI: construction performance. Report by the comptroller and auditor general, HC 700, Session 2002–2003
26. Succar B, Sher W, Williams A (2012) Measuring BIM performance: five metrics. Architectural Eng Des Manag 8(2):120–142
27. Hoang GV, Vu DKT, Le NH, Nguyen TP (2020) Benefits and challenges of BIM implementation for facility management in operation and maintenance face of buildings in Vietnam. IOP Conf Ser: Mater Sci Eng 869(2):22–32. IOP Publishing
28. Crippa J, Araujo AM, Bem D, Ugaya CM, Scheer S (2020) A systematic review of BIM usage for life cycle impact assessment. Built Environ Proj Asset Manag
29. Karakurt A (2019) Benefits of BIM technology on construction management (Master's thesis, Hasan Kalyoncu Üniversitesi)
30. Ahankoob A, Manley K, Hon C, Drogemuller R (2018) The impact of building information modelling (BIM) maturity and experience on contractor absorptive capacity. Architectural Eng Des Manag 14(5):363–380
31. Eastman CM, Eastman C, Teicholz P, Sacks R, Liston K (2011) BIM handbook: a guide to building information modeling for owners, managers, designers, engineers and contractors. Wiley

32. Dresch A, Lacerda DP, Antunes JAV (2015) Design science research. In: Design science research. Springer, Cham pp 67–102
33. Parvan K (2012) Estimating the impact of building information modeling (BIM) utilization on building project performance. University of Maryland, College Park
34. Shen Z, Issa RR (2010) Quantitative evaluation of the BIM-assisted construction detailed cost estimates
35. Ilozor BD, Kelly DJ (2012) Building information modeling and integrated project delivery in the commercial construction industry: a conceptual study. J Eng, Proj, Prod Manag 2(1):23–36
36. Jin R, Hancock CM, Tang L, Wanatowski D (2017) BIM investment, returns, and risks in China's AEC industries. J Constr Eng Manag 143(12)
37. Azhar S (2011) Building information modeling (BIM): trends, benefits, risks, and challenges for the AEC industry. Leadersh Manag Eng 11(3):241–252
38. Becerik-Gerber B, Rice S (2010) The perceived value of building information modeling in the US building industry. J Inf Technol Constr 15(15):185–201
39. Liu R, Issa RR (2014) Design for maintenance accessibility using BIM tools. Facilities 32(3/4):153–159
40. Mom M, Tsai MH, Hsieh SH (2014) Developing critical success factors for the assessment of BIM technology adoption: Part II. Analysis and results. J Chin Inst Eng 37(7):859–868
41. Mesároš P, Mandičák T, Behúnová A (2020) Use of BIM technology and impact on productivity in construction project management. Wireless Netw 1–8
42. Souza L, Amorim S, Magalhães A (2009) Impact from the use of BIM in architectural design offices: real estate market opportunities. Des Manag Technol 26–52
43. Kymmell W (2008) Building information modeling: planning and managing construction projects with 4D CAD and simulations (McGraw-Hill Construction Series). McGraw-Hill Education
44. Zaini N, Zaini AA, Tamjehi SD, Razali AW, Gui HC (2020) Implementation of building information modeling (BIM) in Sarawak construction industry: a review. IOP Conf Ser: Earth Environ Sci 498(1). IOP Publishing
45. Poirier EA, Forgues D, Staub-French S (2017) Understanding the impact of BIM on collaboration: a Canadian case study. Build Res Inf 45(6):681–695
46. Benjaoran V, Bhokha S (2010) An integrated safety management with construction management using 4D CAD model. Saf Sci 48(3):395–403

The Adoption of 3D Printing as a Construction Methodology in South Africa: What Are the Barriers?

Abimbola Windapo, Msindisi Yende, Onele Kala, Lawrence Maboreke, and Lindelani Matshidze

Abstract 3D Printing is the production of 3-dimensional objects using an additive process. It has revolutionized industries such as manufacturing, aerospace, medicine and now commonly construction. This technology has been successfully implemented in the construction industries of countries such as China, United Arab Emirates and Russia. It has improved productivity on site, produced sustainable buildings and lowered costs of production. However, there is limited use of the 3D construction methodology in South Africa. Therefore, this research examines whether there are barriers to implementing 3D printing technology within the South African construction industry. The research employs a qualitative research approach that involves interviews of 12 construction industry professionals in South Africa. The data obtained from interviews and secondary data sources were analyzed using thematic analysis. The findings revealed that critical barriers to the implementation of 3D printing within the construction industry were Regulations, the dominant labour-intensive production process and the traditional methods of procurement used in construction project delivery in South Africa. Based on these findings, the study concluded that 3D printing technology would make the project production process more efficient and effective from a cost, performance, and quality perspective; however, the South African construction industry is currently not conducive to the implementation of an innovative solution in construction such as 3D printing. Further research should be conducted on how 3D printing could be implemented in the current state of the construction industry.

Keywords 3D Printing · Conservative · Construction methodology · Labour-intensive production process · Procurement · Regulations

---

A. Windapo (✉) · M. Yende · O. Kala · L. Maboreke
Department of Construction Economics and Management, Faculty of Engineering and the Built Environment, University of Cape Town, Cape Town, South Africa
e-mail: abimbola.windpo@uct.ac.za

L. Matshidze
School of Construction Economics and Management, University of the Witwatersrand, Johannesburg, South Africa

T. C. Haupt et al. (eds.), *Construction in 5D: Deconstruction, Digitalization, Disruption, Disaster, Development*, Lecture Notes in Civil Engineering 245,
https://doi.org/10.1007/978-3-030-97748-1_7

1 Introduction

The construction industry has experienced problems that have led to a decline in its effective functioning. These problems include labour productivity, high accident and fatality rate, adverse negative environmental impacts and increased waste production. To address these challenges, professionals within the construction industry have begun to look at innovative solutions such as 3D printing technology to improve the industry's performance.

According to Ashraf et al. [3], 3D printing, also known as Additive Manufacturing (AM), is an advanced manufacturing process that automatically produces complex geometric shapes from a 3D computer-aided design model and is originally developed as an automated method of producing prototypes. 3D printed products can be printed for review by the designer or engineers, and revisions printed equally with the same level of ease. Furthermore, the items produced are an original or perfect copy [15]. The technology has been successfully implemented worldwide in countries such as China, United Arab Emirates and Russia. However, there is limited use of this innovative construction methodology in South Africa.

This paper examines whether there are barriers to the use of 3D printing in construction in the South African construction industry. The research is significant because it aids the construction industry to understand how it can overcome the barriers to utilizing innovative technologies such as 3D printing on construction projects towards achieving efficient and effective construction processes. The following section provides an overview of 3D Printing Technology and the barriers to its use. Thereafter, the paper presents the Research Methodology used in the study, the findings, discussion, and conclusion and recommendations.

2 Overview of 3D Printing Technology and Barriers to Its Implementation on Projects

2.1 3D Printing Technology

According to Ashraf et al. [3], 3D printing is an advanced manufacturing process that produces complex shape geometries automatically from a computer-aided design (CAD) model, through the layer-by-layer addition of materials until a three-dimensional product is created [15]. There are six main 3D printing methods, namely: Fused deposition modelling (FDM), Laminated object manufacturing (LOM), Polyjet Photopolymer, Selective Laser Sintering (SLS), Stereolithography (SLA) and Syringe Extrusion [19].

According to Hager et al. [11], contour crafting (CC) is the most promising 3D printing technology used in the building industry, and it involves the manufacture of big elements based on printing the building layer by layer using special material ejection. CC constructs objects layer by layer using robotics,it is used for small scale industrial parts and was identified as the only method capable of delivering components large enough for building structures [9]. The contour crafting machine combines a robotic extruding system that takes its orders from 3D CAD software and pours concrete layers through a nozzle. Furthermore, special hardeners and fibers are used in the concrete mix to make each layer hard enough to carry the next layer.

2.2 Barriers to the Implementation of 3D Printing Technology in the Construction Industry

2.2.1 Procurement Systems

Procurement is an integral aspect of a construction project as it includes the sourcing, purchasing and all activities related to providing knowledge, labour, construction equipment, materials, supplies, supervision, and management services required to complete the project. Procurement systems in the construction industry place a premium on speed and urgency and competition based on price alone. The establishment of rigid role responsibilities within a construction project in addition to promoting adversarial and self-preservation behaviour are also aspects that are emphasized within the procurement systems used in the construction industry, and it is these procurement systems that are most delirious to innovation and that discourage contractors from pursuing novel building solutions [13].

2.2.2 Regulations

The term regulation refers to the political and legal actions a government may consider necessary to oversee the activity in the construction industry and the behaviour of private sectors in the economy [5]. Government regulatory policies and legislations generally exert a significant influence on demand and play an important role in directing the path and the rate of technological change, as well as encouraging or discouraging innovative activity [10, 8].

2.2.3 Fragmentation and Structure of Construction Production

Fragmentation is defined in terms of the number of subcontractors/specialists involved in construction projects and its effects on the multiple processes in construction projects. Therefore, regarding the construction industry, fragmentation is the division resulting from the increasing number of participants involved in all processes of a building project. Furthermore, this division has come about due to the growing demand for differentiation and specialization as building projects increase in both size and complexity [1]. Fragmentation leads to eliminating the opportunity for learning and innovating [7], It decreases the intention to invest in innovation and blocks the mutual sharing of information and knowledge [22].

2.2.4 Risk Allocation

The desire to avoid the use of innovation because of its uncertainty is present in the clients' actions, who invest capital into the project; consultants who advise on the procurement, design and production methods and contractors who are in charge of the production methods. Furthermore, the very nature of the contractual risk allocation between the parties involved in a construction project discourages innovation, especially when there is a possibility that claims, disputes and litigation could arise from the slightest deviation from what is contractually agreed upon [17]. The result is that contractors and sub-contractors choose to avoid innovation in projects in favour of utilizing tried and tested methods to mitigate the risk.

2.2.5 The Cost

According to Attaran [4], the cost is one barrier to entry for the 3D printing technology because of the product's availability. The equipment required to 3D print is very costly, and the materials used in the printer as an implication of the novelty of the product is expensive, while there are few producers of the materials in the industry yet [4].

2.2.6 Culture/Attitude of the Construction Industry

Compared to other industries, the construction industry's research, development, and adoption of new technological innovations are slow, leading to the industry's reputation of preferring conservative, well-known practices to innovative construction methods. The culture of the construction industry is noted as conservative.

2.2.7 Knowledge and Awareness

The nature of the projects within the construction industry is that they are mostly one-off projects, which hinders the innovation from applying to other circumstances. The review found that 3D Printing is an innovative technology that improves the efficiency and effectiveness of construction production processes. Barriers to the use of the 3D Printing technologies found globally are the construction procurement systems, regulations, fragmentation and structure of construction production, risk allocation, cost, culture/attitude of the construction industry, knowledge and awareness that have been identified holistically to limit the use of 3D printing technologies on construction projects. However, it is not known whether these barriers will also be applicable in the South African construction industry.

3 Research Methodology

The study employs a qualitative research approach involving semi-structured interviews of 12 construction industry professionals in South Africa with a minimum of 3 years of experience. The respondents included contractors, civil engineers, construction managers, project managers and architects who were purposively selected because of their experience and exposure to the South African construction industry.

The profiles of participants interviewed are listed in Table 1.

Table 1 shows that the 12 interviewees have a minimum of three years of experience and a combined experience of 191 years (16 years average), and therefore, the respondents can provide reliable perspectives on the industry's receptiveness to implementing 3D printing technology in construction. The research made use of

Table 1 Summary of participants

Code	Designation	Years of experience	Location	Projects handled
P1	Project manager	7	Gauteng	Residential and health care
P2	Project manager	3	Gauteng	Healthcare, commercial and hospitality
P3	Project manager	3	Gauteng	Residential, retail and commercial
P4	Project manager	8	Gauteng	Retail and health care
P5	Architect	20	Western Cape	Commercial, residential and health care
P6	Site agent and construction manager	47	Western Cape	Low-cost housing
P7	Contracts manager	27	Western Cape	Commercial, health care and government
P8	Architectural technologist	25	Western Cape	Industrial, commercial and residential
P9	Project manager and architect	35	Western Cape	High-end residential
P10	Architect	4	Western Cape	Educational facilities
P11	Civil engineer	3	Mpumalanga	Power station
P12	Civil engineer	9	Eastern Cape	Low-cost housing

possible, plausible and preferable futures in which 3D printing technology would have been implemented in the construction research because there is limited use of the technology in project delivery in South Africa.

The research requires qualifiable data to highlight the knowledge-based perceptions of the interviewed professionals on the barriers to the adoption and implementation of 3D printing technology in the South African construction industry. This was necessary because qualitative data cannot be easily reduced to numbers and is related to people's concepts, opinions, and behaviours in a particular context. Furthermore, a qualitative research approach was ideally suited to enable the categorization, analysis and drawing out themes from the information garnered from the respondents.

According to Rossman and Rallis [20], qualitative research is obtained from the differing information garnered due to exposure to various places, activities and events. The data is analyzed and discussed using thematic data techniques. Braun and Clarke [7] describes thematic analysis as an approach that analyzes, identifies and reports patterns within data and translates the data to achieve the outlines objectives of the research study.

The questions presented in the semi-structured interviews were formulated from the literature review, which served as a basis and guideline for the questions. The interviews were designed to address the question and objective set for the research. Most of the questions used in the semi-structured interviews were deliberately open-ended to limit the respondents' answers to each question.

The semi-structured interview questions were also designed to establish the type of organization the respondent worked for, their role within the firm, and their perceptions of the potential impact of 3D printing technology in the South African construction industry. The data collected were transcribed and, after that, analyzed using thematic data analysis. The thematic analysis is an approach that analyses, identifies and reports patterns within data and translates the data to achieve the outlined objectives of the research study.

4 Findings and Discussion

The study found that although the construction design incorporates advanced construction solutions, construction companies are not open to implementing innovative construction solutions during the construction production phase. Their reluctance to implement innovative construction solutions was also evident in the limited number of professionals who provide training for on-site employees. The barriers identified to the implementation of 3D Printing technology include high capital investments, loss of labour, the unwillingness of industry professionals, unaccommodating procurement processes and lack of knowledge and awareness of 3D printing technology. It was found that the construction industry is not ready for the implementation of 3D printing technology.

According to P6, Our production methods would be a barrier to the implementation of 3D printing because we use people to do the work, and if this machine does not use people, then where will our people work? "They will not Have work".

P10 noted that "the construction process involves a lot of parties and a lot of money and if any form of change comes up, people fear it because of the risks of the large amounts of money that are involved".

In the case of awareness and willingness, it was found that construction professionals know about 3D printing conceptually. It was also found that these professionals are not open to staying up to date with the trends while very few professionals are willing to change production processes which is why the limited understanding of 3D printing is expected and was found to be true.

According to P8, "Not currently, there's always innovative technology that is available out there that we could use, we do not always know all of it. I do not think there is anything available out there that could help me greatly right now".

Six themes were drawn from data collected which were in line with the literature. These themes included a High chance of project and budget overrun, the predominant use of the traditional procurement system and traditional production process, limited use of the technical plant, increased incorporation of computer software, limited awareness of future construction technologies and a high level of awareness of 3D printing technology but common understanding. Table 2 summarises the barriers identified in the literature and the participants' responses regarding whether they agree with any of the identified barriers to the use of 3D Printing Technology on their projects.

It can be seen from Table 2 that there is consensus amongst the respondents that Regulations such as the National Building Regulations, SANS10400 and R158 for the health sector is a key barrier to the implementation of 3D printing technology in the South African construction industry. According to P8, "we use regulations, but sometimes, there are too many regulations". P1 notes that they must follow regulations pertaining to how they conduct their production for the health sector,

Table 2 Barriers identified by respondents

Barriers	Respondents											
	P1	P2	P3	P4	P5	P6	P7	P8	P9	P10	P11	P12
Regulations	x	x	x	x	x	x	x	x	x	x	x	x
Production process	x	x	x	x	x	x	x		x	x	x	x
Procurement systems (Traditional)	x	x	x	x	x	x	x	x	x			x
Labour-intensive production process	x		x	x		x	x		x			
Unwillingness to adopt innovation			x	x	x	x				x		
Lack of knowledge and awareness	x					x	x		x			
Cultural aspect as a barrier							x		x			x
Cost of technology as a barrier	x							x				

which creates barriers to the use of innovative technologies such as 3D Printing in construction project delivery.

The production process is also considered a barrier to implementing 3D printing technology by all interviewees except P8. P8 expresses doubt that there is currently any innovative technology available that would improve their production processes significantly, while P4 states that their organization's processes and methods are centred around their current production processes, which are traditional. According to P4, "having their logistics centralized on traditional production methods does not provide much room for innovation to be incorporated", which is why they consider the production methods used as a barrier to implementing 3D Printing technology. P3 buttresses the view held by P4 that satisfaction with the current production methods is a barrier because regardless of the challenges that those production methods have, their delivery process is efficient and effective.

All the respondents except P10 and P11 identify the prevailing procurement methods—the traditional methods with a separation of construction design, used in the South African construction industry as a barrier to implementing the 3D printing technology in project delivery. P10 and P11 use mainly Design and Build, sometimes Turnkey (P10), as procurement strategies in project delivery. P2 states that not much 3D printing is occurring in the construction industry right now and that it will only be implemented if the client requests its use explicitly on their projects. This is buttressed by P12, who observes that "in terms of social housing, the government is the main barrier to the adoption of 3D printing technology, and they are the financiers of social housing projects, therefore, they need convincing first".

Closely tied to the production process is the labour-intensive production methods currently used in construction project delivery. The findings revealed that very few projects utilized technological equipment in their production processes, suggesting that their production processes are labour-intensive. P1, P6 and P7 state that the labour-intensive nature of the projects they are involved in is the main barrier to adopting new technology as they do not want to retrench any of their employees. P4 states that the logistics of their organization are centred around the current methods, which allow for recruiting workers in their target locations; additionally, their employees are trained for the current construction processes employed.

Currently, P6 shows a lack of knowledge of technological trends within the construction industry but is willing to learn and switch to new production methods. The lack of knowledge and understanding in P6 is further substantiated by the fact that they do not provide technical training and educational awareness for their employees, which is viewed as a barrier to adopting new technologies in construction project delivery.

Some respondents view that the barriers to the use of 3D printing technology act in isolation. P3 considers a combination of the capital required to implement 3D printing technology, the current production methods being labour-intensive and keeping them competitive as barriers to implementing 3D printing technology. At the same time, P10 states that the large capital requirements and numerous stakeholders of the current construction processes result in fear of the risks associated with switching to new technology.

The literature reviewed corroborated these findings because scholars have established that many projects in developing countries encounter considerable time and cost overruns, fail to realize their intended benefit or are even completely terminated and abandoned before or after their completion [2, 14, 21]. Support for the findings of the predominant use of traditional practices during the construction phase of a project has been established by Kulatilake [16]. This was aligned to earlier studies by Nam and Tatum [19], who found the conservative nature of the construction industry, in which professionals cling to the accepted methods and practices when fulfilling a client's requirements and do not actively seek to self-educate themselves on the newest developments in the industry, which also aligned with the limited awareness of future technologies found in the construction professionals studied.

The findings are also aligned with the conclusions of Oke et al. [18] regarding automated innovation having the effect of taking over the roles of the labour force and Kulatilake [12]'s discussion on the conservative nature of the construction industry.

5 Conclusion and Further Research

The research examined whether there are barriers to implementing 3D printing technology within the South African construction industry. It emerged that there are barriers to implementing 3D printing technology within the South African construction industry. These barriers include Regulations, the dominant labour-intensive production process used on construction projects in South Africa and the traditional method of project procurement in use.

It can be concluded that the South African construction industry is not ready for 3D printing. The high capital investments of implementing the technology, which includes training individuals on operating the machine, is a trending issue among industry professionals unless the client in their specifications requires the design to be implemented using 3D printing. Knowledge of 3D printing in South Africa is indistinct, and the benefits it brings are not clearly understood. It will take time and much convincing to switch to new ways because the industry is set and rooted in traditional ways which are familiar and have worked in the past regardless of the problems those ways have brought to the industry. There are, however, a minority of professionals who are open and willing to implement 3D printing, but without support from the industry, South Africa will continue with the labour-intensive methods of construction indefinitely.

The potential for 3D printing technology to positively impact the South African construction industry is yet to be fully realized. There is a noticeable lack of awareness and understanding of the technology and regulations, production, and procurement processes that hinder the technology's adoption and implementation in the South African construction industry. Therefore, the study recommends that the government, relevant regulatory institutions should educate construction professionals and other stakeholders on industry construction regulations and procurement processes.

Additionally, the certificate holder of 3D printing technology should produce a prototype of the completed building for the client to review and opportunity for training construction companies in the knowledge and use of the technology. Further research should be conducted on how 3D printing could be implemented in the current state of the construction industry.

Acknowledgements This work is supported by NRF (Grant Number-120843). However, the opinions and conclusions in the study are those of the authors and are not necessarily attributable to the NRF.

References

1. Abadi M (2005) Issues and challenges in communication within design teams in the construction industry: investigation into the use of virtual teams and information and communication technologies (ICTs) in the UK construction industry. Doctoral dissertation. The University of Manchester
2. Adeyemi LA, Idoko M (2008) Developing local capacity for project management: key to social and business transformation in developing countries. Paper presented at PMI® Global Congress 2008—EMEA. 19 May. St. Julian's, Malta. [online] Available at: https://www.pmi.org/learning/library/developing-project-management-developing-countries-8355 [Accessed 18 Oct 2019]
3. Ashraf M, Gibson I, Rashed MG (31 Jan–2 Feb 2018) Challenges and prospects of 3d printing in structural engineering. In: 13th international conference on steel, space and composite structures. Perth, WA
4. Attaran M (2017) The rise of 3-D printing: the advantages of additive manufacturing over traditional manufacturing. Bus Horiz 60(5):677–688
5. Benmansour C, Hogg K (2–4 Sep 2002) An investigation into the barriers to innovation and their relevance within the construction sector. 18th annual ARCOM conference. University of Northumbria
6. Braun V, Clarke V (2006) Using thematic analysis in psychology. Qual Res Psychol 3(2):77–101
7. den Hertog P, Bergman E, Charles D (2001) Innovative clusters: drivers of national innovation policy. OECD Publishing
8. Dubois A, Gadde LE (2002) The construction industry as a loosely coupled system: implications for productivity and innovation. Constr Manag Econ 20(7):621–631
9. Fischer F, Wild F, Sutherland R, Zirn L (2014) Grand challenges in technology-enhanced learning: outcomes of the 3rd Alpine Rendez-Vous. Springer International Publishing, New York
10. Gann DM, Salter AJ (2000) Innovation in project-based, service-enhanced firms: the construction of complex products and systems. Res Policy 29(7–8):955–972
11. Hager I, Golonka A, Putanowicz R (2016) 3D printing of buildings and building components as the future of sustainable construction? Procedia Eng 151:292–299
12. Kulatilake P (2016) Innovations in the construction industry: problems and potentials. Built-Environ Sri Lanka 1(2):2–9
13. Kumaraswamy M, Dulaimi M (2001) Empowering innovative improvements through creative construction procurement. Eng Constr Archit Manag 8(5/6):325–334
14. Larsen JK, Shen GQ, Lindhard SM, Brunoe TD (2016) Factors affecting schedule delay, cost overrun, and quality level in public construction projects. J Manag Eng 32(1):04015032

15. Manners-Bell J, Lyon K (2012) The implications of 3D printing for the global logistics industry. Transport intelligence, pp 1–5
16. Nam CH, Tatum CB (1997) Leaders and champions for construction innovation. Constr Manag Econ 15(3):259–270
17. Naoum S, Lock K, Fong D (09–11 Jun 2010) Is fragmentation of the UK construction industry the main barrier to innovation? The architects' view. The 6th international conference on innovation in architecture, engineering and construction. Pennsylvania, U.S
18. Oke A, Aigbavboa C, Mabena S (2017) Effects of automation on construction industry performance. Second international conference on mechanics, materials and structural engineering (ICMMSE 2017). April. Atlantis Press
19. Prince JD (2014) 3D printing: an industrial revolution. J Electron Resour Med Libr 11(1):39–45
20. Rossman GB, Rallis SF (2011) Learning in the field: an introduction to qualitative research. Sage
21. Samarghandi H, Mousavi S, Taabayan P, Mir Hashemi A, Willoughby K (2016) Studying the reasons for delay and cost overrun in construction projects: The case of Iran
22. Vock P (2001) An anatomy of the Swiss construction cluster. Innov Clusters 3:229

Disruption

Utilisation of Remote Monitoring Systems in Construction Project Management

Nomusa Makhathini, Innocent Musonda, and Adetayo Onososen

Abstract Traditional project monitoring systems are increasingly linked with high financial losses, project delays and poor project performance. Emerging digital technologies such as remote monitoring systems offer immense potential in resolving these challenges. This study identifies the challenges faced in managing multiple projects in the construction industry and how adopting remote monitoring systems can mitigate the challenges presented by these. Case study approach was adopted to contrast practical usage of remote monitoring and on-site monitoring systems. By using similar projects, with the same organisation and client, a comparative analysis between traditional on-site monitoring and remote monitoring systems on project efficiency, resource optimization and project outcome revealed that remote monitored multiple construction projects had better project performance on its goals and objectives compared to traditional monitored systems. Given the urgent need for improved productivity in the built sector, it is therefore imperative that remote monitoring systems adoption is adopted at national and organizational levels in the built environment. Furthermore, its need is imperative for further studies and research with regards to scaling its use for both big construction conglomerates and small and medium-sized construction firms.

Keywords Remote monitoring system · Construction multiple projects · Resource optimisation · Traditional monitoring system

1 Introduction

The need for infrastructure to drive economic growth has seen flag off of diverse construction projects sometimes being managed by a construction organisation thereby leading to distribution of resources to monitor execution of work on these projects [3]. This need raises many challenges for the project managers of which

N. Makhathini (✉) · I. Musonda · A. Onososen
Center of Applied Research and Innovation in the Built Environment (CARINBE), Faculty of Engineering and the Built Environment, University of Johannesburg, Johannesburg, South Africa
e-mail: nomuso.makhathini@yahoo.com

T. C. Haupt et al. (eds.), *Construction in 5D: Deconstruction, Digitalization, Disruption, Disaster, Development*, Lecture Notes in Civil Engineering 245,
https://doi.org/10.1007/978-3-030-97748-1_8

the biggest challenge is the capacity to deliver multiple projects efficiently [8, 1]. Since project monitoring is pivotal to ensuring quality, managing resources and maximizing productivity, adopting construction remote monitoring systems offers an opportunity to manage multi-projects more efficiently, while traditional physical on-site monitoring is incapable [2].

The key thrust of this research was to explore how remote monitoring tools could aid in reducing some of the challenges being faced in managing multiple projects. The study sample was limited to a project management consulting company who is using two different methods to run multiple projects. Using the two sub-cases, a comparative analysis was utilised to study the multiple projects performance and explored how the use of remote monitoring systems enhanced the capacity of individuals to accomplish the project goals.

2 Literature Review

2.1 Managing Multiple Construction Projects and Their Challenges

Multiple projects are defined as an integration of several projects using a common pool of resources with an aim to maximise productivity and deliver projects in alignment with the business purpose [7, 4, 6]. However, multiple projects do not only come with benefits but an array of challenges as each of the projects may have different goals and objectives. Due to the need to ensure quality, maximise resources and increase productivity on the construction sites, monitoring of multiple projects becomes challenging giving sparse human resources, unforeseen hazards and difficulty of terrains which inevitably necessitates the need to adopt remote sensing for managing multiple construction projects.

2.2 Remote Monitoring

Remote monitoring systems offer benefits such as vision-based identification, tracking and analyses of project activities, construction equipment and personnel [5]. The adoption of this emerging digital tool would enable clear visualisation of the activities on-site and make it easy for comparing the planned work against the executed work.

Emerging remote sensing digital tools can: (a) gather data through sensing technologies utilised in recording the as-built scenes, (b) retrieve information through the processing of extracted data required from the as-built data [5]. Furthermore, cloud computing is being integrated with laser scanning and digital models to improve the overall remote monitoring experience for construction site usage [9].

3 Research Methodology

To establish the contrast in utilisation of remote monitoring systems compared to the traditional monitoring systems, two projects were examined as case studies. The projects identified were similar in terms of design, execution period, budget and design use. One project was executed using traditional monitoring systems while the other was executed using remote monitoring systems. A research questionnaire as research instrument was distributed to the project managers on the two different projects.

The questionnaire was generic and only request participation on project manager in multiple projects as with categories shown on Fig. 1.

The analysis of the impact of remote monitoring systems on multiple projects is presented by means of a cross comparison of the two sub-cases. The researcher was able to draw conclusions from the findings, which provided information to make recommendations.

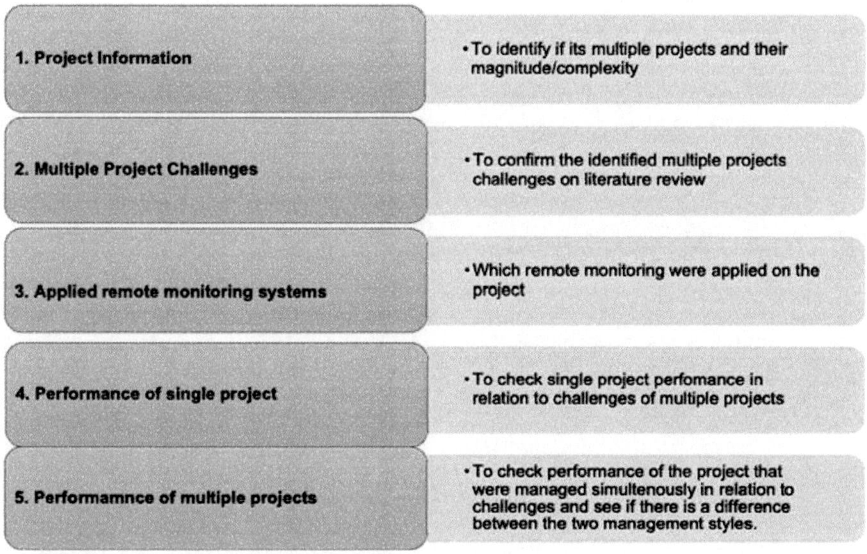

Fig. 1 Questionnaire categories

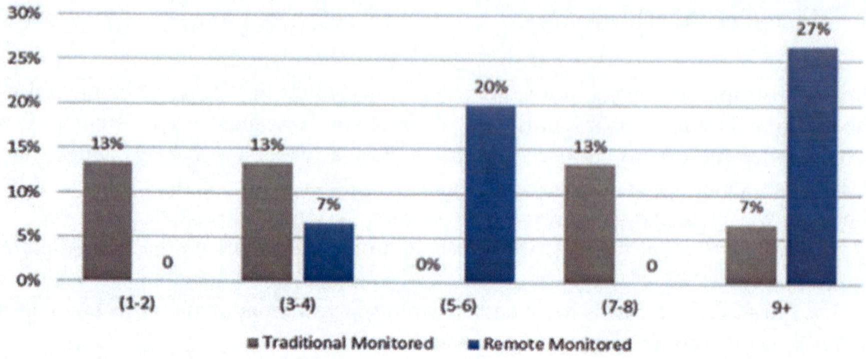

Fig. 2 Number of projects managed by respondents

4 Data Analysis and Findings

4.1 Number of Projects Monitored

The respondents were asked to indicate the number of projects engaged in the results are shown on Fig. 2.

4.2 Project Monetary Value

The respondents were asked to indicate the monetary value of the projects they managed. Table 4.2 below illustrates the spread of project values across the two management methods.

Figure 3 indicates that the majority of projects managed under traditional managed

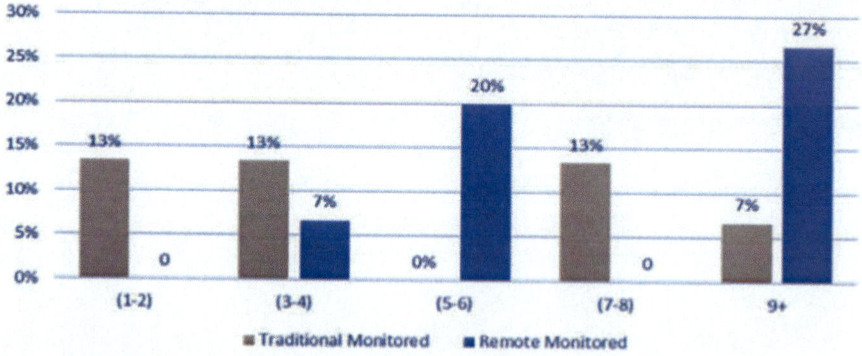

Fig. 3 Project monetary value range of multiple project

simultaneously are between 60 ≥ 200 Million rand. While the remote monitored method manages, a majority of project on a much lesser value bracket 10 – ≥20 Million peaking at 33%.

4.3 Multiple Projects

As stated by Ireland (2002) and Jerbrant [4], many project managers are responsible to manage and achieve the goals of the multiple projects simultaneously, with each project having its own challenges that are different from the others. In the questionnaire, the respondents were asked to indicate the frequency in monitoring of multiple projects simultaneously. Table 1 shows that most of the respondents are experienced in managing multiple projects.

The findings demonstrated that Case 1, the traditional monitored method projects face a higher number of challenges, such as project control, delays, and costs, compared to the remote monitored projects. Furthermore, it was observed that the traditional management was not delivered on time (Table 2).

Table 1 Frequency of managing multiple projects

P1:	Respondents results					Weighted mean
Likert scale	Always	Often	Sometimes	Rarely	Never	
Traditional monitored	5	2	0	0	0	4.71
Remote monitored	6	2	0	0	0	4.75
Average mean						4.73

Table 2 Cross comparison-ranking analysis on multiple project challenges

Challenges	Traditional monitored projects		Remote monitored projects	
	Weighted mean	Ranking	Weighted mean	Ranking
Projects on similar stage of progression	3.86	1	2.88	2
Projects experience similar challenges	3.71	2	2.75	3
Project prone to delays	3.71	2	2.5	5
Projects prone to cost over run	3.57	4	2.5	5
Projects prone to contractual claims	3.57	4	3	1
Projects prone to quality issues	2.86	6	2.63	4
Projects prone to safety issues	2.43	7	2	7

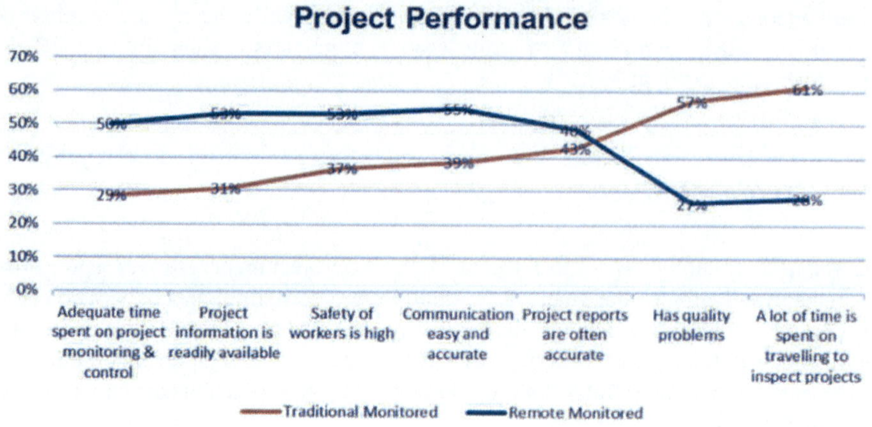

Fig. 4 Project performance

4.4 *Project Performance*

The intention of the research was to find out how the project performed under different management methods. The respondents were asked how agreeable the statements in relation to the projects' performance. Each respondent used a score: 5- strongly agree to 1 strongly disagree.

The comparison of the project performance between the two cases is shown in Fig. 4 revealed that the traditional monitoring approach was encumbered by several challenges as compared to the remote monitoring systems.

4.5 *Remote Monitoring System Use and Benefits*

As with Fig. 5 shown:

- E1—Was aimed to check whether automation or manually was used to undertake the project site inspections. The traditional method has a higher percentage at 67%, which implies that most of the site inspections monitoring progress, are done manually. The question of remote monitored project scored 33%, meaning that there are less manual site inspections done using this management method.
- Traditional monitored projects scored a weighted average of 14%, which means 100% of the respondents have not used any remote monitoring systems of those listed on Table 4.9 (E2 to E9)
- Remote monitored projects show an exposure to all the listed available remote monitoring systems. The results show E2 peaked at 86%, which is the use of remote digital real time cameras, E3 at 86% the use of augmented reality tools, followed by the use of drones at 86% and BIM at 86%.

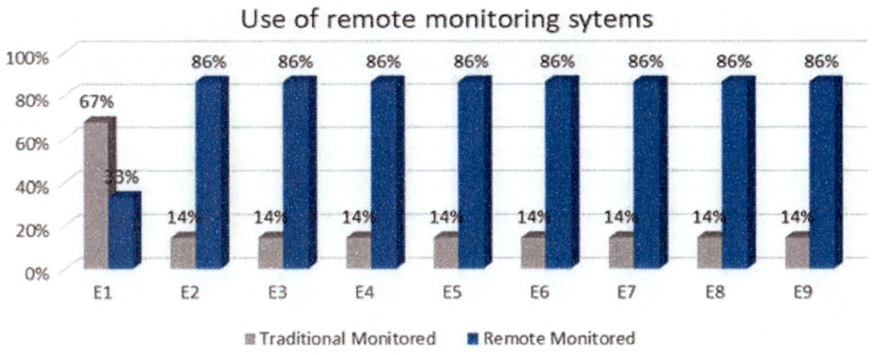

Fig. 5 Use of remoting monitoring systems

5 Conclusion and Recommendations

5.1 Conclusion

The findings clearly demonstrated that on Case 1 (the traditional monitored projects) faced a higher number of challenges such as; project control, delays and cost compared to the remote monitored projects (Case 2). The study outlined that; traditional management results into inability to deliver project at appropriate time with consequential effect resulting into cost overruns and dissatisfaction of clients. It is deemed as an insufficient method of management, and it doesn't provide good quality control which leads to delays and decreases productivity.

Remote monitored projects peaked with positive results on the project performance even though they are all subjected to similar challenges. It is evident as literature suggests that multiple project management needs sophisticated methods such as remote monitoring systems to manage and control challenges, also to aid the project managers enabling them to withstand the poised challenges, brought forth by a multi-project environment to ensure project success. Literature showed that multiple projects face many challenges under one project manager supervision, which negatively affects the project performance and delays the project to meet its goals. However, with the use of remote monitoring systems, the capabilities of the project managers are enhanced, and the projects perform better under this management method.

5.2 Recommendations

It is recommended that further research includes an in-depth analysis of the relationship between the project values, as mentioned above that the bigger the project value the less number of projects could be managed at each time. These might have an impact on why the smaller projects perform better, even though they are prone to similar challenges.

For future studies, it is recommended to focus on the cost implication related to change in project performance courtesy of the remote monitoring systems. This would be a way of quantifying the benefits of this management style making it easier to make a stakeholder decision for investment purposes.

References

1. Aarseth W, Ahola T, Aaltonen K, Økland A, Andersen B (2017) Project sustainability strategies: a systematic literature review. Int J Proj Manag 35(6):1071–1083
2. Chou JS, Yang JG (2012) Project management knowledge and effects on construction project outcomes: an empirical study. Proj Manag J 43(5):47–67
3. Cleland D, Ireland L (2010) Project management's portable handbook, 3rd edn. McGraw-Hill, New York
4. Jerbrant A (2014) A maturation model for project-based organisations with uncertainty management as an ever-present multi-project management focus. S Afr, J Econ Manage Sci 17:33–51
5. Kopsida M, Brilakis I, Vela PA (2015, October) A review of automated construction progress monitoring and inspection methods. In: Proceedings of the 32nd CIB W78 conference, vol 2015, pp 421–431
6. Miyang AF (2018) Multiple project environments and management challenges, higher national institute of Siantou, cameroon, 38th international academic conference, Prague, 11 June 2018
7. Patanakul P (2010) Key drivers to the effectiveness in managing multiple projects: an empirical investigation in an IT organization of a large financial institution. Paper presented at PMI® research conference: defining the future of project management, Washington, DC. Newtown Square, PA: Project Management Institute
8. Porter M, Siggelkow N (2008) Contextuality within activity systems and sustainability of competitive advantage. Acad Manag Perspect 22(2):34–56
9. Wang X, Truijens M, Hou L, Wang Y, Zhou Y (2014) Integrating augmented reality with building information modelling: onsite construction process controlling for liquefied natural gas industry. Autom Constr 40:96–105

Factors Influencing the Adoption of Disruptive Construction Procurement Models in State Universities

Jacob Mhlanga, Theo C. Haupt, and Mariam Akinlolu

Abstract Construction procurement models determine how projects are executed. For a long period, the design-bid-build model dominated government institutions and departments. However, the model proved inefficient in most situations within Zimbabwean state universities, resulting in the emergence of disruptive models such as design-bid, direct labor, labor only, and Public Private Partnerships (PPPs). Even so, as disruptive models gain traction, there is a scarcity of literature which identifies the factors that influenced their adoption. For that reason, this study explored the factors that influenced the selection of disruptive models in state universities. Premised on the interpretivism paradigm, the study employed a multi-case study approach. A purposive sample of five state universities was drawn constituting a total of 15 key informants. These informants were interviewed via online platform. The study findings show that most state universities had fully migrated to disruptive models. Furthermore, client expectations, financial, technological, social, and legal factors were influential in the selection of disruptive construction procurement models. It means that, in addition to the project quadruple parameters (cost, time, scope, and quality), state universities considered external factors. The study recommends that the government continue to support the state universities whereas on the other hand, the lagging universities are urged to explore adopting innovative disruptive models.

Keywords Disruptive · Construction procurement model · Projects · State universities · Design-bid-build

J. Mhlanga (✉)
Department of Landscape Architecture and Quantity Surveying, BA ISAGO University, Gaborone, Botswana

T. C. Haupt · M. Akinlolu
Department ofEngineering, Mangosuthu University of Technology, Umlazi Durban, South Africa

T. C. Haupt et al. (eds.), *Construction in 5D: Deconstruction, Digitalization, Disruption, Disaster, Development*, Lecture Notes in Civil Engineering 245,
https://doi.org/10.1007/978-3-030-97748-1_9

1 Introduction

In Zimbabwe, the state universities (quasi-government institutions) have long struggled with a plethora of issues. At the heart of these issues is a lack of infrastructure to keep up with the growing student population [18]. The Ministry of Higher and Tertiary Education, Innovation, Science and Technology Development [MHTESTD] has expressed its dissatisfaction with the lack of facilities and attributes that to the construction procurement model employed [19]. The procedure by which a client solicits goods and services is referred to as a construction procurement model [13]. As the main client, the government through the MHTESTD [19] points out that state university projects are predominantly solicited and implemented through the Public Service Investment Program (PSIP) system. The PSIP system is based on the design-bid-build (DBB) model. Notable, the DBB model takes a sequential approach, with design, bid, and construction stages separated [9]. Furthermore, Ali and Badinelli, [3] note that the DBB model has dominated government projects worldwide since the eighteenth century. Despite its prevalence, perhaps, its sequential methodology affects its effectiveness in project delivery. Tirivanga [30] reports that out of the 700,000 students enrolled in universities across Zimbabwe, only 200,000 had secured campus accommodation. The remaining students resorted to renting apartments within the proximity of the universities, a situation that has made them susceptible to squalid living conditions [5]. Apart from accommodation woes, several state universities across the country are characterised by lack of libraries, office spaces, lecture theatres, while in some instances they are embroiled in contractual disputes with contractors or consultants leading to project stalling, incomplete buildings or project abandonment [10], Tirivanga, [30]. Nonetheless, in the midst of the aforementioned challenges, understanding the factors informing the choice of construction procurement models is essential in addressing the facilities problems faced by state universities.

The construction procurement models are classified into two broad categories, namely traditional and disruptive models. In this study, the traditional model refers to the DBB while disruptive models include the Private Public Partnerships (PPPs), design and build (DB), direct labour (DL), labour only (LO) [9, 30, 33, 21]. There are various factors that influence the choice of a construction procurement model. Of late, the MHTESTD has been canvassing for the adoption of disruptive models [21]. Such a policy direction within government seeks to address the challenges that resulted from the PSIP approach. Arguable, Emmanuel [9] notes that the DBB model is deemed the easiest within governments hence its widespread adoption. However, Cynthia [8] suggests that as a result of DBB model, most public projects in Zimbabwe were incomplete while others were abandoned. A trend that is common in most state universities [15]. Therefore, project failures affected the universities. For that reason, some universities have resorted to adopting disruptive construction procurement models. In South Africa, Wits University abandoned the DBB and developed an in-house construction team [12, 16]. While in Ghana, Badu, et al. [4] noted that finances influenced universities to transition from DBB to disruptive models. In Zimbabwe, the Great Zimbabwe University (GZU) instead of constructing facilities, resorted to purchasing buildings around its campuses and converted them to lecture theatres, offices and libraries [21]. Whereas in some cases, universities established in-house construction units. For instance, the Midlands State University (MSU) acquired state of the art construction earthmoving equipment for its in-house construction unit to deliver its projects a move which disrupts the established DBB model [20]. As a result, disruptive construction procurement models have gained momentum and for that reason, the study explored the factors that influenced their adoption within the state universities. The remainder of the article is arranged as follows, Sect. 2 covers the literature review, and Sect. 3 details the methods employed while Sect. 4 and 5 highlights the results and discussions respectively. Section 6 concludes the study.

2 Literature Review

2.1 Construction Procurement Challenges in Universities

Construction procurement concerns have been the order of the day in most universities. These challenges have been evidenced by a plethora of symptoms that are prevalent in these places of higher learning [18]. Predominantly, Fitzgerald and Hodgson, [12] cement this observation by highlighting that most universities in the developing countries still had a long way to go in realizing the infrastructural necessities relative to the level of educational brilliance they seek to pursue. Universities have been experiencing contract disputes, project cost overruns, unprofessional project teams, clientele dissatisfaction, project abandonment, time overruns, rampant corruption [30, 15]. These problems have consequently led to a failure of attaining the quadruple constraints (time cost, scope, and quality) in most projects. Laryea and Watermeyer [16] contend that that failure in construction projects at universities to meet the project constraints was correlated to the antediluvian construction procurement models used. The key implication drawn from this assertion is that failure to innovate and use disruptive procurement strategies will cause universities to continuously fail in their project delivery. Another sombre construction procurement challenge faced by universities is linked to poor project planning. Ewa [10] postulates that project abandonment at universities was closely connected to "rushed projects" that circumvent the planning phases which had dire consequences on the decisions made about the construction procurement model choices. The study seems to attribute such an approach as the ultimate reason of failure by these institutions in providing the much needed infrastructure. The lack of infrastructure has led to the overpopulation in scant facilities within campuses.

2.2 Disruptive Construction Procurement Models

A disruptive construction procurement model is any model besides the DBB model. These models are perceived to address the DBB shortcomings. The DB model is whereby a client approaches a single organization to carry out both the designing and construction of the project [17]. The approach engages contractor at the earliest possible time in the project development stage. The contractor absorbs the risks associated with the project which indemnifies the client from any losses in the process. The DL model is when the client either engages the professionals or not for designs but employs their in-house tradesmen to carry out the works [1]. In this model the client retains total control of the project delivery by eliminating the services of an external contractor. On the contrary, Adegoke, [1] disputes the effectiveness of this model and concludes it is suitable for small and non-complex projects. Such an argument is well-founded because with project complexity emanates the challenge of experience and proficiency. The LO model is whereby the client purchases materials and solicits the services of an experienced contractor for project execution [2]. The model is flawed in that the client has no control over the efficient use of materials [24]. By and large, state universities rely on government for funding. Arguable, the government budgetary deficits have resulted in them pursuing alternative funding for their infrastructural projects. This has given birth to the PPPs which are joint ventures or cooperation between the government (public sector) and the private sector in project delivery [23]. The main reason behind this approach is that the government seeks to tap on the resources of the private sector. Of late, the state universities in Zimbabwe are pursuing this strategy [21]. Individually, these disruptive models have limitations but perhaps combining different models is likely to give yield to an effective hybrid model.

2.3 Factors Influencing Construction Procurement Model Selection

The choice of a construction procurement model involves a diversity of factors that need to be considered. This implies that there are intervening variables that tend to influence the selection of the construction procurement model apart for the project success parameters. Khattak [14] seems to claim that selection of the procurement route is largely dependent on the three aspects namely, client, contractor, and the supply chain of services. This suggests that how these relate and coordinate the activities of the project determines the project delivery. It dovetails with the findings by Wahaj et al. [31] that postulate that the success of the project is more inclined to the reasons that influence the selection criteria of the construction procurement model. The study identifies project objectives, experience, time, and quality issues as some of the core decision matrices that are used in arriving at the choice. These factors seem to address the long-lasting dispute that has been faced in the selection of the construction

Table 1 Factors influencing construction procurement selection

Environment	Factors
Internal	(1) Project characteristics (2) Client characteristics (3) Client requirements (cost, time, scope and quality)
External	(1) Economic or finance (2) Political or legal (3) Technological (4) Social (5) Natural disasters

procurement models. Primarily, studies focused mainly on the quadruple constraints to measure project success while turning a blind eye on external forces [31]. It seems a divergent view that addresses other external factors has emerged. Muhammad et al. [22] allude that the selection of a construction procurement model should be informed by both the internal and external considerations. It contradicts paying attention to project constraints only. Along the same vein, Ratnasabapathy et al. [29] and Naoum and Egbu [25] identified some of the key factors affecting construction procurement models selection (see Table 1). The factors seem to address the project from both the inward and outward perspective.

The internal environment represents the platform that the client controls. Therefore, within the organization, apart from the client characteristics and requirements, the project characteristics play a crucial role. Debatable, the complexity of the projects in state universities determine which model would likely deliver the intended results [16]. Therefore, the approach is at variance with clients that are fixated on the success parameters only [28]. The external environment on the other hand, represents an arena where the client has limited control. Within universities factors such as finance, legality, and technology are critical. Investments, market conditions, and cost at completion are also of paramount importance [31]. As for technology, e-procurement systems address rampant corruption in government departments [11, 4]. Notable, the advent of COVID 19 pandemic, disrupted construction procurement plans and hence the sustainability of the universities is determined by how they respond to such natural disasters [26]. The pandemic affected collaboration and supply chains, arguable, in-house construction units perhaps proved handy in ensuring that project disputes resulting from time or cost overruns were minimised.

3 Methods

The study employed an interpretivism paradigm. As a result, an exploratory research design was used [27]. The reason behind selecting this type of research design was informed by the need to explore the factors that influenced the adoption of disruptive construction procurement models in state universities [7]. Again, to further gain in-depth perceptions and attitudes of different key informants, a multi-case study approach was adopted [32], [6]. Out of the 12 state universities in Zimbabwe, 5 were purposively sampled. These universities represented Bulawayo, Manicaland, Midlands, Mashonaland West and Masvingo provinces respectively. Each state

Table 2 Study demographics

Department	Interviewee name	Profession/Qualification	Years of experience
Works and estates	A	Plumber and drain layer	41
	B	Architect	12
	C	Architect	9
	D	Quantity surveyor	2
	E	Engineer	15
Procurement	F	Procurement	12
	G	Purchasing and supply	7
	H	Tenders and contracts	17
	I	Purchasing and supply	10
	J	Logistics	5
Bursar	K	Accountant	14
	L	Finance	5
	M	Accountant	21
	N	Accountant	3
	O	Accountant	6

university provided 3 key informants drawn from the works and estates/projects, procurement, and finance/bursar departments respectively. These departments constituted the professionals that are actively involved in the decision-making processes of project implementation at universities. The key informants were interviewed via Google Meet for approximately 45–60 minutes with the researcher transcribing the responses. The researcher clarified any vague questions which enabled interviewees to provide rich responses. The data was cleaned and then coded prior to categorizing factors into themes. The analysis was done using Atlas.ti software.

4 Results

4.1 Demographics

All the interviewees that were sampled for the study availed themselves for the interview sections. Table 2 shows the demographics of these interviewees.

4.2 Description of Case Studies

To gain insight about the factors influencing the construction procurement models selection, the researcher used parameters as shown in Table 3.

The capital expenditure on projects in 3 universities (UA, UB and UE) fully adopting the disruptive models is averaging ZW$30.7 million, ZW$89.6 million, and ZW$450 million respectively. These are sizeable amounts in project investments.

Table 3 Description of case studies

University code	Projects	Estimated cost in millions (ZWL)	Dominant procurement model	Source of funds
UA	A1 A2 A3	$46.7 $38.2 $7.2	Disruptive	Government
UB	B1 B2 B3	$126.8 $107.8 $34.2	Disruptive	Government internal resources
UC	C1 C2 C3	$93 $65 $125	DBB Disruptive	Government internal resources
UD	D1 D2	$25.5 $68	DBB Disruptive	Government internal resources
UE	E1 E2	$800 $100	Disruptive	Government internal resources

Note: Disruptive either comprises of DB, DL, LO, PPPs or combination of any of these models

4.3 Factors Influencing Selection of Disruptive Procurement Models

The interviewees provided several factors that they consider when choosing to adopt their construction models. The researcher grouped these into five themes (Table 4).

5 Discussion

The study revealed that UA, UB and UE had fully migrated to disruptive procurement models with the remainder still employing DBB model. Those still adopting DBB probably have not formulated in-house construction units to tackle complex

Table 4 Factors considered in disruptive construction procurement models selection in state universities

Themes	Factors
Client expectations	Provision of infrastructural needs, quadruple constraints, composition of project teams
Financial	Source of funds, availability of funds
Technological	Availability of equipment and machinery
Social	Employment creation, collaboration within the supply chain
Legal	Compliance to Procurement Regulatory Authority of Zimbabwe (PRAZ) regulations, tendering for specialist works only

structures in line with Fitzgerald and Hodgson, [12] study findings. Nonetheless, UA remains reliant on government as its sole source of funds because the university is fairly new [4]. However, other universities have devised means to complement government funds through their internal resources. It corroborates findings by Laryea and Watermeyer [16] who advocated for innovativeness within universities in order to reduce the burden on the government fiscus. The estimated costs at completion for UA, UB and UE projects are indicative of how the universities have become innovative in project delivery instead of relying on DBB model (see Table 2). In line with the study by Ratnasabapathy et al. [29] and Naoum and Egbu [25] this study revealed the following factors:

5.1 Client Expectations

The introduction of disruptive construction procurement model saw the universities being the clients, designers, and contractors at the same time. Although such a setup has its own challenges, majority of the interviewees from UA, UB and UE hailed the approach and cited that it reduced some bottlenecks synonymous to the DBB model. Furthermore, *interviewee A* and *B* acknowledged that their keen desire to meet the needs of their students was demonstrated by projects being completed in the shortest possible time, within budget and required quality it agrees with findings by Naoum and Egbu [25]. Inversely, observations indicated that UA and UB were characterised by the construction of relatively simple designs. Questioned about that, *interviewee A, B* and *D* highlighted that the project team capabilities and construction team composition informed the nature of projects they carried out [12]. The interviewees also attested that simple structures had attracted a negative connotation among students and other stakeholders. Nonetheless, disruptive procurement models had delivered projects. Whether these are aesthetically appealing or not, that is open for debate.

5.2 Financial Considerations

Construction projects require a sizeable amount of funding. Wahaj et al. [31] study in India identified the source of funds as the determinant in model adoption. *Interviewee F, G* and *M* confirmed that at UA, UB and UC the choice of the construction procurement model was influenced by both the source and availability of finances. However, besides UA all other universities seem they are no longer reliant on government funds for their projects. This supports findings of Badu et al. [4] which claimed that the future of university projects lied in financial freedom. The interviewees revealed that these funds came from third stream income generating projects. UB formed a subsidiary company which is wholly owned by the university which ventures into farming, brick molding, tailoring of personal protective equipment and graduation regalia. *Interviewee B, K and M* hailed this approach and noted that it made the university liquid and hence with the available resources generated its projects are bound to be completed. It is worth noting that the five universities complemented government's effort of providing the infrastructure at state universities.

5.3 Technological Factors

Interviewee A lamented that the absence of construction technologies in universities influenced the adoption of their construction procurement model. As a result, the in-house construction teams used archaic systems with occasional hiring of plant and equipment in the process [21]. However, UB purchased its own equipment and plant which provided the construction with fully fledged construction equipment [20]. It confirmed the claim that technology was at the heart of any choosing a construction procurement model. Furthermore, *interviewee H* and *J* noted that the procurement of equipment allowed the university to deliver projects efficiently. It also demonstrated the confidence that universities bestowed on their project teams. *Interviewee K* argued that *the purchase of machinery reduced substantially the equipment hiring costs incurred by the universities.* It is also important to note that during the downtime period, when the university had no running projects, the equipment can be hired out and it becomes a third income stream. UD also confirmed that technological advancements drove their approach in terms of adopting procurement models just like in Ghana [4]. However, innovative systems such as e-procurement remain scarce in these institutions contrary to Fatima and Ahmad, [11] suggestions.

5.4 Social Factors

Employment creation is at the centre of state universities projects. *Interviewee D* noted that the social aspects of increasing employment opportunities had greatly influenced their construction model choice. Mostly, the disruptive procurement model enhanced creation of local employment. *Interviewee F* and *G* highlighted that university projects were now employment hubs which complement findings of Laryea and Watermeyer [16]. Contextually, having both the design and construction team within the university presented an opportunity for the universities to control the employment process [19]. For the design team, construction professionals and supervisory personnel constituted of permanent employees while the construction team was made up of contract workers. In that light, *Interviewee C* added that, *the disruptive construction procurement has seen the universities moving from just being sources of academic excellence but creating employment for the communities they are situated in.* On the other hand, collaboration with suppliers had increased because of the in-house construction units. Notable, Naoum and Egbu [25] echoed that social factors were essential when determining the factors of procuring services. *Interviewee K–O* further suggested that the engagement of local manufacturers enabled material acquisition on credit which further eased pressure on the university finances. Contrary, *interviewee A* and *G* bemoaned how universities oftenly failed to fulfill their obligations.

5.5 Legal Factors

Quasi-government institutions are regulated by the public procurement regulations. In that vein, *interviewee D* and *F* stated that the PRAZ regulations stipulated threshold for purchasing materials without going for a competitive bidding process. According to these regulations, public institutions have a threshold of $US10,000.00. Disputable, since construction projects require hefty amount of materials, the universities have resorted to purchasing these in tranches. It could be regarded as an adapted LO method as cited by Akinkunmi et al. [2]. However, arguable the approach is not cost-effective especially in an economy that is characterised by fluctuating prices like Zimbabwe. *Interviewee A* and *E* postulated that the purchasing of materials had in some cases led to project delays because of market shortages [24]. In combatting this potential challenge, UB and UE entered into contracts with manufacturers who produced goods upon demand. Regrettably, these arrangements were affected by payment delays by state universities. In areas of specialization such as roof installation, the universities outsourced these services through competitive bidding using DBB [9]. Nevertheless, the winning bidder is subcontracted by the in-house construction unit which shows the combination of construction procurement models.

6 Conclusion

The state universities adopted disruptive construction procurement models at varying levels. However, most of these universities had totally abandoned the DBB model and devised disruptive models. The migration was largely influenced by different factors which were summarized into five themes: clientele expectations, financial, technological, social, and legal considerations. The study further revealed that within these factors, it seemed the disruptive models were yielding positive results as compared to the DBB model. However, state universities that established in-house construction units, having both the design and construction teams under the same roof is likely to present sustainability challenges. Therefore, the study recommends for the government to continue supporting the state universities while those that are lagging to consider adopting disruptive models. The major limitation of the study is that it adopted a qualitative approach characterised by a small sample size. To address this limitation, the researcher recommends that a further study be carried out to explore quantitatively whether these hybrid models do yield value for money.

References

1. Adegoke B (2010) Problems of direct labour projects procured in Lagos State. Munich, GRIN Verlag, pp 1–46
2. Akinkunmi OG, Aghimien DO, Awodele OA (2018) Appraising the use of labour-only procurement system for building construction in Nigeria. Organ, Technol Manage Constr 1719–1726
3. Ali AK, Badinelli R (2016) Novel integration of sustainable and construction decisions into the design bid build project delivery method using BPMN. Procedia Eng 145:164–171
4. Badu E, Kissi E, Boateng EB, Antwi-Afari MF (2018) Tertiary educational infrastructural development in Ghana: financing, challenges and strategies, Africa education review, pp 65–81
5. Chiguvi D, Ndoma JT (2018) The effects of shortage of accommodation on students' performance in private tertiary institutions in Botswana. J Econ Sustain Dev 97–104
6. Cook L, Kamalodeen V (2019) An introduction to mixed methods case study research. In: MMIRA Webinar
7. Creswell JW (2014) Research design: qualitative, quantitative and mixed methods approaches, 4th edn. Sage Publications, CA
8. Cynthia T (2014) An investigation into the problems faced in construction project completion: case of department of public works Midlands province, Bachelor's thesis, Midlands state university, pp 1–68
9. Emmanuel OO (2013) Comparisons of performance of traditional and labour-only procurement in construction projects in Nigeria. DBA Afr Manag Rev 1–15
10. Ewa UE (2013) Root causes of project abandonment in tertiary institutions in Nigeria, Cross River Univ Technol 1–20
11. Fatima S, Ahmad A (2018) A review paper on analysis of procurement methods used in construction industry using sampling techniques. Int J Eng Techn 489–492
12. Fitzgerald P, Hodgson S (2010) Higher education infrastructure expansion: the case of wits based on the presentation by Prof Fitzgerald at the SARUA executive programme, 05 August. University of Dares Salaam, Tanzania pp 1–13
13. International Standards Organization ISO 10845–1:2010 (2010) International standard for construction procurement—Part 1: processes, methods and procedures. Geneva, Switzerland
14. Khattak IK (2015) Construction management, department of engineering and built environment. Anglia Ruskin University, pp 1–63
15. Laiton C (2019) MSU ordered to pay over US$3,2 m for engineering services. Newsday
16. Laryea S, Watermeyer R (2014) Innovative construction procurement at wits university. Proceedings of institution of civil engineers: management, procurement and law, pp 220–231
17. Lee JH (2018) Changes in quality management approaches for design-build highway projects. Georgia institute of technology, pp 1–84
18. Majoni C (2014) Challenges facing university education in Zimbabwe. Greener J Educ Training 20–24
19. MHTESTD (2019) Strategic plan 2019–2023, pp 1–88
20. Midlands State University Homepage, https://ww5.msu.ac.zw/blog/2017/09/12/vc-commissions-construction-equipment/. Last accessed 18 Feb 2021
21. Mswazie W (2018) Spotlight on students' housing. The Herald
22. Muhammad BA, Adamu T, Ladi BD (2015) Appraisal of construction project procurement policies in Nigeria. Am J Eng Res (AJER) 19–24
23. Mungkasa O (2018) The role of government in PPP, world urban forum 9 community collected data and partnerships to attract and assure inclusive private sector investment human city coalitions (HCC) and Slum Dwellers international (SDI) Kuala Lumpur
24. Munyoro F (2020) Consultant loses US$294 m suit. The Herald
25. Naoum S, Egbu C (2015) Critical review of procurement method research in construction journals. Procedia Econ Financ 21(5):6–13
26. Paul SK, Chowdhury P, Moktadir MA, Lau KH (2021) Supply chain recovery challenges in the wake of COVID-19 pandemic. J Bus Res 316–329

27. Plano Clark VL, Foote LA, Walton JB (2018) Intersecting mixed methods and case study research: design possibilities and challenges. Int J Multiple Res Approaches 14–29
28. PMBOK Guide (2017) A guide to the project management body of knowledge, 5th edn. Project Management Institute, USA
29. Ratnasabapathy S, Rameezdeen R, Gamage I (2006) Macro level factors affecting the construction procurement selection: a multi criteria model. The joint international conference on construction culture, innovation and management (CCIM), pp 581–591
30. Tirivanga A (2018) Higher and tertiary education infrastructure investment conference…-lessons for students. The Patriot
31. Wahaj M, Deep S, Dixit RB, Khan MB (March 2017) A study of project success and procurement frameworks in Indian construction industry. Int J Civil Eng Technol (IJCIET) 8(3):167–174
32. Yin RY (2018) Case study research and applications: design and methods 6th edn. Sage
33. Zinyama T, Nhema AG (2015) Public-private partnerships: critical review and lessons for Zimbabwe, pp 39–44

The Effect of Disruptive Technologies on Facilities Management: A Case Study of the Industrial Sector

Kathy Michell, Nick Brown, Jason Terblanche, and James Tucker

Abstract Disruptive technology is rooted in the fourth industrial revolution. Technological innovation and advancements have benefited a multitude of global industries. However, the benefits from disruptive technology are yet to be realised in the real estate industry. As such, the emergence of the global e-commerce industry positions the industrial sector ideally for technological adoption, specifically within the distributive warehousing and logistics sub-markets. Facilities management (FM) itself is further ideally placed within this framework to necessitate this change given its services and operational functions. This paper documents the findings of an investigation into the effect of disruptive technologies on FM within the context of the industrial sector. The findings show that the impact of disruptive technology can be disaggregated into two stances, namely: the cautious stance where little impact is occurring; and the open stance where significant impact is occuring. The nature of the different stakeholders gives rise to a variety of perspectives on the implementation of technology. Property-owning companies were more inclined to choose labour as a solution over technology as in South Africa labour proves to be a cheaper alternative than the capital required to incorporate and maintain technology. The paper concludes that although the industrial sector is experiencing the implementation of disruptive technologies the industry is still lagging behind the developed world. Not much will change in this regard as long as labour costs are cheaper than technological installations, or property owners are forced to adopt technology in order to remain competitive.

Keywords Facilities management · Disruptive technology · Industrial sector

K. Michell (✉) · N. Brown · J. Terblanche · J. Tucker
Sustainability Oriented + Cyber Research Unit for the Built Environment, Department of Construction Economics and Management, University of Cape Town, Cape Town, South Africa
e-mail: kathy.michell@uct.ac.za

T. C. Haupt et al. (eds.), *Construction in 5D: Deconstruction, Digitalization, Disruption, Disaster, Development*, Lecture Notes in Civil Engineering 245,
https://doi.org/10.1007/978-3-030-97748-1_10

1 Introduction

The benefits from disruptive technology are yet to be realised in the real estate industry, and therefore there are gains to be realised in this market. As such, the emergence of the global e-commerce industry positions the industrial sector ideally for technological adoption, specifically within the distributive warehousing and logistics sub-markets. The patterns of technological advancement is growing exponentially in relation to their size and cost, giving rise to the need for an awareness of this rate of technological change [34]. Progressive projects should infer the influence of these discoveries both presently and in the future allowing institutional alternatives to be created to maximise efficiency and productivity through the potential that such disruptive technology affords.

In this context, the industrial sector is ideally placed within the real estate space to adopt these technologies. Specifically, this market is disaggregated into logistics and distributive warehousing. The emergence of technologies introduced in developed countries especially by logistics companies such as Amazon point towards future opportunities for developing countries. Facilities Management (FM) itself is further ideally placed within the sector to facilitate this technological change given its strategic and operational function.

2 Literature Review

2.1 Facilities Management (FM)

The International Facilities Management Association (IFMA) defines FM as "a profession that encompasses multiple disciplines to ensure functionality, comfort, safety and efficiency of the built environment by integrating people, place, process and technology" [22]. This definition incorporates many attributes of definitions by other authors (see [4, 7, 9, 38]. Central to all these definitons is that of a theme of performance and an integrated approach to FM. A key aspect of the IFMA [22] definition is that it aligns with the core competencies mentioned by IFMA [23]. More importantly, it recognises the importance of technology in FM in leveraging value from the real estate asset. In this regard, it is important to note that the actual practice of FM has had to advance exponentially in order to keep up with a changing world [40]. In this regard, contemporary FM services are considered to include property/real estate; financial, change and human resource management; technology management; operations and maintenance; and lastly, occupational health and safety management [2, 6, 37]. The concept of leveraging value underpins strategic FM in terms of improved asset management; lower occupancy costs; higher productivity in employees; and operational FM by either improving services for the same cost or delivering the same service at a decreased cost [23, 31].

In terms of strategic FM, [4] and [8] argue that it entails developing a strategy that is oriented towards planning for continuous improvement by identifying the business needs. In addition, strategic FM incorporates the identification of future trends and risks that need to be considered and planned around in order to ensure that the organisation is able to constantly improve or remain competitive. The emphasis here on strategic planning emphasises the need for organisations to develop a strategically orientated approach and utilise operational FM as a tool to implement the decisions made based on future trends [11, 19, 23]. Operational FM deals with the day-to-day activities of an organisation regarding general upkeep of the facilities to allow the organisation to run in a functioning environment [6, 9, 30, 38]. The activities that fall within the scope of operational FM are considered to be 'non-core services' as they do not form a primary function of the objectives in the organisation but need to be aligned in order to optimise efficiency [8, 13]. These activities are routine as they are primarily linked to the caretaking role and require no thought process solely at this level to deviate from the current method of operation [27]. Therefore, operational FM alone will not satisfy organisational needs and will struggle to provide a competitive advantage for the organisation as it functions as a mechanism to deliver services to the organisation. In order to leverage value for the organisations, the FM needs to think strategically and act at the operational level.

2.2 *Disruptive Technology*

Disruptive technology is defined as a "new technology which has a rapid and major effect on technologies that existed before" [14]. These modern technologies are varied and in terms of their application to FM could include: the Internet of Things (IoT), Smart sensors, Radio Frequency Identification (RFID) technology; cloud computing; Artifical Intelligence (AI); Fifth Generation (5G) cellular networks; and robotics. Xia et al. ([43]: 1101) define the IoT as "the networked interconnection of everyday objects, which are often equipped with ubiquitous intelligence". This interconnectedness enables the development of new ways of doing things by connecting physical objects, coupled with intelligent decision making, and communication and coordinationbetween physical objects [3]. IoT provides seamless connectivity between heterogeneous networks. The eventual aim of IoT is to introduce the plug and play technology providing the end-user, ease of operation, remotely access control and configurability [35]. Coupled to the IoT are smart sensors and RFID technology.

In any smart application, sensors are very important. They detect any physical/chemical change and after processing the collected data, sensors automate the application/devices to make them smart. IoT integrates various types of sensors, devices and nodes having the capability to communicate with each other without intervention of humans [5]. In any IoT application, sensors bring the physical world very close to the digital world that can be implemented by leveraging cloud computing. Everything including managing of facilities are getting smarter and data generated at all levels of the process are used to improve efficiency, flexibility, and productivity. This would not be possible without smart sensors, which generate the data and allow further functionality from self-monitoring and self-configuration to condition monitoring of complex processes [16]. RFID technology essentially allows physical objects to be data-coded digitally which can then be interpreted by a reader via radio waves [1]. The National Institute of Standard and Technologies (NIST) further describe Cloud Computing (CC) as "a model for enabling ubiquitous, convenient, on-demand network access to a shared pool of configurable computing resources that can be rapidly provisioned and released with minimal management effort or service provider interaction" ([10]: 686).

Fifth-Generation (5G) cellular network is considered the latest cellular communication network technology that is aimed at improving internet connectivity and therefore the quality of service levels [18]. 5G is intended to support the connectivity of a large amount of devices, including the real-time transmission of data from information inputs, such as smart sensors and RFID tags. [18]. It is also intended to facilitate the upload of large amounts of information to cloud based technology, the inter-communication between things; and creating the network for accessing and locating sensors [12, 17, 25].

The term Artificial Intelligence (AI) covers the notion of an intelligent machine in terms of both operational and social consequences. It is commonly referred to the area of science concerned with machines functioning learning, perception, reasoning and logic processes [32]. Robotics is a branch of technology that concerns design, construction, operation, and application of robots [28]. Robotics fundamentally upgrades traditional logistics management to meet the demands of exponential growth in e-commerce business [20]. Robotics are able to improve efficiency, accuracy, and cost-competitiveness while accommodating the speed with which e-commerce orders need to be processed and delivered. In addition, robotics technology boosts efficiency and cost competitiveness, including automated guided vehicles, goods-to-person robotics picking, and automated racking and shelving solutions [39]. Compared with conventional robots, advanced robotics have a superior perception, flexibility, integratability, and mobility. These improvements permit faster setup, commissioning, and reconfiguration, as well as more efficient and stable operations [26]. Robots enhance a business' brand value by serving as a visible testament to smart operations, improved productivity and overall innovation. The installation of robots into industrial property/real estate would have significant implications for the FM who would likely be responsible for the maintenance, and would need to plan for upgrade and/or replacement.

2.3 Industrial Property Context

This research specifically examined the disaggregated market of industrial property that is known as logistics and distributive warehousing. An overview of the industrial property sector is provided in order to provide context to the discussion. In 2016, the SA property sector was valued at around R5.8 trillion [24].

The industrial/warehouse sector was valued at R188 billion, and the office and commercial sector at R1.8 trillion [24, 42]. The manufacturing sector in South Africa is experiencing a declining trend, which is consistent with Stats SA's report indicating a falling contribution to the economy since 1993 [36]. The increase in property value for the industrial sector can be attributed primarily to the warehousing and logistics sectors [15]. The industrial sector is experiencing expansionary growth over the medium to long term globally. The top two property types identified in the *Emerging Trends Europe*'s sector rankings were data centres and logistics facilities [41]. In this regard, Africa is considered an untapped market for growth in data centres [29]. In addition, South Africa specifically is experiencing sustained growth in the demand in the logistics and warehousing sector [33].

The role of FM is crucial in terms of leveraging value from these assets, and the adoption of technology in this regard is likely to provide both financial and economic benefits to the investor. How these technologies are implemented, and the extent of the impact brought about by these technologies are yet to be ascertained. However, this ambiguity will become clearer the more these technologies are properly understood and utilised. It is likely that FM practice will inevitably change to accommodate this technological disruption. The purpose of the paper is to report on research that was aimed at developing an understanding of the impact of disruptive technologies on the strategic and operational FM practices in the industrial sector.

3 Methodology

The research utilised a qualitative research methodology in the form of a case study of the industrial sector with the aim of identifying the impact of disruptive technology on FM practices within the industrial sector. Key stakeholders within the industrial sector were interviewed using a semi-structured interview outline together with site visits to industrial properties in the greater Cape Town metropole. The approach taken to select interview candidates was a non-random or purposeful sampling technique where interviewees were specifically selected according to their position in the industry. The stakeholder positions that were interviewed included property-owners, logistics tenants and a technology supplier for industrial purposes. The data was analysed using thematic analysis and the *NVivo 12* software package.

4 Findings

4.1 Emergent Theme 1: Changing Scope of FM Practice

It was clear from the data that the role of the facilities manager in the industrial sector has evolved to be more than just that of a maintenance manager or building manager. The modern role of an FM has evolved significantly with interviewees describing how the job has become more technical and intricate (INT03, INT04). Interviewees identified the need now for all-round skills and substantial years of experience to successfully manage industrial properties. In addition, their role needs to incorporate strategic thinking as the field in South Africa has become more professional and pushes the FM to be one step ahead of the game to mitigate as much exposure to risk. These strategies are mainly focussed around controlling the design of the facility in order to be compatible with the FM function; lifecycle costing and planning; lease agreements; and to some degree technology implementation.

A number of key roles were identified by the interviewees. Interviewees highlighted the importance of FM in the design process in industrial buildings in order to ensure more efficient and cost effective FM once the facility is completed: "*It's something we have been talking to our development guys a lot about because we told them that when you are planning to build a new building, they design the build, after 18 months they hand the build to us. They move on. We sit with the baby*" (INT03). The need for the FM to be involved in the decision-making at the design stages of a new build points to the strategic nature of the function and the integration of people, place, process and technology in leveraging best value from a real estate asset.

Another role identified was that of needing to be "*a step ahead of the game*" with life-cycle costing and planning (INT03). Life cycle costing and planning was considered an important part of the modern-day FM's role as it ensures that equipment is replaced timeously: "*Thing is, what you can't do in facility management is you can't cut corners. If this thing needs to be serviced, it needs to be serviced*" (INT04). The role identified here in terms of active financial forward planning further reinforces the strategic role played by FM.

It was further noted in the data that the "traditional" role of the FM has not disappeared with interviewees acknowledging that cleaning, maintenance, security, fire safety compliance, and occupational health and safety compliance still forms a significant part of everyday operations. Interviewees did not explicitly describe many challenges they face in terms of their evolving role, but did acknowledge that in a resource constrained environment, FM is the first place where companies look to apply budget constraints (INT01). This leads to FM's having to do their job with a smaller budget, yet the same level of service and compliance is expected of them.

In terms of implementing new technology, some interviewees disapproved with the way their operations are currently run and property owner(s) being not inclined to try new technology. However, one interviewee explained how improving the operation constantly is essential to the successful management of the facility and that an FM should constantly be testing new equipment and technology. A commitment to continuous improvement further reinforces the notion of thinking strategically, but acting locally.

The leasing structure adopted by those interviewed is that of a triple-net lease i.e., the tenant/lessee pays the base rental plus property taxes, building insurance and all maintenance and operating costs in their allocated space: "*I prefer the lease where we look after the external structure ... roof, gutters ... and then anything exclusive the tenant looks after*" (INT03). Therefore, the property-owner is responsible for only the structure and providing a suitable space for tenants. There is a risk associated with investing capital into technology to customise spaces for tenants. The nature of the leasing agreements could be a reason for the relatively slow rate of implementation of disruptive technology in that many of these systems would need to be integral to the 'smart' building structure/shell in order to realise the benefits of improved efficiency, flexibility, and productivity. If this is the case, then a significant portion of the capital expenditure in adoption and implementation will be carried by the landlord/property owner.

Lastly, relationships were considered critically important in FM where services are outsourced (INT01). Interviewees emphasised the importance of strong relationships with all parties and thus being able to get the best out of service providers.

4.2 Emergent Theme 2: Technology Adoption and Implementation

Technology is disaggregated between disruptive technology and established technology.

Disruptive technology

The findings indicate that RFID technology is seen as disruptive to barcode systems. Moreover, it would allow for greater automation in operations and has application in the health and safety compliance. INT04 and INT05 mentioned the use of an imported system by SICK manufacturers in their operations: "*RFID overcomes quite a bit of that because it's a radio based thing. Even if it's an invisible object, it may maybe visible with a radio frequency type of technology. And the same is true in parcels and logistics..... the progression of the technology would be towards the RFID stuff because it is seen as the sort of the 'future technology*'" (INT05).

Smart sensors also emerged as a key disruptive technology finding application in robotics, health and safety systems, and conveyor systems that aid in cost reductions in the operational costs of the facility (INT05, INT01).

Cloud technology in distributive warehousing also found application amongst interviewees. Cloud technology has a high degree of application in the logistics space and helps with warehouse management and inventory systems. INT04 indicated the use of a cloud-based solution that provides an integrated warehouse, inventory and workforce management platform that easily scales with increased volume and complexity (INT04).

The Internet of Things (IoT) enables intelligence in and between machinery by manipulating and self-interpreting data. INT05 stated the IoT opens possibilities for preventative maintenance thereby making the duties of the facilities manager more accurate and efficient: "*I'm reaching 12,000 h of operation, so you probably need to think about preventative maintenance. That level of intelligence is now becoming quite common and it's being reported across the machine and up the machine as well*". Apart from preventative maintenance, interviewees also stated that the IoT is useful along the logistics chain, particularly with regards to pre-emptive stock control.

Programmable Logic Controller (PLC) technology and Machine-to-Machine (M2M) technology allows for coordination of machines in a warehousing environment and therefore increases efficiency along the production line. INT05 mentioned their use of Profibus which is an industrial communication protocol that allows machines to communicate with one another: "*...there might be... different machines and they need to all be put in line and made to talk to each other to move one thing from one to the other... or have robots doing processing things all the way down a production line... So machine to machine communications is already very much every day. So you've got Profibus or whatever the communication methods are communicating between them, reporting maybe up to a high level control system (which) is quite common.*"

Robotics in terms of the findings was limited, despite rich literature around its applications in logistics. This finding is ultimately owing to South Africa and its third world nature. INT05 indicated a growing interest in the adoption of this form of technology in the industrial sector in South Africa.

Lastly, sorter technology applications were considered by interviewees to become an integral function of the logistics process as they have the potential to result in better lead times, customer delivery time and manufacturing: "*We've actually just implemented a new sorter in Joburg, which automatically sorts all this stuff in dispatch. So that whole manual process in dispatch from which parcel goes to which customer and to which house and everything, that is now happening automatically*" (INT04).

Established Technology

Established technology identified during the data collection were: security technology; camera technology; barcode readers; mobile applications; electric and fire safety technology; renewable energy and automated conveyor systems. Security technology such as electric fencing, vehicle and people scanning devices, CCTV and global monitoring systems emerged as the greatest concern and as a result the adoption level of this form of technology is high.

Interviewees felt that camera technology is useful in operations, especially for scanning barcodes and RFID tags. Barcode readers help with accuracy and efficiency but are limited in their potential owing to its requirements of optical technology (INT05). Mobile application is considered a useful established technology for the FM, particularly in monitoring construction productivity and building maintenance

(INT01). Mobile application, however, was considered limited by interviewees in that the applications discussed still require physical inspection and logging of the issue.

It was clear from the data that in making decisions to implement technology in the logistics and warehousing sector, FM's weight up the challenges of implementation against the need for technology in logistics processes, comparative advantage economics and production efficiency. The key challenges identified in implementation lies in the incentives to deviate from business as usual, the FM professionals' adaptivity to change and, predominantly cost concerns.

5 Discussion and Conclusion

The impact of disruptive technology can be disaggregated into two stances, namely: the cautious stance where little impact is occurring and the open stance where significant impact occurs. The nature of the stance taken by the property owner and/or facilities manager gives rise to a variety of perspectives on the implementation of technology and the resultant effect on FM. In the cautious stance, it was identified that those interviewees from the property-owner side were reluctant to implement technology into their FM approaches. They were more inclined to choose labour as a solution over technology as in South Africa labour proves to be a much cheaper expense than the capital required to incorporate and maintain technology. The open stance interviewees don't necessarily see labour and technology as a trade-off but prefer to have the view that the implementation of technology provides the labour with the necessary tools to allow the business to function. Therefore, the two are not substitutes but rather complement each other. With little implementation there is little impact and with significant implementation, significant impact can be expected. The conditions of the market and economy are significant contributing factors to this.

It is notable that the nature of the property-owners relationship or lease agreement with the property also plays a role in their stance. Many property-owning companies have a lease contract with industrial tenants that operate in the spaces provided i.e., they do not occupy these spaces themselves. This allows for situations where the tenant has little control over perfecting the space for themselves for their operations and are unlikely to spend large capital amounts to retrofit an industrial space to be 'smart'. The combination of the nature of leases where tenants are happy for the owning companies to provide a 'white box' for them combined with labour which is seen as a more desirable solution over technology gives rise to the cautious approach adopted by the property-owning companies. However, the companies have not ruled out the implementation of technologies but only in very extreme conditions when turbulent changes in the market are occurring will they be considered. It must reach the level where it becomes a requirement to adapt to stay competitive.

In conclusion, the industrial sector is experiencing the implementation of disruptive technologies, however the industry is still lagging behind the developed world. Not much will change in this regard as long as labour costs are cheaper than the cost of technological installation and property-owners need to adapt their buildings in order to remain competitive.

References

1. AB&R (2021) What is RFID and how does RFID work? https://www.abr.com/what-is-rfid-how-does-rfid-work/. Accessed 14 Sept 2021
2. Adama UJ (2019) A study of the impact of technological innovations on the social sustainability of facilities management employees in South Africa. Ph.D. Thesis, University of Cape Town
3. Al-Fuqaha A, Guizani M, Mohammadi M, Aledhari M, Ayyash M (2015) Internet of things: a survey on enabling technologies, protocols, and applications. IEEE Commun Surv Tutor 17(4):2347–2376
4. Alexander K (1994) A strategy for facilities management. Facilities 12(11):6–10
5. Altohami A, Haron N, Law T (2021) Investigating approaches of integrating BIM, IoT, and facility management for renovating existing buildings: a review. Sustainability 13(7):3930
6. Atkin B, Brooks A (2015) Background to facilities management. In: Atkin B, Brooks A (eds) Total facilities management, 3rd edn. Wiley-Blackwell, Oxford, pp 1–11
7. Barrett P (1995) Facilities management: towards best practice, 1st edn. Blackwell, Oxford
8. Barrett P (2000) Achieving strategic facilities management through strong relationships. Facilities 18(10/11/12), 421–426
9. Barrett P, Baldry D (2003) Facilities management: towards best practice, 2nd edn. Blackwell, Oxford
10. Botta A, Donato WD, Persico V, Pescapé A (2016) Integration of cloud computing and internet of things: a survey. Futur Gener Comput Syst 56:684–700
11. Boyle L (2016) Urban facilities management as a systemic process to achieve urban sustainability in South Africa. Master's Thesis, University of Cape Town
12. Cheng J, Chen W, Tao F, Lin C-L (2018) Industrial IoT in 5G environment towards smart manufacturing. J Ind Inf Integr 10:10–19
13. Chotipanich S, Nutt B (2008) Positioning and repositioning FM. Facilities 26(9/10):374–388
14. Collins Dictionary: Definition of 'disruptive technology', https://www.collinsdictionary.com/dictionary/english/disruptive-technology. Accessed 13 Sept 2021
15. Commercial Property News (2021) What to expect from the industrial property market in 2021, https://eprop.co.za/commercial-property-news/item/22265-what-to-expect-from-the-industrial-property-market-in-2021. Accessed 13 Sept 2021
16. Fairchild A (2019) Twenty-first-century smart facilities management: ambient networking in intelligent office buildings. In: Mahmood Z (ed) Guide to ambient intelligence in the IoT environment. Springer, Switzerland, pp 271–289
17. Fettweis GP (2016) 5G and the future of IoT. In: Proceedings of ESSCIRC conference: 42nd European solid-state circuits conference. IEEE, Lausanne, Switzerland, pp. 21–24
18. Goudos SK, Dallas PI, Chatziefthymiou S, Kyriazakos S (2017) A Survey of IoT key enabling and future technologies: 5G, mobile IoT, sematic web and applications. Wirel Pers Commun Int J 97(2):1645–1675
19. Grimshaw B, Cairns G (2000) Chasing the mirage: managing facilities in a virtual world. Facilities 18:392–401
20. Huang GQ, Chen MZQ, Pan J (2015) Robotics in ecommerce logistics. HKIE. Transactions 22(2):68–77
21. International Facilities Management Association (IFMA): What is Facility Management? (2021a). https://www.ifma.org/about/what-is-facility-management. Accessed on 13 Sept 2021
22. International Facilities Management Association (IFMA): What do facility managers do? (2021b). https://www.ifma.org/about/what-is-facility-management. Accessed on 13 Sept 2021
23. Jack J (1994) Strategic facilities management. Prop Manag 12(4):40–43

24. Kilian A (2016) Property sector contributes significantly to South Africa's economy. https://www.tysonprop.co.za/news/property-sector-contributes-significantly-to-south-africas-economy/. Accessed on 16 Sept 2021
25. Kumar NM, Dash A (2017) The internet of things: an opportunity for transportation and logistics. In: Proceedings of 2017 IEEE international conference on inventive computing and informatics (ICICI). IEEE, pp 194–197
26. Küpper D, Lorenz M, Knizek C, Kuhlmann K, Maue A, Lässig R, Buchner T (2019) Advanced robotics in the factory of the future. Boston Consulting Group, Boston
27. Langston C, Lauge-Kristensen R (2002) Strategic management of built facilities. Butterworth-Heinemann, Oxford
28. Lexico (2021) Definition of Robotics. https://www.lexico.com/definition/robotics. Accessed on 13 Sept 2021
29. Mhlanga D (2021) Data centres and energy usage. Asset Magazine 96:100–107
30. Noor NMM, Pitt M (2009) A critical review on innovation in facilities management service delivery. Facilities 27(5/6):211–228
31. Nutt B (2000) Four competing futures for facility management. Facilities 18(3/4):124–132
32. Perez JA, Deligianni F, Ravì D, Yang G-Z (2017) Artificial intelligence and robotics. In: UK-RAS network, pp 1–56
33. Property Wheel (2020) Africa's logistics and warehousing sector to weather pandemic storm. https://propertywheel.co.za/2020/06/africas-logistics-and-warehousing-sector-to-weather-pandemic-storm/. Accessed on 13 Sept 2021
34. Rivers R (2018) Simplified law of accelerating returns. https://www.ronrivers.com/2018/09/04/simplified-law-of-accelerating-returns/. Accessed on 13 Sept 2021
35. Shafique K, Khawaja BA, Sabir F, Qazi S, Mustaqim M (2020) Internet of things (IoT) for next-generation smart systems: a review of current challenges, future trends and prospects for emerging 5G-IoT scenarios. IEEE Access 8:23022–23040
36. Statistics South Africa (2017) Three facts about the ICT sector. Statistics South Africa, South Africa. http://www.statssa.gov.za/?p=9852. Accessed on 13 Sept 2021
37. Tammo M, Nelson M (2012) A critical review of the concept of facilities management in community-based contexts. In: The 28th annual ARCOM conference. ARCOM, Edinburgh, pp 1379–1388
38. Tay L, Ooi JTL (2001) Facilities management: a "Jack of all trades"? Facilities 19(10):357–363
39. Teoman S (2020) Achieving the customized "rights" of logistics by adopting novel technologies: a conceptual approach and literature review. UTMS J Econ 11(2)
40. Tobi S, Amaratunga D (2013) Social enterprise applications in an urban facilities management setting. Facilities 31(5/6):238–254
41. Urban Land Institute (ULI) (2021) Emerging trends in real estate: an uncertain impact Europe 2021. Urban Land Institute and PwC, Washington
42. Warehouse Finder (2019) SA Property sector values at R4.9 trillion. https://www.warehousefinder.co.za/news/sa-property-sector-values-at-r49trillion/. Accessed on 15 Sept 2021
43. Xia F, Yang LT, Wang L, Vinel A (2012) Internet of things. Int J Commun Syst 25(9):1101–1102

Application of Artificial Intelligence for Construction Workers Wellbeing in South Africa

Mohlomi Terah Raliile and Theo C. Haupt

Abstract Globally, the construction industry is known for its complexity, where a multi-array of interdependent activities all take place at the same time. Construction job sites are increasingly dangerous with historically low levels of innovations and adaptation to change. Construction workers are required to perform repetitive manual tasks involving heavy lifting, bending, twisting, reaching overhead or away from the body and working under generally unfavorable conditions. Construction companies have a moral and legal responsibility to ensure working environments for their workers that do not present a threat to their health, safety and general wellbeing. The purpose of this study is to explore how the application of disruptive AI technologies could improve construction workers' wellbeing and safety on the job site. This paper reviewed previous studies on the application of AI on construction workers' wellbeing such as, for example, Weak AI, Strong AI, Machine Learning, Deep Learning and Big Data. Furthermore, keywords such as, for example, Construction, Workers, Health and Safety, Artificial Intelligence and Wellbeing were used to search online databases. The findings of this study indicated that the integration of AI on construction job sites may be used to prevent negative and health and safety prejudicial occurrences from happening by monitoring and tracking the workforce and their construction activities using, for example, applications (apps), wearables or emotional wellbeing tools. Moreover, the complex industry could benefit from machine learning to bank an unlimited amount of data for better overall health and safety performance. Based on the findings, the benefits of AI technologies imbued with machine learning, have a far-reaching impact than the negatives as traditional health and safety systems continue to fail to eliminate construction accidents. Therefore, the adoption of AI could help improve construction workers wellbeing.

M. T. Raliile (✉)
Department of Quantity Surveying and Construction Management, University of the Free State, 205 Nelson Mandela Drive Park West, Bloemfontein, Free State, South Africa

T. C. Haupt
Mangosuthu University of Technology, 511 Mangosuthu Highway Umlazi, Durban 4031, South Africa

T. C. Haupt et al. (eds.), *Construction in 5D: Deconstruction, Digitalization, Disruption, Disaster, Development*, Lecture Notes in Civil Engineering 245,
https://doi.org/10.1007/978-3-030-97748-1_11

Keywords Artificial intelligence · Construction workers · Health and safety · Machine learning · Wellbeing

1 Introduction

The construction industry, when compared with other industries is evidently one of the most difficult and challenging [1]. Workers are generally required to perform repetitive manual tasks involving heavy lifting, bending, twisting, and turning in awkward and cramped positions, reaching overhead or away from the body and working under generally unfavorable conditions [1]. The industry is heavily reliant on manpower. However, there is a great cause for concern for wellbeing as workers are often considered invaluable to the construction processes [20]. Furthermore, the industry does not attract young people and is characterized by an ageing workforce [5, 7]. The negatives outweigh the positives in this regard as construction projects are carried out within client-driven short durations and rely on high levels of productivity [5].

Human error, defective equipment, and dangerous work environment among others, are cited as the common causes of construction accidents with their associated negative consequences [11, 16]. Despite numerous efforts to improve health and safety on construction sites, traditional approaches have reached saturation and continue to fail, and innovations are required [8]. Zhou et al. [25] attested that human error is inevitable and further argued that it is impossible to eliminate all errors. In this light, innovative technologies for timely detection of symptoms, leading decisions and events, triggers that might result in accidents if left unaddressed, are a means to prevent accidents, and the construction industry is not immune to disruptive technologies that could bring change and opportunities [19, 25].

The adoption of Artificial intelligence (AI) provides suitable and advanced decision-making tools that have been shown to effectively analyse and provide solutions to construction problems [21]. With the assistance of AI, managers can be more specific in their assessments, help improve workers' health and safety, and recognize latent risk factors before they become dangerous [14]. AI can assist in the identification of indicators of safety risk in photos by observing activities such as workers executing construction activities at height or potential outcomes of slips, trips, and fall hazards (ibid.). These platforms can review thousands of images and project data points daily without getting exhausted and identify risk factors and make observations that would never have been identified by humans [3, 14].

Artificial Intelligence (AI) is gradually evolving as a multifaceted technology that could have far-reaching effects in several sectors and cause disruptive changes [4]. This paper aims to explore various applications of AI on improving the well-being of construction workers. A comprehensive review of previous studies is conducted to establish how AI can be adopted in construction. This study identifies historical and current trends in AI technologies, their benefits and the extent of their applications and limitations. The next section explores the methodology employed in this study.

2 Methodology

This study is an initial overview of the AI landscape in the construction industry. It is a literature review study which forms part of ongoing empirical research. The study is based on AI research across all fields and its applicability in the construction industry. Relevant studies and research were primarily obtained from position papers, policy documents and recent scientific publications from online research databases such as Scopus and Web of Science. Database search was conducted using keywords such as Construction, Workers, Health and Safety, Artificial Intelligence and Wellbeing. The initial search revealed many irrelevant documents across all subject areas as the word construction is a common term. The initial search was filtered to reveal only research from 2008 to 2019 in the area of engineering. A total of 59 papers were obtained. Of the total papers obtained, 30 specifically investigated occupational health and safety in construction and machine learning was more prevalent than AI. Only 7 papers specifically addressed some of the applications of AI especially in machine learning and NLP. Articles reviewed were not sufficient to solely conduct literature finding and led to the adaptation of studies published in other fields such as the financial sector, healthcare sector, legal sector and information technology. Publications cited in this study include publications from reputable scholarly journals and reports.

3 Literature

Artificial Intelligence (AI) is currently regarded as an important topic across all industrial sectors for both economical and societal development [12]. AI is creating an ever-increasing role in global economies and creating a new industrial revolution [19]. Recent developments in computer power, new algorithms and big data have led to major breakthroughs in the advancement of AI and machine learning (ML) [9]. The applications of AI and ML such as machine translation and image recognition are increasingly being used [9]. Therefore, there are many opportunities, many of which are yet to be realised.

3.1 What is AI?

The term Artificial Intelligence was born in the late 1950s from a group of computer scientists and most notably, John McCarthy and was defined as ‘making a machine behave in ways that would be called intelligent if a human was so behaving’ [17]. However, the main underlying ideas and practical computer programs had previously been established in the late 1940s by Alan Turing and other pioneers of computer science [4]. There are many competing AI definitions and some of the most referred to are: “non-biological intelligence (intelligence being the ability to

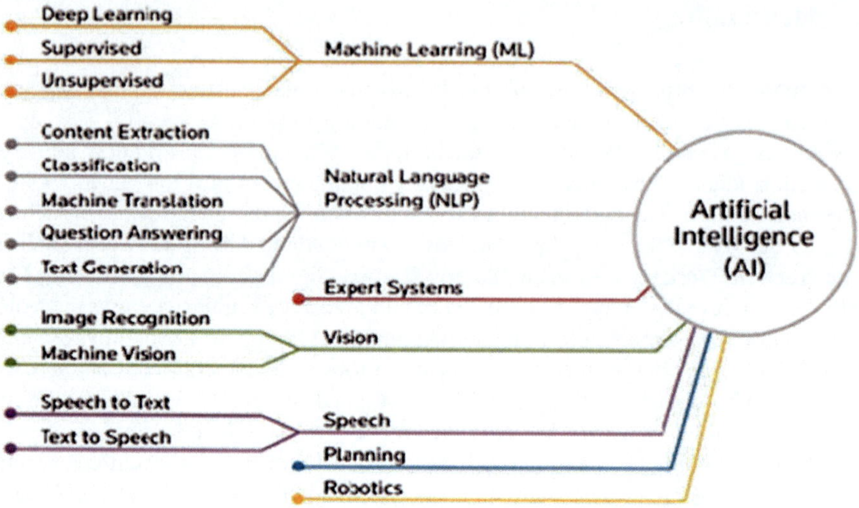

Fig. 1 AI and its dimensions (*Source* Thomson Reuters cited in Richardson 2018)

accomplish complex goals)" [23]. The European Commission (2018) referred to 'AI' as systems that exhibit intelligent behaviour by analysing their environment and taking actions with some degree of autonomy to achieve specific goals.

Artificial Intelligence is an umbrella term often used to describe techniques linked with data analysis and pattern recognition such as machine learning and natural language processing [4]. The growth of AI systems has been made possible by advances in computer memory, availability of data and the development of new machine learning methods, especially by neuro-inspired learning algorithms, notably deep learning algorithms [23]. "Artificial Intelligent-based systems can be purely software-based, such as, for example, conversational assistants, image analysis software, search engines, speech and face recognition systems) or can be integrated into hardware devices such as, for example, autonomous cars, drones, medical devices, advanced robots)" [4]. Figure 1 shows dimensions of AI technology:

3.2 *Levels of AI*

There are different levels of AI and scientists have classified them into Weak AI and Strong AI. Weak AI also known as Narrow Intelligence is the ability of a machine to accomplish a specific set of intelligent tasks using software to guide investigation and responses [23]. Hetter highlighted that this type of AI does not reach a level of consciousness or full sentience as such but acts as a problem-solver within a specific field of application. Strong AI also linked with Artificial General Intelligence (AGI) and Superintelligence is when a machine can demonstrate cognitive functions at the

Table 1 AI application fields and AI data sources [15]

AI application fields	AI data sources
Cyber security	Unstructured and structured data
Industry 4.0/production	Images
Medicine	Videos
Internet/social media	Text
Connected cars	Audio
Autonomous vehicles	Environment
Digital identity (recognition of biometric features)	Speech
Smart city	Internet of things and complex systems
Telecommunications networks	Network data
Robots	Log-files of software applications
Smart farming	QoS parameters of network protocols (nework traffic)
Customer service (bots)	Databases
Requirement-based code generation	Canvas drawing files (CVS)

level of humans or acquire general intelligence far beyond the level of humans [23]. Currently, AI levels of are described as narrow AI [4].

3.3 Applications of AI

Artificial Intelligence is now proving probable and can complete automation tasks, where robots are becoming competitive as humans without demonstrating human-like characteristics such as tiredness or sickness [18]. Moreover, computer memory capacity is increasing and programmes are becoming more sophisticated (ibid). Recently, AI has made prompt advances in a wide range of application areas such as machine translation, autonomous vehicle navigation, control of assisted driving as well as voice, video and image recognition [4]. Table 1 illustrates different AI application fields and their data sources:

3.4 AI in Construction Health and Safety

Artificial Intelligence is expected to have a major impact in addressing many social challenges such as ageing, transport, and the environment [9]. It is also expected that it will help improve the quality of life of people both at home and at work [9]. Unhealthy workplaces can take a toll on workers' well-being and productivity both physically and mentally [3]. Adopting AI in construction could bring meditative practices into the workplace, employee support, and improvements to communication [24]. AI brings the ability to sift through thousands of data points, automatically and instantly categorise and classify them, and highlight vital information on spend and risk to management [24]. Construction companies can, therefore, bank on as much data as possible for every project, giving AI as much information as possible to learn from [10].

The integration of live videos and image detection with machine learning can be used to learn and identify construction risks in real-time, such as workers not wearing suitable health and safety gear on-site, or hazardous gaps in scaffolding or buildings [10]. AI can identify hazards and alert workers in real-time to ensure a safer work environment [10]. Furthermore, the applications of machine learning are not only transforming the way construction companies execute projects but also have a positive impact on the employees working conditions daily.

Future applications of AI with 5G connectivity (5th generation wireless technology for digital cellular networks) could eventually provide the platform for remote operation of construction machinery thousands of kilometres away and operation of machinery underground without the network latency issues suffered previously [10]. Construction workers will be able to operate machinery from a single remote base, speeding up projects and avoiding health and safety hazards that could cause delays on-site [10].

AI applications are currently under scrutiny on how they can be used to help the architecture, engineering, and construction (AEC) industry. A 2016 Engineering News-Record ENR photo competition employed AI agent that reviewed photo submissions to look for potential job-site risks. The system processed 1080 images in less than 5 min, while it took more than 5 h to do a similar review by human experts. Construction companies such as Skanska have invested in AI image processing technologies like those used during the 2016 ENR competition to reduce job-site risks. Skanska reviewed how to expand the types of risk indicators identified by AI through BIM 360 Field software; a software platform used for the collection of site imagery. They further developed a pilot project where the technology is being expanded to handle additional types of visual information, including videos with voice narration and real-time image streaming from job-site cameras, to prevent accidents [5, 14].

Cortexica and Cisco also developed a video analysis system that will be able to monitor all worker's PPE equipment where any worker who is flagged as non-compliant is alerted by sound or, access to the site could be denied. The system reinforces one of the responsibilities of a supervisor and the other staff. Apart from the apparently visible hard hats and high visibility jackets, it is improbable that a supervisor could monitor all workers always wearing the correct PPE especially on large construction sites. However, it is unclear whether AI could cover large areas economically, especially with regularly changing working locations [2].

4 Concerns

Despite the benefits of AI in the construction sector, there are concerns about how it will negatively affect the industry. Stress, musculoskeletal difficulties, discrimination, heightened precarisation, job losses and possibly work intensification have been cited as risks to psychosocial violence and physical violence in digitalised workplaces [18a]. Public debates have spawned panic on potential impacts of AI. However, theories about the negatives and positives of AI are currently inconclusive [9]. PwC [22] reported that in the next 20 years, AI will create as many jobs as it displaces in the UK. However, Moore argued that it is not the technology in isolation that creates benefits or risks for workers, but rather its implementation that creates negative or positive conditions.

Although the net effects across industries at this stage are almost neutral, it was estimated that AI in the construction industry could displace 15% jobs and create 12% jobs resulting in a net effect of negative 3% [22]. Newton [19] attested that although AI could lead to fewer jobs in construction, there is a general trend nonetheless towards fewer people wanting to enter manual construction trades in any case. The implications could also mean better safety and working conditions, with shorter periods on site, less waste and, critically reducing accidents. Although there will be fewer jobs on-site, they will be performed by higher-skilled labour who will be better protected (ibid). Figure 1 identified some key barriers preventing faster deployment of AI structures in European Countries (Total vs. SMEs) (Fig. 2).

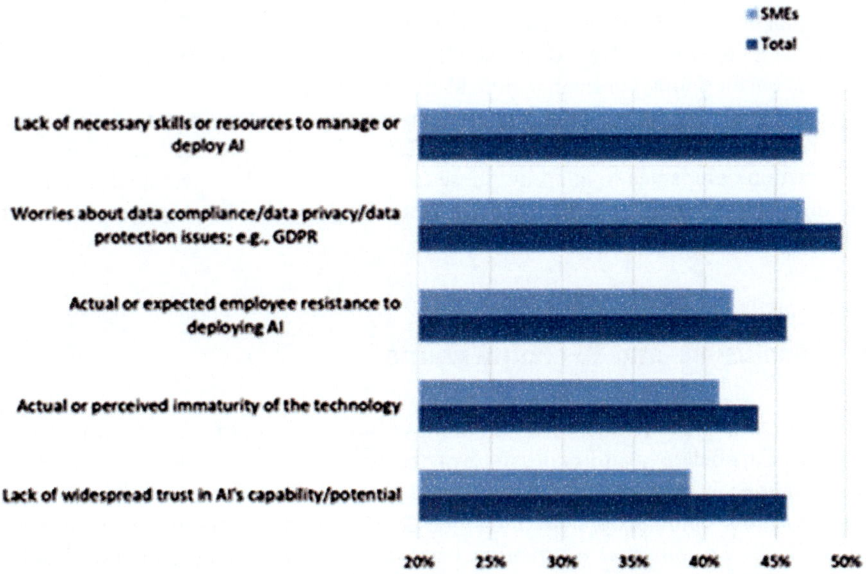

Fig. 2 Areas preventing faster AI deployment in Europe (Source: IDC's Western Europe AI/Cognitive Solutions Survey, Ju. 2018 ($n = 350$))

The 2018 IDC survey carried out among European industry and SMEs identified three primary types of barriers to AI adoption as (i) those related to the internal technical capacity, (ii) those concerned with AI policy and regulatory risks, and (iii) those related to the early stage of development of AI applications in business and social acceptance of AI [4].

5 Findings and Discussions

Artificial Intelligence is not a new technology and has been in existence since the 1950s. Recent advances in computer algorithms, big data analytics and computer hardware have led to rapid advances in this field. However, the development of AI technologies is still in its infancy. Current AI applications are described as Weak or Narrow AI. It is blurry what the future of AI will be as there are various concerns around AI safety, security, social acceptance, lack of necessary skills and whether it will displace the current workforce. The construction industry is not immune to these concerns. It is also speculated that the rise of AI will take over humans. The fears and concerns construed could retard the growth of AI in construction especially in developing countries where construction work is heavily reliant on manpower. However, the benefits have a far-reaching impact than the negatives as traditional health and safety systems continue to fail to eliminate construction accidents.

The applications of AI are proving probable in assisting managers and workers to improve on-site health and safety. There is a need for skills retraining as the development of AI is highly reliant on an educated workforce thereof. Although some sectors are more advanced in AI adaption, the construction industry still lacks behinds. Furthermore, political, technical and legal factors could hinder or accelerate progress of AI. Companies and government must plan on how best to help people acquire new skills to prepare them for the future. Some of the challenges faced by construction companies in adopting these technologies could be persuading clients and legislatures both of whom can be resistant to change with a tendency to resort to familiar tried-and-tested methods.

6 Conclusions and Recommendations

Proper use of AI in construction could improve working conditions and most importantly lead to better health and safety of the workers. Human error is subject to most fatalities in the workplace and AI could greatly improve standards by reducing mistakes and keeping people out of danger. One of the greatest components of AI is the ability to take unlimited information, learn from algorithms and make autonomous decisions. Construction companies could bank on as much data as possible to eliminate construction accidents and help improve workers wellbeing. Although the construction industry is under-digitised, AI systems have the potential to drive growth and improve construction health and safety. Some construction companies are slowly realising the benefits of this AI technology even though its applications, especially in construction health and safety are still limited. Implementing AI in construction could significantly disrupt the way companies operate.

Although AI systems could significantly improve the wellbeing of the construction workforce, they also bring about own challenges such as job losses and how the disruption will affect current roles. However, it is argued that the benefits far outweigh the negatives. The construction industry does not attract young people, is poised with a shortage of skilled workforce and is also one of the most difficult industries in which to practised adequate health and safety. Applications of AI in construction could significantly improve the current health and safety landscape. There is a need for more education and training in the field of AI, and moreover, the need to embrace these technologies which are shaping the future of business and major economic sectors globally.

To effectively implement AI in construction health and safety, recommendations must be made to maximise its benefits and to recognise how current challenges could be overcome. The implementation of AI systems should not be driven by unrealistic expectations but should instead address social values. AI safety research should be conducted, and AI systems must be tested prior to their implementation. Construction companies and government must allocate funding to ensure that the applications of AI systems prioritise and support construction workers' efforts with respect to making sound and impartial decisions that address workplace challenges. Furthermore, ethical codes of conduct should promote and protect workers values and their wellbeing by using these technologies responsibly. There is also a need for policy reforms to protect the dignity of construction workers, autonomy and responsibility from threats posed by autonomous AI decision and actions. Furthermore, ethical concerns poised by AI should be extensively analysed and correct ethical policies should be developed.

There is a need for governments to invest more in 'STEM' (science, technology, engineering, and mathematics) skills or subjects that are most beneficiary to people in the increasingly automated world. STEM skills and Arts could greatly benefit workers. Construction companies and governments should encourage workers to acquire new skills to remain relevant and to meet the current demands of AI technologies and systems in the workplace.

The applications of AI for construction workers wellbeing are relatively new and there are currently few studies in the general field of AI applications in construction health and safety. The benefits are yet to be realised and it is important for construction companies to leapfrog and take on the disruptive AI technologies as the industry is faced with ever-increasing challenges. Despite numerous challenges and fears around AI systems, it is important to realise that the benefits could greatly improve construction health and safety and furthermore save lives of many construction workers working in unsafe environments.

This paper has provided a comprehensive and traditional review of previous literature on the applications of AI for construction workers wellbeing. Artificial Intelligence used as an umbrella term and encompassed Machine Learning (ML), Deep

Learning and Natural Language Processing (NLP). The importance of construction workers' health and safety continues to grow in the construction industry. It is discovered that the application of AI can effectively improve construction workers' health and safety. However, despite the benefits of AI, there some concerns on the job security of the general workforce. The adoption of AI imbued with machine learning could significantly improve the wellbeing of construction workers on construction sites. Applying AI in construction health and safety may provide a practical means for safety personnel and practitioners to track and monitor interactions of construction workers on construction sites to realise the full potential of workers.

Acknowledgements The authors much acknowledge the support of the National Research Fund (NRF).

References

1. Asanka W, Ranasinghe M (2015) Study on the impact of accidents on construction projects. [online] Available at: http://www.civil.mrt.ac.lk/conference/ICSECM_2015/volume_4/Extract/SECM-15-088.pdf. Accessed 17 July 2018
2. Conlan S (2019) AI in health and safety—SHEilds health and safety blog and news. [online] SHEilds health and safety training. Available at: https://www.sheilds.org/ai-health-safety/. Accessed 5 Nov 2019
3. Croman J (2018) How artificial intelligence can improve your health and productivity. [online] entrepreneur. Available at: https://www.entrepreneur.com/article/317047. Accessed 5 Nov 2019
4. Delponte L (2018) European artificial intelligence leadership, the path for an integrated vision. Brussels, policy department for economic, scientific and quality of life policies, European Parliament
5. Eaves S, Gyi D, Gibb A (2015) Facilitating healthy ageing in construction: stakeholder views. Procedia Manuf 3:4681–4688
6. Enr.com (2016) ENR's 2016 year in construction photo contest winners. [online] Available at: https://www.enr.com/media/photos/81-enr-s-2016-year-in-construction-photo-contest-winners. Accessed 5 Jan 2020
7. Eppenberger M, Haupt T (2009) Construction worker injuries and costs—a comparative study of older and younger workers. [online] Accessed 17 Sept 2018
8. Esmaeili B, Hallowell M (2012) Diffusion of safety innovations in the construction industry. J Constr Eng Manage 138(8):955–963
9. Commission E (2018) Communication on artificial intelligence for Europe. European Commission, Brussels
10. Hepton J (2019) The future of AI in the construction industry. [online] ITProPortal. Available at: https://www.itproportal.com/features/the-future-of-ai-in-the-construction-industry/. Accessed 8 Nov 2019]
11. Hinze J, Teizer J (2011) Visibility-related fatalities related to construction equipment. Saf Sci 49:709–718
12. Holdren J, Smith M (2016) Reparing for the future of artificial intelligence national science and technology council preparing for the future of artificial intelligence. [online] Obamawhitehouse.archives.gov. Available at: https://obamawhitehouse.archives.gov/sites/default/files/whitehouse_files/microsites/ostp/NSTC/preparing_for_the_future_of_ai.pdf. Accessed 3 Nov 2019

13. Hutter M (2012) One decade of universal artificial intelligence. Theor Found Artif Gen Intell 4:67–88
14. Kanner J (2019) How artificial intelligence can drive construction safety. [online] Blogs.oracle.com. Available at: https://blogs.oracle.com/construction-engineering/how-artificial-intelligence-can-drive-construction-safety. Accessed 4 Nov 2019
15. Leopold H, Krenn W, King R, Mateis C (2019) Artificial intelligence landscape—an introduction in technology fields and research areas. [online] p.7. Available at: https://www.researchgate.net/publication/332471378_Artificial_Intelligence_Landscape_-_An_Introduction_in_Technology_Fields_Research_Areas/stats. Accessed 4 Dec 2019
16. Li RY (2015) Generation x and y's demand for homeownership in Hong Kong. Pacific Rim Real Estate J 21:15–36
17. John M, Minsky Martin L, Rochester N, Shannon C.E (1955) A proposal for the Dartmouth summer research project on artificial intelligence. Online: http://www-formal.stanford.edu/jmc/history/dartmouth/dartmouth.html
18. Moore, Phoebe V (2018a) The threat of physical and psychosocial violence and harassment in digitalized work. Geneva, Switzerland: International Labour Organization, ACTRAV
19. Newton J (2018) Artificial intelligence in the construction industry. [online] Curriebrown.com. Available at: https://www.curriebrown.com/media/1666/thought-leadership-artificial-intelligence-by-jeremy-newton.pdf. Accessed 5 Nov 2019
20. Okoro C, Mudonda I, Agumba J (2016) Assessing safety performance of construction workers in Gauteng, South Africa. ASOCSA. [online] Available at: https://ujcontent.uj.ac.za/vital/access/services/Download/uj:22482/SOURCE1. Accessed 19 Aug 2018
21. Patil R (2017) Artificial intelligence for the prediction of safe work behavior in construction projects: a survey. Int J Innov Res Sci Eng Technol 6(1):1418–1423
22. Price Waterhouse Cooper (2018b) AI will create as many jobs as it displaces by boosting economic growth. Online: https://www.pwc.co.uk/press-room/pressreleases/AI-will-create-as-many-jobs-as-it-displaces-by-boosting-economicgrowth.html
23. Tegmark M (2017) Life 3.0. Being human in the age of artificial Intelligence
24. Venugopal M (2018) How AI and machine learning are helping construction reduce risk and improve margins. Retrieved from Construction Executive
25. Zhou Z, Irizarry J, Li Q (2013) Applying advanced technology to improve safety management in the construction industry: a literature review. Constr Manage Econ 31:606–622

Artificial Intelligence (AI) in Sustainable Construction Management: A Scientometric Review

Motheo Meta Tjebane, Innocent Musonda, Chioma Okoro, and Adetayo Onososen

Abstract The possibilities and potentials of Artificial Intelligence (AI) in enhancing the effectiveness of sustainable construction management is gaining momentum in research. While this area is in need of further studies, it is imperative to identify the trends, research clusters, emerging thematic areas and future directions of AI in sustainable construction management. Using scientometric analysis, this study examines AI in sustainable construction management research published primarily in the Scopus database from 2011 to 2021. A critical study was used to define the research trends in AI sustainable construction management research, while a scientometric analysis critically visualised the research status quo. From analysis China, the United States of America and the United Kingdom are the top contributing countries. The top co-occurring keyword is "Artificial Intelligence". Four research themes were identified with the combination of cluster analysis and critical review: "GA Decision Support System (GA-DSS)", "High-Performance Building Design and Decision-Making", "Intelligent Concrete Prediction" and "Geographic Information Model." This analysis examines current research and identifies emerging trends in this field.

Keywords Artificial intelligence · Construction management · Sustainable scientometric · VOSviewer

1 Introduction

Artificial Intelligence (AI) is fast becoming the most influential tool in the construction industry. It is the backbone of 4IR, which establishes a link between the physical and virtual (digital) worlds and uses intelligent systems and technologies [1].

M. M. Tjebane (✉) · I. Musonda · A. Onososen
Center of Applied Research and Innovation in the Built Environment, Faculty of Engineering and the Built Environment, University of Johannesburg, Johannesburg, South Africa

C. Okoro
Department of Finance and Investment Management, College of Business and Economics, University of Johannesburg, Johannesburg, South Africa

T. C. Haupt et al. (eds.), *Construction in 5D: Deconstruction, Digitalization, Disruption, Disaster, Development*, Lecture Notes in Civil Engineering 245,
https://doi.org/10.1007/978-3-030-97748-1_12

AI is increasingly being adopted worldwide to promote the efficiency of sustainable construction management [2]. It also alleviates the social and environmental effects of construction projects on society. Construction organisations are increasingly aware that achieving customer satisfaction based on traditional methods is not enough to ensure a competitive advantage [3]. It is also achieved through the usage of AI, which is a decision support tool. This is because AI offers new opportunities for the development of new applications in the construction industry. It stimulates a new wave of highly dependable, available, reliable, safe, secure and performing smart applications. It also has several advantages in the project management environment, such as improved logistics, safety, and security [4].

Artificial intelligence (AI) is a specialised system that can recognise intelligent entities, construct them, and make decision-making easy, fast, and effective. Artificial intelligence is concerned with automating intelligent behaviour that thinks and acts similarly to humans [5]. These traits allow AI to bring immense sustainability opportunities while also posing an ethical and socioeconomic challenge in the construction industry [6]. Sustainability relates to social, economic, and environmental factors and the usage of AI to optimise it in any setting [7].

The construction industry's adoption of sustainable development has been a source of widespread concern [8]. Industry players using AI as a decision-making tool. This ranges from costing, scheduling right up to programming. This helps in the sustainable development of the construction industry as it uses programmed data. Furthermore, AI is being welcomed by all countries around the world. However, no previous study has conducted a Scientometric review of AI in sustainable construction management. In addition, no previous studies have examined its research corpus in the detail required by a scientometric analysis, considering aspects such as co-citation clusters, keywords, and research clusters [9]. This research explores the relation between AI, construction management and sustainability, and sustainable development in a bibliometric literature review. The study avails previous research endeavours in the thematic area of AI and emerging areas to guide further studies in improving innovation, development and awareness of AI in sustainable construction management.

2 Research Method

A systematic approach is essential for conducting a comprehensive review. This research used a science mapping approach to create a bibliometric map based on Scopus database sources for AI in sustainable construction management between 2011 and 2021. Bibliometrics evaluates and tracks the progress of disciplines by sorting data such as citations, author affiliations, keywords, topics explored, and methods used for published studies in the disciplines using basic/advanced statistical techniques [10]. Several studies have used bibliometric methods to shed light on methodologies, prominent publications and scholars, geographic locations, and disciplines' science [11]. This method aids in visualising significant patterns and

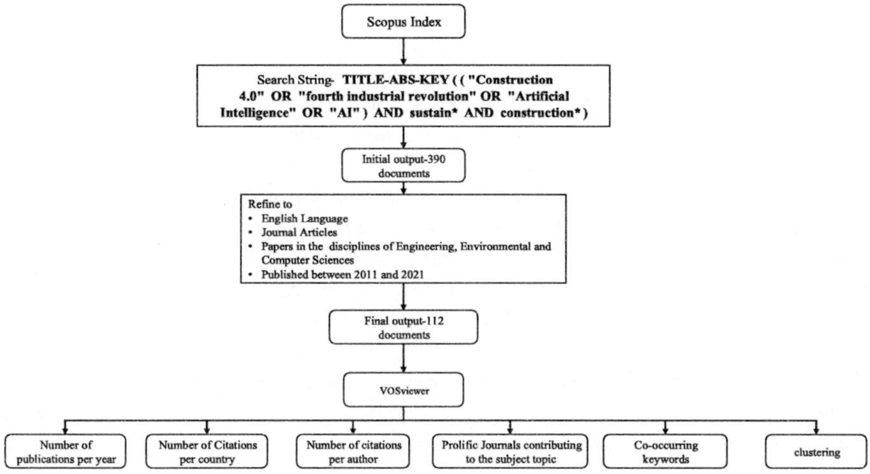

Fig. 1 Research methodology

trends in large amounts of literature and bibliographic data. It enables researchers to make observations about literature that would be impossible to make using other approaches. In order to obtain quantitative scientific knowledge of current research publications and contributions made in journals and regions about sustainable AI research in the construction industry, a bibliometric analysis was used in this study. The review structure and processes are depicted in Fig. 1, including a bibliometric search, Scientometric data analysis, and a detailed discussion of the analysis results.

2.1 Data Selection

For bibliometric research, Scopus was used to identify existing and relevant literature on the topic. Scopus one of the largest curated abstract and citation databases contains millions of documents from journals, books, and conference proceedings. Scopus was chosen over other databases because it contains Engineering, Environmental Science, and Computer Science publications that other databases do not [12]. Further, Scopus uses rigorous content filtering and re-evaluation to ensure that only the highest quality data is indexed by a broad Content Collection and Advisory Board [13]. Web of Science, EBSCOhost, and ProQuest have smaller coverage of sustainability, construction project management (PM), and construction IT journals than Scopus. There are also some more recent publications [14].

2.2 Data Collection

A search string was performed after having selected Scopus as a database for the bibliometric analysis. This search aimed to find journal articles using keywords related to AI and sustainability in the construction industry in the title, abstract, and keywords of the study. Keywords were used to find journal articles. However, it is possible that some of the articles do not have any of the selected keywords. The search used a database-specific operator OR to avoid overlaps and asterisks. Before starting the hunt, the length of the observation had to be determined. Initially, there were 390 documents. A study using bibliometric methods for relevance is recommended to be limited to ten years [11]. Journal articles written between 2011 and the 13th of May, 2021, were used for this analysis. Engineering, environmental, and computer science disciplines published in English contributed to these journals resulting in 112 documents that satisfied this criterion.

2.3 Data Analysis Tool

The data was analysed using Vosviewer. Vosviewer (Visualisation of Similarities), an internationally free bibliometric program, was used to analyse and visualise the relationships among the authors, countries, journals, co-citations and terms [15]. Similar studies have used Vosviewer (version 1.6.15), which allows for the importation of Scopus data sets. It is an easy to use software that employs a distance-based approach in its visualisation. Therefore, the relatedness of the nodes is expressed by the distance between them in the visualisation. Normalisation is performed for the differences between the nodes by default using association strength normalisation. Vosviewer uses an intelligent local moving algorithm to cluster the network, and objects in the same cluster are coloured the same. Visualising the network also provides a complete counting methodology and a fractional counting methodology [16]. Science mapping is part of the scientometric research method. It evaluates research policies and processes vast amounts of bibliometric data. It also displays the dynamic and structural aspects of a research domain. Science mapping, in particular, uses a spatial approach to illustrate the relationships between disciplines, fields, and individual publications [17].

3 Scientometric Analysis Results

3.1 Annual Publications Per Year (2011–2021)

The output of journal publications is shown in Fig. 2, which shows the annual number of publications in Scopus. In the year 2011, Scopus has three publications. The years

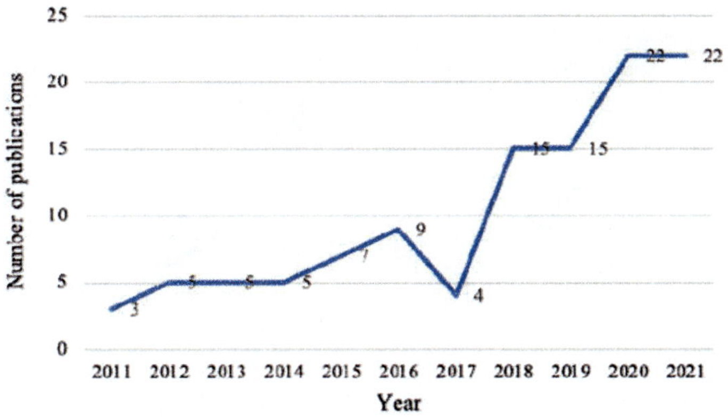

Fig. 2 Annual publications from 2011 to 2021

2012, 2013 and 2014 saw an increase as each had five publications. The upward trend continued as 2015 had five publications, 2016 with nine publications. A dip occurred in 2017 where there were only four publications. 2018 saw an increase with 15 publications, and consistency was kept in 2019 with the same number of 15 publications. The year 2020 saw more growth of 22 publications relating to AI in sustainable construction management, notably, until the 13th of May 2021.

3.2 *Citations Per Country*

Table 1 shows the top 10 countries contributing to this research field. Figuer 3 shows a world map of countries showing the distribution of the 112 documents included in the study. In Table 2, China ranked first as it accounts for 23 documents, contributing

Table 1 Top ten contributing countries

Country	Number	Documents	Citations	Contributing percentage (%)
China	1	23	196	20.54
United States of America	2	17	372	15.18
United Kingdom	3	16	110	14.29
Spain	4	13	247	11.61
Greece	5	6	57	5.36
Sweden	6	5	58	4.46
Vietnam	7	5	50	4.46
Malaysia	8	5	27	4.46
Canada	9	5	22	4.46
Nigeria	10	5	0	4.46

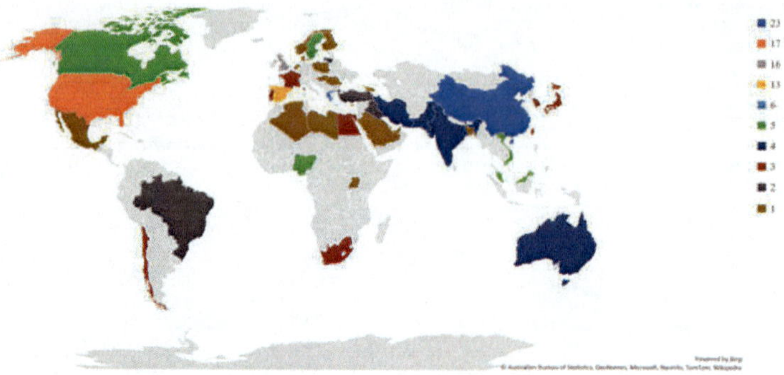

Fig. 3 World map of all documents distributed per country

Table 2 Top ten contributing authors

Author	Number	Documents	Citations	Contributing percentage (%)
Gómez D	1	3	132	2.7
Javed M. F	2	3	4	2.7
De La Cruz M. P	3	2	118	1.8
Del Caño A	4	2	61	1.8
Cerezo V	5	2	33	1.8
Asteris P. G	6	2	25	1.8
Xu J	7	2	17	1.8
Ali Z. H	8	2	16	1.8
Guo J	9	2	16	1.8
Salih S. Q	10	2	16	1.8

to 20.54% of the analysis. In second place in the United States of America, with 17 documents contributing. The United Kingdom followed in third place with 16 documents making up 14.29%. Spain ranked fourth, 11.61%, and Greece in fifth place, with six documents contributing 5.36%. Sweden, Vietnam, Malaysia, Canada and Nigeria were ranked sixth, seventh, eighth, ninth and tenth, respectively, with five documents each, contributing 4.46% each.

3.3 Citations Per Author

Table 2 presents the top ten contributing authors. Gómez was ranked first with three documents and 132 citations, making a 2.7% contribution. This is followed by Javed. with three publications and four citations, contributing to 2.7% of the analysis. De La Cruz. ranked second with two publications and 118 citations, making up 1.8%. Del Caño. ranked third with two documents (1.8%), Cerezo fourth (1.8%) and Asteris

Table 3 Top ten contributing journals

Source	Number	Documents	Citations	Contributing percentage (%)
Sustainability	1	10	40	8.9
Journal of cleaner production	2	6	101	5.4
Automation in construction	3	6	11	5.4
Engineering, construction and architectural management	4	4	26	3.6
Journal of construction engineering and management	5	3	135	2.7
Sustainable cities and society	6	3	52	2.7
Buildings	7	3	0	2.7
Civil-comp proceedings	8	3	0	2.7
Engineering structures	9	2	24	1.8
Journal of civil engineering and management	10	2	20	1.8

(1.8%) sixth. Xu, Ali, Guo, and Salih, were ranked seventh, eighth, ninth and tenth, respectively, with two documents and 16 citations each. They make up 1.8% of the study.

3.4 Prolific Journals

As identified in Table 3, prolific jounnals contributing studies to the thematic areas are: Sustainability, Journals of Cleaner Production, Automation in Construction, Engineering, Construction and Architectural Management, Journal of Construction Engineering and Management, Sustainable Cities and Society, Buildings, Civil-comp Proceeding, Engineering Structures and Journal of Civil Engineering and Management respectively. This is imperative in directing researchers to top outlets of significant publications in AI in sustainable construction management.

3.5 Co-Occurring Keywords

Table 4 shows the top ten co-occurring keywords, and Fig. 4 shows the keyword co-occurrence network. The word 'Artificial Intelligence' occurs 82 times in the 112 documents, with 361 link strength ranking first. This is followed by Sustainable development with 50 occurrences and 272 link strength. Decision Support Systems ranked third with 39 occurrences and 277 link strength. Sustainability ranked fourth with 30 (130 links) occurrences and Decision Support System fifth with 20(128 links)

Table 4 Top ten co-occurring keywords

Keyword	Number	Occurrences	Total Link Strength
Artificial intelligence	1	82	361
Sustainable development	2	50	272
Decision support systems	3	39	227
Sustainability	4	30	130
Decision support system	5	20	128
Construction industry	6	21	109
Optimization	7	14	100
Construction	8	13	57
Forecasting	9	12	60
Architectural design	10	9	74

occurrences. The remainder of the contributors is shown in Table 4. Figure 5 shows Visualising information of the largest and the most recent keywords.

4 Cluster Analysis and Critical Review

Cluster analysis is a common statistical data analysis and information discovery technique for detecting hidden semantic themes in textual data [18]. Cluster analysis groups words into different units which help in identifying research themes, trends and interconnection within the relevant research domain [9]. This scientometric literature review identified four clusters. A large node size suggests that the objects are frequently encountered, and a thick link line indicates a close relationship between two items. Nodes are divided into clusters by different colours [2]. Figure 4 shows the keywords network according to the clusters, Fig. 5 shows the overlay visualisation of when the keywords occurred. Table 5 shows the keywords sorted according to the clusters.

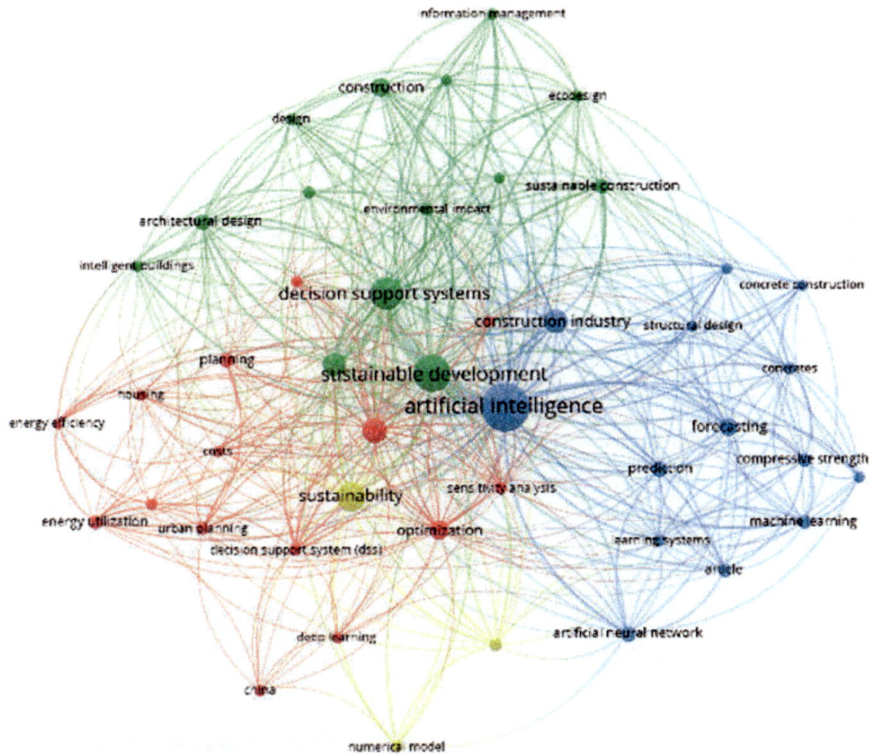

Fig. 4 Keyword co-occurrence network

4.1 Cluster 1—GA (Genetic Algorithm) Decision Support System (GA-DSS)

GA Decision Support System (GA-DSS) represents one of the major clusters observed in Fig. 4 in red. The cluster comprises of 15 items. GA Decision support system for construction management is essential for construction professionals when setting sustainability goals and ensuring that the objectives are met throughout the design process [19]. The GA-DSS's key characteristics can be defined as being decision-oriented, flexible and adaptive, controlled by the user, and supporting decision-making using AI algorithms. This Intelligent decision support system is frequently chosen for intelligent decision-making independently without human interference. The prediction algorithms such as artificial neural networks and support vector machines are required [20].

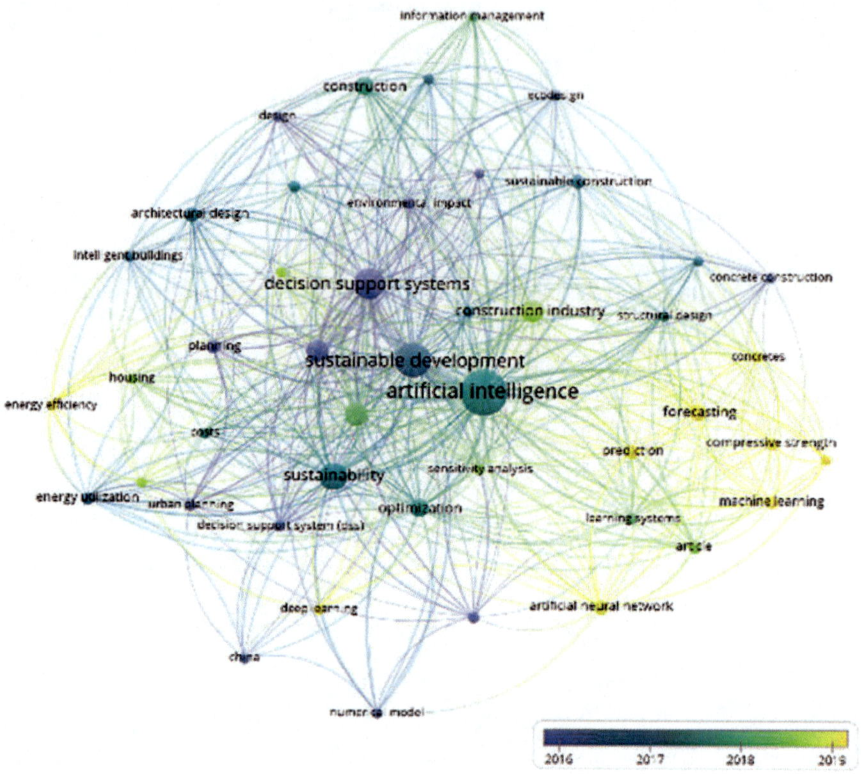

Fig. 5 Visualising information of the largest and the most recent keywords

4.2 Cluster 2—High-Performance Building Design and Decision-Making

The Green colour represents High-Performance Building Design and Decision-Making. The cluster comprises 14 items. The proposed approach enables what-if scenarios to better promote sustainability in design decisions and evaluate sustainability during the infrastructure project's design process [21]. According to [22], the integration of complex design aspects during the early design phases is extremely complex, time-consuming and requires a high level of expertise and decision making that are not available. Achieving high-performance building design requires the application of the optimal variable combinations through AI. This can be achieved if multidisciplinary optimisation AI tools are integrated with the building design platforms to search for the optimal solution automatically [23].

Table 5 Research clusters

Cluster 1 (15 items) red	Cluster 2 (14 items) Green	Cluster 3 (14 items) blue	Cluster 4 (3 items) yellow
Automation	Architectural design	Article	Geographic information
China	Construction	Artificial intelligence	Numerical model
Costs	Construction projects	Artificial neural network	Sustainability
Decision making	Decision support system	Compressive strength	
Decision support system (DSS)	Decision support systems	Concrete construction	
Deep learning	Decision supports	Concretes	
Energy efficiency	Design	Construction industry	
Energy utilization	Ecodesign	Forecasting	
Genetic algorithms	Environmental impact	Forestry	
Housing	Information management	Learning systems	
Life Cycle	Intelligent buildings	Machine learning	
Optimisation	Project management	Neural networks	
Planning	Sustainable construction	Prediction	
Sensitivity analysis	Sustainable development	Structural design	
Urban planning			

4.3 Cluster 3—Intelligent Concrete Prediction

Intelligent Concrete Prediction is represented by blue colour. The cluster comprises of 14 items. Intelligent Concrete Prediction provides in-place concrete strength knowledge to help with concrete shape removal and construction scheduling [24]. Artificial Intelligence is a reliable and useful simple prediction tool for the unconfined compressive strength of compacted soils [25]. The concrete prediction can be used as a crack monitoring device on the original concrete and form the intelligent concrete. Great signs of progress, such as optimising the design of concrete, will be made in civil engineering and the construction industry at large [26].

4.4 Cluster 4—Geographic Information Model

The yellow colour represents the Geographic Information Model with 13 items. The Geographic Information Model (GIM) brings together spatial knowledge and methodology from a variety of disciplines. They include geography, cartography, remote sensing, landscape design, landscape ecology, urban planning, geodesy and survey science, computer science, civil engineering and environmental science, decision science, network science, and many others [27]. GIM provides a solution for interoperability in the construction industry by coordinating resources, sharing information, and ensuring efficient communication between construction personnel and utility owners. This technique could enhance information extraction from the virtual world to the real world and improve the access and utilisation of information [28].

5 Conclusion

Over the past decade, the AI sustainable construction management issues have attracted a sharp increase in research attention. This is evident in the number of publications annually. These topics span a variety of fields, including decision support, construction engineering, and sustainability. This is because it can significantly promote sustainable construction management as it is more convenient, precise and efficient than the traditional tools. Existing literature were classified and discussed through a bibliometric metric analysis. The bibliometric analysis tool VOSviewer is used to analyse the co-occurrence keywords activeness and top authors. It also analysed from the collected articles, publication years, and contributing countries'. This review will provide researchers with a road map for developing integrated construction management of spatial knowledge through data analysis and visualization, AI technology in conventional fields. These includes construction and environmental engineering, more effective planning methods, and promoting sustainable construction development. Furthermore, the study guides on emerging thematic areas for further studies and recommends more research collaboration across institutions to deepen exchange of knowledge in increasing awareness and knowledge on AI in sustainable construction management.

1. References

1. Darko A, Chan APC, Adabre MA et al (2020) Artificial intelligence in the AEC industry: scientometric analysis and visualisation of research activities. Autom Constr 112:103081. https://doi.org/10.1016/j.autcon.2020.103081
2. Li CZ, Zhao Y, Xiao B et al (2020) Research trend of the application of information technologies in construction and demolition waste management. J Cleaner Prod 263:121458. https://doi.org/10.1016/j.jclepro.2020.121458

3. Lima L, Trindade E, Alencar L et al (2021) Sustainability in the construction industry: a systematic review of the literature. J Cleaner Prod 289:125730. https://doi.org/10.1016/j.jclepro.2020.125730
4. Chen X, Chen Q, Zhang Z et al (2018) BIM+ robot creates a new era of building construction. J Phys Conf Ser 1069:012142. https://doi.org/10.1088/1742-6596/1069/1/012142
5. Patil A, Patted L, Tenagi M, et al (2017) Artificial intelligence as a tool in civil engineering—a review. J Comput Eng 36–39
6. Patil G (2019) Applications of artificial intelligence in construction management. Int J Res Eng IT Social Sci 9:21–28
7. Olawumi TO, Chan DWM (2019) Critical success factors for implementing building information modeling and sustainability practices in construction projects: a Delphi survey. Sustain Dev 1–16. https://doi.org/10.1002/sd.1925
8. Wang W, Zhang S, Su Y, Deng X (2018) Key factors to green building technologies adoption in developing countries: the perspective of Chinese designers. Sustain 10:4135. https://doi.org/10.3390/su10114135
9. Olawumi TO, Chan DWM (2018) A scientometric review of global research on sustainability and sustainable development. J Cleaner Prod 183:231–250. https://doi.org/10.1016/j.jclepro.2018.02.162
10. Zupic I, Čater T (2014) Bibliometric methods in management and organization. Organ Res Methods 18:429–472. https://doi.org/10.1177/1094428114562629
11. Ellegaard O, Wallin JA (2015) The bibliometric analysis of scholarly production: how great is the impact? Scientometrics 105:1809–1831. https://doi.org/10.1007/s11192-015-1645-z
12. Cavacini A (2014) What is the best database for computer science journal articles? Scientometrics 102:2059–2071. https://doi.org/10.1007/s11192-014-1506-1
13. Baas J, Schotten M, Plume A et al (2020) Scopus as a curated, high-quality bibliometric data source for academic research in quantitative science studies. Quant Sci Stud 1:377–386. https://doi.org/10.1162/qss_a_00019
14. Chadegani AA, Salehi H, Yunus MM, et al (2013) A Comparison between two main academic literature collections: web of science and scopus databases. Asian Social Sci 9. https://doi.org/10.5539/ass.v9n5p18
15. Guo Y-M, Huang Z-L, Guo J, et al (2019) Bibliometric analysis on smart cities research. sustain 11:3606. https://doi.org/10.3390/su11133606
16. Saka AB, Chan DWM (2019) A scientometric review and metasynthesis of building information modelling (BIM) research in Africa. Buildings 9:85. https://doi.org/10.3390/buildings9040085
17. Jin R, Zou Y, Gidado K et al (2019) Scientometric analysis of BIM-based research in construction engineering and management. Eng Constr Archit Manage 26:1750–1776. https://doi.org/10.1108/ecam-08-2018-0350
18. Zhong B, Wu H, Li H et al (2019) A scientometric analysis and critical review of construction related ontology research. Autom Constr 101:17–31. https://doi.org/10.1016/j.autcon.2018.12.013
19. Nielsen AN, Jensen RL, Larsen TS, Nissen SB (2016) Early stage decision support for sustainable building renovation—a review. Build Environ 103:165–181. https://doi.org/10.1016/j.buildenv.2016.04.009
20. Ateş KT, Şahin C, Kuvvetli Y, Küren BA, Uysal A (2021) Sustainable production in cement via artificial intelligence based decision support system: case study. Case Stud Constr Mater 15:e00628. https://doi.org/10.1016/j.cscm.2021.e00628
21. Liu Y, van Nederveen S, Wu C, Hertogh M (2018) sustainable infrastructure design framework through integration of rating systems and building information modeling. Adv Civ Eng 2018:1–13. https://doi.org/10.1155/2018/8183536
22. Attia S, Gratia E, De Herde A, Hensen JLM (2012) Simulation-based decision support tool for early stages of zero-energy building design. Energ Build 49:2–15. https://doi.org/10.1016/j.enbuild.2012.01.028
23. Rahmani Asl M, Stoupine A, Zarrinmehr S, Yan W, Optimo (2021) A BIM-based multi-objective optimisation tool utilising visual programming for high performance building design, 2015. Accessed 01 Sept 2021 [Online]

24. Naderpour H, Rafiean AH, Fakharian P (2018) Compressive strength prediction of environmentally friendly concrete using artificial neural networks. J Build Eng 16:213–219. https://doi.org/10.1016/j.jobe.2018.01.007
25. Besalatpour A, Hajabbasi MA, Ayoubi S, Afyuni M, Jalalian A, Schulin R (2012) Soil shear strength prediction using intelligent systems: artificial neural networks and an adaptive neuro-fuzzy inference system. Soil Sci Plant Nutr 58(2):149–160. https://doi.org/10.1080/00380768.2012.661078
26. Liu B, Wang Q, Yin K, Wang L (2019) An analytical model for crack monitoring of the shape memory alloy intelligent concrete. J Intell Mater Syst Struct 31(1):100–116. https://doi.org/10.1177/1045389x19880010
27. Yuan M (2020) Geographical information science for the United Nations' 2030 agenda for sustainable development. Int J Geogr Inf Sci 1–8. https://doi.org/10.1080/13658816.2020.1766244
28. Fenais A, Ariaratnam ST, Ayer SK, Smilovsky N (2019) Integrating geographic information systems and augmented reality for mapping underground utilities. Infrastruct 4(4):60. https://doi.org/10.3390/infrastructures4040060

Disaster

Sustainable Approach to the Replacement of Water Mains: Environmental, Social and Economic Considerations

Bert Ediale Young, Rafiu Deji Seidu, A. Kelani, Herbert Robinson, Obas John Ebohon, and Upeksha Madanayake

Abstract Water mains in many areas of London and some parts of the Thames Valley are still thus the oldest and in need of replacement. The mains have been susceptible to corrosion, leaks and subsequent breakage, causing water wastage, continuous maintenance cost, social and environmental impact because of disruptions to water supply to local communities, road congestion and damage to the road infrastructure. The aim of this study is to examine the factors, which determines the methods adopted for the replacement of water mains and the economic, environmental, and social considerations that underpins the decision-making process. The research method adopted is a mixture of quantitative and qualitative approaches using surveys and interviews. These trenchless techniques are preferred due to lower cost, speed and productivity. Directional drilling was selected due to its cost, speed and productivity but also because it is the most customer driven method to ensure a constant supply of water. There is a need for a sustainable procurement approach incorporating social and environmental factors which affects productivity such as ground conditions, unknown utilities, the impact of water disruption on residents/schools and delays

B. E. Young (✉)
Department of Construction, Property and Surveying, London South Bank University, London, UK
e-mail: youngb6@lsbu.ac.uk

R. D. Seidu
Department of Construction, Property and Surveying, London South Bank University, London, UK

A. Kelani
J Browne Construction Co. Ltd, London, England, UK

H. Robinson
Knowledge & Learning Department, The African Capacity Building Foundation, Harare, Zimbabwe

O. J. Ebohon
School of the Built Environment and Architecture, London South Bank University, London, UK

U. Madanayake
Department of Construction, Property and Surveying, London South Bank University, London, UK

T. C. Haupt et al. (eds.), *Construction in 5D: Deconstruction, Digitalization, Disruption, Disaster, Development*, Lecture Notes in Civil Engineering 245,
https://doi.org/10.1007/978-3-030-97748-1_13

due to obtaining road access permits. Involvement of Local authorities in the design and accessibility discussion can help speed up the process and increase productivity.

Keywords Water mains · Replacement methods · Environmental · Social and economic factors · Sustainability

1 Introduction

Despite major investment over the past decades within the UK's water industry, water mains in many areas of London and some parts of the Thames Valley are still thus the oldest and in need of replacement [21]. Historically water mains installed in the twentieth century were made from cast and ductile iron which were expected to hold a life expectancy of 50–100 years of trouble-free services [17]. Unfortunately, the mains have been susceptible to corrosion, leaks and subsequent breakage causing overall water wastage, continuous maintenance cost, disruptions of water distribution to local communities, road congestion and damage to road infrastructure [21]. Ofwat (2016), the water industry watchdog annual report highlighted 20% of water lost is occurring before it reaches homes and there is no sign of it declining. For example, Thames Water the biggest water company in the UK was reported as the leakiest, with 20,500 L escaping everyday per kilometre of main [9]. A Thames Water spokesman highlighted difficulties in the rehabilitation of water mains within London and argued that "Large scale mains replacements are disruptive, especially with two-thirds of our network running under the busiest and hardest to reach roads in London [9].

With many thousands of miles of water mains still to be replaced in the UK, it is anticipated that the costs of infrastructure renewal will be high with significant disruption to customers, households, road users and other stakeholders. Since the early 1990's the construction industry has been increasingly pressured and challenged to improve its efficiency and effectiveness. The focus of this research is on the replacement of water mains. Considerations will be given to the current methods used to replace water mains such as trenching/ open cut (OC) method, trenchless water main replacement technology, and the factors which are considered when selecting the most efficient and cost-effective methods of rehabilitation. There is also a need to balance social, economic and environmental factors to ensure sustainable procurement (see Fig. 1) for illustration.

In determining the most appropriate method, it will be crucial to consider the associated costs involved in replacing water mains, as well as social and environmental factors such as service disruption to customers and road users, accessibility, and other challenges/ restrictions. Hence, the aim of this research is to examine the factors, which determine the methods adopted for the replacement of water mains and the role of environmental and social considerations.

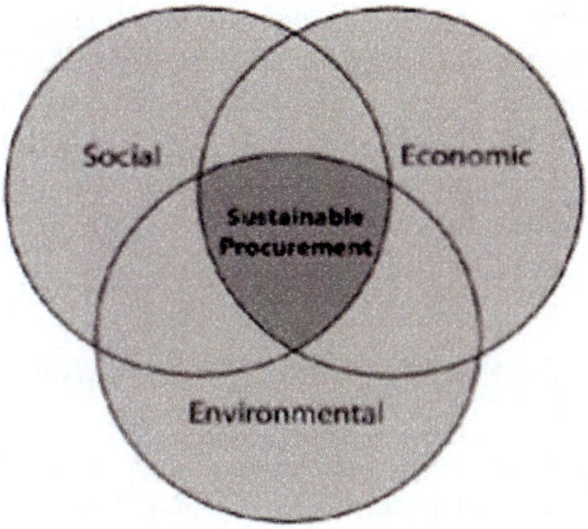

Fig. 1 Balance of sustainable procurement [5, p. 11]

2 Literature Review

2.1 Structure of the Water Industry

The UK has regional statutory water and wastewater companies responsible for public water supply and wastewater networks in the entire country. For example, Thames Water is responsible for providing water in the Greater London and the Thames Valley. Thames is the largest water and wastewater company in the UK and every day it treats and supplies 9.5 million customers with treated drinking water and removes and treats wastewater from 14.9 million customers [21]. All the regional water companies are regulated by the Water Services Regulation Authority ("Ofwat"), the Environment Agency and the Drinking Water Inspectorate ("DWI").

1. Ofwat is the economic regulator for all appointed water and wastewater companies and water-only companies in England and Wales. Ofwat is responsible for price control in a process known as Periodic Review ("PR") [21].
2. The Environment Agency (EA) seeks to maintain and improve the quality of 'raw' water in England and Wales and is responsible for issuing water companies with abstraction licences and discharge consents [21].
3. The DWI regulates all appointed water companies in England and Wales. It acts on behalf of the Secretary of State for the Department of Environment, Food and Rural Affairs ("Defra") and undertakes technical audits of water suppliers to examine all aspects of water quality, treatment and monitoring [21].

In an attempt to improve the delivery of water in the UK, water companies forge partnerships with major contractors and subcontractors. For example, in April 2017, steps were taken by Thames Water to launch the largest alliance in the water sector named Eight20. The partners involved are Thames Water and two designs and built joint ventures; made up of Costain, Atkins, IBM; and Skanska, MWH and Balfour

Beatty (SMB). The aim is for the alliance to carry out £1.75 billion of capital investment work during the AMP6 period (2015–2020), with the potential to extend the contract to 2025 [21]. Such partnerships by water companies are designed to put their best people, practices, and techniques in water projects to deliver innovative, sustainable solutions to generate greater value for money and benefits to water companies including a stable return.

2.2 Water Mains Replacement Methods

In many developed countries, the urban water supply infrastructure is in crisis due to various factors, such as increasing urban populations, insufficient attention to maintenance and replacement planning 23. There has been rapid innovation in mains rehabilitation techniques in the water industry within the UK, where existing practices are commonly categorised as "Trench" and "Trenchless" methods. Over the past 30–40 years a suite of "trenchless" technologies has been developed for rehabilitation of water main and wastewater assets that no longer require full ground excavation and replacement 20. The various methods are discussed below.

Open Cut" Trench Methods

Traditionally, pipe rehabilitation is undertaken using an open cut method, as shown in Fig. 2, in which the ground where the old pipe is situated is completely excavated and the old pipe is totally removed and replaced with new pipe. Based on the type of work, this method is also called dig- and—install, dig-and-repair or dig- and- replace [19].

It is often described as more time-consuming and does not always yield the most cost-effective method of pipe installation and renewal [15]. The social costs include cost to public, environmental impacts, damage to pavement existing utilities and

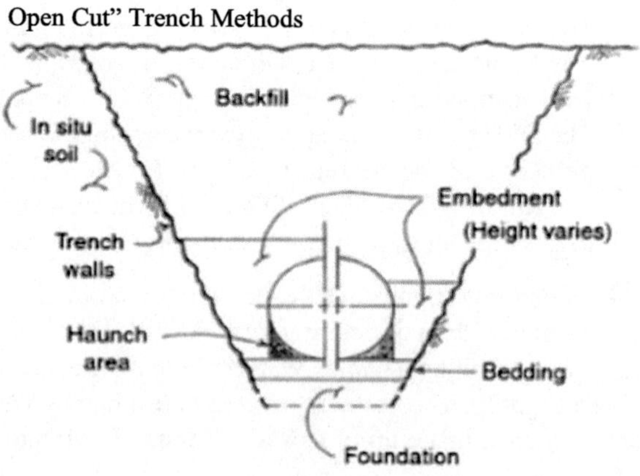

Fig. 2 Open-Cut installation [11]

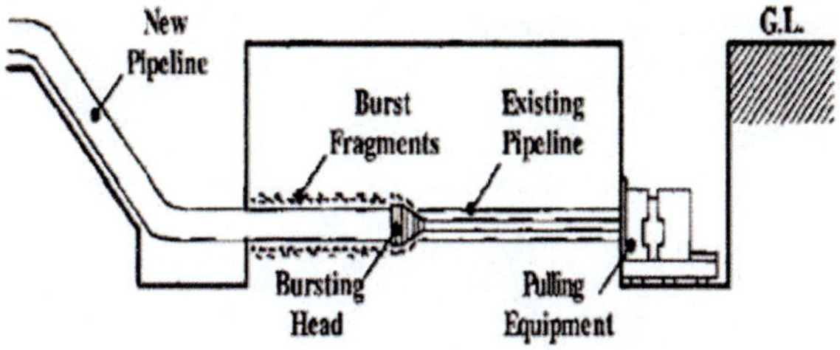

Fig. 3 Configuration for pipe bursting [12]

structures in addition, loss of access to businesses and homes and undesirable noise and sight pollution [8].

Figure 3 shows another method called pipe bursting, a trenchless rehabilitation technique which involves installing a new pipe by pulling or pushing a device as a bursting head through the existing pipe [13]. This method allows the installation of larger diameter pipes, increasing the water pipeline capacity and addressing increased urban water consumption. However, a major limitation of this process includes the need to disconnect and reconnect existing service connections from the surface, so an element of the conventional open-cut method is still required. Additionally, the technique cannot negotiate bends in the existing pipe.

Lueke and Ariaratnam 14 outline the three main systems currently used in the pipe bursting industry as: Static, Pneumatic, and hydraulic. The main difference between each method is the way force is generated and transferred to the original pipe during the bursting operation (See Fig. 4).

2.2.1 Horizontal Directional Drilling (HDD) (Trenchless)

Horizontal Directional Drilling (HDD) which is shown in Fig. 5 is the most rapidly growing method in the range of trenchless technology and techniques available [2]. This technique was originally developed by the oil industry in the United States but is now widely used for installing all pressure pipes under obstacles such as motorway, large rivers and airport runways [6]. The HDD equipment consists of five group components, (1) Drill rigs, (2) Bore drilling, (3) Drilling, fluid system, (4) Tracking system and (5) Accessories. The method involves the pipeline being bored under the crossing to emerge at a target point on the opposite side. Figure 5 opposite illustrates the process of Horizontal Directional Drilling (HDD).

Allouche et al. [1] further identified the advantages of the HDD technique over other trenchless technologies as not requiring vertical shafts as drilling starts from the surface, short installation and setup time, flexibility of borehole elevation alignment and manoeuvrability around the existing underground services and one single drive installation length is longer than any other non-man entry trenchless method.

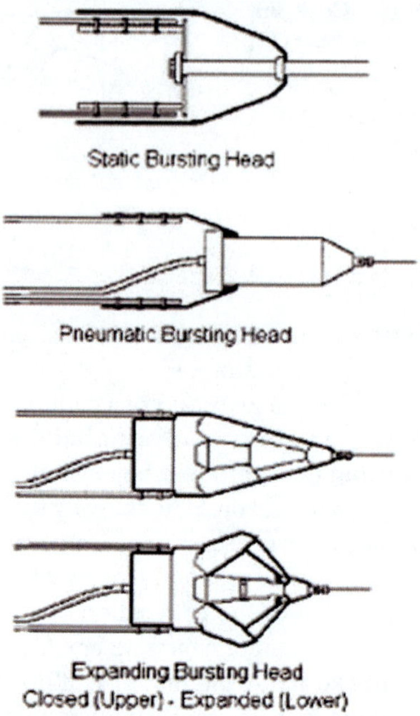

Fig. 4 Schematic different bursting head [12]

2.3 *Environmental and Social Considerations*

Allouche et al. [1] noted that trenchless methods are preferred due to the high cost incurred in tackling numerous environmental and social factors. [3] argued that the HDD method gradually evolved to a preferred method due to the high costs associated with the open cut method in crowded urban areas, consideration of social costs such as traffic delays, distraction of business activities and environmental issues such as placement of pipelines across rivers, and other environmentally sensitive areas. Lueke and Ariaratnam [14] further noted that accessibility constraints within urban areas and increased underground congestion has resulted in making the traditional open cut method a more expensive technique and even impossible in some situations. Jung and Sinha [12] highlighted how negative social and environmental impacts influence the effectiveness of pipe laying methods.

McKim [16] argued that disruptive open-cut methods are often not acceptable when working with underground infrastructure systems due to the increase in traffic congestion which causes major inconvenience to the public, and decreased road lane widths which can make road accidents more likely to occur. Myers et al. [18] identified the key considerations for local businesses and residents due to congested construction sites which are likely to result in loss of customers due to traffic disruption or loss of access when the open-cut method is used. This may also result in significant loss of sales for businesses and tax revenue for the local government. Additionally, major inconveniences such as traffic congestion and delays are often imposed

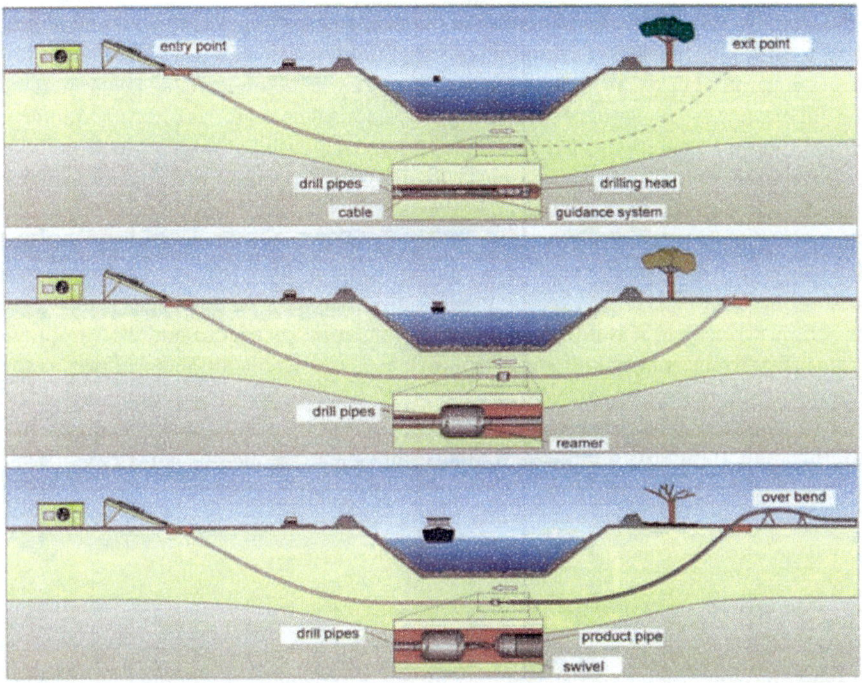

Fig. 5 The stages in horizontal directional drilling [4]

on neighbourhoods which can make commuting strenuous. Jung and Sinha [12] also argued that the open-cut method often requires removal of pavements followed by subsequent restoration, which significantly reduces pavement life. Surface subsidence of the pavement from the cutting and patching process in the open-cut method can reduce pavement life by about 40%. Additionally, Jung and Sinha [12] further argued that the open-cut method frequently causes environmental damage to grass, trees, and other landscaping features and can have negative effects on conservation.

McKim [16] discusses how pollution in the form of air, noise and water can be caused through trench excavating. Firstly, air pollution in the form of fine soil particles may become airborne in the form of dust due to wind blowing over the soil stockpiles created during the process with the open-cut method. Secondly, rain or water encountered during construction using open-cut methods can cause soil erosion and run off of contaminated solids into streams, rivers, and sewers. Thirdly, mains replacement techniques require the use of heavy equipment that produces high levels of noise causing a great deal of disturbance, especially to established communities and residents. Allouche et al. (2001) highlighted the importance of geotechnical investigation and the need for awareness of the soil type. Geotechnical investigations are used to define the existing soil types and conditions to enable the contractor to make the best arrangements and to choose the most suitable equipment for maximum productivity. The quality and quantity of the available geological information during design and bidding phases is very important in estimating production rates, shaft design and maximum drive length for any construction method. Hegab [10] reiterated the importance of knowledge of the soil type as contaminated soil is often encountered during pipeline construction. This is particularly the case within open-cut methods as it requires removing large volumes of soil. The disposal of this material, which requires specialized equipment and labour can be costly. Hegab [10] noted further that unexpected soil conditions may cause a loss of connection with the drilling head and can delay the whole pipe installation process. HDD drilling bits are used according to soil type and pipe length. Ariaratnam and Allouche [3] argued that prior to job initiation, work field should be visited for a visual inspection to address important issues that affect quality and speed of work (i.e., sufficient room for entrance and exit pits; equipment; support vehicles; and fusion machines). In addition, it is noted that weather conditions have a major effect on any form of trenchless technique. Temperature, humidity, rainfall, and snow might cause an obvious delay in work due to their direct effect on machine, soil and worker productivity.

3 Research Method

A case study approach was selected as the authors wanted to conduct an in-depth analysis to understand the social, economic, and environmental considerations of replacement of water mains. After the privatisation of regional water authorities in England, maintenance became the responsibility of private companies, rather than the state which led to the first Asset Management Period (AMP) [22]. The sixth asset management period which is known as AMP6 commenced from year 2015. AMP6 water main replacement project in Reading was selected as the single case study due to its typical nature. Mixed method of both quantitative and qualitative research techniques was used within the case study. Quantitative data was collected through questionnaire survey in order to capture the main replacement methods and key considerations associated with environmental and social factors. Qualitative data was derived through interviews with six (6) key decision-making personnel in the case study project. The interviews provided greater insight on tendering and operational issues and the findings are summarised below.

4 Findings and Discussion

4.1 Main Replacement Methods and Key Considerations

80 web-based questionnaires were distributed via e-mail to those related to the selected case study project. 27 respondents completed the questionnaire, making

the response rate as 33.75%. From the completed questionnaires, the respondents range from Quantity surveyors (25.9%), Directors (18.55%) and Site Supervisors (22.2%) and others involved with the delivery of mains replacement projects within the UK Construction Industry. Horizontal Directional drilling is overwhelmingly the most preferred method favoured by 70.4% of the participants. Sliplining is the second preferred method with 18.5%, followed by the open cut method (11.1%). The trenchless method of pipe bursting was not selected. Most common factors leading to a change in technique is utility obstructions (34.2%) and ground conditions (27.6%). Others included value engineering (9.2%), space restriction (17.1%) and inadequate designing (11.8%). When deciding on the most appropriate method, labour force and quality of equipment was not considered a major factor. However, cost and level of disruption were selected by 21.4% and 23.2% respectively, suggesting that these are the two major influences when deciding on the most appropriate method. Of the 11 subcontractors who participated, 8 selected '*cost*' as key factor and 3 out of the 4 client/ local authority selected '*customer satisfaction*' as a key factor, clearly indicating that subcontractors are more driven by cost whilst clients/local authorities are more concerned with customer satisfaction. On external factors impacting projects, road restrictions (29.6%) and permit issues are the most cited, though utility services (25.9%) is also noted as a prominent factor. However, environmental impacts on wildlife, environmental regulations and pollution were not highlighted as having any impact by participants. The notion that it is harder to achieve productivity in London in comparison to the regions in the UK was widely acknowledge with 92.6% of participants agreeing to that. On other environmental or social factors which impacted productivity, the additional comments made by participants are shown in Table 1.

Table 1 Additional comments on environmental and social factors

Environmental and social factors that impacted productivity
• Technology—Regarding ground conditions, as laid surveys, utility drawings and CAT (cable avoidance tool) and Genny (signal generator) equipment are improving all the time with technology, making it easier to pre-determine ground conditions.'
• Coordination—The coordination between Highways Agency, client, sub-contractor and customer are misaligned. Permit restrictions and limited working length will impact on productivity, cost per metre, which will result in slowing the process, delays and low productivity on site, ultimately causing more customer dissatisfaction and complaints
• Awareness of the project—The sooner the local authority is informed about proposed work, the more likely disruptions will be reduced
• Location—is a key factor which is influenced by density of housing/urbanisation, parked cars, restricted roads, built up areas and busy road due to large volume of vehicles
• Other factors—such as surrounding environment, existing network construction and conditions

Table 2 Profile of the respondents

Participant reference	Sector (job title)	Years of experience in construction industry	Years of experience with water sector
A	Client (Customer Relation Department)	18	8
B	Highways Agency	7	3
C	Main contractor (Construction Manager)	35	35
D	Sub-contractor (Project Delivery Manager)	33	30
E	Client (Contracts Manager)	25	18
F	Sub-contractor (Site Supervisor)	12	10

4.2 The Case Study Project

The project is AMP6 Water Mains Replacement Project in Reading. The contract commenced in June 2017 and finished in April 2018. The project involved the replacement of a total of 12,449 m (m) Cast Iron main to Plastic Polystyrene new mains as well as renewing services. The original design drawings had 11,186 m (89.85%) of the 12,449 m scheduled being replaced using a trenchless (no excavation) technique and the balance 1,263 m using the open cut (OC) method. The contract value was £2,002,435.90 million and it was implemented using an Early Contractor Involvement (ECI) approach. The profile of the interview respondents within the case study are given in Table 2.

4.2.1 Main Factors Influencing Selection Process

Trenchless methods such as directional drilling and slip lining are preferred in general to trench method due to several factors such as they are quick, cheap, efficient, and productive. For example, one of the participants mentioned that "…. *slip lining is cheaper, productive and most importantly it enables customers not to be out of water for more than the 4 h window*" (Participant C). Participant F also stated that "*As we work on price our main objective is to get as much new main in the ground as quick as possible, so speed and productivity is the main factor for me*".

According to the interviews, sliplining can be efficient but the main factor preventing it from being used is the fact the water mains that needed to be two ways fed for customer supply should not be affected. Further due to urbanisation and the demand for water increasing all the time most of the 90 mm existing main had to be replaced by 125 mm and 180 mm so slipping lining was not appropriate as "*Sliplining requires a smaller main to go in larger one*" (Participant C). The element of open cut is guaranteed, for roads with poor ground conditions, restricting accessibility, for launch pits, pipe connection, bends, services, valves, hydrants, and wash outs.

In addition, risk was considered as one of the main factors influencing the selection of appropriate water replacement methods. Participant D supports this by stating as "*we don't like to take risks especially with drilling as we are liable to any utility or cable strikes, which can cost anything from three thousand to five thousand pounds to repair as well as delay the programme*".

4.2.2 Environment, Social and Economic Factors

Environmental and social issues are a major concern in water mains replacement project and are key factors to be considered as early as possible. Participants expressed strong views on environmental issues, for example, Participant A noted "*Environment is a very sensitive factor, especially wildlife issues when working in the provinces. I've experienced problems with nesting, Japanese's knotweed which causes much panic and suspends works, asbestos, TPO on trees (which means they cannot be knocked down). Bushes are not allowed to rip down anymore so works need to be adapted around it*". Contamination within excavations is a major issue now with the directional drilling method, as sometimes the ground conditions are too difficult to drill, so to avoid switching to an expensive method like Open Cut, adding a fluid called bentonite into the drilling rig makes it easier to drill. However, the environmental issue occurs when the bentonite remain that can contaminate the water is left in the excavations. In order to avoid tankers are required to suck the slurry out of the excavation. Whether this counts as a compensation event as such has been a debate as subcontractors argue they are having to incur an additional cost they would not have allowed for in their original price. Noise pollution is also a major environmental concern. The project has taken adequate measures to minimise the impact of noise pollution as acoustic barriers are commonly used. To confirm, Participant B stated, "*the noise from machine has led to strict guidelines that no machine can be turned before 8am and after 6 pm. Night works in residential areas are very minimal too.*"

Accessibility is a general issue rather than one related to a replacement method itself. Participant D commented as *"within this job we had numerous issues in obtaining road closure approval and digging permits on time. This is a continuous problem I've seen working for the Main Contractor"*. When there is a space restriction such as working on tight roads and the trenchless machines are unable to access, an Open Cut method is usually the preferred method as it allows for hand digging in such space restricted areas.

Further, having a good relationship with efficient communication with the local authorities and communities was highlighted as an important social factor. To minimise any social issues, it is essential to liaise with residents, schools, and local businesses to ensure that they are kept aware about the planned work. Participant A from the Customer Relation Department emphasised this point by stating *"Before a construction programme is agreed we must meet with the designers and operation team in order identify any "red flags" which will impact the locals. I then meet with the local council to discover whether it will be signed off .... this needs to be conducted 6 months sometimes 12 months before the proposed start date"*. To minimise the impact of the project on the locals, project programmes that can cause most disruptions are usually scheduled around school holidays or outside school hours. When the water supply needs to be turned off residents and businesses are given 48 and 72 h' notices respectively. The lack of notice commonly delays the works as the Client usually refuses to turn off water with no notices. At public buildings such as schools and hospitals, water supply is never turned off.

The actions taken to mitigating the cost of the environmental and social disruptions were also discussed during the case study interviews. As per the project delivery manager (Participant D), cable strikes can cost anything from three to five thousand pounds (£3000–£5000) to repair every time they are hit. This is more likely to occur when there are a lot of utility services present. The cost of full road closure varies from two to four thousand pounds (£2000–£4000) a week which is a cost for the Main contractor. It can cost up £1250 to gain early road access as such it is key to plan early and adequately to ensure road permits are in order.

Road restrictions/permits, and utility services have a major impact on productivity too. This is echoed by the participants by emphasising the significance of delays in obtaining road permits, the need to conduct trial holes to ascertain the ground conditions and the utility services frequency of the working areas. With the amount of time being spent on conducting trial holes to establish ground conditions, the question posed on the significance of pre-investigation to improve productivity. Pre-investigation is absolutely necessary as it enables the contractor to make the best arrangements and to choose the most suitable equipment for maximum productivity. Participant D stated that "*trial holes are supposed to be done every 25 m when drilling but this is commonly not done to save time and money, but this is a main catalyst for reduction in productivity'*. It has been suggested that the traditional culture of having a short-term view in time and money saving is a reason why thorough pre-investigation is not being conducted which is ironically leading to further loss in time and costs. The delay that occurred in the project due to not obtaining notices and having to conduct trial holes amounted to £31,931.21 which is about 8% of the overall cost of compensation events.

Looking at the processes currently in place to deal with change in technique, submitting Technical Queries (TQ's) has been seen to have received mixed views. The case study highlighted that Participant C (Construction Manager/Main-contractor) requested for a change in technique within 24 h. However, Participant D (Project Delivery Manager—Sub-contractor) argued this is hardly ever achieved within 24 h and that the site supervision should have more delegated powers due to the productivity caused by waiting for protocols to be completed. Changing methods amounted to costs of £131,232.37 which represents about 34% of overall compensation events.

5 Conclusions

Trenchless techniques are methods which are preferred due to lower cost, speed and productivity. Directional drilling was not only selected due to its cost, speed and productivity qualities but it is the most customer driven method as keeping customers with a constant supply of water has been the most determining factor. It has also been established that the drive to be more customer focus is coming at a cost, as additional work is now required to keep customers in constant supply by the installation of under pressure tees and riders during connections from old to new mains. Environmental and social issues are a major concern in water mains replacement project. Several actions such as work time restrictions, use of acoustic barriers, adequate notices on the possible disruptions, choosing appropriate methods according to the site conditions etc. were taken to minimise the social and environmental impact of water main replacement projects. Good relationship and efficient communication with local authorities and communities are key factors as their early involvement in the discussions relating to design and accessibility issues can speed up the process and eliminate additional costs associated with delays and extra work. Despite, the higher initial costs, pre-investigation are strongly recommended as it enables the contractor to make the best arrangements and to choose the most suitable equipment for maximum productivity and will reduce the costs on the long run.

6 Data Availability Statement

Please note, that no data, models, or code were generated or used during this study.

References

1. Allouche EN, Parhami A (2003) A Simplified Approach for Selecting a Trenchless Construction Method for Municipal Sewers. In New Pipeline Technologies, Security, and Safety, pp.1438–1448
2. Allouche E, Ariaratnam S, Lueke J (2000) Horizontal directional drilling: profile of an emerging industry. J Constr Eng Manag 126(1):68–76
3. Ariaratnam ST, Allouche EN (2000) Suggested practices for installations using horizontal directional drilling. Practice Periodical on Structural Design and Construction, 5(4):142–149
4. Bayer H (2005) HDD-practice handbook. Vulkan-Verlag, Essen
5. Berry C, McCarthy S (2011) Guide to sustainable procurement in construction. CIRIA, London, pp. 11,12
6. Blayse AM, Manley K (2004) Key influences on construction innovation. Constr Innov Inf Process Manage 4(3):143–154
7. Deller D, Hviid M. (2016). Ofwat: Review of retail household markets in the water and wastewater sector–Call for evidence
8. Fröling M, Holmgren C, Svanström M (2004) Life cycle assessment of the district heat distribution system. Int J Life Cycle Assess 9(2):130–136

9. Guardian (2017) Water companies losing vast amounts through leakage as drought fear rise 2017. [Online] Available at: https://www.theguardian.com/environment/2017/may/11/water-companies-losing-vast-amounts-through-leakage-raising-drought-fears. Accessed 20 Nov 2020
10. Hegab M (2003) Productivity modeling of microtunneling projects. North Dakota State University of Agriculture and Applied Science
11. Howard G, Teuton J, Luyima P, Odongo R (2002) Water usage patterns in low-income urban communities in Uganda: implications for water supply surveillance. International Journal of Environmental Health Research 12(1):63–73
12. Jung YJ, Sinha SK (2007) Evaluation of trenchless technology methods for municipal infrastructure system. J Infrastruct Syst 13(2):144–156
13. Kramer S, McDonald W, Thomson J (1992) An introduction to trenchless technology. Springer, US, Boston, MA
14. Lueke J, Ariaratnam S (2001) Rehabilitation of underground infrastructure utilizing trenchless pipe replacement. Pract Period Struct Des Constr 6(1):25–34
15. Ma B, Najafi M. (2015). Special Issue on International Conference on Pipelines and Trenchless Technology 2011. Journal of Pipeline Systems Engineering and Practice, 6(3),A2014001.
16. McKim R (1997) Selection method for trenchless technologies. J Infrastruct Syst 3(3):119–125
17. Mirza S (2006) Durability and sustainability of infrastructure—a state-of-the-art report. Can J Civ Eng 33(6):639–649
18. Myers MB, Stickrod TW, Abraham DM, Iseley T (1999) Microtunneling technology for conduit construction. Pract Period Struct Des Constr 4(2):56–63
19. Najafi M. (Ed. 2007). Pipe bursting projects. American Society of Civil Engineers
20. Selvakumar A, Kampbell E, Downey D, Condit W (2012) Quality assurance and quality control practices for rehabilitation of sewer and water mains. Urban Water J 9(4):211–222
21. Thames Water (2016) Replacing London Victorian water mains 2016. [Online] Available at: https://corporate.thameswater.co.uk/about-us/investing-in-our-network/victorian-mains-replacement. Accessed 20 Nov 2020
22. Thomas S (2015) AMP 6 is coming: But what does it mean for you? [Online]. Available at: https://wwtonline.co.uk/features/amp-6-is-coming-but-what-does-it-mean-for-you. Accessed 10 Apr 2021
23. VanBriesen JM, Dzombak DA, Zhang L (2014) Sustainable urban water supply infrastructure. In Ahuja S (ed) Comprehensive water quality and purification. Elsevier, pp 427–449

Learning Online During a Pandemic: Student Perspectives

Theo C. Haupt and Ayesha Mall

Abstract This study explores student experiences of online learning during the global COVID19 pandemic. The paper assesses whether the outcomes of constructivist inquiry-based learning can still be achieved online. A qualitative research approach was used to survey online learning experiences of students. A total of 27 registered Construction Studies university students conveniently sampled across all levels of study completed a questionnaire with close ended questions. The data collection was computed and analysed using SPSS version 27. Mean values, standard deviations and reliability values were computed. Results of the survey indicated that most students spent more time doing work during online learning. Students indicated that they were able to think more critically and discuss concepts and collaborate with other students. There was a handful of students that were more stressed, socially inactive and lacked outdoor activity due to online learning. The sample is drawn from a single university in South Africa and findings cannot be generalized across all student populations.

Keywords Constructivist · COVID-19 · Inquiry based learning · Online learning

1 Introduction

The novel coronavirus disease COVID-19 was detected in Wuhan City, China in the latter part of 2019. It rapidly spread around the world within a couple of months and on March 11th 2020, COVID-19 was declared a global pandemic by the World Health Organization (WHO) [30]. Due to its incredibly fast spread rate and the shocking effects on human lives, researchers believed the virus is transmitted in just minutes

T. C. Haupt
Mangosuthu University of Technology, Umlazi, South Africa
e-mail: theo.haupt@mut.ac.za

A. Mall (✉)
Department of Construction Management and Quantity Surveying, Durban University of Technology, Durban, South Africa
e-mail: ayesha.mall14@yahoo.com

T. C. Haupt et al. (eds.), *Construction in 5D: Deconstruction, Digitalization, Disruption, Disaster, Development*, Lecture Notes in Civil Engineering 245,
https://doi.org/10.1007/978-3-030-97748-1_14

from droplets or by touching surface metals or other materials that has been infected from a person who has the virus [4]. Therefore globally, governments employed a nationwide lockdown to curb the spread of the virus [3].

In South Africa the pandemic was declared a national state of disaster in terms of the Disaster Management Act and President Cyril Ramaphosa instituted a national lockdown that commenced on March 29th, 2020. The lockdown served as a method to limit person to person transmission and decrease the infection rate. As of 16th July 2021, the number of confirmed cases in South Africa had risen to 2,269,179 with a total of 66,385 deaths and 5,046,267 vaccines administered. However higher education institutions (HEI) in South Africa remain closed with online learning as a method of academic course delivery [8]. Although the transition from traditional face-to-face learning was inevitable, COVID-19 drastically revolutionized education. Therefore, this study aims to assess student experiences of online learning during a global pandemic. The paper assesses whether the outcomes of constructivist inquiry-based learning are still achievable online.

1.1 *Impact of COVID-19 on Education*

It is imperative to acknowledge the impact of online learning on the effects of education, and the social consequence of retaining this method of course delivery. Globally there have been extensive studies on the impact of online learning on education and it has been identified that online learning has many advantages such as reducing the costs associated with traditional face-to-face learning, ensuring the continuity of education [27] and increased critical thinking and technical skills for students [1, 22].

Additionally, there are certain challenges with online learning that may have an impact on the online learning experience of students when it comes to the accessibility of study materials. For instance, most universities may not have literature available as a soft copy online but rather as a hard copy in the university library [3] (Santos, 2020). A recent study in Turkey on the views of students and teachers on online learning, it has been identified that students faced challenges such as restricted interaction with fellow classmates and as a result found difficulty collaborating for group projects. Further, universities faced infrastructure problems and lack of equipment [11].

In a different study on online learning in Iran it was identified that the advantages of online learning outweighed the limitations as online learning allowed for studying anytime at any location, students have more time as commuting to university is eliminated and learning online is more affordable [23]. However limitations to online learning include, limited or no social interaction, and increased chances of distraction or boredom [20, 23].

1.2 Learning Online in South Africa

The unprecedented COVID-19 pandemic forced students and instructors to transition from traditional face-to-face classes to an online instructional environment. However, in South Africa shifting the education delivery mode online prematurely presented students and instructors with great challenges [16, 17].

Even though some HEI's made use of online platforms such as BLACKBOARD and MOODLE, the rise in digital technologies were slowly evolving in a pre-COVID-19 era. The pandemic has exposed the world to a 'new normal' by catalyzing digital education. There seems to be a general consensus that the return to traditional teaching and learning may not be possible post pandemic but rather hybrid and blended models of teaching and learning will be employed [31, 32].

Online teaching and learning provides students with the opportunity to be actively involved in their learning and to discuss and reflect on theories with other students [12]. The 'new normal' has given instructors a new perspective in designing good learning experiences. This can be achieved by reevaluating traditional methods of assessment and innovatively creating interactive course content. Therefore, opportunities are presented for students to collaborate with each other in a constructive approach in an out of the classroom [6, 7, 22]. Studies have identified that discussion forums created online allowed students to actively collaborate and reflect with others. Further interaction online is more reflective and personal as opposed to traditional classroom discussions [10, 19, 26, 29].

Online learning in education tends to make students undergo procrastination, contemplation and lack of interaction [23]. Therefore, students who spend their time online tend to experience signs of social isolation, due to the absence of human interaction [20]. Social isolation can lead to several mental health issues such as anxiety, depression or increased stress. In a study on online learning during the COVID-19 pandemic in Saudi Arabia, [13] concluded that students felt that online learning shredded their social life. However, social isolation can be reduced by the use of blended learning environments and promoting increased interaction and engagement between online students and instructors [14, 15, 20, 21]. In a study on the impact of COVID-19 and the impact on students wellbeing [9] revealed that normal routines have been disrupted by COVID-19 by preventing free movement. Subsequently students felt as if they were not doing anything meaningful resulting in anger, lack of motivation, sadness and loneliness.

2 Methodology

The study adopted a quantitative research approach. A questionnaire was developed from literature and an exploratory study was presented to 103 registered Construction Studies students at a university in KwaZulu-Natal. A total of 27 students responded to the survey questionnaire indicating a 26% response rate. A data set of descriptive

Table 1 Cronbach alpha's reliability statistics for online learning

Description	Cronbach's Alpha	Cronbach's Alpha based on standardized items	N of Items
Learning in an online environment	0.814	0.827	10
Health and wellbeing during online learning	0.735	0.750	6

statistics was captured and computed with the use of SPSS version 27. Internal validity tests were conducted of the items using a set of scaled responses reporting the Cronbach's alpha co-efficient for validity. Table 1 indicates reliability analysis. The results suggested that the items in the survey instrument requiring scaled responses had a good internal consistency with Cronbach's coefficient alpha values greater than 0.70. This is supported by Pallant [18] who indicated that the acceptable Cronbach alpha cut-off value of 0.70 was adequate for further processing of the data.

3 Data Analysis

3.1 Respondents' Profile

A total of 27 students responded to the questionnaire survey (male = 66.7%, female = 33.3%). The study sampled a cross section of students across different levels of study enrolled in Construction Studies. The sample was selected from classes that comprised of full-time students only. The questionnaire survey was hosted online on Google Forms and a link was emailed to students.

3.2 Respondents' Demographics

Table 2 presents the demographic information of students who participated in the study. A total of 62.96% (17) of the participants were aged at 21 and below and 37.04% (10) were above the age of 21. Primarily, participants were 1st year students 51.9% (14), 2nd year 11.1% (3) and 3rd year 37.0% (10). Majority of the respondents reside in urban areas 55.6 (15), while 33.33 (9) in rural areas and 11.1% (3) in periurban areas. 48.1% (13) of respondents used a smartphone for online learning and 51.9% (14) used a computer/laptop.

Table 2 Demographics

Characteristics	Discipline	No. of Participants	%
Age group	21 and below	17	62.96
	Above 21	10	37.04
Level of study	1st year	14	51.9
	2nd year	3	11.1
	3rd year	10	37.0
Residential location	Rural area	9	33.30
	Urban area	15	55.60
	Peri-urban area	3	11.1
Device used for online learning	Smartphone	13	48.1
	Computer/Laptop	14	51.9

3.3 Learning in an Online Environment

This section sought to assess student experiences during online learning. Students rated their experience using a 5-point Likert scale, where 1 = strongly agree, 2 = agree, 3 = neutral, 4 = disagree and 5 = strongly disagree. Table 3 indicated that more students spent time doing work (mean score = 2.33), online learning allowed a student to think more critically (mean score = 2.26), students were able to discuss the theories and concepts related to the modules with other students (mean score = 2.26), online learning provided students with the opportunity to participate in more activities (mean score = 2.22), students were able to connect new concepts and theories learnt with knowledge that was already known (mean score = 2.11) were ranked as the most frequent responses about experiences during online learning by the respondents. With a mean score of 1.66 students indicated that they least experienced that online learning increased their technical and computer skills. The study confirmed students that used a computer/laptop increased their technical and computer skills as well as the ability to troubleshoot problems.

3.4 Mental Health and Wellbeing During Online Learning

This section sought to assess the mental health and wellbeing of students during online learning. Table 4 shows that online learning had a minimal impact on students' mental health (mean score = 3.22), experienced restlessness during online lessons (mean score = 3.07) and depression during online learning (mean score = 3.03). With a mean score of (2.67) students indicated that online learning may have reduced their time spent outdoors.

Table 3 Online learning experience

Description	1	2	3	4	5	T	M	SD	R
More time spent doing work during online learning	18.5	37.0	37.0	7.4	0	27	2.33	0.87	1
Online learning allows me to think critically	25.9	25.9	44.4	3.7	0	27	2.26	0.90	2
Students able to discuss the theories and concepts related to the modules with other students	22.2	48.1	14.8	11.1	3.7	27	2.26	1.06	3
Online learning gives me the opportunity to participate more in activities	18.5	44.4	33.3	3.7	0	27	2.22	0.80	4
Students are able to connect new concepts and theories learnt with knowledge that I already know	25.9	44.4	22.2	7.4	0	27	2.11	0.89	5
Students are given the opportunity to discuss and reflect concepts and theories with other students	14.8	63.0	22.2	0	0	27	2.07	0.62	6
Instructor provides an opportunity to share my experiences	37.0	29.6	25.9	7.4	0	27	2.04	0.98	7
Students are given tasks that allows them to learn new concepts their own	37.0	44.4	18.5	0	0	27	1.81	0.74	8
Students are given scenarios based on real life industry problems and can connect to existing knowledge	37.0	44.4	18.5	0	0	27	1.81	0.74	9
Online learning has increased technical and computer skills	51.9	37.0	3.7	7.4	0	27	1.66	0.88	10

Table 4 Mental health and wellbeing during online learning

Description	1	2	3	4	5	T	M	SD	R
Online learning affected my mental health	3.7	14.8	40.7	37.0	3.7	27	3.22	0.89	1
I get restless during online learning	0	33.3	29.6	33.3	3.7	27	3.07	0.92	2
I feel depressed during online learning	0	29.6	37.0	33.33	0	27	3.03	0.81	3
I feel more stressed during online learning	7.4	37.0	25.9	29.6	0	27	2.78	0.97	4
Online learning has made me socially inactive	18.5	29.6	14.4	33.3	3.7	27	2.70	1.23	5
Online learning has reduced my time outdoors	22.22	29.6	11.11	33.33	3.7	27	2.67	1.27	6

3.5 Bivariate Correlation Between Online Learning Experience and Mental Health and Wellbeing During Online Learning.

Table 5 presents bivariate correlation analysis result for the items in the self-efficacy scale. Correlation values between the items ranged from 0.368 to 0.864, indicating that the items were related to each other. Therefore, the result suggested discriminant validity of the instrument as correlation of the factors were less than 0.90.

4 Discussion

4.1 Learning in an Online Environment

Time spent doing work during online learning

The survey showed that students spend more time doing work during online learning. Students feel online lessons were overloaded with activities and therefore required more time, further students spend many hours navigating and learning online learning platforms. In a study at the university of Malaysia, students expressed added pressure due to the overwhelming number of assignments instructors issue during online learning [28].

Online learning allows me to think critically

It is apparent from the study that online learning enabled a student to develop critical thinking skills. The aim of a virtual classroom is to challenge students to think differently and to shift the pedagogical framework to a student-centered approach. For example a constructivist-based approach online can be adopted by involving

Table 5 Bivariate correlation between online learning experience and mental health and wellbeing during online learning

	1	2	3	4	5	6	7	8	9	10	11	12	13	14	15	16
1	1															
2	0.708**	1														
3	0.432*	0.255	1													
4	0.266	0.309	0.458*	1												
5	0.086	0.048	0.751**	0.591	1											
6	0.386*	0.275	0.186	0.068	0.187	1										
7	−0.488**	−0.127	0.315	0.197	0.367	0.173	1									
8	−0.056	−0.050	0.291	0.054	0.125	0.041	0.343	1								
9	0.061	0.106	0.115	0.090	0.082	−0.113	−0.110	0.381*	1							
10	−0.183	−0.102	−0.054	−0.185	−0.060	0.067	0.400*	0.398*	0.383*	1						
11	−0.254	−0.220	−0.192	−0.129	−0.152	0.068	0.393*	0.347	0.395*	0.864**	1					
12	−0.140	−0.053	0.067	0.007	−0.321	−0.192	0.338	0.277	0.595**	0.409*	0.619**	1				
13	0.237	0.039	0.068	0.021	−0.071	0.212	0.032	0.380	0.662**	0.410*	0.441*	0.480*	1			
14	−0.262	−0.308	0.045	0.000	−0.112	−0.311	0.200	0.328	0.548**	0.469*	0.443*	0.676**	0.453*	1		
15	−0.126	−0.083	−0.228	0.039	−0.146	−0.115	−0.237	− 0.005	0.277	0.618**	0.545**	0.456*	0.172	0.475*	1	
16	−0.353	−0.376	−0.221	−0.052	−0.150	−0.233	0.159	0.117	0.203	0.311	0.502**	0.503**	− 0.041	− 0.377	0.456*	1

**Correlation is significant at the 0.01 level (2-tailed)
*Correlation is significant at the 0.05 level (2-tailed)

analysis, evaluation as well as by students formulating their own knowledge based on previous knowledge and experience in a way that might not be achieved in a traditional classroom setting [6, 7, 22].

Students can discuss the theories and concepts related to the course modules with other students

Students learn by investigating and developing questions through an active participation in an activity. As students processed the new information, connections with various previous information and material could be made therefore creating new knowledge [24]. Students collaborate and discuss the new formulated information with other students and link ideas together, sharing and comparing knowledge on various online platforms as well as social media handles [2].

4.2 Mental Health and Wellbeing During Online Learning

Findings from the study showed that students' response to their mental health and wellbeing were as follows:

Online learning affects mental health

Majority of the students confirmed that online learning had not affected their mental health. [5] identified that burnout syndrome, depression and anxiety has decreased after transitioning from traditional learning to online learning.

Restlessness during online learning

The study findings suggest that restlessness did not affect most students during online learning. Although during online learning some students undergo procrastination, boredom and lack of interaction [23].

Depression during online learning

Some students confirmed that they had been depressed during online learning. [21] indicated that depression can be caused due to overwhelming amounts of work and reduced social interaction.

5 Conclusion

Findings from this study established that most university students positively adapted to online learning. Students spent more time on their work during online learning. However, students were also able to think more critically arguably due to the pedagogical shift to a student-centered approach. Students were able to collaborate with other students on online discussions forums or for assignments. This was probably because most students were Gen Zers and therefore connecting and collaborating with the virtual world was natural, unlike the other generations that grew up without technology. There is still much room available for the improvement of online learning systems. Universities need to be geared to create conducive online learning environments to ensure the possible outcomes of constructivist inquiry-based learning are achieved and ensure that students develop key skills to commence their career potentially under the 'new normal'.

References

1. Al-Husban NA (2020) Critical thinking skills in asynchronous discussion forums: a case study. Int J Technol Educ 3(2):82–91
2. Archer-Kuhn B, Lee Y, Finnessey S, Liu J (2020) Inquiry-based learning as a facilitator to student engagement in undergraduate and graduate social work programs. Teach Learn Inquiry 8(1):187–207
3. Bao W (2020) COVID-19 and online teaching in higher education: a case study of Peking University. Human Behav Emerg Technol 2(2):113–115
4. Bender L (2020) Interim guidance for COVID-19 prevention and control in schools. UNICEF
5. Bolatov AK, Seisembekov TZ, Askarova AZ, Baikanova RK, Smailova DS, Fabbro E (2021) Online-learning due to COVID-19 improved mental health among medical students. Med Sci Educ 31(1):183–192
6. Bryant J, Bates AJ (2015) Creating a constructivist online instructional environment. TechTrends 59(2):17–22
7. Colasante M, Oboler A, Walsh L, Civitella M (2018) Striving for authentic social constructivism in online learning; examples from postgraduate law & humanities. Open Oceans: Learning without borders Proceedings ASCILITE 2018:324–329
8. Dube B (2020) Rural online learning in the context of COVID 19 in South Africa: evoking an inclusive education approach. Multidisc J Educ Res 10(2):135–157
9. Duraku ZH, Hoxha N (2020) The impact of COVID-19, school closure, and social isolation on gifted students' wellbeing and attitudes toward remote (online) learning
10. Garrison DR (2003) Cognitive presence for effective asynchronous online learning: The role of reflective inquiry, self-direction and metacognition. Elem Qual Online Educ Pract Dir 4(1):47–58
11. Hebebci MT, Bertiz Y, Alan S (2020) Investigation of views of students and teachers on distance education practices during the Coronavirus (COVID-19) Pandemic. Int J Technol Educ Sci (IJTES) 4(4):267–282
12. Hughes JE, Read MF (2018) Student experiences of technology integration in school subjects: a comparison across four middle schools. Middle Grades Rev 4(1):n1
13. Khalil R, Mansour AE, Fadda WA, Almisnid K, Aldamegh M, Al-Nafeesah A, Alkhalifah A, Al-Wutayd O (2020) The sudden transition to synchronized online learning during the COVID-19 pandemic in Saudi Arabia: a qualitative study exploring medical students' perspectives. BMC Med Educ 20(1):1–10
14. Kwon K, Han D, Bang E-J, Armstrong S (2010) Feelings of isolation and coping mechanism in online learning environments: a case study of Asian international students. Int J Learn 17(2)
15. Ma Z, Idris S, Zhang Y, Zewen L, Wali A, Ji Y, Pan Q, Baloch Z (2021) The impact of COVID-19 pandemic outbreak on education and mental health of Chinese children aged 7–15 years: an online survey. BMC Pediatr 21(1):1–8
16. Mhlanga D, Moloi T (2020) COVID-19 and the digital transformation of education: what are we learning on 4IR in South Africa? Educ Sci 10(7):180
17. Mpungose CB (2020) Emergent transition from face-to-face to online learning in a South African University in the context of the Coronavirus pandemic. Humanit Soc Sci Commun 7(1):1–9

18. Pallant J (2020) SPSS survival manual: a step by step guide to data analysis using IBM SPSS. Routledge
19. Picciano AG (2017) Theories and frameworks for online education: seeking an integrated model. Online Learn 21(3):166–190
20. Raza SA, Qazi W, Khan KA, Salam J (2021) Social isolation and acceptance of the learning management system (LMS) in the time of COVID-19 pandemic: an expansion of the UTAUT model. J Educ Comput Res 59(2):183–208
21. Razai MS, Oakeshott P, Kankam H, Galea S, Stokes-Lampard H (2020) Mitigating the psychological effects of social isolation during the covid-19 pandemic. bmj 369
22. Reid-Martinez K, Grooms LD (2018) Online learning propelled by constructivism. In: Encyclopedia of information science and technology, 4th edn. IGI Global, pp 2588–2598
23. Sadeghi M (2019) A shift from classroom to distance learning: advantages and limitations. Int J Res English Educ 4(1):80–88
24. Sangkala NR, Doorman L (2019) The influence of inquiry-based learning on Indonesian students' attitude towards science. J Phys Conf Ser
25. Santos AP (2020) In the Philippines, distance learning reveals the digital divide. The Heinrich-Böll-Stiftung European Union
26. Secore S (2017) Social constructivism in online learning: andragogical influence and the effectual educator. e-mentor 70(3):4–9
27. Serhan D (2019) Web-based homework systems: students' perceptions of course interaction and learning in mathematics. Int J Soc Educ Sci 1(2):57–62
28. Sundarasen S, Chinna K, Kamaludin K, Nurunnabi M, Baloch GM, Khoshaim HB, Hossain SFA, Sukayt A (2020) Psychological impact of COVID-19 and lockdown among university students in Malaysia: implications and policy recommendations. Int J Environ Res Public Health 17(17):6206
29. Swan K (2005) A constructivist model for thinking about learning online. Elem Qual Online Educ Engaging Commun 6:13–31
30. Toquero CM (2020) Challenges and opportunities for higher education amid the COVID19 pandemic: the Philippine context. Pedagogical Res 5(4)
31. Wangenge-Ouma G, Kupe T (2020) Uncertain times: re-imagining universities for new, sustainable futures
32. Witze A (2020) Universities will never be the same after the coronavirus crisis. Nature 582(7811):162–165

Development

A Bibliometric Analysis of Sustainable Construction Practices—Implication on Construction Productivity

Oluseyi Julius Adebowale and Justus Ngala Agumba

Abstract Damages being done to the ecosystem have continued to foster the clamor for sustainable development practices across sectors, including the construction industry. The construction industry's operations account for about 33% of greenhouse gas (GHG) emissions on the earth, thereby making sustainable construction practices (SCPs) central to the global sustainable development goals. A bibliometric analysis was conducted to map the existing literature in SCPs research field. The analysis was utilized to ascertain the growth trajectory of publications in SCPs research domain, most productive and influential authors, collaboration among authors, geographical distribution of publications, and ultimately the emerging knowledge areas (EKAs) in the research field. The EKAs include: lean construction, procurement, energy, recycling, construction ecology, lifecycle costing, construction waste, and sustainable design. These knowledge areas have varying degree of implication on construction productivity. However, some of the knowledge areas, which include lean construction, energy, construction ecology, and sustainable designs have more notable implications on construction productivity. Documents published in the Scopus database were considered for analysis due to the wider coverage of the database. Besides, the study is limited to journal articles published from 2002–2021. Construction stakeholders can adopt the research findings to develop a framework that promotes SCPs and contributes to improving construction productivity. The reported knowledge areas provide directions for future research, whereas, factors with more notable implications on construction productivity should be investigated more extensively to determine their extent of implication.

Keywords Bibliometric analysis · Construction · Literature review · Productivity · Sustainable construction practices

O. J. Adebowale (✉) · J. N. Agumba
Department of Building Sciences, Tshwane University of Technology, Pretoria, South Africa
e-mail: adebowaleoluseyi@gmail.com

T. C. Haupt et al. (eds.), *Construction in 5D: Deconstruction, Digitalization, Disruption, Disaster, Development*, Lecture Notes in Civil Engineering 245,
https://doi.org/10.1007/978-3-030-97748-1_15

1 Introduction

The construction industry provides the facilities and infrastructure that people need to function, and it is therefore the fundamental foundation on which humanity exists, develops, and survives. Although the roles play by the industry is indispensable for achieving a better society, it is still characterized by negative environmental impacts on humankind, accounting for about 40% of the global energy consumption, 40% of the global resources; and around 30% of GHG emissions on the earth [14]. These statistics make the construction industry one of the largest users of energy, which significantly contributes to global warming [5, 8, 15]. Developed and developing countries alike share a common concern on the environmental impacts of construction operations. Arising from the concern, experts have begun to call for SCPs, as its level of implementation is currently low in practice [3]. Therefore, more than ever, it has become necessary to promote stakeholders' awareness and responsiveness through training and workshops [25].

SCP is important as the subject is inextricably linked with several key global and societal issues such as social, economic, and environmental discourse [44]. Rapidly developing construction, particularly in developing countries, contributes to environmental pollution, high energy consumption, and natural resources [35]. To promote SCPs, deliberate actions are required from the industry's practitioners to imbibe its culture and practice in construction [4]. Energy performance is an important criterion for evaluating environmental sustainability in construction [10]. These are embodied energy—energy used in production, transportation, assembly, and demolition. Likewise, there are operative energy—energy needed to control the building in progressions, such as lighting and operative alternative building appliances [45]. In recent time, multi-stakeholder interactions have continued to foster and promote diverse concepts of SCPs. They believe that SCPs will mitigate the negative impacts of construction operations on the environment, improve sustainability, promote cost effectiveness, and increase productivity [25]. The primary purpose of this study is to determine the EKAs in SCPs and ultimately discuss the implication of the knowledge areas on construction productivity. A bibliometric analysis was preferred to map the existing literature in SCPs research field. The analysis was utilized to ascertain the growth trajectory of publications in SCPs research domain, most productive and influential authors, collaboration among authors, geographical distribution of publications, and ultimately the EKAs in the research field. The reported knowledge areas provide directions for future research, whereas, factors with more notable implications on construction productivity can further be studied extensively to determine their extent of implication. Santos and colleagues [31] used bibliometric methods to review research on building information modelling (BIM), a methodology that leveraged information technology in response to demands for sustainable construction. Zuo and Zhao [48] used bibliometric methods to identify relevant clusters as well as hot topics concentrated in recent years, which identified social sustainability as a gap in the literature. Udomsap and Hallinger [39] adopted a broad perspective to sustainable construction, which extends prior reviews. In contrast, this study focuses on SCPs in relation to construction productivity.

2 Methodology

This section describes the procedures used in the identification of documents for the analysis as well as the methods of data analysis.

2.1 Identification of Sources

The study summarized 2002–2021 publications in the field of SCPs. The Scopus database was preferred to search for the documents. The search was conducted on the July 1, 2021. Scopus database was preferred in this study because it contains a comprehensive publication coverage from numerous field of study [6, 42]. The review adopted a broad scope in terms of document types with the inclusion of journal articles, conference papers, book chapters, and conference review. PRISMA (Preferred Reporting Items for Systematic Reviews and Meta-Analyses) guidelines for conducting systematic reviews of research was adopted in the study. "Sustainable" AND "Construction" AND "Practices" OR "Green" OR "Building" OR "Energy Efficiency" are keywords adopted for the search within "article title" of the Scopus database. The initial search yielded 103 documents. An article by Riedler et al. [30] was filtered out because it was written in German. Additional documents were excluded based upon documents type and publication stage (see Fig. 1). Review and data articles were also eliminated at this stage. After these filtering, the remaining 94 documents were considered for the bibliometric analysis.

2.2 Data Analysis

The data analysed for this review consisted of bibliographic information describing features of the 94 Scopus-indexed documents. Microsoft Excel was used to analyze growth trajectory of the SCPs publications, whereas VOSviewer software was used to analyze and present network visualization maps, which include co-authorship network and co-occurring keywords [40]. Co-authorship network was useful to determine scholars' collaboration in the SCPs research field. The network computes the number of times that two authors have been cited together in the reference lists of documents contained in the review database. The analysis scanned the reference lists and its results reflect patterns of scholarly influence in the broader literature. Thus, to some extent, co-citation overcomes a limitation of traditional citation analysis which is limited to the analysis of documents from a particular document repository [39].

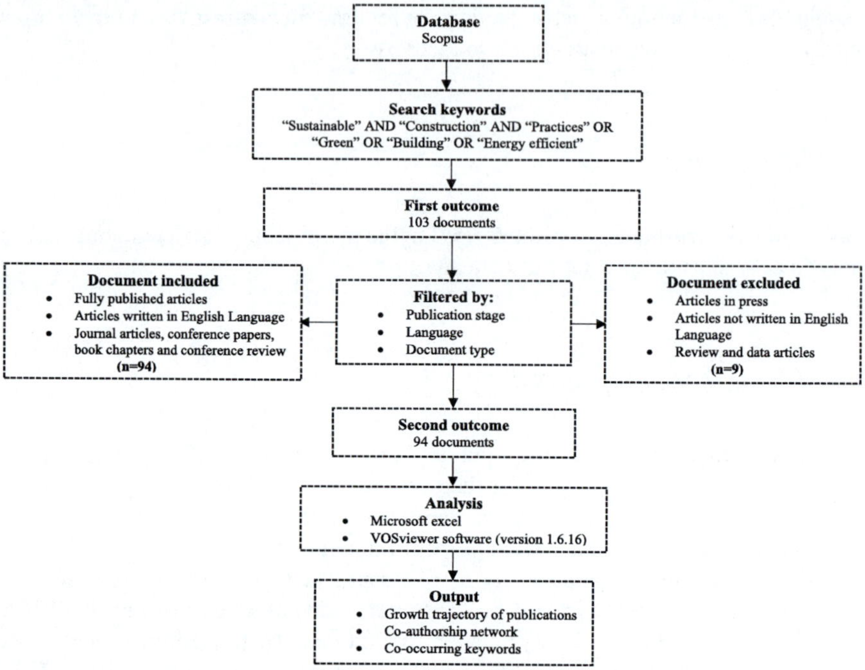

Fig. 1 Research process

The analysis has also been used as the basis for the visualization of similarities, a powerful approach to network mapping [40]. It assumes that when two scholars are frequently cited together by other authors, they tend to share a similarity in theoretical perspective. Author co-citation analysis in VOSviewer transforms patterns of author co-citation into a social network map that visualizes similarities among the authors in a particular literature [40]. Co-citation network has been used to determine authors' collaboration in different research domains, including delays in construction [32], health and safety [7], Building Information Modelling [28], and green building [47]. EKAs in SCPs research field was determined through the application of keyword co-occurring analysis in VOSviewer. Key contents of published documents and the areas studied within a specific field are denoted by keywords [42], hence, keywords constitute the EKAs in any research domain. Implications of the EKAs on construction productivity are finally discussed.

3 Findings

This section describes the results of the bibliometric analysis of the 94 SCPs documents published from 2001 through 2021.

3.1 Descriptive Trends in SCPs Research Domain

The 94 Scopus-indexed documents published over the past 20 years represent a large and rapidly growing knowledge base in SCPs. Over the years, a consistent growth of publications in SCPs research field is lacking. The trend has rather been inconsistent in nature. There is a growth from 2008 to 2010, whereas the domain records a decline in publication from 2010 to 2012. The nature of growth from 2012 to 2016 was haphazard in nature. Publication in the domain also began to experience growth from 2017 to 2019, while there was a decline from 2020 to 2021. The search was conducted at the beginning of the second half of 2021, therefore, there is a possibility that more articles will be published before the end of 2021 (Fig. 2).

The regions that contributed most to SCPs research were identified in VOSViewer. Table 1 presents the distribution of SCPs publications from regions having minimum of 5 publications from 2001 to the end of first half of 2021. The 94 documents were distributed across 33 regions. The United Kingdom achieved the most active region that contributed to SCPs research. Following United Kingdom are United States, South Africa, and Malaysia. Although Australia and Hong Kong have fewer

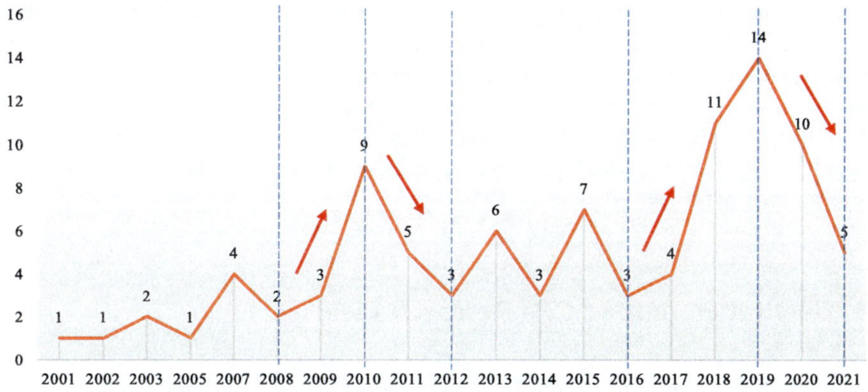

Fig. 2 The growth trajectory of SCPs publication, 2001–2021 (n = 94)

Table 1 Distribution of SCPs publications

Region	Total publication	Total citation	Average citation
United Kingdom	10	89	8.90
United States	8	234	29.25
South Africa	8	41	5.13
Malaysia	8	21	2.63
Australia	7	295	42.14
Hong Kong	6	423	79.50

Table 2 Most productive authors

Author	Total publication	Total citation	Average citation
Aigbavboa, C	4	30	7.50
Shen, I	3	261	87.00
Fortune C	3	21	7.00
Oke A. E	3	11	3.67

publications, the regions have more citations than the regions with more publications. In fact the citations from the regions are more than the other four regions put together. Therefore, Australia and Hong Kong are high performing in terms of total citations of articles from these regions.

4 The Most Productive Authors

The most productive authors in the SCPs research field is presented in Table 2. Minimum number of document of an author was set at 3. Of the 197 authors, 4 met the threshold. Aigbavboa, C. having four published articles is the most productive author in terms of total articles published in SCPs research field. The other three authors have three publications each, however, the influence of their articles in terms of citations varies significantly. For example, Shen, I. though having three publications recorded 261 citations, which is over four times more than the citations of the other three authors put together.

5 Co-citation in the SCPs Research Field

Collaboration is common among researchers to exchange ideas and enhance their research productivity [18]. VOSviewer was used to create and visualize co-authorship bibliometric network (see Fig. 3) to help determine collaboration of authors and citations of author's documents. Minimum number of document of an author was set at 3. Of the 197 authors, 14 met the threshold. Authors that are most influential in terms of their contributions to SCPs research domain are presented in Fig. 3. From the network, some collaborations consist of three authors, whereas some involve two authors. The cluster of authors with three authors collaboration include: (Aigbavboa, C. O., Mashwama, N. X. and Thwala, D.); (Aigbavboa, C. O, Oke, A. E. and Mosenga C.), and (Swan, A. J., Lovegrove, G. and Rteil, A). The cluster of two authors who have collaborated in SCPs research field include (Opoku, A. and Fortune, C.) and (Shafli, H. and Yassin, A. M).

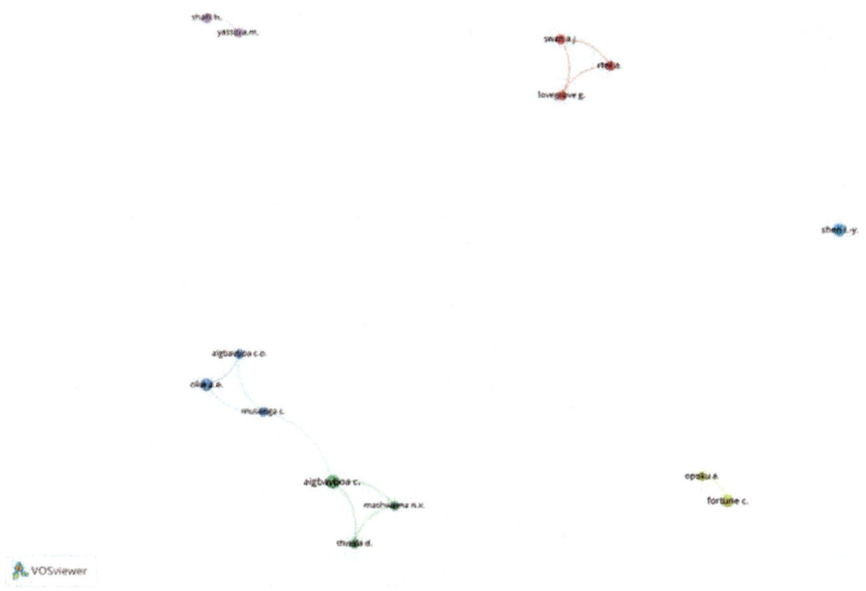

Fig. 3 Co-authorship bibliographic network

In terms of citations, the most influential articles and their information which include: authors' name, year of publication, title of article, and total citation of articles are presented in Table 3. Minimum number of citation of a document was set as 25. Of 94 documents, 6 met the threshold. Shen et al. [34], which explored the key to successful implementation of sustainable and socially responsible construction management practice was the most influential article. Following this article in their order of influence is [37] and Ruparathna and Hewage [31] having 162 and 71 citations respectively. Tan et al. [37] conducted a preliminary study on SCPs and contractor's competitiveness, whereas Ruparathna and Hewage [31] investigated current practices, drivers and opportunities in sustainable procurement in the Canadian construction industry.

6 Emerging Knowledge Areas in SCPs

Key contents of published documents and the areas studied within a specific field are denoted by the keyword [42]. Different researchers might understand keywords differently, however their co-occurrence with other keywords explains their meaning, especially when its usage occurs in many research papers [20]. The co-occurrence analysis forms the basis for deriving clusters and, therefore, research sub-fields. The inter-closeness among the keywords in Fig. 4 emphasizes its co-occurrence [42,

Table 3 The most influential publications in SCPs research

Author	Year	Title	Total citation
Shen, Tam, Tam and Ji	2010	Project feasibility study: the key to successful implementation of sustainable and socially responsible construction management practice	214
Tan, Shen and Yao	2011	Sustainable construction practice and contractor's competitiveness: a priliminary study	162
Ruparathna and Hewage	2015	Sustainable procurement in the Canadian construction industry: current practices, drivers and opportunities	71
Shen, Tam and Li	2009	Benefit analysis on replacing insitu concrete with precasts slab for temporary construction work in pursing sustainable construction practice	34
Akadiri and Fadiya	2013	Empirical analysis of the determinant of environmentally sustainable practices in the UK construction industry	30
Yu, Shi, Zuo and Chen	2018	Critical factors for implementing sustainable construction practice in Hopsca projects: a case study in china	27

41]. Hosseini et al. [18] recommended the use of "Author Keywords" and "Fractional Counting" in the analysis of VOSViewer, which is adopted in this study. Co-occurring keywords analysis was conducted using authors' keyword and fractional counting method in VOSviewer. Minimum number of occurrence of a keyword was set as 2. Of the 212 keywords in the retrieved articles, 28 met the threshold. Some keywords that are general were excluded before the analysis (see Table 4). The network of the remaining keywords is presented in Fig. 4. Connection lines indicate the link between a pair of keywords.

The nodes colour express the cluster of keywords and divides keywords into several clusters [42]. With the colour of the nodes, there are three visible clusters, with each cluster containing keywords that have internal relationships with it. Keywords within the same cluster imply that internal relationships exist among the keywords. One of the clusters in the network indicates that lifecycle costing, construction ecology, and lean construction are closely associated. Besides, sustainable design and procurement exist in the same cluster, which explain the internal relationship of these keywords. These keywords indicate EKAs in the SCPs research domain, which are discussed in relation to construction productivity in the subsequent section.

Fig. 4 Bibliometric network of co-occurring keywords. *Note* The missing keywords in VOSViewer presentation are captured in Table 4

7 Discussion

This section discusses the EKAs in SCPs and their implication on construction productivity. Focusing on key areas of SCPs would contribute to promoting sustainability in construction [25]. Despite introducing new advanced tools and materials to overcome cost overruns in construction, compared with other industries such as manufacturing, the construction industry has failed to achieve major changes in its production systems. The term "lean" refers to the optimal use of construction resources to save costs, time, and ensure quality, thereby contributes to achieving stakeholders' satisfaction [27]. Lean construction eliminates waste in the construction process, while maximizing projects' productivity [38]. Being one of the lean techniques, time studies are of utmost importance in terms of measuring productivity and construction project success. Regrettably, the construction industry is still confronted with time related waste in the form of idle time and non-value adding activities [13]. Among the nine principles of the lean construction philosophy determined by Bajjou and Chafi [9] include: continuous improvement, elimination of waste, personnel participation, planning and scheduling. These principles correlate with productivity, and contractors can imbibe them for their organisations' multi-factor productivity improvement.

Table 4 Influential keywords in SCPs research domain

S/N	Keywords exempted	Keywords included	Occurrence
1	Sustainability	Lean construction	4
2	Sustainable construction	Procurement	3
3	Green building	Energy	2
4	Construction industry	Recycling	2
5	Construction	Construction ecology	2
6	Sustainable	Lifecycle costing	2
7	Sustainable construction practices	Construction waste	2
8	Sustainable practices	Sustainable design	2
9	United Kingdom		
10	China		
11	Construction project		
12	Practice		
13	Zambian construction industry		
14	Sustainable construction practice		
15	Sustainable construction project		
16	Sustainable development		
17	Management		
18	Sustainable performance		
19	Concrete		
20	Leadership		

Sustainable procurement is recently becoming an important discourse in driving sustainability in the construction industry. Due to the environmental impact of procurement, sustainable procurement helps in achieving the desired sustainable development [26]. Traditional procurement practices give credence to time, quality, and cost as performance indicators. Sustainable procurement, however, extends traditional procurement to incorporate other factors, such as environmental and social considerations. Consequently, sustainability performance of suppliers has become an important factor in bid selection criteria [1]. An essential knowledge area in SCPs is construction ecology, which relates to pollution and energy efficiency [44]. Global warming, which is partly due to embodied energy in construction has brought more frequent and severe heat waves, which have severely affected vulnerable groups such as construction workers [46]. Excessive heat stress can have a profound impact on physiological responses, leading to occupational injuries and low productivity. Construction workers are particularly susceptible to heat stress because physical labour and humid working conditions can cause body heat. Construction project deliveries usually involve the operation of mechanized equipment. Operations of the equipment contributes to noise pollution and significant GHG emissions including carbon and diesel exhaust emissions [36]. Noise pollution can interfere with social strategy in terms of quality of site instructions, while GHG emission contributes to global warming. Notwithstanding threats pose by mechanized equipment to the ecological system, they are essential for achieving high productivity, efficiency and quality of work [43]. Consequently, GHG emission emanating from site equipment and machineries can be reduced by eliminating engine idle, task planning, operation planning, and activity planning [36].

Life cycle cost (LCC) analysis helps to compare design options of an asset in order to achieve value-for-money [23]. [16] defined LCC as "*the total discounted cost of owning, operating, maintaining, and disposing of a building or building system over a period of time*". The concept is an economic evaluation technique used to determine the total cost of owning and operating a facility over a period of time. The practice is valuable for evaluating and comparing the initial cost increase and long-term operating cost benefits of different building designs. Some clients conduct LCC analysis to help their decision making, whereas some avoid life cycle costing due to its cost and time implications. The main motivation for applying LCC analysis is to increase the possibility of reducing costs during the operation phase, even if the initial investment needs to be further increased. Natural resources, including construction materials have continued to deplete and the unsustainable use of the depleted resources generate adverse environmental impacts [29]. Recovery, recycling, and reuse of construction materials must be ensured to promote sustainable practices [12]. According to Hendriks and Janssen [17], about 90% of construction and demolition waste processed by the recycling system is generally made up of masonry and concrete, which makes waste management critical to achieving sustainable development [29]. In some developed countries, including the United States, the United Kingdom, and Australia, construction waste management is becoming increasingly popular as an important tool for reducing the negative impact of construction waste on the environment. An effective waste management plan requires the cooperation of construction stakeholders, including customers, contractors, designers, engineers, subcontractors, workers, and even suppliers. Insufficient information on waste management and its relevance to sustainable practices by construction stakeholders is a major issue that makes it difficult for construction waste management practice. The widely acceptance and practice of waste management will promote SCPs, further save construction cost and largely address productivity challenges arising from scarcity of construction materials. The government should play a leading role through appropriate institutions, emphasizing the use of processed construction and demolition waste as raw material for new construction projects [17].

Towards entrenching SCPs, [29] advocate the need for legislation that promotes the recycling and reuse of construction and demolition waste. According to [2], there is insufficient knowledge of sustainable design among design professionals, which contribute to delaying the practice of sustainable design and construction. The assessment of alternative systems and design along with green materials can be more complicated than the conventional building designs. Although green buildings provide significant future benefits, the ability of designers in sustainable design is similarly essential. The construction industry requires participation of many design professionals. Design related issues have been widely reported as some of the challenges that undermine productivity in conventional constructions [11, 19, 22, 24]. Considering the low level of sustainable design expertise among design professionals, especially in developing countries, the practice of sustainable design could further compound design related issues affecting the industry's productivity. This necessitates the need for design professionals and contractors to prioritize capacity building, otherwise constructability challenge will be prevalent in practice [21].

8 Conclusion

The need for sustainable practices has continued to gain popularity in different sectors, including the construction industry. The study conducted a bibliometric analysis of the existing literature in SCPs to determine the EKAs in the research field and also to ascertain the implication of the EKAs on construction productivity. The study determined that publication growth in the research domain over the past twenty years has not been consistent. The United States, United Kingdom, and South Africa represent the three countries that contribute most to publications in the SCPs knowledge area. Further analysis reveals that Aigbavboa C. is the most productive author in terms of number of publications in the research field, whereas, Shen et al. [34] published the most influential documents in terms of citations. Finally, EKAs in the SCPs research domain include: lean construction, procurement, energy, recycling, construction ecology, lifecycle costing, construction waste, and sustainable design. Although these SCPs have varying degree of implications on construction productivity, lean construction, energy, construction ecology, and sustainable designs have more notable implications on construction productivity.

References

1. Agbesi K, Kissi E, Adjei-Kumi T (2020) Implementation of sustainable procurement practices in public sector construction organisa tion in Ghana In: Scott L, Neilson CJ (eds) Proceedings of the 36th annual ARCOM conference, 7–8 Sept 2020, UK, Association of Researchers in Construction Management, pp 526–535
2. Ahn YH, Pearce AR, Wang Y, Wang G (2013) Drivers and barriers of sustainable design and construction: The perception of green building experience. Int J Sustain Build Technol Urban Dev 4(1):35–45
3. Aigbavboa C, Ohiomah I, Zwane T (2017) Sustainable construction practices: "a lazy view" of construction professionals in the South Africa construction industry. Energy Procedia 105:3003–3010
4. Akadiri PO, Chinyio EA, Olomolaiye PO (2012) Design of a sus tainable building: a conceptual framework for implementing sustain ability in the building sector. Buildings 2:126–152
5. Akeiber H, Nejat P, Majid MZA, Wahid MA, Jomehzadeh F, Famileh IZ, Calautit JK, Hughes BR, Zaki SA (2016) A review on phase change material (PCM) for sustainable passive cooling in building envelopes. Renew Sustain Energy Rev 60:1470–1497

6. Akinlolu M, Haupt TC, Edwards DJ, Simpeh F (2020) A bible ometric review of the status and emerging research trends in con struction safety management technologies. Int J Constr Manage, pp 1–13
7. Alruqi WM, Hallowell MR, Techera U (2018) Safety climate dimensions and their relationship to construction safety performance: a meta-analytic review. Saf Sci 109:165–173
8. Asdrubali F, D'Alessandro F, Schiavoni S (2015) A review of un conventional sustainable building insulation materials. Sustain Mater Technol 4:1–17
9. Bajjou MS, Chafi A (2018) A conceptual model of lean con struction: a theoretical framework. Malays Constr Res J 26(3):67–86
10. Berardi U (2012) Sustainability assessment in the construction sector: rating systems and rated buildings. Sustain Dev 20:411–424
11. Chan PW, Kaka A (2007) Productivity improvements: under stand the workforce perceptions of productivity first. Pers Rev 36(4):564–584
12. Defra (2009) Site waste management plans: non-statutory guidance for site waste management plans. Department for Environment, Food and Rural Affairs. Retrieved from www.defra.gov.uk. Accessed 5th April 2021
13. Demirkesen S, Sadikoglu E, Jayamanne E (2020) Investigating effectiveness of time studies in lean construction projects: case of Transbay Block 8. In: Production planning & control, pp 1–21
14. European Commission, Directorate-General for Energy and Transport (2020) Vision: Saving Our Energy. Office for official Publications of the European Communities, Luxembourg. Available online at: http://www.energy.eu/publications/2007_eeap_en.pdf. Accessed 2 July 2021
15. Family R, Menguc MP (2018) Analysis of sustainable materials for radiative cooling potential of building surface es. Sustainability 10(9):3049
16. Fuller S, Petersen S (1995) Life-cycle costing manual for the federal energy management program. In: National Institute of Standards and Technology (NIST) Handbook, p 135. Washington, DC
17. Hendriks CF, Janssen GMT (2001) Reuse of construction and demolition waste in The Netherlands for road constructions. Heron. https://doi.org/10.1016/j.wasman.2004.09.005
18. Hosseini MR, Martek I, Zavadskas EK, Arashpour M, Chileshe N, Aibinu AA (2018) Critical evaluation of o ff -site construction research: a Scientometric analysis. Autom Constr 87:235–247. https://doi.org/10.1016/j.autcon.2017.12.002
19. Hwang BG, Zhu L, Ming JTT (2017) Factors affecting productivity in green building construction projects: The case of Singapore. J Manag Eng 33(3):04016052
20. Isenberg P, Isenberg T, Sedlmair M, Chen J, Moller T (2016) Visualization as seen through its research paper keywords. IEEE Trans Visual Comput Graphics 23(1):771–780
21. Jarkas AM (2015) Factors influencing labour productivity in Bahrain's construction industry. Int J Constr Manag 15(1):94–108
22. Jarkas AM, Al Balushi RA, Raveendranath PK (2015) Determinants of construction labour productivity in Oman. Int J Constr Manage 15(4):332–344
23. Kadarova J, Kobulnicky J, Teplicka K (2015) Product life cycle costing. In: Applied mechanics and materials, vol 816. Trans Tech Publications Ltd, pp 547–554
24. Kazaz A, Ulubeyli S (2006) Physical factors affect ing productivity of Turkish construction workers. In: Proceedings of the 22nd annual ARCOM conference
25. Mashwama NX, Thwala D, Aigbavboa C, Hamma-Adama M (2020) A stakeholder perspective on the implementation of sus tainable construction practices. Proc Int Struct Eng Constr 7(1)
26. Meehan J, Bryde DJ (2015) A field-level examination of the adoption of sustainable procurement in the social housing sector. Int J Oper Prod Manag 35(7):982–1004
27. Natasya A, Wahab A, Mukhtar M, Sulaiman R (2013) A conceptual model of lean manufacturing dimensions. Procedia Technol 11:1292–1298. https://doi.org/10.1016/j.protcy.2013.12.327
28. Noor BA, Yi S (2018) Review of BIM literature in construction industry and transportation: meta-analysis. Constr Innov

29. Papargyropoulou EFFIE, Preece C, Padfield R, Abdullah AA (2011) Sustainable construction waste management in Malaysia: a contractor's perspective. In: Management and innovation for a sustainable built environment MISBE 2011, Amsterdam, The Netherlands, 20–23 June 2011. CIB, Working Commissions W55, W65, W89, W112; ENHR and AESP
30. Riedler J, Wimmer R, Frisch J, Abend T, Flinterhoff A, Van Treeck C (2016) Sustainable construction materials for temporary trade fair structures. 2016. Analysis and recommendations for action in practice [Article@Nachhaltige Baumaterialien für temporäre Messebauten Analyse und Handlungsempfehlungen für die Praxis] Bauingenieur, 91 (July–August), pp 322–325
31. Ruparathna R, Hewage K (2015) Sustainable procurement in the Canadian construction industry: current practices, drivers and opportunities. J Clean Prod 109:305–314
32. Sanni-Anibire MO, Mohamad Zin R, Olatusnji SO (2020) Causes of delay in the global construction industry: a meta analytical review. Int J Constr Manage, pp 1–13
33. Santos R, Costa AA, Grilo A (2017) Bibliometric analysis and review of building information modelling literature published between 2005 and 2015. Autom Constr 80:118–136
34. Shen LY, Tam VW, Tam L, Ji YB (2010) Project feasibility study: the key to successful implementation of sustainable and so cially responsible construction management practice. J Clean Prod 18(3):254–259
35. Stanaszek-Tomal E (2020) Bacterial concrete as a sustainable build ing material? Sustainability 12(2):696
36. Szamocki N, Kim MK, Ahn CR, Brilakis I (2019) Reducing greenhouse gas emission of construction equipment at construction sites: field study approach. J Constr Eng Manag 145(9):05019012
37. Tan Y, Shen L, Yao H (2011) Sustainable construction practice and contractors' competitiveness: a preliminary study. Habitat Int 35(2):225–230
38. Tauriainen M, Marttinen P, Dave B, Koskela L (2016) The effects of BIM and lean construction on design management practises. ProcediaEng 164:567–574
39. Udomsap AD, Hallinger P (2020) A bibliometric review of re search on sustainable construction, 1994–2018. J Cleaner Prod 254:120073
40. Van Eck NJ, Waltman L (2014) Visualizing bibliometric networks. In: Measuring scholarly impact. Springer, pp 285–320. https://doi.org/10.1007/978-3-319-10377-8_13
41. Van Eck NJ, Waltman L (2017) Citation-based clustering of publications using CitNetExplorer and VOSviewer. Scientometrics 111(2):1053–1070
42. Vigneshkumar C, Salve UR (2020) A scientometric analysis and review of fall from height research in construction. Constr Econ Build 20(1):17
43. Waris MSLM, Arazi I (2014) Reducing the greenhouse gas emissions from onsite mechanized construction. Aust J Basic Appl Sci 8(5):391–398
44. Windapo A, Ogunsanmi O (2014) Construction sector views of sustainable building materials. In: Proceedings of the institution of civil engineers-engineering sustainability, vol 167, no 2, pp 64–75. Thomas Telford Ltd.
45. Yadav M, Agarwal M (2021) Biobased building materials for sustainable future: an overview. Mater Today Proc
46. Yi W, Chan AP (2017) Effects of heat stress on construction labor productivity in Hong Kong: a case study of rebar work ers. Int J Environ Res Public Health 14(9):1055
47. Zhao X, Zuo J, Wu G, Huang C (2019) A bibliometric review of green building research 2000–2016. Architectural Sci Rev 62(1):74–88
48. Zuo J, Zhao ZY (2014) Green building research–current status and future agenda: a review. Renew Sustain Energy Rev 30:271–281

Integrated Project Delivery (IPD): An Innovative Approach for Achieving Sustainability in Construction Projects

Ghada Adel, Ayman Ahmed Ezzat Othman, and Nishani Harinarain

Abstract In spite of the positive impact construction can have on the economy and society in general, the industry is being blamed for the negative impact it has on the environment. This problem can be attributed to traditional procurement approaches adopted in construction industry and the inability to handle the different challenges regarding waste generation and pollution of the environment. This paper focused on the role of construction procurement as a method to achieving sustainability. This paper aims to investigate the role of integrated project delivery (IPD) in achieving sustainability in construction projects. In order to achieve this, aim the research method consisted of an in-depth literature review and case studies. Firstly, the literature review was used to identify and categorise the principles of integrated project delivery and their importance in achieving sustainability in construction. Secondly, two case studies were analysed to investigate the benefits of the using of integrated project delivery in construction projects. It was found that the usage of the integrated project delivery results in meeting the goals of sustainability. The two case studies examined showed that IPD had a positive outcome in regards to economic, social and environmental aspects because it reduced the cost and time of the construction of the project, as well as having a positive effect on the community.

Keywords Sustainability · Traditional procurement · Integrated project delivery (IPD) · Construction industry

G. Adel · A. A. E. Othman
Architectural Engineering Department, Faculty of Engineering, The British University in Egypt (BUE), El Shorouk City—Cairo Suez Desert Road, Cairo, Egypt
e-mail: Pg.ghada91830010@bue.edu.eg

A. A. E. Othman
e-mail: ayman.othman@bue.edu.eg

N. Harinarain (✉)
University of KwaZulu-Natal Durban, Durban, South Africa
e-mail: harinarain@ukzn.ac.za

T. C. Haupt et al. (eds.), *Construction in 5D: Deconstruction, Digitalization, Disruption, Disaster, Development*, Lecture Notes in Civil Engineering 245,
https://doi.org/10.1007/978-3-030-97748-1_16

1 Introduction

The construction industry is one of the main sectors affecting the economy of any country. The influence of the construction industry is not only on the economy but also on the social aspects of life, through creating, modifying, and improving the living environment. Simultaneously, the construction industry has a noteworthy impact on the environment throughout the whole life cycle of a building, starting with the mining of raw materials until the end of life of the building, i.e., its demolition [1]. Nowadays, the construction industry is facing some major challenges in maintaining the balance between environmental, social, and economic aspects, and the manner in which the construction process is realized [2]. Knowing that the choice of procurement selected affects the whole construction process, it is important to note that the traditional procurement approaches have proven to be great sources of waste, disputes and time consumption due to the separation between its parties. Various nontraditional approaches have been introduced to solve these issues. One of these approaches is the integrated project delivery (IPD). Therefore, this research aims to investigate the role of IPD in achieving sustainability in construction projects.

2 The Nature of the Construction Industry

The construction industry is one of the largest industries worldwide. It provides the community with high living standards through socio-economic projects and infrastructure facilities such as roads, hospitals, and schools. It also plays key role in generating income in both formal and informal sector. It supplements the foreign exchange earnings derived from trade in construction material and engineering services [3]. Construction industries in developing countries are a major stakeholder in the economy. It is also a source of employment at various levels of skills from manual labour to semi-skilled, skilled and specialist workforce. Experience shows that it is one of the foremost industries in any developing country whose upward activity is related to the economy of the country. In addition, any crash in the construction industry could potentially lead to the fall of the economy [4]. In spite of the flourishing of construction and its positive impact on the economy, the industry is being blamed for the negative impact it has on the environment. This problem could be attributed to traditional procurement approaches adopted in construction industry and the inability to handle the different challenges regarding waste generation and pollution of the environment. This is a call for the construction industry to be more innovative and to consider sustainable solutions [5].

3 Sustainability

The principle of sustainability was inspired from the Brundtland Report of 1987, when the United Nations appointed Gro Harlem Brundtland the former prime mister of Norway to head the new World Commission on Environment and Development. It has found that countries were suffering from extreme poverty despite all the devoted efforts to increase the living standards. This led to the need of balancing resources and consumption. The Brundtland Report defined the sustainable development as development that meets the needs of the present without compromising the ability of future generations to meet their own needs [6]. This definition has developed over the years. Raynsford defined the sustainability, to be:

- Reducing material consumption;
- Minimising waste;
- Using recycled materials;
- Energy efficiency; and
- Avoiding pollution [7].

Sustainability is also defined as the distribution of resources efficiently and equitably through the generations with consideration of the finite eco-system [8]. In 2015, Ben-Eli sees sustainability as a dynamic equilibrium between the people and the capacity of the environment where the people achieve their full potential without negatively affecting their environment [9]. Moreover, Mensah & Enu-Kwesi define sustainability as an application of a concept of improving and nourishing a strong social, economic and ecological system [10].

Sustainability, aims to allow the existence of upcoming generations through protecting the natural and built environment, in addition to the natural resources and human beings [11]. Sustainability is a multi-dimensional system that concentrates on the quality of life of all people through making bonds between the social aspects and economic besides the environmental aspect [12, 13]. There are 3 main dimensions for sustainability showed in Fig. 1, which are environmental, social and

Fig. 1 Sustainability dimensions

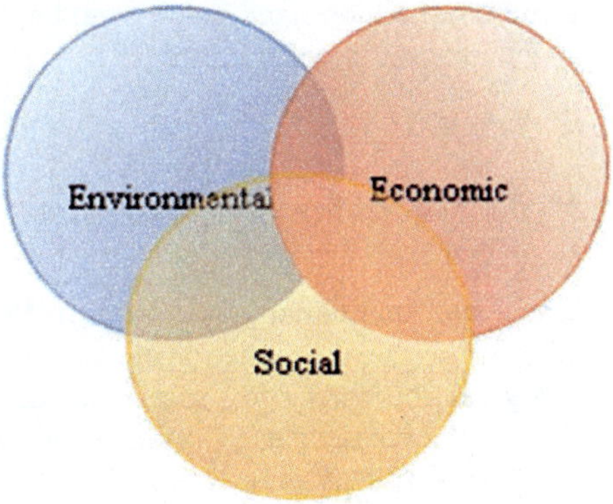

economic [14].

The environmental aspect is described as reducing the waste impact and emissions in the long term with caution in the natural resource usage. The economic aspect is defined as "*an economic assessment of competing design alternatives, considering all significant costs over the economic life of each alternative, expressed in equivalent dollars*" [14]. Basically, it is related to the efficient use of resources leading to an increase in profit [15]. Besides, the consideration of the life cycle costs, the social aspect is also incorporated, which takes into consideration all the needs of the workforce and the people throughout the life cycle of the project [15].

4 Construction Procurement

Procurement is derived from the word procure, which means "*to get possession of something*" and "*to obtain by care or effort*". This term has various definitions within literature. One of which is that it is a process of providing a service from different parties such as contractor, consultant and designer in order to ensure that the project is delivered to the client [16]. Procurement also defined as a process of obtaining services and goods from others with return [16]. Construction procurement is seen as a complex network of relationships between contractor, consultant and client in order to form a building project [17].

4.1 Traditional Procurement Methods

The traditional procurement method is a method in which the design phase occurs before assigning a contractor to the project. This method results in a separation between the design and construction professionals which leads to difficulty in communication and does not foster innovation and creativity, resulting in reducing the value of the project [18]. Types of contracts that are usually used under traditional procurement include fixed price contract, unit price contract, cost reimbursement contract and cost target contract [19].

4.2 Non-traditional Procurement Methods

Different non-traditional procurement methods have been invented since the early 1960s, focusing on the reduction of consumed time in the design and tender processes which gives the ability to proceed to the construction work sooner. In addition, it also emphasizes engaging with the contractor in the early stages of the project (design phase). This cooperation is in the interest of the client which reduces conflicts occurring during the construction work. Moreover, it helps in sharing experiences which increases the quality of the product [20]. Types of contracts that are usually used under the non-traditional procurement include design and build contract, management-based contracts and partnering contracts [19].

5 Integrated Project Delivery (IPD)

The IPD is defined by the American institute of Architects (2007) as an approach that integrates the people, business structures, systems and practices into a method that connects talents and visions of all parties in order to boost the project outcome, increase value for the client and reduce the waste through all the building processes (design, fabrication, construction). Integrated project delivery has offered a great deal of improvement by maximising value and reducing waste through the early collaboration of cross-functional teams to align goals and share risks and rewards through a relational contracting approach [21]. The IPD agreements set rules to limit liability and improve the information and creativity flow. Therefore, the integrated project delivery is considered a whole building design [20] as IPD aims to reduce waste in construction projects which leads to optimal improvement in schedule, cost and quality [21, 22]. In this approach all parties are working together as a team with one goal which is in the best interest of the project. In addition, the profits and losses are shared between the project parties [21].

IPD was introduced to overcome the inadequacies encountered in the traditional procurement approach. Though it has been increasingly adopted in the USA and other parts of the world, its application in the Middle East and Egypt has not commenced yet. Despite the numerous advantages that this new method provides, its application in the Middle East is still marginal because of the required perquisites for its implementation (i.e. technological, legal, and cultural); therefore the number of projects adopting IPD remains relatively small [23, 24].

5.1 Principles of Integrated Project Delivery (IPD)

The main concern of IPD is to eliminate waste through early collaboration between parties, risk sharing and planning. In order to achieve the benefits of IPD, all parties need to follow the IPD principles. The American Institution of Architects guide book in addition to the AEC have set the principles of the IPD in the recent years as follows:

- Mutual respect and trust: All parties understand and respect the value of partnership and are devoted to team work for the best interest of the project.
- Mutual benefit and reward: The benefit from IPD is reflected on all the team members as the process entails the involvement of all parties at early stages. The compensation in IPD is related to the value added to the interest of the project.

- Collaborative innovation and decision making: The innovation is fostered by free exchange of ideas among parties. In the IPD, the ideas are rated by their impact on the project not by the originator of the idea.
- Early involvement of key participants: In the IPD all parties are involved in the early stages of the project which enriches the flow of information and the sharing of experiences and creative ideas.
- Early goal definition: The goal of the project is set in early stages, which promotes the innovation and outstanding performance.
- Intensified planning: IPD believes that the more effort exerted in planning, the more efficient the execution process will be.
- Open communication: The focus of the IPD on team work is achieved through honest, open and direct communication between all parties. The responsibilities are defined and all parties are concerned with finding solutions and resolving problems rather than determining the liability and blaming each other.
- Appropriate technology: "*Integrated projects often rely on cutting-edge technologies*". These technologies are specified at the start of the project in order to maximise functionality, generality and interoperability. Open and interoperable data exchange is essential to support IPD because open standards best enable communication among all participants. This means that technology that complies with open standards should used whenever available.
- Organisation and leadership: In IPD, the project team is considered an organisation in its own and all team members are dedicated to the goals and values of the project.

5.2 Difference Between Traditional Procurement Methods and IPD

Table 1 shows the differences between the IPD and traditional procurement methods.

As shown in Table 1, the traditional procurement methods are transactional contracts that requires each party to finish its role and then the other party would start. Whereas in the IPD, the contract is relational and all parties are involved in the project from early phases which increases the feeling of ownership.

The team structure in the traditional procurement is hierarchal. Meanwhile, in the IPD, it is flat as all parties are partners. The contracts in the traditional procurement are separate as each party has its separate contract with the owner. However, in the IPD, the contracts are interlocking and all parties work as partners after signing the Integrated Form of Agreement.

The risk in the traditional procurement is allocated to the parties and the reward is for the owner only. On the other hand, in IPD, the risk is shared and there is no concept of blaming or disputes. Besides the reward is distributed to the team.

The decision making in the traditional procurement is through the top management only and from the concerned party only. Meanwhile, in the IPD, the decisions are made jointly between the parties and the major decisions are taken by the primary members (core team).

Table 1 Difference between traditional procurement and IPD

	Traditional procurement	IPD
Type	Transactional	Relational
Team structure	Hierarchical	Flat
Contracts	Separate contracts Setup detailing how to sue and claim for compensation	Interlocking contracts Legally: direct and blind team activity Eliminate and reduce the ability to sue and claim compensation
Risk and reward	Risk is primary allocated Reward payment is based on delivery	Risk is primary shared Reward: pooled profit distribution is based on team success in achieving project goal Owner pays for additional extras
Decision control	Hierarchical	Joint decision making Major project decisions are taken by primary team members
Process	Linear information	Cross sharing of information "Best for project" mentality

The process in the traditional procurement is linear and the flow of information is linear which hinders innovation and flow of ideas and creativity. In contrast, in IPD, there is flow of information between all parties sharing experiences and finding what is best for the project.

5.3 IPD and Sustainability

IPD principles serve the sustainability dimensions as shown in Table 2. The mutual respect and trust between parties helps in achieving the project's goals, fosters creativity, and allows the ideas to flow. This, in turn, will reflect in the economic dimension of the project and help in exploring new ideas for energy saving and reducing pollution in order to save the environment. It will also allow for the existence of more projects, thereby socially helping people.

The mutual benefit and reward help increase the feeling that the project is owned by all parties, which, in turn, will make all parties concentrate on early delivery of the project, cost reduction and increase of quality. Moreover, all parties will be dedicated to increase the value of the project through considering the environmental impact of this construction.

The collaborative innovation and decision-making results in all parties making decisions together and agree on risk taking, which allows for the consideration of all ideas on table. That will result in permanent decisions which were discussed by the experts in their respective fields, preventing rework and conflicts. As a result, this will decrease the project duration and cost, as well as, decrease the demolition and the rework affecting the environment.

Table 2 IPD principles and the sustainability dimensions

IPD principle	Sustainability		
	Social dimension	Economic dimension	Environmental dimension
Mutual respect and trust	✓	✓	
Mutual benefit and reward	✓	✓	✓
Collaborative Innovation and decision making	✓	✓	✓
Early involvement of key participants	✓	✓	✓
Early goal definition	✓	✓	✓
Intensified planning	✓	✓	✓
Open communication	✓	✓	
Appropriate technology	✓		✓

The early involvement of key participants and the early goal definition are related because the parties are involved in early stages and define the project goals from early stages, resulting in a project that will be easier to direct easier and the decision making process will be simpler. This could result in the project being handed over sooner and ensures that the quality of the project is according to its specifications, thereby, decreasing the costs.

The intensified planning is an important principle that affects all the dimensions of the sustainability because a well-planned project has minor conflicts and its construction phase has fewer negative effects on the environment.

Open communication allows for the flow of ideas between parties and the sharing of experiences to prevent past mistakes which, in turn, will reflect on the duration needed to finalise the project and help in reducing the costs of construction.

The appropriate technology helps in studying the project better and having a better overview on the construction process. Besides, it helps in studying the environmental impacts of the project and its emissions. Moreover, it gives a visual on how each phase is going to be held and how it will be mobilised because this gives an overview of all the project details. As a result of this study, the project conflicts and the time of construction will be reduced which, in turn, affects the cost of the project, as well as, the environmental analysis.

6 Research and Method

This paper aimed to investigate the role of IPD in achieving sustainability in construction projects. In order to meet this aim, the research method involved a detailed literature review and the investigation of two case studies, that of the Cardinal Glennon

Children's Hospital Expansion and Sutter Medical Centre Castro Valley. The literature review was designed to build a comprehensive background on the research topic, where the sustainability dimensions, the procurement methods and the integrated project delivery was studied to gain better knowledge and understanding. The presentation and analysis of the two case studies show the role of IPD implementation in achieving sustainability.

7 Case Study

This section presents two case studies of health care projects that used the integrated project delivery in order to reduce the time consumed in construction. In these cases, all parties acted as partners. Both projects were renovation and extension of the hospitals. Therefore, in one of the case-studies the application of IPD was not implemented in the early stage of the project, which did not allow the parties to identify goals and apply IPD in early stages.

7.1 Case Study 1: Cardinal Glennon Children's Hospital Expansion

This is located in St. Louis, Missouri.

Project Description

The project was to build a children's hospital expansion on 138,000 square feet with a budget of $45.5 million. The expansion consists of 10 surgical suites, an intensive care unit to hold 60 beds, a central sterile unit, 10 recovery rooms and a shell space for future radiology and laboratories.

This project was the first time that all the parties experienced using IPD. The decision to use IPD was made in the design phase. The architect, the builder and the MEP engineer had worked together previously in phase 1 of this project (Fig. 2).

Shared Risk/Reward

The parties were at the design phase when the decision was made to work under the IPD. The architect had its contract and the builder had the CM at risk arrangement.

The owner held a lean seminar from the lean construction institute and some partners from the design and construction community. The project parties challenged each other to try to work under the IPD in this project and all agreed.

The owner's attorney negotiated an integrated form of agreement based on the project with a plan to make it a model for the future work. As the project had already started, it was to late to involve early application of IPD.

Fig. 2 Cardinal Glennon Children's Hospital Expansion—St. Louis, Missouri

Financially, about $400,000 were saved with approximately $1 million contingency. The incentives were distributed among parties as follows: 40% owner, 20% design team, 40% builder and lean partners.

The Contract

The integrated form of agreement was a four-way contract including the owner the MEP engineer, the architect and the builder. All parties were accountable to each other as equal partners. The builder, along with the architect, had combined contingencies and were jointly responsible for the construction errors and design.

Decision Making

The integrated form of agreement designed two teams, the field team and core team. The field team consisted of mid-level partners and was responsible for resolving routine issues. While the core team consisted of the owner, architect, engineer and builder as well as the lean partners, who set a weekly meeting to resolves issues and make decisions.

The core team was highly cooperative and refused the idea of "*one party win and the other lose*". As an example of the cooperation between the parties, the builder suggested that they use the concrete maturity test (CMT) to measure the strength of concrete which was not used in the traditional method. That resulted in reducing the time needed for the concrete to cure.

Challenges

An example of a challenge experienced was a conflict between rebar in the flat slab and the plumbing sleeves that needed to penetrate the slab. After carefully studying, this could be solved by shifting the entire plan 3.5″ with respect to the column grid. As a result of the full dedication of the design team, they made it work which resulted in early delivery of the project with 6 weeks earlier than planned [25].

7.2 *Case Study 2: Sutter Medical Centre Castro Valley*

The vision of this was to be a landmark medical centre that includes advanced technology and integrates it with the quality of service required. The owner is considered one of the leading non-profit funders and set a budget of $320 million for the project. The project did not only include hospital rooms and a building but also included the demolition of the old hospital, construction of a new parking area and improvement to the campus (Fig. 3).

There were new laws passed that compressed the time frame for construction, which encouraged Sutter to reduce time and budget overrun by using the IPD. The schedule was required to be accelerated by 30% without exceeding the cost target. Early in the planning phase the core team of the contract was gathered signing the Integrated Form of Agreement (IFOA). There were 10 parties in that team, namely, owner, architect, general contractor, mechanical and plumbing designer, electrical designer, structural designer, mechanical designer, electrical designer, fire protection and lean/BIM project integration team.

The IPD approach encouraged the participating firms to change their normal work process and devote people full-time to this project, as opposed to the traditional process where they typically have people working on 3–4 projects at the same time. In addition to the core IPD team, there were over 25 additional firms in the expanded project team, who are also supporting the IPD approach and adjusting their traditional processes to work with it.

Given the challenges of the site, the complicated shape of the design, and the schedule and budget constraints, the use of BIM on this project was indispensable. That helped in generating the 3D model before starting the project in addition to discovering conflicts before facing them in reality. Remarkably, the time for construction was reduced from an expected 15 months to 8 months, and was informed by far more information from other disciplines than what is usually available which led to better design quality.

Fig. 3 Sutter medical centre castro valley

Result

This project can be considered as a landmark project in the US architectural, engineering and construction industry as it was the first one to show that IPD is not just a utopian vision but a practical reality that can actually be implemented on large, as well as, small projects. First and foremost, though, it takes the initiative and leadership of an owner such as Sutter Health, which had the confidence to embark upon unchartered territory and put together a team that could actually deliver on this project as envisaged by the IPD concept.

- The delivery of the project was 6 weeks earlier than the original intention of compressing it by 30%.
- The project cost $2 M less that the budget.
- The steel delivery was 6 months earlier which saved $1 M.
- 97% of the inspections passed on the first time [26].

7.3 *Summary for Case Studies*

The IPD helped in reducing the time of construction which affected the environmental aspect of sustainability as well as the economic aspect. Besides, it reduced the cost of the project affecting the economic aspect of the sustainability dimension. Since the project helps the community it in turn, affects and benefits the social aspect of the sustainability dimension, which makes the application of IPD in these projects an achievement in meeting the sustainability goals. Table 3 summaries the case studies against the sustainability dimensions. In the first case there was no early involvement of the parties as a result of implementing the IPD later in the project. While in the second case study they implemented the IPD in early stages and all the parties were involved in early stage which was reflected on the sustainability dimensions as shown in the table.

8 Conclusion and Recommendations

The construction industry has proven to have a negative effect on sustainability. This paper focused on the role of the construction procurement method in impacting sustainability. Literature shows that the usage of the traditional procurement methods negatively affects sustainability as it consumes more resources; either human, natural or economic. The non-traditional methods for procurement provides a potential solution for these resource wastage problems. One of the non-traditional procurement methods is the integrated project delivery (IPD), which requires the early involvement of all parties and defines their relation as partners. The usage of the IPD results in meeting the goals of sustainability. This was verified using two case studies that of "Cardinal Glennon Children's Hospital Expansion" and "Sutter Medical

Table 3 Case studies implementing IPD principles and their effect on sustainability

IPD principle	Sustainability dimensions					
	Case 1			Case 2		
	Social	Economic	Environmental	Social	Economic	Environmental
Mutual respect and trust	✓	✓		✓	✓	
Mutual benefit and reward	✓	✓	✓	✓	✓	✓
Collaborative innovation and decision making	✓	✓	✓	✓	✓	✓
Early involvement of key participants				✓	✓	✓
Early goal definition				✓	✓	✓
Intensified planning	✓	✓	✓	✓	✓	✓
Open communication	✓	✓		✓	✓	
Appropriate technology	✓		✓	✓		✓

Centre Castro Valley". The usage of IPD in these two cases contributed positively to economic, environmental and social aspects of sustainability, as it reduced the cost and time of the construction of the project, as well as had a positive effect on the community. It is recommended that future projects consider the IPD to reduce disputes, time consumed in construction and project cost, as the IPD application enhances the feeling of ownership and encourages trust between all parties of the project resulting in increasing the value of project.

References

1. Yılmaz M, Bakı¸s A (2015) Sustainability in construction sector. Procedia Soc Behav Sci 195:2253–2262
2. Wibowo MA, Sholeh MN, Adji HS (2017) Supply chain management strategy for recycled materials to support sustainable construction. Procedia Eng 171:185–190
3. Abdullahi AH (2014) Review of subcontracting practice in Nigeria construction industry. J Environ Sci Resour Manage 6(1):23–33
4. Datta M (2000) Challenges facing the construction industry in developing countries. In: Proceedings: the 2nd international conference on construction in developing Countries. Gaborone, Botswana, 15–17 Nov
5. Daoud AO, Othman A, Robinson H, Bayyati A (2018) Exploring the relationship between materials procurement and waste minimization in the construction industry: the case of Egypt. In: International conference on sustainability, green buildings, environmental engineering and renewable energy (SGER 2018), 27 Jan 2018, GTower Hotel, Kuala Lumpur, Malaysia, pp 76–85
6. World Commission on Environment and Development (WCED) (1987) Our common future. Oxford University Press, New York, NY, USA
7. Raynsford N (2000) Sustainable construction: the Government's role. Proc Inst Civ Eng Civ Eng 138(S2):16–22
8. Stoddart H, Schneeberger K, Dodds F, Shaw A, Bottero M, Cornforth J, White R (2011) A pocket guide to sustainable development governance. In: Stakeholder forum 2011. [Google Scholar]
9. Ben-Eli M (2015) Sustainability: definition and five core principles a new framework the sustainability laboratory New York
10. NY info@sustainabilitylabs.org I www.sustainabilitylabs. [Google Scholar]
11. Mensah J, Enu-Kwesi F (2018) Implication of environmental sanitation management in the catchment area of Benya Lagoon, Ghana. J Integr Environ Sci.https://doi.org/10.1080/1943815x.2018.1554591
12. Feil A, Schreiber D (2017) Sustainability and sustainable development: unraveling overlays and scope of their meanings. Cadernos EBAPE.BR 15(3):667–681. https://doi.org/10.1590/1679-395157473
13. Hoúkara E (2007) Ülkesel Koúullara Uygun Sürdürülebilir YapÕm øçin Stratejik Yönetim Modeli, (Doktora Tezi). østanbul Teknik Üniversitesi Fen Bilimleri Enstitüsü, østanbul
14. Oktay B (2005. A model for measuring the sustainability of historic urban quarters: comparative case studies of Kyrenia and Famagusta in North Cyprus Ph.D. Thesis, Eastern Mediterranean University, Famagusta
15. Tahon A, Elyamany A (2017) Sustainability concept and knowledge analysis in construction industry, pp 307–315
16. Constructing Excellence (2004) Sustainable construction: an introduction. Retrieved 10 Jan 2021, from www.constructingexcellence.org.uk

17. Molenaar et al. (2009). Sustainable, high performance projects and project delivery methods: a state-of-practice report. Ph.D.
18. Nasid A, Nordiana W, Ali W, Zainordin ZM (2006) Effect of procurement systems on the performance of construction projects, vol 1996, pp 1–13
19. Clamp H, Cox S, Lupton S, Udom K (2012) Which contract? Choosing the appropriate building contract, 5th edn. RIBA, Riba Publishing
20. Oshungade O, Kruger D (2015) A comparative study on the procurement methods used in the construction industry of South Africa. In: 3rd IPMA conference. https://doi.org/10.13140/RG.2.1.3756.9520
21. Hinze J (2013) Construction contracts, 3rd edn. MC Graw Hill India. ISBN10: 9351340058
22. Othman A, Youssef L (2020) A framework for implementing integrated project delivery in architecture design firms in Egypt. J Eng Des Technol
23. Matthews O, Howell G (2005) Integrated project delivery: an example of relational contracting. Lean Constr J 2
24. Hamzeh F, Rached F, Hraoui Y, Karam AJ, Malaeb Z, El Asmar M, Abbas Y (2019) Integrated project delivery as an enabler for collaboration: a Middle East perspective. Built Environ Project Asset Manage 9(3):334–347
25. Singleton M, Hamzeh FR (2011) Implementing integrated project delivery on department of the navy construction projects. Lean Constr J USA, pp 17–31
26. Dpr.com (2021) [online] Available at: https://www.dpr.com/assets/casestudies/IPD-Whitepaper_2018-2.pdf. Accessed 13 Apr 2021

The Hand-Arm Vibration Syndrome: A Bibliometric Co-occurrence Analysis of a Neurological Disorder in Construction Workers

Mariam Akinlolu and Theo C. Haupt

Abstract Hand arm vibration syndrome has become one of the most prevalent diseases in the construction industry. This paper provides a bibliometric review of research topics on HAVs in construction workers. A bibliometric review was conducted to compile relevant materials from the Scopus database. A total of 240 related papers were examined. VOSviewer was used to develop a co-occurrence network based on the bibliographic data obtained. Emerging trends were prevalence of HAVS, disability in workers diagnosed with HAVS, HAVS related diseases, machine vibration exposure and disease control measures for HAVS.

Keywords Construction workers · HAVS · Musculoskeletal disorder · Neurosensory · VosViewer

1 Introduction

Hand-Arm Vibration Syndrome (HAVS) was first discovered in the early 1900's by Alice Hamilton [9, 17] and the first reports of vibration injury to operators date back to 1946 [22]. HAVS has become one of the commonest occupational diseases of the twentieth century [17]. It is the most common compensated occupation related disorder in the United Kingdom [11].

In the context of the construction industry, hand-arm vibration is the term used to describe vibration that is transmitted into the hands and arms of construction workers that are using mechanical hand-held power tools or processes while carrying out construction work activities [2, 7, 10, 11, 16, 23]. Hand-arm vibration syndrome (HAVS) is a health condition that has vascular, neurologic, and musculoskeletal features [23]. Its vascular component, also known as vibration-induced white finger (VWF), is a type of secondary Raynaud phenomenon because of attacks of well-demarcated finger blanching [8] and the most well established manifestation of HAVS

M. Akinlolu (✉) · T. C. Haupt
Faculty of Engineering, Mangosuthu University of Technology, Durban, South Africa
e-mail: akinlolu.mariam@mut.ac.za

T. C. Haupt et al. (eds.), *Construction in 5D: Deconstruction, Digitalization, Disruption, Disaster, Development*, Lecture Notes in Civil Engineering 245,
https://doi.org/10.1007/978-3-030-97748-1_17

[18, 21]. During an attack the fingers might feel 'numb' and a sensation of 'pins-and-needles' may also be experienced. A more progressed state of VWF can be triggered by a less significant reduction in temperature and if left to progress even further through ongoing HAV exposure and/or no medical intervention the condition will ultimately lead to increased numbness, increased tingling and a significant decrease in manual dexterity of the hands [14]. In the most severe cases, damage to blood circulation may become permanent [15] causing fingers to turn 'blue-black' in colour. In the most exceptional of cases, gangrene may result [16].

Carpal Tunnel Syndrome (CTS) is another, separate kind of disorder, related to HAVS. It is compression and irritation of the median nerve as it passes under the transverse carpal ligament in the wrist [5]. It can cause tingling, numbness and weakness, particularly in the wrist of the hand. Pain and night waking resulting in disturbed sleep are also CTS symptoms. Patients with HAVS can easily be misdiagnosed with CTS and undergo unnecessary surgery [11].

The general lack of proper and timeous diagnosis of HAVS and referral are major reasons for delays in treatment [3]. Further, the lack of awareness by workers and their employers and fear of repercussions by their employers are also barriers [23]. Early recognition and management of this condition are crucial for preventing progression and improving prognosis. Little is known about the effects that vibration exposure to handheld tools can give on the ability to perform activities of daily living [4].

Notably, HAVS has a high prevalence in industries such as construction. The prevalence among exposed workers has been estimated at 50% and varies by intensity and duration of vibration exposure. HAVS is a recognisable problem globally and studies in several countries have confirmed its prevalence such as Malaysia [24], Sri Lanka [19], Canada [23], United Kingdom [7], Norway [2], and the United States [11].

Although several studies on HAVS in construction have been conducted, there is no evidence of studies that determine the trends in publications and collaborations among countries and authors, i.e. bibliometric and scientometric reviews, in this area of studies. This paper presents a bibliometric analysis to elevate the level of awareness of HAVS among construction employers and workers to improve their quality of life.

2 Methodology

2.1 *Material Comprehensive Retrieval*

The Scopus database was selected as a data source for this study because of its comprehensiveness and coverage of publications from diverse areas [12, 13, 20]. Publications related to HAVS were retrieved from the Scopus database using specific keywords such as "HAVS" OR "Construction" AND "Workers" AND "Musculoskeletal". The search was limited to documents published between 2011 and 2021. After the online retrieval was conducted on 19,977 documents were obtained.

2.2 *Literature Selection and Extraction*

To filter the output, books, trade publications, reports and book series were excluded from the data set, limiting the literature selection to 5808 journal articles and conference papers. Vuksic et al. [26] argued that journals and conference papers are considered more reliable sources of literature review and provide detailed information compared to other sources. For a further selection and elimination of irrelevant publications, the abstracts, titles and keywords of the papers were further reviewed, and three filter criteria were applied to eliminate unnecessary papers and select publications that aligned with the research subject and theme.

The filter criteria were as follows:

(1) Articles that did not present studies conducted in the engineering and health industry and were not construction related.
(2) Articles that did not present studies in the English language.
(3) Articles that focused on HAVS in other fields but did not directly address issues of HAVS in construction.

After filtration using these three criteria, a total of 204 bibliographic data records were obtained. A bibliometric analysis was conducted to analyze the final documents obtained. The bibliometric analysis utilises science mapping tools for visualising physical aspects of scientific research and domains and describing the structure of their disciplines [1, 25]. It is the mathematical analyses of publications and their properties such as, for example, authorship, publication source, type, and timelines [6].

A Keyword co-occurrence analyses were then conducted to identify the trend of research topics. Gaps and trends for future research were discussed to guide research directions on HAVS among construction workers.

The VOSviewer software (version 1.6.17) was used for the bibliometric review to analyze statistical evidence obtained from the data source in terms of a cluster network and develop a visualization map of the network. Data is imported from the Scopus database unto the VOSviewer software for the construction of distance-based visualization networks. The software offers the functionality required to display large visualization networks in an easily interpretable way [25].

Items in the visualization network are presented by labels and circles. Therefore, the distance between each item in the visualization network approximately indicates the relatedness of the items (ibid). In general, the closer two items are to each other, the stronger their relatedness. Items in the network are connected with lines that represent links. In the cluster network, the smart local moving algorithm and the color of an item are determined by the cluster to which the items belong to.

3 Co-occuring Keyword Analysis

Numerous topics and themes have evolved in research concentrated on the application of digital technologies for construction safety in the last decade. Data from the Scopus database was evaluated to develop a visualisation map for co-occurring keywords and an overlaying network of research topics in digitalisation for construction safety.

Keyword Clusters and Keyword Trends. A keyword co-occurrence analysis was conducted to determine the fundamental structure and clusters in digital technologies for construction safety research. Keywords are significant words which serve as a point of reference and aid the description of contents and understanding of concepts in research articles [20].

A co-occurrence network was derived from a total of 10,699 keywords using the VOSviewer software. With a minimum number of 5 co-occurrence of keywords, 168 keywords co-occurred, and 5 significant keyword clusters were identified. The proximity of keywords to one another and similarity of keywords determines the degree of co-occurrence. Figure 1 presents a network visualization map of the 5 co-occurring keyword clusters.

Cluster #1 labelled in purple had 4 members with keywords such as prevalence, risk factors, hand arm vibration and epidemiology, and can be summarized as 'Prevalence of HAVS' (PV).

Cluster #2 represented in yellow had 4 members with keywords such as disability, disability evaluation, daily life activity, sensation disorders, Keywords in this cluster can be associated with 'Disability in Workers Diagnosed with HAVS' (DWDH).

Cluster #3 shown in green had 7 keywords such as machine vibrations, vibrations analysis, vibration exposure, hand-arm vibration, exposure estimation, hand tools and upper extremity, and can be summarized as 'Machine Vibration Exposure'.

Cluster #4 represented in red had 5 members paresthesia, vasospasm, nerve conduction, Raynaud phenomenon and disease severity. Keywords in this cluster can be associated with 'HAVS-related diseases' (DWDH).

Cluster #5 shown in blue had 3 keywords such as pathophysiology, pathology and vascularization. Keywords in this cluster can be associated with 'Disease Control Measures' (DCM).

Findings from the co-occurrence of keywords analysis indicates that studies on HAVS among construction workers have majorly foused on the prevalence of HAVS, disability in workers diagnosed with HAVS, HAVS related diseases, machine vibration exposure and disease control measures for HAVS.

4 Conclusion

Given the prevalence of HAVS in construction, where it is misunderstood and ignored despite the negative health impacts of the occupational disease this papers sought to by means of a bibliometric co-occurrence analysis, elevate the level of awareness of

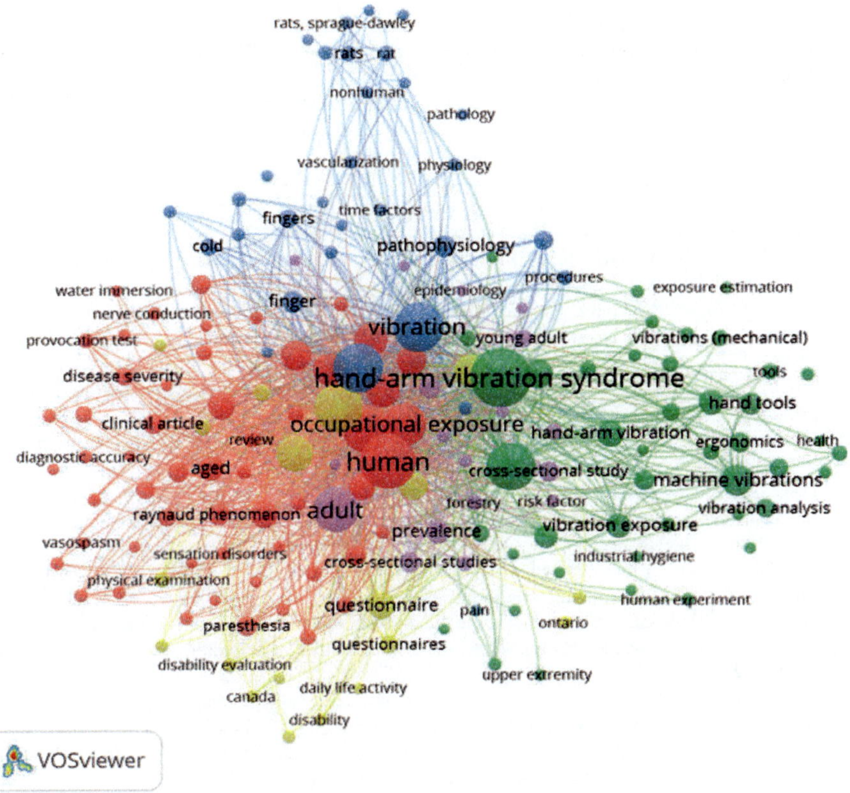

Fig. 1 Cluster visualization map for co-occurring keywords

HAVS among construction employers and workers to improve their quality of life. The study presented the evolution of research topics through a cluster of existing topics, providing researchers with a trend of existing literature directions to guide future research on HAVS among construction workers.

References

1. Akram R, Thaheemb MJ, Nasira AR, Alic TH, Khana S (2009) Exploring the role of building information modeling in construction safety through science mapping. Saf Sci 120:456–470
2. Buhaug K, Moen BE, Irgens Å (2014) Upper limb disability in Norwegian workers with hand-arm vibration syndrome. J Occup Med Toxicol 9:5
3. Bodley T, Nurmohamed S, Holness D, House R, Thompson A (2015) Health-care barriers for workers with HAVS in Ontario, Canada. Occup Med 65(2):154–156
4. Cederlund U, Nordenskiold G, Lundborg R (2001) Hand-arm vibration exposure influences performance of daily activities. Disabil Rehabil 23(13):570–577
5. Davis C (2021) Medical definition of carpal tunnel syndrome. MedicineNet

6. de Rezende LB, Blackwell P, Gon alves MDP (2018) Research focuses, trends, and major findings on project complexity: a bibliometric network analysis of 50 years of project complexity research. Project Manage J 49(1):42–56
7. Edwards D, Holt G (2007) A guide to hard-arm vibration. The Off-highway Plant and Equipment Research Centre (OPERC), Loughborough. ISBN:978-0-947974-54-1
8. Griffin MJ, Bovenzi M (2002) The diagnosis of disorders caused by hand-transmitted vibration: southampton workshop 2000. Int Arch Occup Environ Health 75(1):1–5
9. Hamilton A (1918) A study of spastic anaemia of the hands of stonecutters. United States Bureau of Labor Statistics, Bulletin (part 19), pp 53–66
10. Harada N, Mahbub MH (2018) Diagnosis of vascular injuries caused by hand-transmitted vibration. Int Arch Occup Environ Health 81:507–518
11. Heaver C, Goonetilleke KS, Ferguson H, Shiralkar S (2011) Hand–arm vibration syndrome: a common occupational hazard in industrialized countries. J Hand Surgery (Euro Vol) 36(5):354–363
12. Hong Y, Chan DWM, Chan APC, Yeung JF (2012) Critical analysis of partnering research trend in construction. J Manage Eng 28:82–95
13. Hosseini MR, Maghrebi M, Akbarnezhad A, Martek I, Arashpour M (2018) Analysis of citation networks in building information modelling research. J Constr Eng Manage 144(8):1–13
14. HSC (2003) Proposals for new control of vibration at work regulations implementing the physical agents (vibration) directive (2002/44/EC) hand-arm vibration. Consultative document. Published by the Health and Safety Executive
15. HSE (2005) Hand-arm vibration. The control of vibration at work regulations 2005. Guidance on regulations. The health and safety executive. Document ref: L140
16. HSE (2003) Hand-arm vibration. The health and safety executive, document ref: HSG88
17. Lawson I (2018) Alice Hamilton, 1869–1970, 'the mother of occupational medicine.' Occup Med 68:224–225
18. Marks J (2021) Medical definition of vibration white finger. MedicineNet
19. Madhushanka J, De Silva G, De Silve G (2015) Investigation of whole body vibration exposures of operators of construction vehicles. In: 6th international conference on structural engineering and construction management, Kandy, Sri Lanka, Dec 11–13, pp 117–122
20. Olawumi TO, Chan DWM (2018) A scientometric review of global research on sustainability and sustainable development. J Clean Prod 183:231–250
21. Olsen N (2002) Diagnostic aspects of vibration-induced white finger. Int Arch Occup Environ Health 75(1–2):6–13
22. Pelmear PL, Taylor W, Wasserman DE (1992) Hand-ann vibration: a comprehensive guide for occupational health professionals. Van Nostrand Reinhold, New York
23. Shen S, House R (2017) Hand-arm vibration syndrome—what family physicians should know. Canadian Family Phys Le Médecin de famille canadien 63:206–210
24. Su T, Hoe V, Masilamani R et al (2011) Hand-arm vibration syndrome among a group of construction workers in Malaysia. Occup Environ Med 68:58–63
25. Van Eck NJ, Waltman L (2014) Visualizing bibliometric networks. In: Ding Y, Rousseau R, Wolfram D (eds) Measuring scholarly impact: methods and practice, pp 285–320
26. Vuksic VB, Invansic L, Vugec DS (2018) A preliminary literature review on digital transformation case studies. Conf Proc Rome Italy 20(9,Part VIII):952–957

Client Roles in Influencing Project Delivery Outcomes

Progress S. Chigangacha and Theo C. Haupt

Abstract Construction clients are the originators of projects who are central to the construction process and play a crucial role in influencing project delivery outcomes. Consequently, it behoves all role players to support the client and deliver specific client requirements. The construction industry is fraught with several challenges during project delivery and it is a common occurrence that critical project objectives are not usually achieved. It is therefore, in the client's interest and control to ensure that the various challenges, foreseen or unforeseen, which could result in waste, conflict, excessive costs and delays do not occur or are well managed to minimize their impact. This paper examines the role of the client and the client's team structure in influencing project delivery outcomes, which was explored quantitatively. Data was collected through the use of semi-structured questionnaires. Study samples were selected through purposive convenience sampling. The study demonstrated a link between client roles and their influence on project delivery outcomes. Between public and private clients, the latter was considered by the majority of respondents as being more involved in their projects and significantly performed their roles, hence the high success rate of private projects.

Keywords Clients · Client roles · Construction industry · Project/s · Procurement

1 Introduction

The construction industry is made up numerous role players consisting of the client team, delivery team (made up of professionals, contractors, sub-contractors, suppliers, etc.) and stakeholders, all of whom, in varying degrees, have a direct impact on project outcomes [28]. It is the objective of this team to ensure that a construction project remains within agreed mandates of budget, schedule, function,

P. S. Chigangacha (✉)
School of Engineering, University of KwaZulu-Natal, Durban, South Africa

T. C. Haupt
Faculty of Engineering, Mangosuthu University of Technology, Durban, South Africa

T. C. Haupt et al. (eds.), *Construction in 5D: Deconstruction, Digitalization, Disruption, Disaster, Development*, Lecture Notes in Civil Engineering 245,
https://doi.org/10.1007/978-3-030-97748-1_18

quality, conflict, health and safety and the environment, and progresses successfully to completion. However, it is a known fact that construction projects are fraught with several challenges during project delivery and it is a common occurrence that critical project objectives are rarely achieved [5]. It is therefore, in the client's interest and control to ensure that the various challenges, foreseen or unforeseen, which could result in waste, conflict, excessive costs and delays do not occur or are well managed to minimize their impact.

Gwaya et al. [11], Cox et al. [8] and Kamara et al. [14] considered construction clients to be originators of projects who are central to the construction process and play a crucial role in influencing project outcomes. Consequently, it behooves all role players to support the client and deliver specific client requirements. In South Africa, the construction industry contributes significantly to national growth through infrastructure delivery, resulting in added benefits to citizens, such as, inter alia, job creation, alleviation of poverty, etc. According to Watermeyer [28] and Thyssen et al. [27] it is imperative that client values on all projects are understood, regardless of size and complexity, as clients' influence on project outcomes is apparent and significant.

2 The Client Team

The onus is on the client to formulate a client team that will develop a business case, procure all services and resources required to deliver the project, ensure all payments to that effect have been made, manage the project and all collaborations, and to provide client direction [28]. Ideally, the client team should be 'in-house", but however in frequent circumstances, the client lacks the technical resources to perform specific roles, and hence outsources these resources. The outsourced individuals become the client's representatives and function within a team environment [6, 23], as technical or professional advisers of the client mandated with executing their duties to fulfil the client's interest [6, 20]. It is important to note the presence of a client team does not eliminate the need for the client on the project, but rather to provide expert advice that will enhance the formulation of a solution to achieve the business case [20].

According to the Latham Report [17], Macaulay and Ramsey [18] the client team is made up of individuals with the requisite technical expertise of construction and is tasked with advising the client on the project and construction related matters throughout all project phases. While SIPDM defines it as the team that provides advice and also manages activities associated with the initiation of projects, the formulation of the client's specific requirements during the initial stages of the project and a range of technical matters. The client team in essence has the mandate to perform two functions, where function (i) is to act as a project sponsor and to provide strategic direction and decision making, and function (ii) is concerned with project delivery which entails the provision of strategic, technical, financial and operational direction in order to realise the client's value proposition [28]. Client teams within the construction industry are in most cases temporary, existing only for the project duration. The structure of the client team directly impacts on project delivery outcomes in that those teams that are structured with the objective of achieving client interests and objectives, frequently successfully meet them [20]. There are variants of the team composition and structure, where team structure refers to the design of the team and the purpose it seeks to serve within the organisation [20].

3 Client Roles

It can never be emphasized enough that the client plays a critical role in a project, and equally important is the role played by the client team leader in guiding the project delivery process and the client team to ensure that each consultant effectively executes their role to facilitate successful project delivery and for the client to attain value for money [20]. Collectively, it is the role of the client team to provide effective leadership and direction to the professional team and the project stakeholders [28]. Client roles are typically structured in such a manner to address overall delivery of a project by tackling the following critical sub-items, delivery management, procurement, portfolio, programme and project management and governance issues [28]. These client roles have evolved from being passive fund providers to an increasingly active participant Alharthi et al. [3], largely due to the increased need for improvement and innovation within the construction industry to achieve better value for money [9].

Alharthi et al. [3] and Aiyetan et al. [2] advised that clients have a pivotal role in construction projects which influences project outcomes and these roles often differ depending on the client's experience, project stage and procurement systems used. Importance is placed on clients to understand their roles to ensure prompt delivery of projects. Despite various scholars proposing the basic functions or roles needed to be performed by each client, the lists tend to be similar. According to Alharthi et al. [3] client roles can be sub-divided into primary and secondary roles in which case it was their recommendation that clients should perform all primary roles, whereas secondary roles were considered optional. On the other hand, Jawahar-Nesan and Price [13] found 12 important client roles for improving the project outcomes while Watermeyer [28] listed 9 roles (Table 1).

According to SIPDM [25] the primary role of the client is to initiate, commission and pay for the construction project On the other hand, Watermeyer [28] argued that the main role of the client is to ensure that business case objectives are achieved. SIPDM [25: 32] pointed out that "*The client owns the business case of the project and needs to provide effective leadership of the project throughout the project life cycle, commencing at a strategic level ending at the close out of a project after the beneficiary of the project has accepted and operates the delivered infrastructure*". Furthermore, the client roles expand into ensuring that the arms of governance, procurement and delivery management practices which are key to successful project delivery are well controlled and addressed. As advised by Watermeyer [28] it is the client's role to ensure that project governance parameters are in place, which will facilitate and guide the decision making process from a strategy perspective.

The common causes of project failures have been attributed to ineffective and/or inadequate implementation of client roles. According to Watermeyer [28], these causes can be classified under, poor governance, poor procurement and poor delivery management practices and were listed by the UK's Office of Government Commerce [21] as,

- Unclear link between the project and the organisation's key strategic priorities, including agreed measures of success
- Unclear senior management and ministerial ownership and leadership
- Lack of effective engagement with stakeholders and lack of understanding of, and contact with, the supply industry at senior levels in the organisation
- Lack of skills and a proven approach to project and risk management
- Poor procurement and delivery management practices
- Too little attention to breaking the development and implementation into manageable steps
- Evaluation of proposals driven by initial price rather lifecycle costs and long term value for money
- Lack of effective project team integration between clients, the supplier team and the supply chain.

4 Procurement and Delivery Management Practices

Procurement within the construction industry can be viewed as an organized process for clients to acquire construction products. It has been defined as "*the process which creates, manages and fulfils contracts.*" [12]. Watermeyer [28] asserts that procurement in construction encompasses the branches of construction, professional services and supply. In the case of construction, in addition to the traditional approach to procurement, there has been a proliferation of alternative procurement methods in recent years [9, 22]. As a result, client roles in relation to procurement have transformed and are increasingly more hands on (Ibid). According to Edum-Fotwe et al. [9], client roles in relation to procurement can be divided into two categories, namely, managerial and technical [9]. The client is for the large part responsible for the managerial role, with the technical roles being assigned to the client's technical team. The managerial roles entail, "*Decision making, Effective critique, Communication, Motivation and drive, Team building, Facilitation, Interviewing, Negotiation skills, Assertiveness and Objective setting*" [9]. As demonstrated in Table 2, the managerial roles can either be primary or secondary, with the former being mandatory roles expected to be performed by every client and the latter being roles that are desirable but not necessarily a requisite.

Under the traditional approach, the client actively performs the roles of decision-making, communication, and objective setting which are critical in the establishment of the client brief and business case. Negotiation is a necessary client attribute to facilitate the engagement of all other role players. The list of primary roles was

Table 1 Client roles

Alharthi et al. [3]	Jawahar-Nesan and Price [13]	Watermeyer [28]
Development of the procurement strategy	Preparing and organising	Initiating and financing projects, including development of project brief
Procurement initiation	Developing project definition	Organising a delivery team
Preparation of the project brief	Procurement	Making payments for all services rendered and works executed
Selection of the procurement method	Organising a joint management team	Providing client leadership and direction to the delivery team
Development of Business case	Design management	Set the delivery team up for successful delivery and remove obstacles to progress
Risk management	Safety management	Ensuring that adequate maintenance plans are put in place
Tendering and award	Measuring and reviewing performance	Ensuring adherence to all legislative provisions
Performance management	Communications	Managing and overseeing the management of scope
Change control	Motivation	Leading engagements with internal and external stakeholders and utilities
	Coordination	
	Documentation	
	Project evaluation	

Table 2 Analysed client roles in procurement [9]

Managerial capability	Traditional Approach		Alternative Approach	
	Primary	Secondary	Primary	Secondary
Decision making	■		■	
Effective critique		■	■	
Communication	■		■	
Motivation and drive		■	■	
Team building			■	
Facilitation		■	■	
Interviewing			■	
Negotiation skills	■		■	
Assertiveness		■	■	
Objective setting	■		■	

expanded and exhaustive in relation to the alternative procurement method largely because these methods require a more hands on approach by the client.

5 Methodology

This study employed a multi-case study research design, where structured questionnaire surveys were used to determine the role of clients in influencing project outcomes. Two Professional Quantity Surveying firms in the KwaZulu-Natal province of South Africa were identified and used as case studies in the study population (n = 25), for the projects they were involved in. The study sought to examine the level of client involvement and their role in influencing project delivery outcomes. From which a sample of 18 clients was used. Purposive convenience sampling which is a non-probability sampling technique was used in sample selection. The researcher intentionally chose the samples, focusing mostly on the respondents that were well known to the researcher in order to maximise the response rate and to collect as much data as possible. Respondents were also selected based on their location, preference being on those in close proximity to the researcher. Snowball sampling was used to a smaller extent, where certain individuals were referrals.

6 Results and Discussions

The study response rate was acceptable, from a population of 25 clients, 18 responses were received, indicating a 72% response rate. The client profile was made up of ten private sector and eight public sector clients; of which Case 1 had six private clients and four public clients, and Case 2 had four clients per type, indicative of a good mix and a fair representation of both private and public clients. Table 3 presents the cases representation of the type of client and the number of related projects.

Procurement methods in use

Table 4 presents the experience that clients had of procurement methods which were currently used in construction projects in SA.

Table 3 Cases representation and number of projects

	Private clients (no. of projects)	Public clients (no. of projects)	Total clients (projects)
Case 1	6 (22)	4 (7)	10 (29)
Case 2	4 (5)	4 (7)	8 (12)
Total clients (projects)	10 (27)	8 (14)	18 (41)

Table 4 Comparison of usage of different procurement methods

Procurement method	Clients	
	Public (%)	Private (%)
Traditional procurement system (architect-led)	88	60
Design-build	40	50
Negotiated	25	50
Construction management	13	0
Cost-plus	0	10
Management contract	0	0
Turnkey	0	0

The respondents concurred that the traditional procurement method was the most dominant method in comparison with alternative procurement methods. It is clear that the public sector utilized the traditional method more than the private sector, with nearly 90% of public clients and 60% of private clients having used this method. Preference in adopting alternative procurement methods by the private sector could be an indication that private sector clients desire to be more involved and want to play a more intensive role in their projects as the traditional method limits their involvement.

6.1 *Client Involvement*

In this study, the group interval coefficient value was calculated as **(5 − 1)/5 = 0.80** and the following intervals were taken as reference values in evaluating the responses obtained through the implementation of the 5-point scale.

The range interpretations as shown in Table 5 are used throughout where 5-point frequency scales were used.

Table 5 Data interpretation range [15]

Range	5-point Likert scale		
	Frequency scale	Importance scale	Agreement scale
4.21–5.00	Always	Most important	Strongly agree
3.41–4.20	Often	Important	Agree
2.61–3.40	Sometimes	Neutral	Neutral
1.81–2.60	Seldom	Somewhat important	Disagree
1.00–1.80	Never	Least important	Strongly disagree

Table 6 Comparative client involvement per project phase

Project phase	Clients	
	Public	Pvt.
Pre-construction	4.73	4.45
Post construction	4.71	4.22
Construction	4.70	4.25

Table 6 presents the mean client involvement per project phase as reported by the respondents. A 5-point scale was used where 1 = very low 2 = low, 3 = moderate, 4 = high and 5 = very high. Table 4 is a list of construct statements.

It is evident that both public and private sector clients had a very high level of involvement throughout all project phases (means > 4.20). Furthermore, the results suggest that public sector clients considered themselves slightly more involved in their projects than private sector clients. In their studies, Sivunen [24] and Thomson [26] advised that clients had the greatest influence very early on in the project phases and their involvement ought to highest during the early phases. Notably, both client types indicated highest involvement during the pre-construction phase. The results reflect the client's awareness and need to be actively involved throughout all project phases.

6.2 *Importance of Client Involvement During Different Project Phases*

Respondents were required to identify the project phases they considered to be the most important for client involvement. A 5-point scale was used where 1 = least important, 2 = somewhat important, 3 = neutral, 4 = important and 5 = most important. The results are shown in Table 7.

From Table 7 it is evident that both types of clients considered their involvement in all three project phases to be very important (mean > 4.20). The pre-construction phase predominated, followed by the construction phase for both sectors respectively. The client views are in line with what Boton et al. [7] indicated namely that the pre-construction phase is a very crucial and significant phase in the construction project life cycle.

Table 7 Importance of construction project phases

Project phase	Public sector	Private sector
	Mean	Mean
Pre-construction	4.94	4.90
Construction	4.57	4.60
Post construction	4.43	4.20

6.3 Implementation of Client Roles

Respondents were presented with a series of statements about the perceived current implementation level and importance of their role in driving project outcomes. A 5-point Likert scale was used as follows, namely.

- Implementation—where 1 = 20% implementation level, 2 = 40% implementation level, 3 = 60% implementation level, 4 = 80% implementation level and 5 = 100% implementation level.
- Importance—1 = not important, 2 = slightly important, 3 = moderately important, 4 = important and 5 = very important. Table 8 is a list of construct statements.

While the respondents considered all client roles as very important (mean > 4.20), with a grand mean of 4.5, the related implementation levels were varied, with a grand mean of 3.9. This apparent gap between the importance and implementation levels could be contributing to the evident poor project deliveries and out comes rampant in South Africa. All but two of the items which were believed to have a very high level of implementation (mean > 4.20), namely, preparation of the project brief, development of business case, organising a client team and tendering and award,

Table 8 Statements about client roles

Statement	Implementation Level	Importance Level
Preparation of the project brief and developing project definition	4.3	5.0
Development of business case	4.2	5.0
Organising a client team	4.2	4.5
Design management	3.8	4.4
Documentation	3.7	4.4
Procurement	3.6	4.3
Tendering and award	4.5	4.2
Risk management	4.0	4.3
Performance management	3.9	4.5
Measuring and reviewing performance	3.8	4.7
Change control	4.2	5.0
Health and safety management	3.8	4.5
Communications	4.0	4.5
Motivation	3.6	4.0
Coordination	3.9	4.7
Project evaluation	3.8	4.3
Governance issues	4.5	5.0
Grand mean	3.9	4.5

are all activities that are conducted within the pre-construction phase. These results are consistent with the findings from this study where both client types indicated highest involvement during the pre-construction phase. Watermeyer [28] advised that a crucial client role is providing effective governance, especially in the case of SA where to a large extent infrastructure failure has been attributed a lack of good governance in many cases. The high score associated with governance issues could be a response to addressing these infrastructural failures.

6.4 *Projects that Were Successfully Completed*

The respondents were asked to indicate the percentage of their projects that were completed successfully in terms of budget, schedule and quality within the last 5 years. The following sections discuss the findings from both Case Study 1 and 2 respectively.

Case Study 1

Case study 1 had ten clients altogether, with the private sector having the majority at six. The combined number of projects was 29, the private sector client had 7 projects and the public sector client with 22. Tables 9 and 10 present the number of projects executed by each client type in terms of value and the respective projects that were completed successfully in terms of cost, time and quality.

The results from Table 9 indicate that in overall, private projects were executed better than public ones in respect to the parameters of cost, time and quality. In both cases, the quality of the projects scored high with 86% of private projects being to the quality standards specified and 59% of public projects respectively. The results

Table 9 Case study 1 project outcomes

	Private projects				Public projects			
Upper limit of project value	No. of projects executed	Cost	Time	Quality	No. of projects executed	Cost	Time	Quality
< 650,000	–	–	–	–	–	–	–	–
2,000,000.00	**1**	1	1	1	**3**	1	1	1
4,000,000.00	**1**	–	1	1	**4**	2	1	2
6,500,000.00	**1**	1	–	1	**5**	2	2	3
13,000,000.00	**2**	1	1	1	**8**	3	2	5
40,000,000.00	**2**	2	1	2	**2**	1	1	2
130,000,000.00	–	–	–	–	–	–	–	–
No limit	–	–	–	–	–	–	–	–
Total	7	5	4	6	22	9	7	13
Percentage (%)		71	57	86		41	32	59

Table 10 Case study 2 project outcomes

	Private projects				Public projects			
Upper limit of project value	No. of projects executed	Cost	Time	Quality	No. of projects executed	Cost	Time	Quality
< 650,000	–	–	–	–	–	–	–	–
2,000,000.00	2	2	1	2	1	1	–	1
4,000,000.00	1	–	–	1	1	–	–	1
6,500,000.00	1	–	1	1	1	–	–	1
13,000,000.00	1	1	1	1	2	1	–	1
40,000,000.00	2	1	1	1	–	–	–	–
130,000,000.00	–	–	–	–	–	–	–	–
No limit	–	–	–	–	–	–	–	–
Total	7	4	4	6	5	2	0	4
Percentage (%)		57	57	86		40	0	80

suggest that the challenge of time overruns persists, especially within the public sector where a mere 32% of the projects were completed within the anticipated schedules. 41% of public projects were completed within budget, compared to 71% for the private sector.

Case Study 2

Case study 2 had eight clients altogether, with both sectors having four clients each. The combined number of projects was 12, the private sector client had 7 projects whereas the public sector client had 5. Table 10 presents Case Study 2 findings.

Case study 2 results also indicate that the private sector projects performed better than the public ones, with more than half of private projects being executed with budget and schedule, and all but one being to the required quality standards. On the other hand, public projects performed dismally with none of its projects being completed on time and a mere 40% being on budget. Performance in relation to the quality aspect was better, with 80% of the projects being to the specified quality standards.

In both case studies, it is evident that the public sector projects are still prone to both cost and time overruns, which according to Gaetsewe et al. [10] is a common phenomenon. These challenges have been linked to the lack of knowledge by clients which directly or indirectly impedes the client from executing their roles adequately. The majority of these projects, especially the public sector projects could be considered as failures. This is in-line with Merrow [19], where it was postulated that a project is a failure if it experiences time and budget overruns, which was the case in both case studies for public sector projects. The leading causes to time overruns especially within the public sector have been attributed to the client and their failure to adequately implement their role, and these roles include, change in scope of work, delayed payments, information gaps and lack of the client knowledge [1, 4, 16].

7 Conclusion and Recommendations

This paper examined the role of the client and the client's team structure in influencing project delivery outcomes. It can be concluded that the client is the most important role player and that effective client involvement, leadership and execution of their roles throughout the project life cycle, commencing at a strategic level and ending at project closeout is critical in fostering successful project outcomes. While the private sector has demonstrated the ability to perform better, the results indicate that the majority of public sector projects are delivered unsuccessfully and are considered failures in relation to the agreed mandates of schedule, budget and quality. The study demonstrated a link between client roles and their influence on project delivery outcomes. Between public and private clients, the latter was considered by the majority of respondents as being more involved in their projects and significantly performed their roles, hence the high success rate of private projects.

The results further suggest that there is an apparent gap between the importance and implementation levels of client roles, and this gap could be contributing to the evident poor project deliveries and out comes rampant in South Africa. As a result, it is a recommendation that clients, especially public clients gain more control of their projects and become more knowledgeable about their roles. This could be done through, inter alia, outsourcing of the required skills, forming partnerships with the private sector, providing frequent continuous skills and professional development trainings.

Poor procurement and poor delivery management practices have been cited as the major contributors of undesirable project outcomes. This study confirmed that the traditional procurement method was the default choice, though not necessarily the best, for 90% of public procurement, and thus possible contributing to the undesirable outcomes. It is therefore a recommendation that the public client equips itself with the requisite knowledge of the alternative procurement methods and ultimately makes informed decisions on the right method to use on a project specific basis.

References

1. Aiyetan AO (2010) Influences on construction project delivery time. Unpublished Ph.D. thesis, Nelson Mandela Metropolitan University, Port Elizabeth
2. Aiyetan AO, Smallwood JJ, Shakantu W (2013) Influence of client understanding of quality design. In: Proceedings of the 5th international conference of engineering, project and production management, PE, 26–28 Nov 2014
3. Alharthi A, Soetanto R, Edum-Fotwe FT (2014) Revisiting client roles and capabilities in construction procurement. In: Proceedings of the international conference on construction in a changing world—CIB W92 procurement systems, Sri Lanka, p 12
4. Apolot R, Alinaitwe H, Tindiwensi D (2013) Investigation into the causes of delays and cost overruns in Uganda's public sector construction projects. J Constr Dev Countries 18(2):33–47
5. Banaitis A, Banaitiene N (2012) Risk management in construction projects in risk management—current issues and challenges. In: Banaitienė N (ed) Chapter 19. InTech, pp 429–448
6. Bougrain F (2009) The role of the client in building site innovations in clients driving innovation
7. Boton C, Kubicki S, Halin G (2011) Understanding pre-construction simulation activities to adapt visualization in 4D CAD collaborative tools. In: International conference on computer aided architectural design, pp 477–492
8. Cox A, Ireland P, Townsend M (2006) Managing in construction supply chains and markets. Reactive and proactive options for improving performance and relationship management. Thomas Telford, London
9. Edum-Fotwe FT, Thorpe A, McCaffer R (2005) Changing client role in emerging construction procurement. In: Kähkönen K, Sexton M (eds) Understanding the construction business and companies in the new millennium. combining forces advancing facilities management and construction through innovation series. VTT—Technical Research Centre of Finland (www.vtt.fi)
10. Gaetsewe R, Monyane TG, Emuze F (2015) Overruns again in public projects: perspective from Northern Cape, South Africa. In: 6th international conference on engineering, project, and production management. Mariott Resort and Spa Hotel, Brisbane Queensland Australia
11. Gwaya A, Wanyona G, Sylvester MM (2014) The need for a structured construction clients' performance assessment in Kenya. Int J Soft Comput Eng 3(6):47–51
12. ISO 10845-1 (2010) Construction procurement: part 1: processes, methods and procedures. International Organisation for Standardisation, Geneva
13. Jawahar-Nesan L, Price ADF (1997) Formulation of best practices for owner's representatives. J Manage Eng 13(1):44–51
14. Kamara JM, Anumba CJ, Evbuomwan NFO (2002) Capturing client requirements in construction projects. Thomas Telford, London
15. Kan A (2009) Effect of scale response format on psychometric properties in teaching self-Efficacy. Eurasian Journal of Educ Res (EJER) 34:215–218
16. Kikwasi GJ (2012) Causes and effects of delays and disruptions in construction projects in Tanzania. Austr J Constr Econ Build Conf Ser 1(2):52–59
17. Latham Report (1994) Constructing the team: joint review of procurement and contractual arrangements in the United Kingdom construction industry. In: Latham M (chair) HMSO, London
18. Macaulay M, Ramsey L (2002) Construction and procurement law. W.Green
19. Merrow E (2012) Why megaprojects fail. Project Manager Magazine, pp 18–21
20. Mosalaesi S, Laryea S (2019) Organisational structure of client teams in two successful infrastructure projects. In: CIB world building congress, Hong Kong SAR, China, 17–21
21. Office of Government Commerce (2005) Common causes of failure—OGC best practice
22. Rashid A, Taib M, Ahmad W, Basiron, W, Nasid M, Ali W et al (2006) Effect of procurement systems on the performance of construction projects. In: Symposium conducted at the meeting of the international conference on construction industry 2006 (ICCI 2006), Padang, Indonesia

23. Senaratne S, Hapuarachchi A (2009) Construction project teams and their development: case studies in Sri Lanka. Archit Eng Des Manage 5(4):215–224. https://doi.org/10.3763/aedm.2008.0075
24. Sivunen S (2014) Clients' role in construction innovation management process. Ph.D. thesis. Aalto University
25. Standard for Infrastructure Procurement and Delivery Management (SIPDM) (2016)
26. Thomson D (2011) A pilot study of client complexity, emergent requirements and stakeholder perceptions of project success. Constr Manage Econ 29:69–82
27. Thyssen MH, Emmitt S, Bonke S, Kirk-Christoffersen A (2010) Facilitating client value creation in the conceptual design phase of construction projects: a workshop approach. Archit Eng Des Manage 6(1):18–30
28. Watermeyer R (2018) Client guide for improving infrastructure project outcomes, 1st edn. School of Construction Economics and Management, University of the Witwatersrand, Johannesburg, Engineers Against Poverty

Challenges to Project Delivery in the Dynamic Human Settlement Environment in South Africa

Abimbola Windapo and Fabio Companie

Abstract Governments worldwide are continually battling to address the housing backlog. In South Africa the backlog stems from previous discriminatory regimes and increasing urbanization. The housing backlog in South Africa has increased significantly and most especially in Cape Town. Housing is seen as a measure to transform and unify the segregated population. Project Leaders (PLs) are saddled with the responsibility of housing delivery and managing diverse stakeholders. Although project management is the driving force behind housing delivery, the project process is not free from dynamics. These housing projects constantly attract social and political attention, resulting in the PL repetitively contending with the inherited social and political dynamics of the Human Settlement Environment (HSE). The numerous challenges facing the PL created by the dynamism and ensuing complexity is unknown. Therefore, this research examines the challenges faced by PLs in the HSE and whether PLs exercise a capacity to transform, maintain, and lead the project organization creatively. A qualitative research approach with inductive-philosophical reasoning that employs interviews for data collection was chosen for the study. A sample of 19 PLs working in the public sector of human settlements in Cape Town was purposefully selected to participate in the study. The collected data were analyzed using thematic analysis to identify appropriate themes. The results identified six challenges: Social, Political, Organizational, Legislative challenges, Multi-stakeholder, and Skills faced by PLs. Following these themes, two key issues showed that social and political influence were the dominant factors affecting the implementation of housing projects. The research found that PLs do not have the authority in housing delivery. This limits the PL's ability to transform, maintain and lead the project organization creatively.

Keywords Cape Town · Housing · Human settlement environment · Project delivery · Project leader · Transformation

A. Windapo (✉) · F. Companie
Department of Construction Economics and Management, University of Cape Town, Cape Town, South Africa
e-mail: abimbola.windapo@uct.ac.za

T. C. Haupt et al. (eds.), *Construction in 5D: Deconstruction, Digitalization, Disruption, Disaster, Development*, Lecture Notes in Civil Engineering 245,
https://doi.org/10.1007/978-3-030-97748-1_19

1 Introduction

Housing is one of man's basic needs. In 1948, the right to housing was realized in Article 25 (1) of the Universal Declaration of Human Rights and was preceded in 1976 by one of the most well-known documents worldwide, namely the Vancouver Declaration on Human Settlements [14: 17–18]. As per the Constitution, this responsibility for housing lies with the public sector, specifically the Department of Human Settlement [24]. According to Mbanga [16], HS has the most dynamic human elements on the earth's surface, with the most visible signs of human culture imposed on the natural world, and directly propels the rate of urbanization. Through the National Department of Human Settlements (NDHS), the public sector fulfils its role and responsibility to the public by building homes for the impoverished inhabitants of the provinces. Municipalities are mandated to consult with all relevant stakeholders in the housing delivery process as part of the project and public legislative requirements.

The fundamental means of achieving a successful housing delivery project is establishing a project management system to deliver projects to the Human Settlements Environment (HSE) [11: 27]. It is the responsibility of the Project Leader (PL) to make sure that the project objective is achieved. The PL, through the utilization of a project management approach in the public sector, is a crucial role player in the delivery of houses for the HSE. The terms "Project Manager" and "Project Leader" are used interchangeably in this paper because of the broadness of the role of a project manager, which is a working condition in the Human Settlement environment.

Developing countries have witnessed massive urbanization, and the ways to solve problems associated with informal settlements are not straightforward for cities. The Cape Metropolitan MSDF of 2018 has projected a housing demand of 500,000 housing units over 20 years (2012–2032). The provision of housing in the public sector has been declining since 2007, while the demand has been rising steadily, and in Cape Town, the proliferation of informal settlements with increasing pent-up demand has increased the demand for housing. However, the resources available to meet the demand for housing are not sufficient. Based on the resources currently available to the City of Cape Town and the traditional housing procurement approach, it is estimated that it will take more than 70 years to eliminate Cape Town's current housing backlog [6: 231].

Since 1994 after the end of the apartheid era, and at the beginning of democracy in South Africa, the National Department of Human Settlements (NDHS) has implemented various housing projects, with 4.3 million houses and housing facilities, with a positive effect on more than 20 million people, according to the Human Settlement Yearbook 2015/2016 [18: 261]. Even with this significant progress, the shortfall is estimated at more than two million housing units [18]. Although the NDHS target is to produce 300,000 social houses per annum, it has failed to reach the target due to challenges and complexities [28].

The public sector has its operational process for each sphere, and with the rules and regulations set to deliver housing, delivery becomes a challenge. What emerges is that the needs of the people and the organization's process and requirements do not match. The HSE is dynamic, and the challenges for a PL within the public sector becomes complex. The role, authority, accountability, and responsibility of a PL are not agile (not on a fast cycle). Furthermore, the housing projects constantly attract social and political attention, resulting in the PL repetitively contending with the inherited social and political dynamics of the Human Settlement Environment (HSE). The numerous challenges facing the PL created by the dynamism and ensuing complexity in the HSE is unknown. Therefore, this research sought to investigate the challenges PLs (as referred to within the human settlement department) face in a dynamic HSE, with the focus being limited to the Western Cape, particularly Cape Town.

The rest of the article is structured as follows: Sect. 2 provides an overview of the concept of HSE, its dynamics and challenges to project leaders. The qualitative research approach used in data collection is presented in Sect. 3. In Sect. 4, the data, the results of the data analysis and findings are presented. Finally, the conclusions and recommendations are presented in Sect. 5.

2 Overview of the Concept of HSE, the Dynamics and Challenges to Project Leaders

2.1 The Human Settlement Environment

The HSE has intrinsic complexities. Housing developments are usually initiated by the private and public sectors, with the public sector mainly concentrating on the poor. Adding to the complexity of the bureaucratic nature of the National Department of Human Settlements (NDHS) is that it forms part of the public sector with a predetermined organizational structure with its responsibilities, authority, and accountability [29]. PLs are thus relied upon when it comes to housing developments to coordinate and synchronize the various project factors whilst managing various multi-stakeholders to achieve service delivery. The expedition of service delivery is directly linked to the quality-of-life aspiration, as people are not looking for a house or place to exist alone, but to be part of the socio-economic environment of which they can form part [34].

Bremner [4], cited by Windapo and Goulding [34], noted that communities within these informal settlements are multifaceted to outsiders due to the dynamics within the settlement. Dynamic in general refers to something energetic, capable of action and change, and forceful. They depend on social networks interlinked to each other to make their occupant's lives easier by using minimal resources at their exposure, making it a very complex social environment [1]. The survival in informal settlements feeds on politics, social network structure and the socio-economics system. Misselhorn [17: 4] argues that ignorance about the dynamics within informal settlements results in inappropriate interventions. These dynamics include the intricate social and survival systems and the practical and communal challenges of housing and infrastructure development.

2.2 Project Management and Leadership in Human Settlements

Thamhain [27] noted that the intricacies of projects and the commercial environs have intensified in terms of labour and the structure and general managerial mechanisms of organizations. The PM is responsible for managing the project delivery process by managing the project team. Gido and Clement (2012: 324) agree with the statement above when highlighting that the PM provides leadership to the project team to accomplish the project objective. Pulmanis [22] notes that governments and organizations usually embark on different projects to create new services or improve the functional efficiency of the existing ones. Tubey et al. [30] note that leaders create cultures within the environment, and their essential role is to influence others whilst sharing the effective role of radical change and transformation with managers. The two constructs overlap. For example, when managers exercise influence over a group, this involves leadership, and when leaders get involved with planning, organizing, staffing, and controlling, they are involved with management [19].

2.3 Challenges Faced by Project Leaders in the HSE

Social and communal challenges: Some housing projects occur at infill sites where the community and its beneficiaries have their expectations about the land parcels to be developed. The nature of housing projects is so unique and specific that it creates conflict among various interest groups such as communities, politicians (ward councillors) and local authorities. The PL stands as a mediator between the beneficiaries, local government, and the project itself. The significance of involvement and consultation of the community at the beginning of the housing project is necessary to develop a successful project. The Housing Delivery process has multiple stakeholders within the project life cycle, with individuals or groups affected by the outcomes of a project.

Complex and dynamic project environment: The NDHS faces challenges through being part of the public sector compared to the private sector. While within conventional projects, the PL, the employer, developer (implementer), and funder can be one organization; on the other hand, in the public sector, organizational operations have a complex framework due to policies and procedures, including the funds/grants that they rely on. Whilst the NDHS as the regulating body is unique as a public sector organization because it implements project management within all HS departments at local, provincial, and national levels, its project environment is very different from the public sector's because it must attend to community/social and political engagement, as well as a plethora of legislative compliance requirements and organizational challenges within specific socio-economic constraints. The HSE is complex and dynamic, which influences both public sector organizational processes and the societal demographics of where the project occurs.

Limited information is available on the challenges faced by the PL in the HSE. Further in-depth research is required in the area to examine the challenges faced by PLs in the HSE.

3 Research Methodology

The research adopted an interpretive philosophy and a qualitative approach. The research focuses on the HSE and nature of project leadership within the context of housing delivery. The research study is undertaken within the Cape Metropolitan area in Cape Town, a vibrant city that attracts people from all walks of life. Cape Town is in the Western Cape and covers an area of approximately 2461 km^2. Its population in 2016 was estimated at 4,004,793, with the number of households being 1,264,849 and an average household size of 3.17 people [7, 8].

In this study, 37 participants were identified and invited to take part in the research. The respondents were drawn from the Provincial and Local Municipality Human Settlement Department. A non-probability purposive sampling technique was used in the study in selecting research participants since they possess features of the required data. The participants were purposefully selected based on their project leadership role in the HSE. An information and consent form were provided to all participants regarding the research. A total of 19 participants responded, of which 12 agreed to be interviewed personally, and seven chose to self-report by filling in an interview protocol. All the respondents were stratified based on their being PLs mainly from the level of PM, up to the level of the director in the public sector both provincial and municipal HSE.

This research adopted the method of self-report and personal interviews in data collection. The questionnaire's content was aligned to the structured interview protocol to allow the participant's freedom to choose the most convenient way of responding, as far as confidentiality and anonymity in participating in the research study were concerned [20]. The interview schedule was developed before the interviews. The items of the interview protocol followed a logical sequence, divided into three categories: Background information of the respondents, Professional experience and affiliated body of the profession; Challenges faced by PLs in a dynamic HSE; and the future of human settlement project delivery.

The personal interview process was clarified with each participant. Interviews were recorded to ensure the quality of information and later transcribed for analysis [5]. Permission for recording the interviews was requested from each participant before each interview. Field notes were also taken during the interview process to ensure that relevant information was captured accurately. The personal interviews were conducted using handwritten notes and audio recordings. Permission for the recordings was requested from the participants before conducting the interviews.

The information obtained from the interviews was transcribed from the recordings. This allowed time to listen to the recording several times and read the transcripts to comprehend what the respondents said and how it tied into the research question and objectives. Transcribing data from the interviews and allocating it to themes guided the ability to highlight patterns of views that the respondents articulated.

To proceed with the topic and research, ethical clearance was obtained from the University of Cape Town, City of Cape Town Municipality and Western Cape Department of Human Settlements. The clearance condition was that the intended participants and supervisors should be willing to participate in the research so that no intimidation could be implied. Care was taken during interviews to ensure that no respondent was subjected to any risk. No compensation of any kind was offered to the participants, and their participation was completely voluntary. All information obtained from the respondents remained strictly anonymous and confidential, and the research outcome was not directly linked to any participant. Pseudonyms or numbers were used in the research to protect the identity of the participants. The specific public sphere was also not revealed in the report for ethical reasons.

4 Data Presentation and Analysis

The information provided by each of the participants was transcribed from an audio recording and notes taken during the interview.

4.1 Background Profile of the Respondents

The study sought to know the background profile of the respondents. The findings revealed that 15 of the participants interviewed were not registered as project management professionals. The majority of the respondents indicated that they have some form of project management education, with only one participant indicated on-the-job training. Furthermore, seven of the respondents specified that they were registered with other professional bodies. Experience amongst participants reflected that 16 of them had more than five years' experience, of which eight deemed themselves at professional project management level, four at senior project management level, and four at an intermediate project management level. When respondents were asked about human settlement projects completed, four indicated they had not completed any, and 15 indicated they had completed between two and 18 projects. The preceding demonstrated that the respondents have the knowledge and experience to lead HS projects.

4.2 Challenges Faced by PLs in a Dynamic HSE

The success of human settlements projects is usually measured according to the traditional project management triple constraints, namely: scope, cost, and time. Out of the 19 participants, the success rate on the housing projects was less than half to none. This low percentage is directly linked to the challenge PLs face during project delivery. During the coding process, the data were categorized into six themes that highlighted challenges the project leaders faced during project delivery. These themes were: social challenges, organizational challenges; political interference; multi-stakeholder management; legislation by-laws; policies, and skills.

Social Challenges

All the interviewed participants noted that social/communal challenges were the most problematic for project delivery. Most participants indicated that a varying degree of social issues had an impact on project delivery. For example,

> Participant 2 mentioned that "*communicating project technical details to the community was very difficult as it was necessary to break it down to a level that they could understand, while they lacked a basic understanding of those technicalities.*"
>
> Participant 4 noted that "*when beneficiaries have to move from one site to another, the relocation process becomes a challenge on its own, which leads to a project within a project. These social dynamics tend to emerge as the project unfolds, and there is no academic knowledge within project management that empowers or prepares a project leader for these practical problems.*" As previously indicated, most participants have a built environment background or experience that is technically related, rather than social studies, to cope with such challenges.

The communities in which these projects often occur have a social and political dynamic stemming from their collective history, which the PL needs to manage. All respondents described the social and political dynamics during the interviews, and they confirmed that these dynamics could drive project implementation to the edge of failure. Various interested parties in communities seek to take advantage of the project which has been initiated, and that is the risky situation that the PL must mitigate against.

> According to Participant 11, "*The buy-in from communities is a challenge; sometimes the solution that we propose does not suit the needs of the community in the contexts of the allowed resources. This requires negotiating with the community and working with them to achieve a solution that fits both parties*".

Political interference

The political interference from ward councillors or MayCo members was highlighted as a major hindrance by project leaders. Most respondents indicated that political members tended to make promises to communities that were not in line with the project objectives or what legislation and policy would allow. The challenges that were created due to these promises caused social and community interference during the roll-out and implementation of the project and negatively affected the project timeframes. The fact that housing delivery takes place within the political domain opens it up to political interference. The findings suggest that politicians believe they have the power of authority and that the PL needs to act on their promises, while the PL is governed by legislation, by-laws, policies and compliance, which are mostly in conflict with the politicians' promises. Although projects are borne out of the political realm, the implementation of the project depends on its legislative and organizational compliance. The decision-making of the PL on public housing projects is always subjected to political interference, which is either local area politics or municipal management politics. The respondents' comments in this regard were as follows:

> Political interference because there will always be a ward councillor or MayCo member who have their own agenda. The overall lack of leadership in human settlements, as the politicians change, so does the priorities of projects. Your project which is planned to happen in the future can shift to this year, or resources like budget can be moved from your current project to the one that management feels is a priority at that particular moment. Internal disputes, internal policies and procedures are forever debated amongst departments. Political interference by ward councillors: for example, the beneficiary criteria because when we need to upgrade an area, people are usually promised that all of them will get houses. Later they find out about the beneficiary qualifying criteria policy and if they don't qualify it leads to all kinds of issues such as protests. Participant 1

> Political interference especially MayCo or ward councillor they always want to tell us what to do and how we need to do it. SCM is also a challenge due to market prices that do not line up with the subsidy, leading to tender cancellations. Community protest and the theft on-site are affecting the projects. The fact that we also don't look at another social service also becomes an issue when you present the layout to communities as for example if there is no school within the development and the kids need to travel far for schooling. Participant 8

> Political interference, the influence of ward councillors and Mayco members is big in the project if they are not informed or even if they are, they make promises to communities which the project cannot provide. Participant 11

Organizational challenges

These as different expressions of the human settlement environment (regarding population growth and urbanization) and organization environment (structure, behaviour, and operations). The respondents generally deal with organizational challenges related to procurement; non-cooperation from line departments; lack of resources, compliance measures; lack of alignment between community needs and the project objectives; and organizational changes that influence project implementation. All participants unanimously stated that fusing the organizational and contract management, namely private sector management techniques and public policies, at the project level was challenging. It required a multi-disciplinary approach by a person who understood the organizational policies, including procedures and contract law.

The findings confirmed that the bureaucratic nature of the public sector organization does not allow PL the authority for decision making over the project. It was revealed that the PL acts from instructions or when instructed on their projects. Only four of the respondents out of 19 indicated they had complete authority over their projects. The findings show that prioritization of housing projects within the organization is a challenge as the infrastructure service departments do not emphasize housing in their allocation of resources, which can often lead to an increase in scope and budget for housing projects.

Legislation, by-laws and policies

Public sector project delivery occurs in a bureaucratic process, and project management is subjected to various legislative policies of compliance. The participants considered the magnitude of legislation and policies that impact housing developments as a burden to project delivery as it affects the way the project is structured and implemented. Most participants indicated that public housing depended on the subsidy, which was not market-related, while consultants and contractors from the private sector charged market-related fees. Several respondents highlighted that PPM, a project management system, was a compliance measure that did not assist them in their work.

It emerged that everything PLs have to do within HS is prescribed by legislation, by-laws or policies. This places the PL in a peculiar position regarding implementation, especially when dealing with revisions on construction contracts, as Supply Chain Management (SCM) overrules the contract law, according to the PMFA and MFMA. The findings indicate that compliance measures such as housing standards, housing financial instruments, MFMA, PFMA, procurement, legislation, standard operating procedure, subsidies and grants all overrule project management in the public sector. This hinders the process of housing project delivery. Municipalities are excellent regulators; therefore, the project is compliance-driven and not production orientated. The challenges of this organizational compliance are that it requires the PL to be organizational process-driven than project-delivery driven. It burdens the PL to perform duties as a manager and leader, which result in project delays. This lack of authority does not give the PL the power to make immediate decisions on projects, which results in scope, cost, time and quality implications.

Multi-Stakeholders

Managing stakeholders in housing are very complex due to the number of internal and external stakeholders. HS projects fall within the public domain and attract various public and private interests, which puts no limitation on the number of stakeholders involved in the projects. The participants considered the number of individuals, groups and organizations both public and private involved in a human settlement project as quite extensive, making the communication unnecessarily complex. HS projects have several stakeholders, and the PL is responsible for managing them. Most respondents indicated that it is becoming a challenge to manage multi-stakeholder interests. In addition, the respondents indicated that not having control over these stakeholders becomes a coordination challenge because it affects the project objectives. Several respondents also agreed that the procurement process is challenging, as the internal infrastructure service departments are not allowing housing projects the importance they demand.

Lack of Soft Skills by Project Managers

A HS project involves managing stakeholders who have different interests. Even the skills that are required to manage these stakeholders differ. The study revealed that the PLs lacked the various soft skills to deal with these stakeholders. Most participants considered technical or hard skills as not a problem on a project. However, their view was that stakeholder management required soft skills when interacting with politicians and community members. The respondents, in general, noted that a lack of soft skills could lead to project disruptions. Their idea of instilling these soft skills was through mentorship and learning on the job.

The study revealed that the lack of project management skills and professionals within the local government sphere was why most PMs in this sector rely on consultants, organizational policies, and procedures to execute their duties instead of professional competency. It also emerged that a PL's lack of public management skills and social skills to deal with community and political issues prohibits them from exerting an ability to transform and lead creatively. The lack of mentorship within the department of human settlements makes it impossible for the project leaders to learn from their successes and failures. The findings suggest that most PLs have the technical competencies but lack the organizational, institutional knowledge and soft skills for delivering HS projects.

Discussion of Findings

The FFC [9] highlights the policy challenges such as increasing building standards, increase unit size and cost. These challenges are the leading causes of a decline in effective project delivery. The participants highlighted organizational, social and political challenges related to project delivery. In essence, it is the combination of elements such as the National Housing Policy and organizational and stakeholder challenges linked to the decline in housing delivery. Van der Waldt [32] and Starling [25] noted that it is difficult for lower-level managers, such as a PM, to exercise authority on public sector projects as the hierarchy and government bureaucracy are complex and the approval processes impact project success, and because the PMs authority is limited.

Mabelebele [13] avers that this SCM process is incompatible with the standard project management role because it involves political principals with authority to make decisions. It appears that the rigidity of a bureaucratic organization prohibits the PL from exercising their creative ability to successfully deliver projects because of the lack of authority to make decisions. The identified project of the participants further revealed that the success of the human settlement project could range from professional services, completion of civil service, top structures (houses), or the entire scope of these activities forming the whole project. All that is contrary to the PMI [21] and APM [3] because understanding the project manager's authority, responsibility and accountability depends on professional competencies within the context of project complexity and differs from situation to situation. Fox et al. [10] highlight these contradictions in a management approach, saying that public and private sector organizations share a common understanding of development techniques, but not so much at the level of political and managerial judgement and decision-making.

The challenge of developing human settlements is much broader than housing; it requires other components that make it socio-economically viable. According to the [31: 1]: these informal settlements are excluded from opportunities, physically, politically, and economically marginalized; and slum dwellers are particularly vulnerable to crime and violence. When it comes to public sector projects, [12: 29] made it clear that these projects involve the community and that the PL is responsible for managing the community and the project team. According to Steyn et al. [26], stakeholder management forms part of the responsibility of project management in general. Therefore, project leaders in housing delivery face an insurmountable task of satisfying different stakeholders with conflicting goals.

Moreover, all respondents agree that the most critical stakeholder in the community requires soft skills to be managed. Radmila et al. [23] and Archer et al. [2] noted that the PL leads the organization through these challenging times by inspiring individuals to participate in the project. On the one hand, the PL relies on soft skills to address social and multi-stakeholder challenges. On the other hand, PLs require enhancing their hard skills to deal with the organization, legislation, and skills. Van der Waldt [33] states that project management requires one to manage technical and non-technical aspects of a project. At the same time, Northouse [19] and Marando [15] support this by averring that a project manager needs to learn and evolve their craft continuously.

5 Conclusion and Recommendations

5.1 A Subsection Sample

This research examines the challenges faced by PLs in the HSE and whether PLs exercise a capacity to transform, maintain, and lead the project organization creatively. The research used a qualitative research approach with inductive-philosophical reasoning that employs interviews for data collection was chosen for the study. The study found that the organizational bureaucratic nature of the public sector revealed that it is difficult for a PL to lead because they lack decision-making authority in that setting, and therefore, may only react to and form instructions. Further, the challenges of compliance within the organization are a burden on the PL. It is expected that the PL should perform their duties as prescribed by legislation, by-laws or policies, but these are factors over which the PL has no control, resulting in project delays.

The prioritization of housing projects within the organization and other internal departments becomes another challenge as all resources are not necessarily aligned to the project objective if it is, for example, a community centre or a bridge which is one specific objective that must be built in the environs of the settlement, not all line departments are affected and have minimal input if any. In contrast, a housing project relates to infrastructure services that include community centre, bridge, water, sewer, electrical, stormwater and landscaping. The houses to be constructed require a multi-disciplinary focus aligned to one objective: achieve a human settlement. The PL has no control over these internal resources and may also lack financial and supporting resources. The social and communal environment in which these projects take place is another challenge as it may bring forth political interference and various community and social dynamics that the PL needs to manage. For a PL to manage these challenges requires soft skills, which PLs did not have, and this poses another challenge, and it is a requirement for a PL to have to lead within such difficult situations.

References

1. Abbott A, Martinez I, Huchzermeyer M (2001) An analysis of informal settlements applicability of visual settlement planning (ViSP) in South Africa. WRC report no. 786/1/01. Water Research Commission, Pretoria. Administration and Law at the University of Zululand, p 101
2. Archer MM, Verster JJP, dan Zulch BG (2010) Leadership in construction project management: ignorance and challenges. In: Proceedings 5th built environment conference, July 18–20. Durban, South Africa, pp 429–440
3. Association for Project Management (APM) (2012) Association for project management body of knowledge, 6th edn. Association for Project Management, Buckinghamshire
4. Bremner L (1994) Development and resistance: the lessons for planners of Phola Park. Urban Forum 7(1):53–67
5. Bryman A (2012) Social research methods, 4th edn. Oxford University Press, New York
6. City of Cape Town (2018) Cape town municipal spatial development framework (MSDF) 2017–2022 review draft
7. City of Cape Town (2019a) Draft housing strategy. City of Cape Town, Cape Town
8. City of Cape Town (2019b) City of Cape Town five-year integrated development plan July 2017–June 2022: as amended for 2019/20. City of Cape Town, Cape Town
9. Financial and Fiscal Commission (FFC) (2013) Exploring alternative finance and policy options for effective and sustainable delivery of housing in South Africa Johannesburg. FFC
10. Fox W, Schwella E, Wissink H (1991) Public management. Juta & Co., Kenwyn
11. Gido J, Clement J (2012) Effective project management, 5th edn. Cengage Learning, Australia
12. Kerzner H (2017) Project management: a systems approach to planning, scheduling, and controlling, 12th edn. Wiley, United States of America
13. Mabelebele JM (2006) Prospects and challenges of implementing projects in public service. Government Communications GCIS, Pretoria
14. Marais L, Wessels J (2005) Housing standards and housing rights: the case of Welkomin the Free State Province. Urban Forum 16(1):17–34
15. Marando A (2012) Balancing project management hard and soft skills: master of science in management of projects and programs. Brandeis University
16. Mbanga L (2018) Human settlement dynamics in the Bamenda III Municipality. North West Region, Cameroon. J Settl Spat Plan 9(1):47–58
17. Misselhorn M (2008) Position paper on informal settlement upgrading, part of a strategy for the Office of the South African Presidency. Urban Landmark, South Africa
18. National Department of Human Settlements (NDHS) (2015) Strategic plan 2015–2020. National Department of Human Settlements, Pretoria
19. Northouse PG (2016) Leadership: theory and practice, 7th edn. Sage Publications, Thousand Oaks, CA

20. Plotnik R (1993) Introduction to psychology, 3rd edn. Brooks-Cole, Pacific Grove, CA
21. Project Management Institute (2017) A guide to the project management body of knowledge (PMBOK), 6th edn. Project Management Institute Inc., Pennsylvania
22. Pūlmanis E (2013) Polycentric development projects initialisation process problems in Latvia. Integrated and sustainable regional development: proceedings of the international scientific conference. Econ Sci Rural Dev 31:220–224
23. Radmila K, Gragana R, Bojana M (2011) Leadership in modern organizations. Int J Soc Sci Stud 1(3):60–69
24. Republic of South Africa (RSA) (1996) Constitution of the Republic of South Africa, act no. 108 of 1996. Government Printer, Pretoria
25. Starling G (1993) Managing the public sector, 4th edn. Wadsworth, Belmont
26. Steyn H, Carruthers M, Dekker A, du Plessis Y, Kuschke B, Sparrius A, van Eck S, Visser K (2012) Project management—a multi-disciplinary approach, 3rd edn. FPM Publishing, Pretoria
27. Thamhain HJ (1991) Developing project management skills. In: Pinto JK, Trailer JW (eds) Leadership skills for project managers. Project Management Institute, 1998
28. The Fuller Centre for Housing Western Cape (2014) Housing delivery South Africa. The Fuller Centre for Housing Western Cape, Cape Town
29. Toit D and Van der Waldt G (1997) Managing for excellence in the public sector. Juta, Kenwyn
30. Tubey RJ, Odenyo M, Chebet WT, Rotich J (2015) Transformational leadership and human capital management in 21st century organizations. Euro J Res Reflection Manage Sci 3(5)
31. UN-HABITAT (2007) Indicators of sustainable of development: guidelines and methodologies. United Nations, New York
32. Van der Waldt G (2007) Project management: a new service delivery paradigm: research Koers. Bull Christian Schol 72(2):239–260
33. Van der Waldt G (2001) Establishing a management by projects approach for service delivery', J Public Administration Manage 36(4):296–311
34. Windapo AO, Goulding J (2013) Value-based perspectives of stakeholders' building requirements in low-cost and government-subsidised housing projects in South Africa. Constr Innov 13(4):424–444

Functionality and Mechanical Properties of Steel Fibre Reinforced Concrete in Stemming Building Collapses in Nigeria

Olubisi Ige, Obas John Ebohon, Theo C. Haupt, and Mariam Akinlolu

Abstract In Nigeria, building collapses is a common occurrence, and several factors have been advanced to explain the incidents, amongst these is the lack of understanding of the functionality and mechanical properties of steel fibre reinforced concrete. Such understandings can assist in stemming building collapses and attendant economic and social effects, particularly the health and safety of construction workers and occupants. This is achieved by understanding the functionality and mechanical properties of steel fibre reinforced concrete. As a way of demonstration, an efficient way to achieve this is to use thin short discrete steel fibres, this will allow, a multi-directional reinforcement, the process modifying and improving properties of the concrete, especially its ductility. Critical to this, is the understanding of the preference and selection of materials to make appropriate mixes for efficient result of steel fibre reinforced concrete. In other words, failure to pay attention to selection and use of materials will produce weak concrete structure and precipitating unwarranted disasters. This study investigates the effects of length and aspect ratio of steel fibres when mixed with different sizes of coarse aggregate on the workability and subsequently, on the mechanical properties of the material. Variables selected for the study were fibre lengths of 50 and 60 mm, aspect ratio of 45, 50 and 60, fibre dosages of 25, 40, 50 and 60 kg/m^3 and maximum aggregate sizes of 10 and 20 mm. Mix proportions for the investigation were kept constant throughout the study. Slump test was performed on fresh concrete while compressive strength was measured using 100 mm cubes and flexural performance assessed through 150 mm $\times$ 150 mm $\times$ 600 mm prism. The experimental results confirm that the combination of geometry and maximum aggregate size in the mix has an important influence on the mechanical properties of hardened concrete. It is hoped that construction practitioners in Nigeria, and by extension, Africa, will evaluate their practices considering this study.

O. Ige (✉) · O. J. Ebohon
School of Built Environment and Architecture, London South Bank University, 103 Borough Road, London SE1 0AA, UK
e-mail: igeo4@lsbu.ac.uk

T. C. Haupt · M. Akinlolu
Faculty of Engineering, Mangosuthu University of Technology, Durban, South Africa

T. C. Haupt et al. (eds.), *Construction in 5D: Deconstruction, Digitalization, Disruption, Disaster, Development*, Lecture Notes in Civil Engineering 245,
https://doi.org/10.1007/978-3-030-97748-1_20

Keywords Steel fibres · Fibre geometry · Aggregate size · Workability · Compressive strength · Flexural properties

1 Introduction

Building collapses, according to experts, owe to failures of buildings to perform the purpose for which they are built [7]. As often as it is the case, building collapses result from three basic scenarios, which are serviceability limit state, ultimate limit state, and durability limit state [16]. Ultimate limit state failure is said to happen when the columns start to wobbles and overturn resulting in ultimate collapse or failed in ultimate limit state. The serviceability limit state relates to a civil engineering structure that fails to meet technical requirements for use despite its relative strength to remain standing. Structures of this nature would have surpassed a defined limit for either of the following properties—excessive deflection, vibration, and local deformation. Durability failure occurs when the components of the structure is weakened beyond reparable limits. Building collapse is a world-wide phenomenon with 181 incidents of building collapse reported across 51 countries, including India with 32 reported building collapse disasters, 13 in China, 13 in Egypt, 12 in Brazil, and 12 in Nigeria. The Bangladeshi reported building collapse incident in 2013 is instructive on the deadly consequences of such an event with the death of 1127 occupants [11].

In Nigeria, incidents of building collapse have become more frequent in recent times and so has been the casualty figures [6], this can be said to owe to the unspring of high-rise buildings in the country amongst other factors. According to the Nigerian Institute of Building (NIOB) in one building collapse in 2014, 115 people died, and in 2019, there were 43 building collapses with the unfortunate loss of many lives. This has forced a critical look at how buildings are procured and regulated in the country, and according to NIOB president, incidents of building collapses in Nigeria "looks like an indictment on building control agencies", indicating the need for a reassessment of the building processes in Nigeria. One strategy is to look at strengthening building structures by paying particular attention to material, process and technological innovation in construction. This research has identified concrete as the major material component of construction for scrutiny. Indeed, concrete has continued to undergo relentless advances through theoretical and experimental research for decades, in order to significantly improve the material properties such as its strength and brittle nature of failure [8, 10]. Although, some of these goals have been achieved, such as enhancement in compressive strength but this made the material to be more brittle and less ductile [1]. In spite of the major milestones already accomplished in the technological enhancement of concrete properties, the challenges of concrete structures being subjected to more loadings as a result of growth in the world population, surge in natural disaster and terrorist attack through explosions experienced in progressive manner call for more technological solutions [9]. Therefore, in order to further mitigate some of the undesirable properties of concrete, steel fibres which are short, thin, and discrete were introduced into the concrete matrix to improve many of its engineering properties such as its ductility and toughness [1, 10]. This is accomplished by the ability of steel fibres to prevent or control the initiation and propagation of micro cracks into macro cracks within the concrete matrix, transferring stress across a crack by bridging it [10].

Previous studies have investigated the effects of the inclusion of steel fibres in concrete and have concluded that some of the engineering properties of steel fibres significantly alter the properties of the resulting steel fibre reinforced concrete (SFRC). The fibre properties include fibre type, dosage, shape, length, and aspect ratio. The magnitude of the improvement in ductility and flexural strength is impacted by these parameters while the distribution and orientation of fibres also influence how the fibres are positioned for the bridging of the cracks, which subsequently affects the post-cracking ability of the matrix [1, 9]. Additionally, the engineering properties and the volume or dosage of the fibres in the matrix affect directly the workability of the fresh concrete. It has been reported that apart from the fibre geometry, lower steel fibre dosages of not more than 1% had negligible effect on the compressive strength, whereas, the flexural strength was significantly enhanced. However, it was also reported that high steel fibre dosages higher than 3% can negatively affect the workability in the fresh state, the compressive and flexural strengths of SFRC [13]. This could be as a result of balling effect causing an increase in voids in the matrix.

The hooked end steel fibres have been found to show best performance by their provision of strong mechanical anchorage within the matrix [10]. Notwithstanding, several research works that have been carried out on how to best explore SFRC in practice, experimental results on investigating the collaborating effects of steel fibres with the concrete components of the matrix and how the relationship between these materials affects the maximal productivity of the SFRC are considered few. Hence, more understanding of how the steel fibres and concrete components collaborate is essential to maximise the contribution of fibres to the improvement of mechanical properties of SFRC and subsequently, it's potential in practice, especially for those in construction industries and academic communities.

This research work studies the effects of fibre geometry, dosage, and coarse aggregate size on the behaviour of fibres within the concrete matrix and the subsequent influence on the workability and the effects on flexural properties of steel fibre reinforced concrete.

2 Experimental Procedures

The experimental programme for this research work was designed to investigate the performance of different geometry of steel fibres and dosages when combined with two maximum sizes of coarse aggregates. The effects of these variables were then investigated on the workability of the fresh steel fibre reinforced concrete and the subsequent effects on the mechanical performance of the steel fibre reinforced concrete (SFRC).

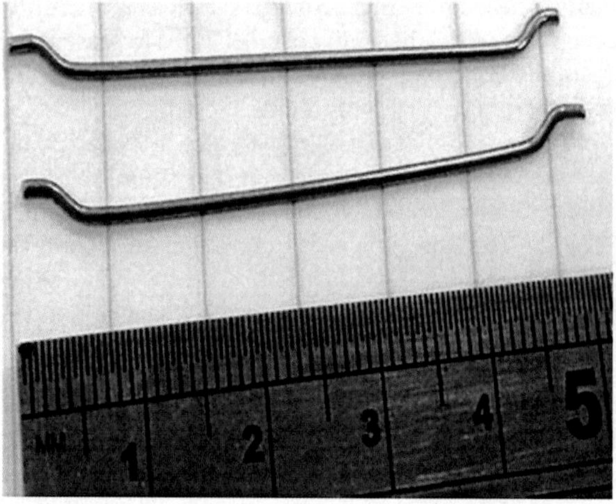

Fig. 1 Hooked end steel fibres

2.1 *Materials*

The steel fibre reinforced concrete mixtures studied in this research were produced with commercially available materials. The main constituents of the concrete were made to be consistent throughout, sustaining the same source so as to allow for consistency and proper comparison. Portland cement conforming with BS EN 197-118 (CEM 1 52.5R) [5], polycarboxylate polymer-based superplasticizer necessary to provide good workability at a low water/cement ratio, tap water and sea-won coarse and fine aggregates were employed in the concrete mixes. The 0.5 water/cement ratio was adopted for this study and was made constant for all the mixtures as well. In order to investigate the effect of materials on fresh and hardened properties of the steel fibres reinforced concrete, the variables selected were fibre lengths of 50 and 60 mm, aspect ratio (ratio of length to diameter of fibre) of 45, 65 and 80, and dosages of 0 kg/m^3 (no fibres), 25, 40, 50 and 60 $^{kg}/m^3$ of steel fibres were employed. The fibres used for this investigation were the hooked end type supplied by Bekaert as shown in Fig. 1. The coarse aggregates with maximum sizes of 10 and 20 mm as shown in Fig. 2 were also chosen for the study. The maximum coarse aggregate size (d_{max}) and the geometry of the steel fibres used are as shown in Table 1.

2.2 *Specimen and Procedures*

A pan mixer of 0.05 m^3 capacity was employed in mixing of the concrete while the same particular procedure was adopted for all the mixes in order to accomplish an even dispersion of fibres in the mix and to avoid balling effects in the mixtures. Gravel, sand and cement were first mixed in the mixer in a dried state for few minutes before water was added, and immediately followed by superplasticizer before being

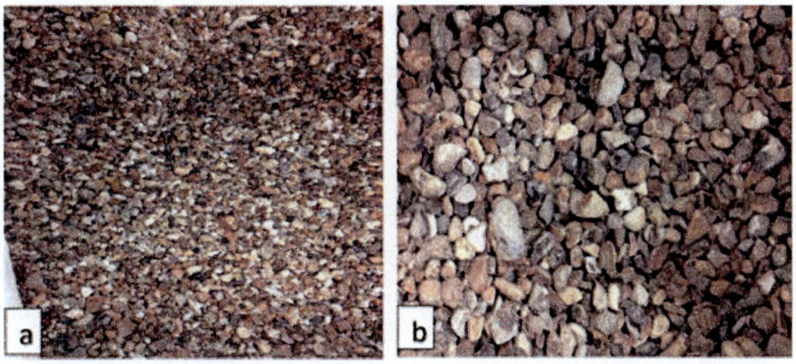

Fig. 2 Gravel used as coarse aggregate with **a** 10 **mm** and **b** 20 mm size variables

allowed to mix for another few minutes. The fibres were added last and then a further one to two minutes mixing was allowed to have the material sufficiently mixed. The workability of each mix was checked through slump test and findings were recorded. The mixtures were accomplished by following the dosages as shown in Table 1 for each maximum coarse aggregate size (d_{max}) and the geometry of the steel fibres.

In each mixture, the workability of the mix was carried out through slump test according to BS EN 12350 Part 2, [3]. Slump test was carried out on both plain concrete without steel fibres and steel fibre reinforced concrete. Subsequently, for each aggregate size/fibre combination, three replicate of 100 mm cubes and 150 mm by 150 mm by 600 mm beam specimens were produced. All the specimens were cured in a water tank at 20 ± 2 °C until testing at 28 days. Compressive strength tests were carried out on 100 mm cubes employing ADR–Autotest machine with 2000 kN maximum capacity while beam specimens were tested for flexural properties using a Zwick/Roell Z250 universal testing machine with maximum capacity of 250 kN. All the tests in this research were carried out at a testing age of 28 days. The tests for beam specimens were accomplished under three point loading, (Fig. 3) with notches introduced in the middle section to a depth of 25 mm using a diamond saw before testing. A crack mouth opening displacement (CMOD) gauge was positioned in the notch to control the test at the CMOD rates specified by BS EN 14651 [2].

3 Results and Discussion

The effects of addition of fibres to concrete on the workability of fresh concrete investigated through slump test can be seen as presented in Table 2. This shows the values of slump tests conducted on fresh plain concrete and steel fibre reinforced concrete with varying fibre dosages, fibre aspect ratio, fibre length and two maximum sizes of coarse aggregate. The results from Table 2 reveal that slump values differ widely between 15 and 140 mm, showing decline in slump values as dosages of fibre increased. It can also be deduced that mixtures with maximum coarse aggregate size

Table 1 Aggregate maximum size and geometry of the steel fibres

d_{max} (mm)	Fibre type	Fibre length (mm)	Fibre aspect ratio (length/diameter)	Fibre dosage (kg/m^3)	% by fibre volume
10	45/50	50	45	0	0
				25	0.31
				40	0.51
				50	0.64
				60	0.76
	65/60	60	65	0	0
				25	0.31
				40	0.51
				50	0.64
				60	0.76
	80/60	60	80	0	0
				25	0.31
				40	0.51
				50	0.64
				60	0.76
20	45/50	50	45	0	0
				25	0.31
				40	0.51
				50	0.64
				60	0.76
	65/60	60	65	0	0
				25	0.31
				40	0.51
				50	0.64
				60	0.76
	80/60	60	80	0	0
				25	0.31
				40	0.51
				50	0.64
				60	0.76

of 20 mm generally show better workability than mixtures with 10 mm maximum aggregate size as displayed by their slump values. Meanwhile, the mixtures of fibre length of 60 mm and aspect ratio (length/diameter) 65 and 80 with 10 mm maximum aggregate size provided the least slump values of 15 mm and 18 mm respectively at 60 kg/m^3 dosages. The decline in workability noticed for mixture of steel fibre with length 60 mm and maximum aggregate size of 10 mm may be attributed to the

Fig. 3 Measuring flexural strength under 3–point loading

Table 2 Slump values of plain concrete and SFRC

Fibre type AR/length	Fibre dosage kg/m^3	Slump, mm	
		10 mm max aggregate size	20 mm max aggregate size
	0	62	120
45/50	25	115	95
	40	53	125
	50	27	88
	60	36	53
65/60	25	55	140
	40	36	70
	50	32	55
	60	15	88
80/60	25	105	115
	40	34	135
	50	32	80
	60	18	65

non-compatibility of higher length of fibres and smaller aggregate size leading to balling effect especially at higher dosages, hampering the flow of cement paste. It is however noticed that the mixture with 25 kg/m^3 dosage of steel fibres for both maximum aggregate sizes demonstrate higher slump values mostly, producing a better workability. This has been credited to the even distribution of fibres within the fresh concrete and hence, influencing the workability of the mixture. Largely, presence of steel fibres reduced the workability of concrete as checked by slump test. This assertion agrees with many other research [13, 15].

The 100 mm cubes produced from different mixtures were employed to check the density of the plain and fibre reinforced concrete according to BS EN 12390-7 [4] with the average results of three replicate cubes presented in Fig. 4. The results depict that the average density of concrete with 10 mm maximum aggregate size having no

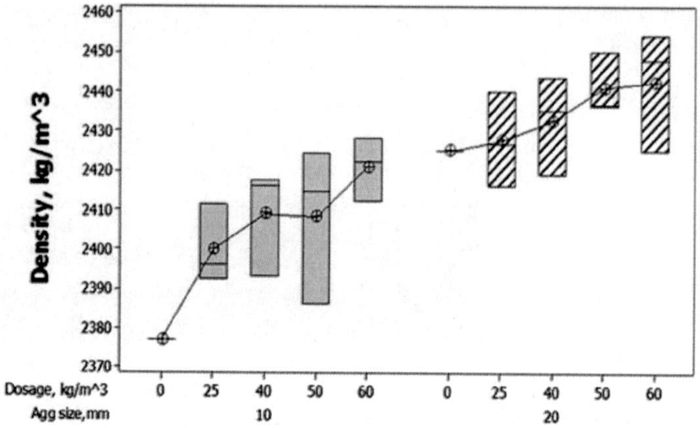

Fig. 4 Summary values of density vs dosage and aggregate size

fibres had the least value of 2377 kg/m^3 whereas, the highest average density value of 2455 $^{kg}/m^3$ was recorded from the concrete mixture of 20 mm maximum aggregate size containing steel fibre 45/50 at 50 kg/m^3 fibre dosage. The experimental values of density from this study have been found to be generally lower when compared to the theoretical values of density calculated, especially for the mixtures containing steel fibres. This could be attributed to the consequences of short period of compaction allowed at casting of concrete to guard against settlement or segregation of steel fibres in the mixture. Therefore, the plunge in the values of density obtained experimentally could have been as a result of air content of the mixture. Moreover, cubes from mixtures of 20 mm maximum aggregate size have higher values of density than their 10 mm maximum aggregate size counterparts. This trend is observed in the values of density recorded for all the mixtures of 20 mm maximum aggregate concrete when compared to that of 10 mm maximum aggregate concrete as clearly seen in Fig. 4 which is the boxplot of summarised values of density as against the dosages of steel fibres and maximum aggregate sizes. This trend of the results was also seen in the workability results where 20 mm maximum aggregate size concrete had higher values of slump in general.

Figure 5 shows the compressive strength results of 100 mm cubes demonstrating the impact of steel fibres and maximum size of aggregate on SFRC. The presented results generally indicate a minor improvement in the compressive strength by means of the inclusion of steel fibres in concrete. The maximum increase in compressive strength initiated by the presence of steel fibres in 10 mm maximum aggregate size mixtures is about 5 MPa, translating to about 9% increase while that of 20 mm maximum aggregate size mixtures had the maximum increase of about 8 MPa, which is 14% increase. The compressive strength of SFRC increases with increase in fibre dosage until optimum compressive strength was achieved at 50 kg/m^3 fibre dosage for both aggregate sizes after which it declined. This may not be unconnected with the compaction effects on the mixtures with highest dosage of steel fibre (60 $^{kg}/m^3$)

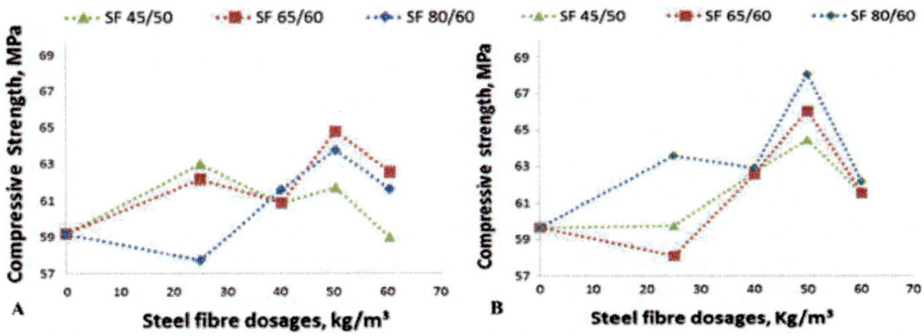

Fig. 5 Relationship between compressive strength and dosages of different fibres, **a** 10 **mm** aggregate size, **b** 20 mm aggregate size

making the compaction of the mixture a bit difficult to accomplish, resulting in lower compressive strengths regardless of coarse aggregate sizes in the concrete mixture. In addition, the lower strengths obtained at the 60 $^{kg}/m^3$ can be attributed to consequences of low workability of fresh mixture achieved at higher fibre dosage. The results achieved are closely related to results from previous findings on compressive strength of SFRC [12, 14].

The typical flexural stress versus CMOD relationship of the notched beams determined by three-point bending tests is shown in Fig. 6. Average values of maximum load, flexural toughness and residual strength of SFRC beams exhibited related trends to those observed in the figure. Figure 6 is presented for both mixes of 10 and 20 mm

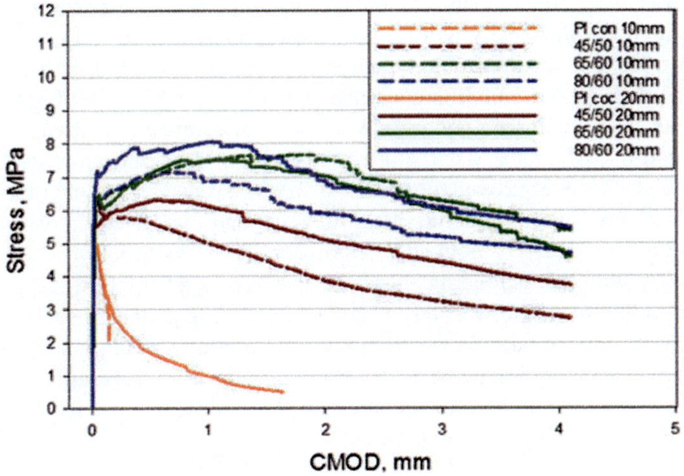

Fig. 6 Flexural behaviour of plain and SFRC with 50 kg/m³ fibre dosage

maximum aggregate sizes summarised at steel fibre dosage of 50 kg/m^3 for ease of comparison. Each concrete mixture is designated by its constituents, the type of steel fibre assigned by 'aspect ratio/corresponding length of the fibre' and the maximum coarse aggregate size. Plain concrete represents concrete mixture without steel fibres. It can be observed from the Figure that plain concrete of both aggregate sizes failed demonstrating brittle nature while other SFRC exhibited ductility behaviour. For each maximum aggregate size, all the flexural strength parameters showed increase as the aspect ratio increased. Moreover, the maximum aggregate size had a significantly effect on the flexural strength of the beam specimens, with the 20 mm maximum aggregate demonstrating higher strength for all fibre types. This is an suggestion that fibre length—coarse aggregate maximum size relationship is a factor to consider when choosing materials for fibre reinforced concrete.

4 Conclusions

Investigation of effects of maximum aggregate size, fibre type and dosage on the fresh and mechanical properties of concrete containing steel fibres has been carried out in this study. The results of the investigation have shown that all the parameters evaluated in this study have significant impact on both the fresh and mechanical performance of hardened SFRC. The workability of the fresh SFRC has been observed to influence the mechanical performance of the SFRC. The workability of fresh steel fibre reinforced concrete declined as the fibre dosage increases with lowest results witnessed at the highest dosage of 60 kg/m^3 steel fibre in concrete. The density of plain concrete and steel fibre reinforced concrete mixes revealed that there was increase in density as the fibre dosage increases in the mixture. Also, mixtures of 20 mm maximum aggregate size generally recorded higher values of density than their 10 mm maximum aggregate size. Meanwhile, the effects of fibres on compressive strength are noticeable, with optimal compressive strength detected at fibre dosage of 50 kg/m^3 and with fibre of 80 l/d ratio. The addition of steel fibres in concrete substantially improved the mechanical properties of concrete transforming it from a brittle to ductile material. The influence and effectiveness of steel fibres in concrete are particularly noticed at right dosages, appropriate combination of fibre geometry with corresponding aggregate size in concrete matrix. It is instructive that countries that disproportionately experience building collapses such as Nigeria should focus on the formation of form works that produce building structures. It is not sufficient to mix aggregates and steel fibre as necessarily determining strengths of structures, it is the combinations of these materials and the particular processes that determines the strengths of structures.

References

1. Boulekbache B, Hamrat M, Chemrouk M, Amziane S (2016) Flexural behaviour of steel fibre-reinforced concrete under cyclic loading. Constr Build Mater 126:253–262
2. BSI (British Standards Institute) (2005) Test method for metallic fibre concrete. Measuring the flexural tensile strength (limit of proportionality (LOP), residual). BS EN 14651:2005. BSI, London
3. BSI (British Standards Institute) (2009a) Testing fresh concrete-part 2: slump test. BS EN 12350-2:2009. BSI, London
4. BSI (British Standards Institute) (2009b) Testing hardened concrete density of hardened concrete. BS EN 12390-7:2009. BSI, London
5. BSI (British Standards Institute) (2011) Composition, specifications and conformity criteria for common cements, in part I: BS EN 197-1:2011. BSI, London
6. Ede A (2010) Building collapse in Nigeria: the trend of casualties in the last decade (2000–2010). Int J Civ Environ Eng IJCEE 10:6–32
7. Ede A (2016) Structural stability in Nigeria and worsening environmental disorder: the way forward. In: The West Africa built environment research conference (WABER), Accra, Ghana. https://doi.org/10.1088/1757-899X/1036/1/012021
8. Ferrara L, Meda A (2006) Relationships between fibre distribution, workability and the mechanical properties of SFRC applied to precast roof elements. Mater Struct 39(4):411–420
9. Ige OA (2017) Key factors affecting distribution and orientation of fibres in steel fibre reinforced concrete and subsequent effects on mechanical properties (Doctoral dissertation, University of Portsmouth)
10. Ige O, Barnett S, Chiverton J, Nassif A, Williams J (2017) Effects of steel fibre-aggregate interaction on mechanical behaviour of steel fibre reinforced concrete. Adv Appl Ceram 116(4):193–198
11. Keim A (2021) When buildings collapse. About disaster doc. https://disasterdoc.org/when-buildings-collapse/. Accessed 06/09/2021
12. Mohammadi Y, Singh SP, Kaushik SK (2008) Properties of steel fibrous concrete containing mixed fibres in fresh and hardened state. Constr Build Mater 22(5):956–965
13. Nehdi ML, Najjar MF, Soliman AM, Azabi TM (2017) Novel steel fibre-reinforced preplaced aggregate concrete with superior mechanical performance. Cement Concr Compos 82:242–251
14. Sebaibi N, Benzerzour M, Abriak NE (2014) Influence of the distribution and orientation of fibres in a reinforced concrete with waste fibres and powders. Constr Build Mater 65:254–263
15. Soulioti DV, Barkoula NM, Paipetis A, Matikas TE (2011) Effects of fibre geometry and volume fraction on the flexural behaviour of steel-fibre reinforced concrete. Strain 47:e535–e541
16. Taiwo AA, Afolami JA (2011) Incessant building collapse: a case of a hotel in Akure, Nigeria. J Build Appr 6(3):241–248. https://doi.org/10.1057/jba.2011.1

Challenges Facing Supply Chain Management on Time Delivery of Construction Projects in Tanzania

Geraldine J. Kikwasi and Nyamagere Gladys Sospeter

Abstract Construction supply chain management (CSCM) is an influencing factor for timely delivery of construction projects. Despite its benefits, the construction industry still faces inefficiency and poor time performance. As a result, project delays and costs overrun are experienced in the industry as a sign of failure to control time in projects implementation. The adoption of SCM is expected to improve time performance of construction project since it has its origin in Just-In-Time (JIT) production and logistics. This study assesses the challenges faced in application of CSCM towards timely delivery of construction projects in Tanzania. Descriptive research design was used and the targeted respondents were from construction firms registered in class I and II located in Dar es Salaam. Data was collected using questionnaires and out of 91 distributed 79 were returned fairly filled for analysis. Data was analyzed using two features of IBM SPSS version 20 which are descriptive statistics-frequencies and Compare Means-One Sample T Test. Findings reveal that: lack of a common standard for collaboration, procurement system deficiencies and low level of commitment between partners are significant challenges facing CSCM on-time project delivery. This study contributes to the body of knowledge on the subject within a previously unexplored context with a focus on-time control to minimize project delays. The study provides insights on the challenges that hinder the construction industry to benefit from adoption of SCM on the aspect of on-time projects delivery.

Keywords Construction · Procurement · Project · SCM · Time

G. J. Kikwasi (✉) · N. G. Sospeter
Department of Building Economics, Ardhi University, Dar es Salaam, Tanzania
e-mail: geraldine.kikwasi@aru.ac.tz

N. G. Sospeter
SBL, University of South Africa, Pretoria, South Africa

T. C. Haupt et al. (eds.), *Construction in 5D: Deconstruction, Digitalization, Disruption, Disaster, Development*, Lecture Notes in Civil Engineering 245,
https://doi.org/10.1007/978-3-030-97748-1_21

1 Introduction

Construction supply chain management (CSCM) has numerous benefits to the industry and considered as influencing factor for on-timely delivery of the projects. It has several responsibilities in the construction industry including coordinating the activities of the participating companies. Supply chain management (SCM) in construction has the task of mixing organizational units along a supply chain and organizing materials, information and financial flows to fulfil customer demand with the aim of improving performance of a project [6, 14]. Players of the supply chain especially contractors must make sure each member to the chain take their responsibility to make it more effective and have a successful project delivery. Some researchers note that CSCM plays a great role to ensure a successful project delivery in different aspects [8, 20]. Some of the benefits of CSCM are reduced real costs, reduced maintenance cost, incentive to remove waste from the process, competitive advantages, greater certainty of out-turn costs, delivery of better underlying value to the client and on-time project delivery [24, 11]. Despite these benefits, the construction industry still faces inefficiency and poor time performance. As a result, project delays and cost overruns are experienced in the industry due to inefficiencies in SCM application.

Successful project delivery is proved by cycle time, on-time completion percentage, time spent, number of adjustments to the schedule and planned hours versus time spent with an efficient SCM [8]. Timely delivery is seen as an important factor in the project undertaking as works, materials, components or goods must be delivered within specified timeframes. More so, client's desire is to receive what is ordered on time and budget (agreed cost) [12]. It requires special attention while undertaking any construction project because it forms large portion of what is delivered in construction projects. On the other hand, time control is very essential for a successful construction project. Despite that, delay is still one of the biggest problems often experienced on construction project sites, resulting to increased costs, loss of productivity and revenue and many lawsuits between owners and contractors and contract termination [13].

Research carried out globally have evaluated SCM and project performance, where time, quality and cost were targeted [8, 12–14]. Ying and Tookey [21] investigated the nature and extent of current CSCM practice in the New Zealand construction industry with focus on the challenges and issues that the industry is confronted with during efficient CSCM implementation. Very few studies have focused on SCM and how it affects timely delivery of projects like Moreira [11] which assessed barriers to the application of CSCM towards timely delivery in building construction projects, Mello [24] which investigated the factors that cause delays, Pekuri et al. [15] who identified challenges of procurement as part of SCM in delivering projects and [22]which explored constraints to the application of SCM in delivering the construction projects on time. These studies focused SCM in other contexts where there are differences in business environment, nature of construction activities, construction methods and dominant procurement systems. Since projects are unique and the nature of supply chain fragmentation differs based on the procurement system adopted, it is important to assess the challenges facing SCM on timely delivery of construction projects in Tanzania.

2 Literature Review

Delays in construction project

Delay is one of the biggest problems often experienced on construction project sites. Delays can instigate negative effects such as increased costs, loss of productivity and revenue many lawsuits between owners and contractors and contract termination [13]. Many studies have been conducted to identify the causes of time overrun hence delays in construction projects. Generally these causes include: poor economic condition, poor site management, slowness in giving instructions, delay of material approval by consultant, delay in bill settlement, unskilled operators, low productivity of labor, lack of maintenance for the equipment, poor site condition, delay of material delivery, insufficient number of staff, unrealistic contract duration imposed by the owner, owner interference, high quality of work required, lack of contractors skills, poor material handling on site, absence of consultants site staff, shortage of material in market, bad contract management by the consultant, equipment and tool shortage on size, contract modification and equipment availability [3, 12, 13]. Few studies have been done on delays related to the poor management of the supply chain. A study by Luvara et al. [9] asserts that the poor financial status of contractor, delay of material delivery, poor supervision, shortage of material in the market, lack of communication and coordination between parties are challenges that affect time delivery of the projects. As described lack of effective communication among the parties involved and lack of adequate information from consultants are limitations to efficient SCM [13]. Others are poor communication and coordination between parties, transportation delays as some modular construction elements take time to get transported to the site [3]. If the machinery and equipment required for construction get delayed to reach the site, the project timelines will eventually be affected [3].

Challenges Facing Construction Supply Chain Management

There are barriers to effective CSCM which have resulted in its failure to accomplish the desired outcome for time control in projects. According to [23], factors limiting CSCM in contributing to timely delivery of construction projects are: many individuals are involved in the supply chain, inappropriate knowledge about SC and CSCM, lack of attentiveness in SCM, benefits in building projects, poor higher authority commitments, systemization and support between the partners involved in the project. Other challenges are: unsuitable organization structure to support supply chain system, deficiency in suitable technology, different languages and procedures when participating with international companies, no trust on partners who get involved in the project, the understanding between the participants [13, 14]. Moreira [11] explained barriers to the application of CSCM towards on-timely delivery in building construction projects as difference of technology used by the actors, especially between the contractor and the supplier, and, difficulty to establish trust-based relationships among actors of the supply chains. Fawcett et al. [4] claim that building trustbased relationships is hard and potentially expensive, both because of demanding investments and to the vulnerability that these relations impose. According to [22] constraints to the application of supply chain management in building construction project delivery are, partners low level of commitment. A lack of trust among supply chain partners often results in inefficient and ineffective performance because commitment is fundamental to all the relational exchanges between the firm and its various partners in the entire chain [16].

Shukla et al. [18] report lack of understanding of SCM concepts, lack of commonly accepted definition of supply chain management and the problems associated with supply chain activities that makes the understanding of supply chain management difficult. They further suggested that participant in the chain should be presented with main activities of supply chain and the step-by-step approach for understanding a complete picture of the supply chain. Lack of the appropriate organizational structure in supporting the system is also a challenge [2]. Construction companies give less attention to supply chain management strategies and are reluctant to employ transparent integrated system to link them with other actors [2, 10]. Lack of organizational set up and resistance to the concept is perceived as weakening of authority as well as inability to share information on project [19]. Traditionally, project information exchange among the client, consultants, main contractor, sub-contractors, and suppliers have mainly been based on paper document system. Sospeter and Rwelamila [19] found that two thirds of the construction problems are a result of an inadequate exchange of information and inefficient information. Deficiency of mutuality and lack of appreciation of other's performance deficiencies of the procurement system purpose are among the challenges [15]. With poor procurement system it may result into budget overruns, schedule delays, poor quality, low customer satisfaction, and weak productivity development [12]. Lack of a common standard for collaboration and ignorance on the contributions and needs of subcontractors and suppliers are see as challenges to SCM. In numerous industries and demographics if an organization fails to integrate with their partners like suppliers and customers (client) it cannot avoid these risks and it will fail to improve their performance [17]. Most of these studies explore challenges and constraints of SCM on project performance without analysing the critical ones and by considering time, cost and quality. The methodology used in previous studies targeted modern procurement systems in developed countries. There is need to research on challenges facing SCM on-time delivery of projects in Tanzania where a traditional procurement system is dominant.

On-time delivery of a project is regarded as a dependent factor which is being affected by several challenges as summarized in Fig. 1.

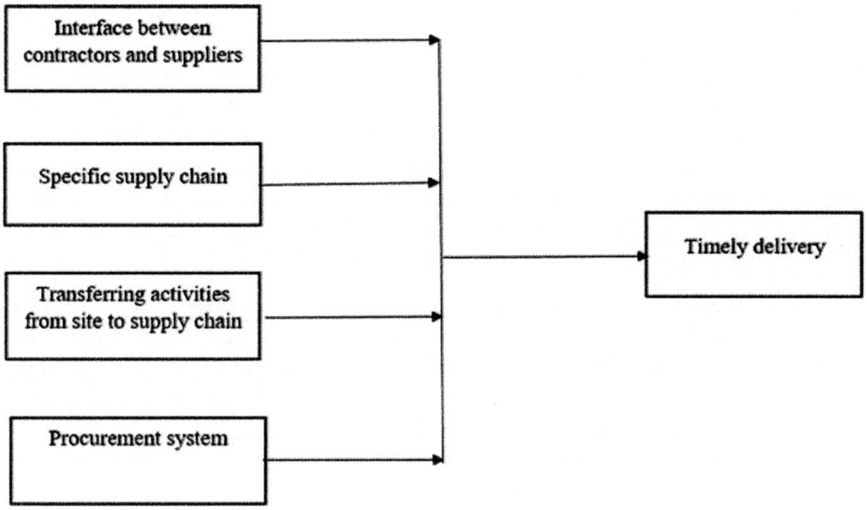

Fig. 1 Conceptual Framework (source: Authors compilation from Shahbaz & Raja 2019 and Naveenkumar & Prabhu 2016)

3 Research Methods

Research design

Descriptive research specifically descriptive survey was adopted to the study as it is type of research that is not only restricted to fact findings but may often result in the formulation of important principles of knowledge and solution to significant problems [7]. The target population for the study is 118 contractors registered in class I (106) and II (45) based on the data obtained from Contractors Registration Board of Tanzania (CRB, 2019). The selection of these contractors considers the mandatory involvement of contractors in supply chain implementation and development because they are the main coordinators of the entire supply chain process. The sample size was selected using stratified sampling technique for building contractors because contractors are categorized in strata. Criteria set to obtain the sample from building contractors were building contractors Class One and Two because in Tanzania these are grouped as large contractors. The rationale behind selecting class I and II only is that the supply chain is more applicable to large building projects due to the complexity nature of the projects undertaken which makes it necessary to have a stable supply chain and ensure that contractors will have suppliers for all required materials through their projects. Using Yamane formula (1967) the sample size for the population of contractors surveyed was 91 made up of 66 class I and 25 class II contractors.

Data Collection

Data was collected using a review of relevant literature and questionnaires. The information gathered from the literature review related to the study guided the design of the research instrument used which was structured questionnaire (Saunders et al. 2016). Questionnaires were used as method of data collection because of the large sample size that was difficult to be reached by using interview method. Furthermore, the questionnaire gives a chance to reach large number of respondents in a short time and convenient. The questionnaire was divided into two distinct sections:

- *Section 1* encompassed the general background information of contractors registered in Class One and Two who were involved in the study and assumed to have played main part of SCM practices. Questionnaires had closed and open questions on challenges for SCM and attribute variables of professional status, experience of respondents and class of contractors and gender. To enable cross comparative analysis as part of a robust data protocol, the responses were nominally coded questions so that they be entered one of various pre-prepared categories.
- *Section 2* comprised of the rating and ranking of the 11 challenges for SCM implementation that affect on time delivery of projects from which critical challenges are determined. SCM challenges for on-timely delivery from the literature formed a list that was included in the questionnaires sent to respondents.
- Questionnaires had mainly closed and open questions on challenges of SCM and attribute variables of profession, gender, and experience of respondents. Eleven challenges extracted from literature were rate for which 5-Likert scale was used. Challenges were rated on 5 = Strongly agreed; 4 = Agreed; 3 = Neutral; 2 = Disagree; and 1 = Strongly disagree. Quantitative data from the field was analyzed by descriptive analysis. Out of 91 questionnaires distributed 79 were returned for analysis. Data was analyzed using two features of IBM SPSS version 20 which are descriptive statistics-frequencies and Compare Means-One Sample T Test.

4 Analysis and Results

The demographic characteristics of 79 respondents obtained from the survey in terms of professional status; most of the respondents were quantity surveyors (39) followed by Engineers (18) and Architects (11). Others were procurement officers (6), clerk of works and construction manager (5). In terms of work experience, majority were between 6 and 10 years (43) followed by 1–5 years (28) above 10 years (6) and less than 1 years (2). In terms of gender, there were more males than females as 58% of respondents were males while 42% were females.

Challenges of CSCM on Timely Delivery of Construction Projects

Table 1 presents one-sample statistics of the challenges of CSCM on-timely delivery of construction projects. Results reveal that lack of a common standard for collaboration (MS = 4.48), deficiencies in the procurement system (MS = 4.46) and lack of

Table 1 One sample statistics and T test

Variable/challenge	N	Mean	Std. deviation	Ranking	Std. error mean	Test value = 3.7		
						t	df	Sig. (2-tailed)
Lack of a common standard for collaboration	79	4.48	0.638	1	0.072	10.885	78	0.000
Deficiencies in the procurement system	79	4.46	0.656	2	0.074	10.235	78	0.000
Lack of top management support	79	3.95	1.049	3	0.118	2.113	78	0.038
Frequent changes of the design in a construction project	79	3.35	4.472	4	0.503	– 0.687	78	0.494
Partners low level of commitment	79	2.90	0.761	5	0.086	-9.357	78	0.000
Poor higher authority commitment	79	2.68	0.708	6	0.080	– 12.762	78	0.000
Lack of good information flow	79	2.61	0.823	7	0.093	– 11.797	78	0.000
Deficiency in suitable technology	79	2.56	0.797	8	0.090	– 12.754	78	0.000
Deficiency of mutuality	79	2.51	0.714	9	0.080	– 14.863	78	0.000
Lack of willingness to invest in supply chain management	79	2.49	0.799	10	0.090	– 13.426	78	0.000
Inappropriate knowledge about SC and CSCM	79	2.27	0.780	11	0.088	– 16.349	78	0.000

top management support (3.95) are CSCM challenges in time delivery of construction projects. Looking closely, Table 1 presents One Sample T Test for the challenges of SCM on time delivery in construction. Results indicate that three challenges are significant with positive t-value. These are: lack of a common standard for collaboration (t = 10.885, sig = 0.000), deficiencies in the procurement system (t = 10.235, sig = 0.000) and lack of top management support (t = 2.113, sig = 0.038) are significant challenges on time delivery in construction projects in Tanzania.

5 Discussion

These challenges are lack of a common standard for collaboration, deficiencies of the procurement system and Lack of top management support.

Lack of a common standard for collaboration

Common standard for collaboration between customers and clients in supply chain is a requirement for delivering projects on time. In Tanzania, most of the projects are procured through a traditional form of procurement. This being an old way of managing construction supply chains using traditional procurement system, it is characterized by severe fragmentation nature of the key supply chain participants namely client, consultant, main contractor, sub-contractor, and suppliers. For organizations to be competitive, they should continuously seek to provide their products and services to customers within agreed budget, time, and quality better than their competitors. While it is not possible to do it alone, they should collaborate with their supply chains and normally through trust for successful delivery of a project. The finding supports the claim that if an organization fails to be integrated with their partners like suppliers and customers (client) it cannot avoid the risks and will fail to improve their performance [17, 22]. The absence of framework for collaboration between construction firms and their counterparts may become open for wastage of time without knowing what to focus on and what to be done. As part of the process, there should be supply chain management database to ensure the smooth flow of information within the different chains which results in increasing the supply chain collaboration within the supply chain partners [20]. Absence of a common standard for collaboration causes significant challenge in CSCM that affect on-time delivery of construction projects.

Deficiencies of the procurement system

Deficiencies of the procurement system is revealed as a significant challenge facing CSCM that affect on-timely delivery of building projects. The finding is supported by [22] that construction firms fail to broaden the supply chain vision beyond procurement or product distribution. The procurement system was also determined by Kikwasi [5] as among the causes of delays in construction projects. Procurement is a backbone in the operation of acquiring suppliers to provide materials needed for construction. Payment delay by the client is also another factor that can hinder the procurement process and hence affect project progress and completion time of the project. In the situation where there is improper procurement system, challenge in CSCM are inevitable and hence project delay. In Tanzania, the Public Procurement Act (PPA) and its regulations are occasionally amended to address procurement challenges. The recent amendment of 2016 of the PPA 2011 and its Regulations 2013 has shortened timelines for responding on procurement matters.

Lack of top management commitment

Lack of top management commitment is amongst the significant challenges facing CSCM on-time delivery of building projects. Top management commitment means to get involved in the process of SCM, spend time about learning as well as supervising the process as part of their support and provide the necessary support/resources for the successful of SCM. It serves time and beneficial to all supply chain players if well managed or committed. Top management should ensure all supply chain team have business processes which are connected to each other, be it procurement, manufacturing, storage, information technology, distribution, and transportation to make it easy to play their roles and deliver products/services timely. The nature of fragmentation in the traditional procurement system is a challenge by itself, if not well managed and aligned can cause time overrun on project delivery. The finding is consistent with Al-Werikat [1] who noted that lack of top management support was the root cause barrier to supply chain management and within organization and with other organization in a project. Absence of all these may lead to lack of top management commitment. Nakiab also conclude that top management could allocate the necessary resources for supply chain endeavors. The study has therefore indicated that lack of top management commitment has the potential to cause time related challenges for CSCM due to failure to provide guidelines or in still motivation policies that aim at accomplishing organizational or inter organizational objectives.

6 Conclusion

The study sought to determine the challenges of CSCM that affect on-time delivery of construction projects. This study concludes that lack of a common standard for collaboration, deficiencies of the procurement system and lack of top management commitment are critical challenges facing CSCM for on-timely delivery of building projects in Tanzania. One of the main contributions of this study lies in the identification of an ordered grouped set of SCM challenges for on-timely delivery of projects in Tanzania. Another significant contribution of this paper is that it sheds light and provides insights on the understanding of the challenges facing SCM that affecting on-time delivery of projects in Tanzania, an area previously under-researched. The findings are important to construction stakeholders as it gives a picture of something to be done in procurement practices to accommodate SCM. Henceforth, a need for guidelines or structured procedure to strengthen SCM in the construction industry. To have a common standard for collaboration in the supply chain management goals should be placed prior to the process while involving all key chain participants at early stages of the project and to have policy that controls and monitors successful and efficient supply chain management in construction projects. For increasing the level of commitment, participants are expected to be responsible to the chain by having a direct contribution in the vital aspect of the chain, because a strong commitment and high level of trust between supply chain partners are the bases for a successful supply chain performance.

Limitation and implication

The findings can be used by the practitioners (project teams) as a basis for providing the foundation to address the challenges during construction stage to improve projects delivery in the future. Secondly, the findings provide insights into how the uniqueness of the time control as one of project success hard criteria have a detrimental effect on cost and quality that may affect not only the process but also outputs of projects. This study has limited its scope to contractor's class I and II in Tanzania traditional method of procurement, hence the results are influenced by the nature of activities, characteristics, and procurement systems of the industry. However, developing countries of similar economic, cultural, and business set up could benefit as well as these insights provide awareness of critical challenges of CSCM that affect on-time delivery of a project as hence limit industry competitive advantage of which SCM belongs.

References

1. Al-Werikat G (2017) Supply chain management in construction: revealed. Int J Sci Ad Technol Res 6(3):106–110
22. Amade B, Akpan EOP, Ubani EC, Amaeshi UF (2016) Supply chain management and construction project delivery: constraints to its application. PM World Journal V(V), 1–19
2. Aziz RF, Hafez SM (2013) Applying lean thinking in construction and performance improvement. Alexandria Eng J 52
23. Battula VR, Namburu SK and Kone V (2020) A study on factors involved in implementation of supply chain management in construction industry. Materials Today: Proceedings, https://doi.org/10.1016/j.matpr.2020.04.900
3. Comrode D (2019) 14 factors that delay a construction project's timelines [online]. Available at https://comrode.com. Accessed 7th Feb 2020
4. Fawcett SE, Jones SL, Fawcett AM (2012) Supply chain trust: the catalyst for collaborative innovation. Bus Horiz 55(2):163–178
5. Kikwasi GJ (2012) Causes and effects of delays and disruptions in construction projects in Tanzania. Austr J Constr Econ Build Conf Ser 1(2):52–59
6. Kolbe H (2010) Aggregated construction supply chains: success factors in implementation of strategic partnerships. Supply Chain Manage Int J 15(5):404–411
7. Kothari C, Gaurav G (2014) Research methodology, methods and technique, 3 edn. New Age International (P) Ltd., New Delhi
8. Leconte P (2020) The 25 most important project management KPIs [online]. Available at https://www.clearpointstrategy.com
9. Luvara VG, Phoya S, Tesha DN, Lyimo KS (2018) Critical factors causing delay and cost overrun in public building projects in Dar es Salaam, Tanzania. Indian J Res 7(7):11–16
10. Mcdermott P, Khalfan M (2012) Achieving supply chain integration within construction industry. J Learn Analytics
24. Mello MH, Strandhagen JO, Alfnes E (2015) Analysing the factors affecting coordination in engineer-to-order supply chain, International Journal of Operations & Production Management, 35(7):1005–1031. https://doi.org/10.1108/IJOPM-

11. Moreira R (2016) The influence of supply chain on the innovation process: a systematic literature review. Supply Chain Manage Int J 21(3):1–35
12. Naveenkumar G, Prabhu V (2016) Factors influencing time and cost overruns in construction projects. Int J Innov Res Sci Eng Technol 5(4):6468–6473
13. Owolad JD, Amusan O, Olanyeni G (2014) Causes and effect of delay on project construction delivery time. Int J Educ Res 2(4):197–208
14. Papadopoulos GA, Zamer N, Gayialis SP, Tatsiopoulos IP (2016) Supply chain improvement in construction industry. Universal J Manage 4(10):528–534
15. Pekuri L, Pekuri A, Aapasalo H (2014) Analysing the problem of procurement in construction. Norway, Researchgate
16. Salaam A (2011) Supply chain commitment and business process integration: the implications of confucian dynamism. Eur J Mark 45(3):358–382
17. Shahbaz M, Raja Z (2019) Impact of supply chain collaboration on operational performance. Revista de Ciencias Sociales y Humanidades 4(17):59–73
18. Shukla RK, Garg D, Agarwal A (2011) Understanding of supply chain. Int J Eng Sci Technol 3(3):2059–2072
19. Sospeter NG, Rwelamila P (2018) Knowledge or practice of the integrated construction supply chain in Tanzanian building projects. Durban, South Africa
20. Sospeter NG, Rwelamila P, Matiku JP (2019) Challenges for improving integrated construction supply chain in tanzanian building projects; the case of design & build competent entities. Dar es salaam, Tanzania
21. Ying F and Tookey J (2015) SCM competencies in construction: issues and challenges in New Zealand. Journal of Engineering Design and Technology 13(4):522–538

Construction Organisation's Planning and Implementation: The Case Between Conceptualization and Implementation Teams

Rashen Maharaj, Innocent Musonda, and Adetayo Onososen

Abstract Project failure linked to poor performance of construction firms has highlighted the effectiveness of the tendering process in selecting capable contractors. Therefore, tender decisions are key to the success of a project from the conception to the completion of the project. The disconnect between decisions made during tendering and the execution of the project also is further challenging to the success of the project and overall productivity of the constructions sector. It is vital that the teams are aligned, as this will lead to project success and ultimately, organizational success. This research paper focuses on literature and research interviews conducted within South African construction organizations to establish the gap or cause of the disconnect between the two teams. The analysed research interviews present findings that indicate the results of the disconnect which are due to a lack of systems or procedures and siloed processes and structures within the organizations interviewed. The researcher recommends that further investigations need to be carried out to develop a solution that would standardize, systemize, integrate, and cross-skill the handover process between the estimators and project teams.

Keywords Construction organisation · Construction planning · Tendering

1 Introduction

Construction companies, both small and large, procure a substantial amount of their work through the conventional tender process [1]. This process requires companies to provide rates in a tender bill of quantities, and through this, a tender price is obtained. Once all the rates are completed, contractors generally use their estimating

R. Maharaj (✉) · A. Onososen
Department of Quantity Surveying and Construction Management, University of Johannesburg, Johannesburg, South Africa
e-mail: maharajras@aforbes.com

I. Musonda
Center of Applied Research and Innovation in the Built Environment (CARINBE), Faculty of Engineering and the Built Environment, University of Johannesburg, Johannesburg, South Africa

T. C. Haupt et al. (eds.), *Construction in 5D: Deconstruction, Digitalization, Disruption, Disaster, Development*, Lecture Notes in Civil Engineering 245,
https://doi.org/10.1007/978-3-030-97748-1_22

or cost management departments to complete these tenders [1]. When completing these tenders, the estimators obtain prices from suppliers and service providers, as well as their internal departments [2]. In finalizing these rates, estimators derive these rates using either their previous experience and/or in discussions with other senior management members of their team. This process more often forces the estimating team to make assumptions and/or allowances [3].

This tender data or assumptions are made and used in the tender estimates. Assumptions are made, at times, due to unclear scope delivery from the client, or slipshod feasibility studies due to time constraints, or unprofessionalism. These are labelled as client factors [1, 4]. The assumptions made have a bearing on the profitability, quality, safety, and schedule of a project [5]. According to Shahhosseinia et al. [4], one of the main causes of project failure is the bidding process. Various factors contribute to project failures such as contractor, client, consultant, and external factors.

To mitigate potential failure, risks, or problems at the bidding stage, suitable solutions need to be analysed and preventative measures need to be developed. Kim and Reinschmidt [1] demonstrate that tender adjudication before a submission is a vital step, and if rushed or handled improperly, pertinent facts of the tender estimate can be missed or misinterpreted in the implementation phase, resulting in a project or even business failure. The disconnect between the tender and the implementation phase occurs because of the unavailability of systems, procedures, and protocols within construction organizations. Therefore, this study investigates existing processes within construction organizations that are in place to ensure that assumptions or allowances initially made at the tender phase are implemented at the construction phase.

Firstly, to establish the extent to which construction companies ensure that the assumptions and allowances made during the compilation of the tenders are carried through to the construction phase, and the subsequent impact of these systems and procedures on the overall profitability, quality, safety, and eventually delivery of the project. Secondly, the study focuses on describing the current systems or procedures within construction companies, and the roles of the estimating and site teams. Thirdly, the research evaluates the causes for project failure and how cost, quality, schedules, and safety are managed and controlled in construction projects. The research endeavours to establish the current guidelines/structured processes that are in place in the handover process of the project from the estimating team to the implementing team.

2 Literature Review

2.1 Tender Processes Within Construction

Good tender processes are key to a successful project. With good planning and a clear roadmap of how the project should be executed, the probability of success in a project is high [2]. The tendering process is complex and can be tiresome [6]. This process requires experts with in-depth knowledge, as the tender documents are complex and the tasks for the tender process are pressurizing [7]. Due to the inconsistent and unstructured tender information, a structured process is required to be followed, which would mitigate a negative result such as incorrect decision making when completing a tender [7]. Most tenders involve the offering of goods, services, utilities, or works for a certain price. Hence, tendering is more strategic and complex than just dealing with the price [8]. "The tendering processes are coined as all the activities related to managing procurement or tender documents including producing, publishing, aggregating, assessing and awarding, performed by awarding authorities to acquire certain products or services while responding and bidding are activities of parties who are interested to win contracts" [35: 283]. Methods of evaluations that clients adopt affect the tender strategy approach, which in turn affects the estimates that contractors produce.

Cao [9], describes three different methods of tender evaluation namely, the Comprehensive Evaluation Method, Lowest bid method, and Average bid method. In comprehensive bid method, the method contributes the least amount of risk to a project and enables project success. However, this method is time-consuming and can be costly to the client. Lowest bid method is based on the lowest price. The project cost will be very low [10]. However, due to the nature of this method, Contractors are forced to be competitive. This method opens room for high risk and project/business failure. An average bid method is based on the average score of all the bidders [11], and the successful bidder is appointed on whichever comes closest to the average.

The strategic approach of a tender leads to the development of estimates. Estimates are generated from historical data and the resources used for the previous implementation. However, resources are not always available as per what the estimator based the estimate/tender on. This then leads to project failure, e.g., the estimator would typically price a front-end loader at a rate of 'x' amount—this does not mean that when implementing, the site team would find the front-end loader at the 'x' rate. The probability is that the rate is now 'x + 10%', and this is due to estimators not factoring in the time of when the project will be awarded [12]. The duration incurs escalation. Thus, the importance of a fully developed handover process from planning to implementation will eliminate the risk of project failure [12].

Figure 1 illustrates the complexity of the pre-construction phase, and if not followed it can lead to project failure. It further illustrates the importance of the handover from the tender to the implementation phase [13].

2.2 *Handover to the Implementation Phase*

The success or failure [14] of the implementation process is identified as an internal measure on the project team, mainly based on efficiency, and the criteria adopted for this is as follows, [15] keeping schedules [16], keeping to a budget, meeting the

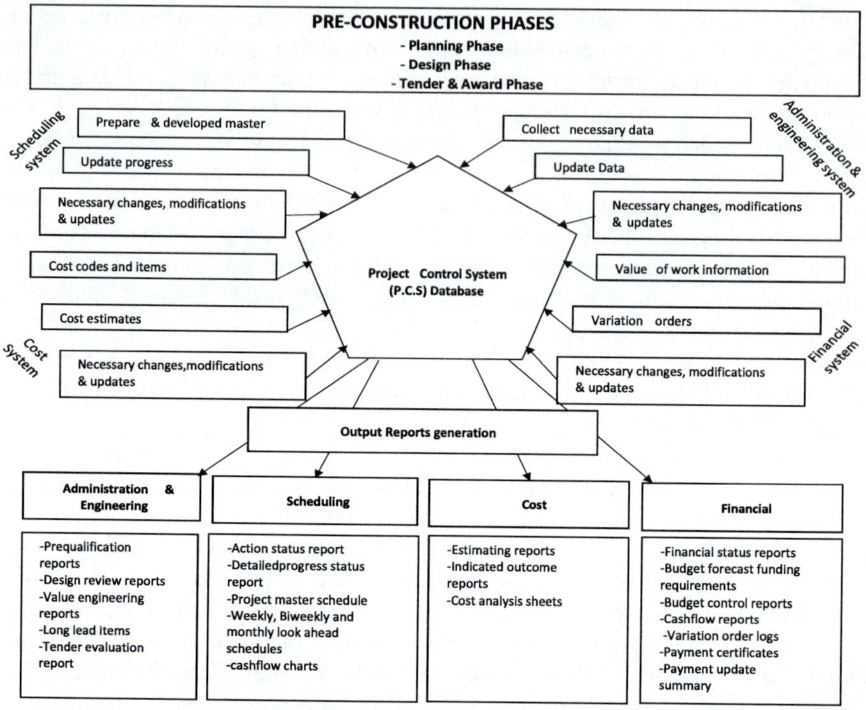

Fig. 1 Pre-construction phases [13]

technical goals of the project, and maintaining smooth working relationships with the team and parent organization.

However, if the above information is vague, it brings uncertainty to the project from the early stages, and this results from informal handover processes [17]. Al-Reshaid et al. [13: 351], clearly state that if no risk mitigation is implemented in the pre-construction phase, it will cause, project delays and budget overruns thus, leading to project failure. Where there are proper systems and processes to handover to the implementing team, the risks can be discussed and mitigated at the handover phase, thus improving the connection between the planning and implementing team [18].

Innovative construction processes eliminate project failure [19] and integrate the management of quality, safety, schedule, and costs [20]. Prierto [21], reported that project success can be improved through better handling of indirect construction costs. There are a variety of process/control mechanisms within the construction organization that can be used. However, many construction projects still overrun time and costs [22, 23]. With the development of clear construction processes, managing a project becomes seamless. Planning a project and implementing the plan is vital to project success [20]. It is critical in developing processes/strategic principles on cost management that must be applied across the construction industry that focuses on

Estimation, Cost Control, and Management of indirect construction costs. The benchmark comparison is between costs incurred to prevent defects, and costs incurred to rectify defects, where the costs to prevent defects trump the costs to rectify [24], as the rectification costs cannot be controlled or managed and thus construction organizations have made total quality management a priority [25, 26]. There are various ways to schedule a project. However, the most common tool used in the construction industry is building information management systems [27].

Building information management systems are used to integrate the information/flow of data developed at the tender stage, which is a smart gateway to work collaboratively and track the entire project with the various moving parts. This plays a vital role in integrating the tender information into the implementation phase, which ultimately results in project success because what was planned is executed and tracked in a systematic approach [27]. Globally, the construction industry output affects economic activities, and despite its importance, the construction industry fatality and incident rates are much higher than any other industry [28]. So, the management of safety is very important to the success of a project [29].

Various aspects contribute to project failure [15]. Project failures according to Ren et al. [36: 25], happens due to factors like deficient planning, poor risk management, lack of change control systems, and poor integration management [30]. The causes of project failure mainly result from lack of information, failure to maintain the information, and poorly executed handover of accurate and integral information [31]. When this insufficient and mishandled information is used to implement the project, and an implementation methodology is selected based on the information provided, the result is usually failure.

It is quite evident that the tender process is dynamic with various aspects to be considered when estimating. The processes vary from organization to organization, however, the principles applied are similar. The handover from planning to implementation is done through a meeting between the two parties, with no substantiating evidence indicating that the handover via a meeting is successful. A positive result can be achieved if the process is followed, however, the handover process needs to be improved for the implementing team to envision what the planning team envisions during the tender stage.

3 Research Methodology

3.1 Research Design

To understand the current South African construction organizational handover protocols and procedures from estimating to implementation, research was needed to provide insight. The research would provide a deeper understanding of the procedures currently adopted or used by construction organizations. Qualitative interviewing methods were used to provide an in-depth view of an individual's experience and provided the researcher with a well-rounded collection of data for this research paper [32]. One-on-one interviews were conducted as part of the qualitative method to establish "the heart of qualitative research and the desire to expose the human part of a story" [33: 68]. This approach also provided an in-depth view and further interrogation of the issue at hand [34]. The framework below was used as a guide to conduct research interviews [35].

The researcher used Rowley's framework on conducting research interviews, which incorporates why interviews are practical in information gathering and the various types of interviews. Rowley's framework also highlights the techniques required in, identifying the right interviewees, formulating interview questions especially question construction, length, order, and delivery. Furthermore, the framework demonstrates ways in which a fitting number of interviews and interview questions can be constructed to fit the desired type of study. Finally, the approach highlights how to begin and end interviews, tips to keep interviewees engaged, and how to analyze and present the obtained data.

For this study, the researcher used the standardized open-ended interview method [32]. This method is ideal for studies that are information-based, as it allows for unlimited responses, new and unexpected insights, more details, sentiment, and opinions, and offers deeper, qualitative data [36]. This method is more time-consuming to answer for interviewees and more so for researchers in terms of data analysis and processing [37]. However, it opens the study to a more comprehensive overview of themes, viewpoints, and facts on the subject of study. Standardized open-ended interviews are key to producing the required output from this research.

3.2 *Interview Questionnaire*

The questionnaire that was developed played a critical part in the interview process, as the questions aimed to allow the interviewees the ability to dig deep into their experience and knowledge [32]. The semi-structured interview approach was used consisting of 14 questions and prompting questions [32] that emphasized the main question to ensure that it was adequately answered. The questionnaire was well-phrased in a set order aimed at answering the research questions of this paper [35].

A profile was developed of an interviewee that had the relevant [32] title, experience, and qualifications. The credibility of the data was substantially based on the interviewee meeting the profile requirements [35]. An interviewee profile criterion was then structured and used as a tool to select interviewees [33]. Practice interviews were carried out, as part of the preparatory method, with senior staff members of the organizations that were going to be interviewed, with the intent of establishing which titles from the organization are best suited for the interviews [32]. The choice of senior staff was that these members would understand the company processes and had adequate knowledge of the projects from key perspectives. From this, it was established that estimators and site managers/agents are best suited to be interviewees.

A pragmatic approach was followed in this research paper, as this approach was based on the availability and willingness of interviewees. Another deciding factor was interviewee experiences and roles within the organization. It was paramount to conduct enough interviews so that the gathered data produced evidence of adequate research from various points of view [38]. A total of 12 interviews, of 15–30 min in duration, were conducted [39, 38]. The 12 interviews were split amongst 4 South African construction organizations, three interviewees per organization.

The researcher prepared the interviewees by sending out structured emails guided by Bolderston and Amanda's [33] approach/framework [40]. The interviews were like casual conversations that flowed [34]. The starting point was to go through the research questions developed to check for jargon, or for something the interviewee might not understand [41]. Consent was also requested before the interview, consent to record and use the interview information for research data [42]. Also, basic questions were asked to gain background on the interviewee [43].

The main aim of the interview was to obtain answers to the questions developed. How the researcher achieved this through employing the technique of prompting questions [35] by repeating the question, using prompting words like why, what, how, and who. The other employed prompt is the use of silence [16]. The researcher also creatively used keywords as prompts [41]. The keyword prompts were, disconnect, handover process, and implementing planned work. In some instances, the interviewee had to make on-the-spot changes to the interview protocol [43]. Through this, the interviewee was further engaged in the topic of discussion and the relevance to their field. This also brought further insight into the interviewees [41].

3.3 Data Analysis Process

After the interviews, the researcher transcribed them [34]. This process took a while. The researcher then started by highlighting key facts and take-outs and looked for commonalities with the various interviews [35]. This was then put into an excel spreadsheet, which made it easier to make sense of [39].

To better explain the above, the following 4 steps were followed to analyse the data [38]. The researcher organized the data set alongside the questions asked, then, got acquainted with the data classified, coded, and interpreted the data then presented the findings in the next chapter. The following key themes were developed through the process [38], checklist and data handover processes, siloed organizational structures, and integration processes. The processes revealed how the process is people dependent, there is no standardization, systems, and documented processes, and how it leads to inconsistency.

In this research paper, the choice of collecting research data was done by face-to-face interviews and in some cases, due to availability, was done via video call. Face-to-face interviews provided an in-depth view of the individual's experience and viewpoints. There were 12 face-to-face interviews conducted, and the interviewees were selected on a specific profile. The profile was based on experience, qualification, and role/position within the construction organization. The interview was conducted through a semi-structured interview method, and a questionnaire was used as a tool to guide the interview. The questionnaire consisted of 14 questions which were at times prompted by further questions to gain a deeper understanding. The interviews were then transcribed so that the data could be analysed, coded, and put into themes for presentation.

4 Findings and Discussion

4.1 Tender Processes in Construction

The interview session was closed out by asking the interviewees what they would say in general about the research topic, "The disconnect between planning and implementation in construction organizations" (Table 1).

The results indicate the following:

- Estimators don't get feedback from the site team.
- Lack of client information provided.
- The site team needs to be involved in the tender process.
- There is a disconnect between the estimating and site team.
- The tender and handover process needs to be improved.

Based on the above, it indicates that there is a disconnect between planning and implementation. The process from estimating to implementation needs to be improved and systemized.

4.2 Handover to the Implementation Phase

The first question asked in the interviews was: Is there a system/protocol in place to handover from estimators to the site team? (Table 2).

The results indicated that the handover process is informal and is based on a checklist that is used to ensure that the project team receives the tender information but does not ensure that the project team understands how to apply theory in practice. Further findings indicate that the estimators also required feedback from the site team after completing the project so that when proposing future tenders, they could be refined according to the feedback provided. The answers varied from organization to organization and even varied from estimator to site manager within the same organization. Based on the data and literature, there seems to be a disparity. The results indicated that there is no importance emphasized in the handover process, and according to Al-Reshaid et al. [13], the handover process is vital in the pre-construction phase.

Table 1 Tender processes in construction

Company	Estimators	Site managers
A	I would appreciate feedback from the site team regarding the tender information provided. I only receive feedback if a mistake is made	In the past, the site team would get to the site and do the job. Nowadays, we must develop a process and plan before we start and there are no proper handovers from the estimating department, this needs to be ironed out There is a disconnect, the estimating team tender to win the project and the site team is expected to deliver. If it's not communicated and understood properly, how we will be able to deliver?
B	For me, the biggest gap is the information provided by the client at the tender stage I realized that the site team involvement provides a holistic approach, and we need to get our site team more involved	I do believe that there is a disconnect, and I think early exposure to a tendering process for a site manager would be ideal. I would like to focus on-site; it currently works well the way it does
C	I think it is a very real situation. Things are done out of habit and are also influenced by the need to protect our company property as far as possible	There is a disconnect. I also think there is a bigger disconnect, not only with the contractor but with the contractor and client. Rates are not understood by our clients
D	There is a disconnect if it's not the same team doing the estimate and implementing, because we just hand over the tender pack. There's a need to improve the handover process	I'd say there is a very big loophole between the tendering department and the site base, because the information is not understood, and that is why when we go to the site, we adapt our methodologies. Time is a challenge, as we are not all able to always get involved in the tender process

Table 2 Handover to the implementation phase

Company	Estimators	Site managers
A	Tender forms are handed to the site manager through a hard and soft copy	An informal handover meeting is held, with a checklist of tender documents that are received by the site manager
B	There is no handover from estimators to the site manager. A quantity surveyor is allocated to the site to manage this	A start-up meeting is held with a high-level discussion. A quantity surveyor is dedicated to the site to ensure the tender estimates are applied
C	No formal system, contract managers are involved in the tender process and the project. They are the link between the estimators and site managers	No system, we have an informal discussion on the scope of work
D	No system, we handover electronic and hard copies of the tender documents	No system, the tender pack is handed to us and we must make sense of the documents

Table 3 Management and implementation of cost, quality, schedule, and safety

Company	Estimators	Site managers
A	At the tender stage, we have a generic quality plan. This is handed over in the tender data pack	We develop a project-specific quality plan and audits are done by the quality department to ensure compliance
B	The tender quality plan is not aligned with the project quality plan. When the project starts, a new quality plan is developed	The tender quality plan is generic, and we develop a project-specific plan
C	The quality plan is generic, and a new plan is developed when the project starts	A new quality plan is developed when the project starts and is aligned to the BOQ
D	Quality is also a generic set-up, and a new quality plan gets developed when the project starts	The quality plan is generic, and a new plan is developed when the project starts

4.3 Management and Implementation of Cost, Quality, Schedule, and Safety

The results show that a cost report is used to manage costs. The cost reports listed below are done by the various stakeholders:

- Cost reports are done by the quantity surveyor with no involvement from the site manager.
- Cost reports are done with no involvement of estimators.
- Daily cost tracking is done in 1 out of the 4 organizations.
- Costs are managed via a monthly meeting (Table 3).

Only 1 out of the 4 organizations align the quality plan to the project costs. From the data set, 3 out of the 4 organizations interviewed have reactive cost processes and are disjointed from estimating and implementation—this results in a disconnect between planning and implementation. In summary, the findings show that the quality plan used at the tender stage is not the same plan used in the implementation phase.

4.4 Schedule

The results indicate that 3 out of the 4 organizations use a generic schedule at the tender stage and develop a new schedule at the start of the project and 1 out of the 4 organizations link the schedule to the cost plan.

The verbatim responses on how the project schedule is managed are listed in Table 4.

In summary, the results indicate that the project schedule developed at the tender stage is not used in the project—a new schedule is developed when a project starts.

Table 4 Schedule

Company	Estimators	Site managers
A	A schedule is developed by a planner and as the estimator. I align my costs to the plan. I handover this information and don't get involved further	The program is managed by daily talks and reports submitted to the head office every week. A new schedule is created to be project specific
B	A new schedule is developed when the project starts and is approved by the client	The tender schedule is generic, and we develop a project-specific plan when the project starts
C	A new project schedule is developed at the start of the project	A new project schedule is developed when the project starts
D	A new schedule is done when the project starts	The tender schedule is generic, and a new plan is developed when the project starts

When comparing literature to the feedback received, it shows a clash between theory and practices. The schedule developed at the tender stage and the schedule used in the project implementation are not the same—this shows a disconnect from planning to implementation. According to the literature the cost and schedule work hand in hand and the planned cost and schedule are applied in the implementation phase to achieve project success [27].

Based on all the data above for implementation and management of project cost, quality, safety, and schedule plans, there is a clash between theory/literature and practice. Within the 4 organizations planned/ tendered quality, safety, and schedule plans are not implemented. A generic version is developed for the tender stage and a new version is then developed on project inception. Within the 4 organizations in practice, the quality, safety, schedule, and cost plans are not integrated, and this does not allow for seamless project management, resulting in a disconnect between planning and implementation, ultimately resulting in project failure.

4.5 Safety

The results indicate that the safety plan used at the tender stage is a generic version, and at the project stage, a new safety plan is developed to be site-specific.

The verbatim responses on how safety is managed are listed in Table 5.

According to literature, globally, the construction industry output affects economic activities and despite its importance, the construction industry fatality and incident rates are much higher than any other industry. So, the management of safety is very important to the success of a project [29]. In summary, the safety tender plan is generic and is not integrated with the other key components such as quality, schedule, and costs. This shows a disconnect from planning to implementation.

Table 5 Safety

Company	Estimators	Site managers
A	At the tender stage, we have a generic safety plan. This is handed over in the tender data pack	A Site safety officer develops a safety plan, and it is audited by the client and by safety officers from head office
B	This is generic, and a new safety plan is developed for the project	The tender safety plan is generic, and we develop a project-specific plan when the project starts
C	The safety plan is project-specific, and we employ a safety consultant to manage the plan	The safety plan is project-specific, and we have a safety consultant looking after the safety plan
D	Safety is a generic setup and its new safety plans get developed when the project starts	The safety plan is generic, and a new plan is developed when the project starts

4.6 *Causes of Project Failure*

In the interview, the following question was asked to further establish how the organizations manage risks: Question from interview questionnaire: Please comment on how effective risks identified at the tender stage are mitigated at project implementation (Tables 6 and 7).

The results show that 1 out of 4 construction organizations discuss project risks every week and develop mitigating actions. From the other 3 companies, it shows that it is either discussed or communicated via the tender information and from then onwards it is informally managed by the site team. The data set further indicates that there were no formal procedures and tools used to manage risks, and site managers captured the risk onto an excel spreadsheet/developed their methods for discussion. Only 1 out of the 4 organizations meet weekly, document the risks, and discuss mitigating actions. Joint risk management is essential to managing and mitigating project risks. Laryea [8], indicates that tender review meetings are a significant part

Table 6 Causes of project failure

Company	Estimators	Site managers
A	Risks are indicated in the tender data and we don't get involved further	It is discussed at the handover meeting and we develop mitigating actions
B	We identify the risk and communicate them to the site team, and it is documented in cost reports	This is managed by the commercial team; I don't get involved
C	Risks are recorded on a sheet and discussed weekly with mitigating actions	Risks are discussed weekly and plans are developed to mitigate them
D	We indicate risks on the tender pack and estimators don't get involved further	Nothing in place

Table 7 Tools to manage risk

Company	Estimators	Site managers
A	I don't get involved	No tools, we come up with methods
B	This is managed directly by the commercial team	It is handled by the commercial department; I don't get involved
C	Weekly meetings and it is recorded on an excel spreadsheet	We have weekly meetings to discuss risks
D	I don't know, I don't get involved	We don't have any We report them on the monthly cost report

of the tendering process and that review meetings are essential. Risk meetings are key to ensure that there is clarity from planning to implementation and that they are correctly implemented [5].

5 Conclusion

The tender process needs to be reviewed from a holistic view encompassing the need for estimators and site managers to be jointly involved in the key milestones of a tender and project implementation. The findings the study contributes is shown in the disconnect from planning to implementation and that the handover process from estimating to the site team needs to be improved. Therefore, there should be a strong emphasis on the handover process which must be open and focused on the further development of project parameters. Integration of proactive management should be further developed, ensuring that assumptions made at the tender stage are applied in the project phase. An approach that embraces cross/multi-skilling must be implemented in both conceptual and implementation.

The theme processes are people dependent and inconsistencies were noted from the data analysis. The current processes in construction organizations are reliant on people and inconsistent practices are applied. There is a need for a well thought and researched handover procedure to be developed. If the processes applied are from individual to individual, it causes a disconnect and inconsistencies. The organizations interviewed indicate that senior managers should manage the handover process. However, from interviewing the team who are part of the process, it is evident that this practice of dedicating a senior manager to oversee the handover or alignment from estimating to the project team is flawed and still results in a disconnect. Continuity of projects becomes challenging, and if there is a churn of staff within the organizations, this could result in project failure. The tender process needs to be reviewed to encompass a holistic approach.

Recommendations

Tender review meetings must take place involving the conceptual and the implementation teams. It is recommended that the above tender review meetings take place and be implemented or integrated into the current process, which can then be further refined to suit the organizations. Implementing these basic steps will initiate the success of a project.

From the findings, it shows that the handover process is informal to none, and very little emphasis is placed on the handover process. The following is advised in achieving success; Keeping to schedule, Keeping to budget, Meeting the technical goals of the project, and Maintaining smooth working relationships with the team and parent organization. Furthermore, it is recommended that a proper system needs to be developed for handover, with a strong emphasis on risk mitigation. Clear strategies with a systematic approach must be adopted within construction to improve handover processes.

Also, it is strongly recommended by the researcher that the process adopted in developing and managing cost, quality, safety, and schedule plans must be integrated to ensure that the process is seamless and becomes a success. It is recommended that early warning measures be developed and implemented in the tender and implementation stages. Implementing a proactive measure with an early warning system will enable early risk mitigation for construction projects, eliminating project failure.

Further Research

When carrying out this study, it was noted that various areas required further research and it is recommended that the following further research take place. The general development of software programs such as information management systems, that can be used to integrate and improve the tender process. The second line for further research could be, the relationship between lack of client information at the tender stage, with project failure. The output of this study was to investigate the current systems/procedures within an organization to ensure that assumptions initially made at the tender stage are implemented and managed in the project stage. Due to the limitation of time, the researcher recommended that the above be further researched.

References

1. Kim H-J, Reinschmidt KF (2011) Effects of contractors risk attitude on competition in construction. J Constr Eng Manage 4:275. Available at http://0-search.ebscohost.com.ujlink.uj.ac.za/login.aspx?direct=true&db=edsgao&AN=edsgcl.255242577&site=eds-live&scope=site. Accessed 24 Feb 2020
2. Baldwin A, Li H, Huang T, Kong CW, Guo HL, Chan N, Wong J (2009) Supporting pre-tender construction planning with virtual prototyping. Eng Constr Archit Manage 16(2):150–161
3. Hassim S, Muniandy R, Alias AH, Abdullah P (2018) Construction tender price estimation standardization (TPES) in Malaysia. Eng Constr Archit Manage 25(3):443–457
4. Shahhosseinia V, Afshara MR, Amirib O (2016) The root causes of construction project failure. Sci Iranica:93–108

5. Laryea S, Hughes W (2009) Commercial reviews in the tender process of contractors. Eng Constr Archit Manage 16(6):558–572
6. Laryea S (2019) Procurement strategy and outcomes of a new universities project in South Africa. Eng Constr Archit Manage 26(9):2060–2083
7. Mohemad, R, Hamdan AR, Othman ZA, Noor NMM (2011) Modelling ontology for supporting construction tender evaluation process. In: 2011 international conference on semantic technology and information retrieval, Putrajaya, pp 282–288
8. Laryea S (2013) Nature of tender review meetings. J Constr Eng Manage 139(8):927–940
9. Cao HL (2011) Conference proceedings of the Xth conference held in Y. IEEE
10. Yu W, Wang K-W (2012) Best value or lowest bid? A quantitative perspective. J Constr Eng Manage 138(1):128–134.https://doi.org/10.1061/(asce)co.1943-7862.0000414
11. Shrestha SK (2014) Average bid method—an alternative to low bid method in public sector construction procurement in Nepal. J Inst Eng 10(1):125–129. https://doi.org/10.3126/jie.v10i1.10888
12. Ren Z, Shen GQ, Xue XL (2013) Failure caused by inappropriate construction methods: an expensive lesson. J Manage Eng 29(1):25–34
13. Al-Reshaid K, Kartam N, Tewari N, Al-Bader H (2005) A project control process in pre-construction phases. Eng Constr Archit Manage 12(4):351–372
14. Firing M, Lædre O, Lohne J (2016) Main challenges found in the handover of a shopping centre in Norway. Procedia Soc Behav Sci 226(1):100–107
15. Attarzadeh I, Ow SH (2008) Project management practices: success versus failure. In: 2008 international symposium on information technology, information technology, 2008. International symposium on TSim 2008, vol 1, pp 1–8. https://doi.org/10.1109/ITSIM.2008.4631634.
16. Bunn R, Marjanovic-Halburd L (2017) Comfort signatures: how long-term studies of occupant satisfaction in office buildings reveal on-going performance. Build Serv Eng Res Technol 38(6):663–690
17. Atkinson R, Crawford L, Ward S (2006) Fundamental uncertainties in projects and the scope of project management. Int J Project Manage 24(8):687–698
18. Investni. (n.d.). Tendering guide—the tender process (PDF). https://www.investni.com/sites/default/files/documents/static/library/invest-ni/documents/tendering-guide-the-tender-process.pdf
19. Imbeah W, Guikema S (2009) Managing construction projects using the advanced programmatic risk analysis and management model. J Constr Eng Manage 135(8):772–781
20. Mizuno Y, Kubota Y (2018) Structural form of bridges reflecting the construction processes. Frattura ed integrità strutturale 13(47):209–220
21. Prierto R (2020) (PDF) Indirect field costs—ResearchGate. ResearchGate. https://www.researchgate.net/publication/343318957_Indirect_Field_Costs
22. Derakhshanalavijeh R, Teixeira JMC (2017) Cost overrun in construction projects in developing countries, gas-oil industry of Iran as a case study. J Civ Eng Manage 23(1):125–136
23. Varadharajan T, Varadharajan T, Senthamilkumar S, Senthamilkumar S (2011) Control of indirect cost in construction. Indian J Appl Res 4(4):75–78. https://doi.org/10.15373/2249555x/apr2014/256
24. Nasir N, Nawi MNM, Radzuan K (2016) Relationship between time management in construction industry and project management performance. AIP Conf Proc 1761(1):1–6
25. Abdelsalam HME, Gad MM (2009) Cost of quality in Dubai: an analytical case study of residential construction projects. Int J Project Manage 27(5):501–511
26. Hamzah N, Ramly A, Salleh H, Tawil NM, Khoiry MA, Che Ani AI (2011) The importance of design process in housing quality. Proc Eng 20(1):483–489
27. Li C, Hong JK, Xue F, Shen GQ, Xu XX, Mok M (2016) Schedule risks in prefabrication housing production in Hong Kong: Kong: a social network analysis. J Clean Prod 134:482–494
28. Mohammadi A, Tavakolan M (2019) Modeling the effects of production pressure on safety performance in construction projects using system dynamics. J Safety Res 71(1):273–284

29. Zou PXW, Sunindijo RY (2013) Skills for managing safety risk, implementing safety task, and developing positive safety climate in construction project. Autom Constr 34:92–100
30. Sage D, Dainty A, Brookes N (2014) A critical argument in favor of theoretical pluralism: project failure and the many and varied limitations of project management. Int J Project Manage 32(4):544–555
31. González P, González V, Molenaar K, Orozco F (2014) Analysis of causes of delay and time performance in construction projects. J Constr Eng Manage 140(1):4013027
32. Turner DW III (2010) Qualitative interview design: a practical guide for novice investigators. Qual Rep 15(3):754
33. Bolderston A, FCAMRT (2012) Conducting a research interview. J Med Imag Radiation Sci 43(1):66–76
34. Rosenthal M (2016) Qualitative research methods: why, when, and how to conduct interviews and focus groups in pharmacy research. Curr Pharm Teach Learn 8(4):509–516
35. Rowley J (2012) Conducting research interviews. Manage Res Rev 35(3/4):260–271
36. Adams WC (2015) Conducting semi-structured interviews. Handbook Pract Progr Eval:492–505.https://doi.org/10.1002/9781119171386.ch19
37. Opdenakker R (2006) (PDF) advantages and disadvantages of four interview techniques in qualitative research [electronic journal]. ResearchGate. https://www.researchgate.net/publication/48666088_Advantages_and_Disadvantages_of_Four_Interview_Techniques_in_Qualitative_Research_Electronic_Journal
38. Rowley J (2014) Designing and using research questionnaires. Manage Res Rev 37(3):308–330
39. Lietz P (2010) Research into questionnaire design: a summary of the literature. Int J Mark Res 52(2):249–272
40. Murray CD, Sixsmith J (1998) E-mail: a qualitative research medium for interviewing? Int J Soc Res Methodol 1(2):103–121. https://doi.org/10.1080/13645579.1998.10846867
41. Qu SQ, Dumay J (2011) The qualitative research of interviews. Qual Res Account Manage 8(3):238–264
42. Fritz RL, Vandermause R (2017) Data collection via in-depth email interviewing: lessons from the field. Qual Health Res 28(10):1640–1649. https://doi.org/10.1177/1049732316689067
43. Hawkins J (2018) The practical utility and suitability of email interviews in qualitative research. Qual Rep. https://doi.org/10.46743/2160-3715/2018.3266

Investigating the Effects of Gentrification in the Creation of Socially Sustainable Urban Precincts: A Case Study of Bo-Kaap, Cape Town, South Africa

Kathy Michell and Jake Le Roux

Abstract African cities are characterised by a variety of challenges in terms of urban degradation. In South African cities the research surrounding gentrification is limited. Therefore the significance of looking at the way gentrification is handled by the various role players of gentrification is imperative in ensuring social sustainability and more importantly social cohesion. This paper documents the findings of a social constructivist single case study that investigated the effects of gentrification on the social sustainability of the Bo-Kaap precinct in Cape Town. The collected data were analysed using thematic analysis and *NVivo 12* software package. The Bo-Kaap precinct is unique because of its status as a heritage protected overlay zone. The residents of Bo-Kaap have a strong social character in the place and a deep attachment to the community. It emerged from the data that there is lack of commonality of understanding between the role players acting within the precinct, which disables any form of effective holistic participation in the regeneration of the area. The cause of this either comes down to an ignorance that separates these role players and a lack of effective communication streams enabling parties to understand different perceptions of development. Stakeholders are influenced by both the fear residents have for displacement, loss of identity and development, and the economic mindset of developers and government. The paper concludes that in mitigating the negative impacts of gentrification on historically marginalised communities, role players need both a level of understanding and a clear direction on how to manage this transition.

Keywords Gentrification · Bo-Kaap · Social sustainability

J. Le Roux
Urban Real Estate Research Unit, Department of Construction Economics and Management, University of Cape Town, Cape Town, South Africa
e-mail: kathy.michell@uct.ac.za

K. Michell
Sustainability Oriented + Cyber Research Unit for the Built Environment, Department of Construction Economics and Management, University of Cape Town, Cape Town, South Africa

T. C. Haupt et al. (eds.), *Construction in 5D: Deconstruction, Digitalization, Disruption, Disaster, Development*, Lecture Notes in Civil Engineering 245,
https://doi.org/10.1007/978-3-030-97748-1_23

1 Introduction

This paper explores the relationship between gentrification and the creation of socially sustainable urban precincts. Sustainability is defined as: "development that meets the needs of the present without compromising the ability of future generations to meet their own needs" [14: 1]. This definition highlights the value of generational stakeholders in the achieving sustainability goals. Pieterse [32] refers to several interlinking fields that comprise sustainability, these include: economic, social, ecological, physical, and political. ODPM [30] further refine social sustainability to a definition that is at a neighbourhood scale, which includes a discussion on meeting diverse needs of all generations, built to be safe, allowing for equality and opportunity for all. Dempsey et al. [13] argue that social sustainability involves the extent of the social interaction that influences the level of social capital achieved within a community. Another dimension to social sustainability is the value of inclusivity. Landman [23] argues that the value of inclusivity lies in a community's potential to be diverse, and diversity is a sought-after aspect of a community in achieving the desired results of social sustainability.

There is an important connection between sustainability and urban regeneration in their core elements. Urban regeneration is widely recognised as being "a comprehensive and integrated vision and plan to solve the multi-faceted problems of urban areas and to improve the economic, physical, social and environmental conditions of deprived areas" [29: 369]. Apart from its connection with sustainability, urban regeneration is widely discussed as having similarities to gentrification. The only commonality between the two concepts highlighted within the literature is that of displacement of the original residents within the urban precinct [31]. Hence, in theory, it possible to have urban regeneration without gentrification.

Gentrification has been depicted as having four different waves which have encouraged its establishment. The changes the waves have brought about include changes in size, location, its drivers, and the protest it receives [19]. Although the definition has continued to grow, certain defining characteristics have remained. These elements include: the influx of higher income participants to the urban space, the appreciation of prices and the displacement of the original residents [4, 22, 31]. Atkinson [1] argues that the role players within a precinct dictate the perceptions with regards to gentrification. Moreover, this includes residents of the area, who fear displacement and a loss of their social cohesion; and the private sector that see financial benefit from the reconstruction of the physical landscape of a precinct [1, 39]. The controversy surrounding gentrification can be attributed to its effects, which have differing impacts on the various role players. One needs to understand the effects to understand who they impact with the primary effect being seen as that of the concept of displacement, which can be both direct and indirect [11]. There are other effects both positive and negative that impact varying role players, but what is needed is an understanding of who drives gentrification in an area, the players that are involved in and make up a specific urban precinct. Since it is the role players who form and are immersed in the social sustainability of a place, the discussion of the relationship between gentrification and social sustainability needs to encompass the effects, relationships, motives, and intentions of these role players.

2 Literature Review

2.1 Sustainability

The Brundtland [7] definition of sustainability refers to development that relies on a strategy that meets current and future generational needs. This strategy includes the dynamics of the synergy between the relevant role players, whilst acknowledging different generations within this process (Hansman et al., 2012). Synergy is also relevant among the respective pillars of sustainability, which Brundtland [7] states as being social, economic, and environmental. Hansman et al. [20] argues that achieving common ground amongst these role players is critical in achieving social sustainability. This encourages researchers to refine their search, to look more specifically to neighbourhoods and their requirements when discussing the concept of urban sustainability. Similar to sustainability, the term neighbourhood has certain complexities in how it is understood. Rohe [33: 210] argues that although there are a variety of definitions, they are all founded on the common understanding that a neighbourhood is an urban area "whose physical and social characteristics distinguishes it from another". Yigitcanlar et al. [40] outlines the value in understanding neighbourhoods as sections of a city and underpins the reliance of a city's sustainability on the neighbourhood's it is comprised of.

Social sustainability is the pillar of sustainability that this research focused on. Glasson and Wood [16] highlight the need for social inclusion and community integration. Bahadure and Kotharkar [3: 76] touch on similar points and define their understanding of social sustainability to be about "the type of development that promotes social interaction and inclusion". Another aspect of social sustainability highlighted in the literature is community resilience, which entails engaging and empowering a community's current resources [28]. Hence, intrinsic to the concept of social sustainability in a neighbourhood is the achievement of a more diverse social group, which is of particular significance within the context of South Africa [21, 23]. Diversity in this context is seen as communities with different, "income levels, races, ethnicities, ages and family types" [23: 53]. Diversity is often dependent on the characteristics of a community and thus needs to be acknowledged in the context of South Africa's history. If this is avoided it can sometimes disrupt specialised neighbourhoods and prevent the benefit they receive from existing as a specialised neighbourhood in the first instance [10].

Segregation is a frequently discussed topic in the context of South Africa. In South Africa there remains strong racial-based spatial segregation, despite the South African governments attempt to address this [18, 21]. Groebel [18] argues that there is a friction of priorities between government budgetary duty and the ability to address the previously marginalised individual's needs. Lohnert et al. [25] further explains the issue, stating an inability to correctly assess the resident's sensitivities and to address them with policies aimed at redressing the imbalances of the past. The current issues South Africa faces is rooted in a "lack of legislative and institutional support from other spheres of government" [21: 675].

2.2 Gentrification

Porter and Shaw [31] explain the concept of gentrification as the movement of middle-income higher-class people into an urban area, which results in an increase in the property values of the area, consequently pushing out lower income bracket original residents who cannot afford these prices. While the concept has evolved since its establishment, the core ideas of gentrification have remained constant. Understanding why gentrification comes about is equally as important as describing it. The drivers of gentrification are spilt into production/supply side and consumption/demand side factors. Smith [35] explains that the production side can be seen as the investment and disinvestment of capital into the urban environment by the private sector. Brown-Saracino [6] explains that the production side is the culmination of both political and economic factors. In contrast the demand side factors are the consumption by middle class or gentrifiers demand for specific residential space, this is critical as the catalyst for gentrification to begin [26: 2003].

The effects will first be looked at from the residents and gentrifiers side. The positives include: "improved housing quality and social service levels" [5: 185]. They also look at "decreased crime, increased social mix, increased property values" [1: 112]. The negative effects include what causes friction or inner network conflicts among role players [34, 37]. Furthermore, they can cause "loss of affordable housing, homelessness, loss of social diversity, increase crime" [1: 112]. The most important effect for the purpose of this research is displacement, depicted as "beyond the household reasonable ability to control or prevent; occur despite the household having met all previously imposed conditions of occupancy; and make continued occupancy by that household impossible, hazardous, or unaffordable" [17: 8].

The effects of gentrification on the private sector include the way in which gentrification supposedly improves the physical makeup of the area, making it more presentable overall. This change then allows for further investment and development in the area, maximising the economic benefit [39]. These effects extend to the improvements to the quality of services, infrastructure as well as retail spaces in a precinct [1]. Vigdor et al. [38] explains how it creates a more culturally rich precinct, with greater accessibility and improved job opportunities. The benefit of fostering this public–private relationship should be considered as this can encourage efficient change in an area that was once derelict, as it can now be rehabilitated [2]. Lemanski [24] explains further that there is a consistent trade-off between a city remaining globally competitive, whilst simultaneously addressing the needs of the poor and vulnerable.

The purpose of this paper is to report on a case study of a Cape Town precinct that is in the process of undergoing a process of gentrification and to develop an understanding of the subsequent manner in which the different, and sometimes apposing, role players are being affected by the process of gentrification. To this end, it was intended to investigate the extent to which gentrification influences the social sustainability of the urban precinct.

3 Methodology

A qualitative research methodology in the form of a single case study was employed in this research study. The case study chosen was the Bo-Kaap precinct, which lies adjacent to the central business district of the City of Cape Town. This case is rich in historical and cultural association and is a declared Heritage Overlay Protection Zone. Semi-structured interviews with key stakeholders within the precinct were undertaken, namely: developers, the resident association, municipality officials and an urbanist. The data was analysed using thematic analysis and the *NVivo 12* software package.

4 Findings and Discussion

4.1 History of Bo-Kaap

This land was first developed in the 1780s when houses were built and leased to both skilled "free" Muslims and white immigrants [12]. The second movement of people into the area was during the emancipation of slaves in 1834, where Muslim people began to gather around this area because of the proximity of mosques [9]. By the late 19th Century Bo-Kaap was home to musicians, tailors, labourers and skilled artisans, who Davis [12] states "effectively built Cape Town". One of the next transformative periods was during apartheid and the implementation of the Group Areas Act, which segregated South African cities by race. This resulted in Bo-Kaap being "designated" strictly a Muslim area and forced removals of residents did occur during this era [12]. Arguably the post-apartheid Bo-Kaap has been in a constant battle with "gentrification" since the mid-1990s. This prolonged battle on the part of residents to retain the Bo-Kaap's historical landscape culminated in the precinct being gazetted as a heritage protection overlay zone in 2019 (Langer 2019).

4.2 Emergent Theme 1: Heritage Protection Overlay Zone (HPOZ)

The HPOZ, in essence, stipulates additional development rules over and above the existing land use zoning requirements [8]. Furthermore, the intention of the HPOZ is to protect the heritage value of the properties located within the precinct and to "protect the Bo-Kaap's long-term sustainability as a cultural asset for future generations". In undertaking the interviews, there were some opposing views with regards to the conception of HPOZ in the Bo-Kaap precinct, it was believed by INT-02 that it came about because of the "*political moves and political factions within the City of Cape Town*". According to INT-02 this then initiated the move to declare Bo-Kaap as an HPOZ very quickly. However, this is not the same view that INT-03 adopts. INT-03 believed that the conception of HPOZ was because of an increase in the intensity of the protests that took place in the area, initiated by the residents. While the outcome of both views was the same, one needs to acknowledge the fact that these are two different streams of thought and the resultant need to consider these opposing views of the HPOZ becomes important in understanding the relationship between gentrification and the social sustainability of the precinct.

The purpose of the HPOZ was something pursued during the interview stages. Different role players were involved in the conception therefore it can be inferred that there were different agendas. INT-03 highlighted the fact that the building fabric of Bo-Kaap would be maintained under the HPOZ. INT-03 also explained how although the majority of the residents were positive about the HPOZ, there remained scepticism of what social benefits it would bring. Understanding the effect of the HPOZ on the community becomes highly relevant in this context. The effects are positive foremost but there are associated risks. INT-01 discussed in detail how the HPOZ can impact the viability of land negatively, through a delayed building period and the chances of legal action as a result. This increases the risk involved with developing the properties in the precinct. Furthermore, according to INT-04 it can also restrict residents' abilities to renovate or build onto their own property.

4.3 Emergent Theme 2: Gentrification

It was clear from the data the perceptions of gentrification differed depending on the role player interviewed. It appeared that this was a result of the manner in which the interviewee had experienced gentrification as well as their position within the Bo-Kaap community. INT-01 felt that if gentrifiers, as demand drivers, have the means to enter into an area they will do so as this is a "*function of their tastes and preferences*". INT-02 felt that the increase in demand and lack of supply will increase the property values which may not necessarily be viewed in a positive light by the community due to affordability issues. INT-01 argued that the increase in property values has resulted in the cost of rates and taxes increasing. The result is an "*improved*" area but also the fact that the marginalised people tend to be displaced in an area exposed to these factors. When discussing displacement, it was clear that INT-02 and INT-01 differed in their perception of displacement. An important point that INT-02 highlighted when defining gentrification was a clear lack in "*strategic intention*".

In a statement made by INT-01 it was evident that there is generally an expectation for "*opposing*" role players to have a holistic understanding of other role players. However, without a long standing discourse between the government, the developer(s) and the residents this holistic expectation for any of the role players is not possible. Consequently, tension exists between the various role players, which remains unresolved. There needs to be a holistic understanding of how role players are impacted on, and further how they react to these impacts. INT-02 importantly states above, that gentrification takes away the ability of a neighbourhood to be able to function how they want, this is because of the influences outside of its control, which are related to the challenges of rapid urban development. According to INT-04, rapid development is however non-existent in Bo-Kaap. This understanding of positions within the realm of gentrification is complex. It is evident that what needs to be understood is the fact that depending on who you are: resident, gentrifier, government or developer you will have a different relationship with gentrification, and this is as a result of how you are affected by it. INT-02 explains that there has been minimal development in Bo-Kaap because of the "*pushback*" it receives from the residents. The residents have taken part in protest action against developers, thereby creating a lack of supply. This is initiated by the group called the Bo-Kaap Rate Payers and Civic Association which all the interviewees agreed are against development in the Bo-Kaap area. The idea of the lack of supply was mentioned by both INT-01 and INT-04. Bo-Kaap has cut off the supply side driver but has not cut off the demand side driver of gentrification, as INT-01 states: "*if supply side is restricted and you haven't dealt with the effects of the demand side, gentrification will continue to occur*". Even without the benefit of developers in Bo-Kaap it has somewhat maintained itself. INT-02 highlighted the occurrence of an organic form of development that has taken place because of its residents. INT-04 explains how the residents have improved their own properties and immediate areas, thereby allowing them to remain somewhat resilient to gentrification as a community.

Bo-Kaap is resistant to development but this doesn't mean it hasn't experienced displacement. South Africa has a history of racially driven forced removals. There appears to be remnants of this remaining within the minds of residents in how they view gentrification. INT-03 explained how the historical injustice flows into the fears experienced by the Bo-Kaap residents, their perceived lack of influence or even participation in this matter only serves to emphasise a feeling of insignificance amongst the residents. Moreover, there is a belief that government has done nothing to protect these people (INT-03). INT-04 highlighted the irony of the situation in that residents protest against development thereby restricting supply, which increases the value of the property, resulting in the rates and taxes increases, and as a result stimulates the historical cycle of displacement.

Since Bo-Kaap is located so close to the central business district of the City of Cape Town, it is unusual for the building fabric to be predominantly freestanding houses, with no high-rise buildings. It is unique in this respect and it also means that forced removal associated with rental apartments generally won't happen. INT-02 explained how it is for this reason that displacement is often a choice by the residents. INT-02 further pointed out that it is an unequal economic circumstance which stimulates the removals and not any form of forced removal because of a landlord. INT-02 and INT-03 also indicated that the houses that are often targeted by "outside developers" to be bought are poorer older people who don't understand the value of their property. This explains the power imbalance that exists in Bo-Kaap, in that developers have a certain amount of financial information which residents do not, the detriment of this is based in how this information is used. However, there is no rapid displacement occurring in Bo-Kaap (INT-04) due to the efforts of the Bo-Kaap Rate Payers and Civic Association who play a significant role in maintaining their understanding of what Bo-Kaap should be. The narrative that emerged in the data here is supported in a statement from Davis [12] that "the Muslim residents of the Bo-Kaap do not mind others moving in. What they do object to is that much of the urban renewal has been conducted without any meaningful consultation with them, with the result that much of what has been done has not been acceptable to the residents."

4.4 Emergent Theme 3: Social Sustainability

Bahadure and Kotharkar [3] discuss several factors that should be consistently reinforced in attempting to achieve social sustainability. These include social networks, inclusivity and encouraging cultures. The struggle with Bo-Kaap in attempting to align with these factors, is that it has a unique community which has been generationally occupied by Muslim people. The focus for Bo-Kaap is therefore on its diversity and relationships, both of which can define its resilience if handled with intent. Diversity has proven to be a complexity for Bo-Kaap. However, diversity has always been a goal for cities as Fainstein [15] points out. Bo-Kaap is 95% Muslim, with a small margin of gentrifiers entering the area. INT-02 emphasised that the slow migration of people into an area is how gentrification should take its course. Furthermore, INT-02 felt that one needs to come into an area open minded and respectfully, that this is how diversity builds a resilient community as it builds both tolerance and acceptance. INT-01 noted that diversity is a part of our country (South Africa) and that neighbourhoods should follow in its footsteps. This is not a simple endeavour since South Africa has distinctly spatially unique cities and within those cities are unique suburbs. Diversity within the context of a unique precinct like Bo-Kaap raises a lot of concerns. INT-02 explained that there is a lot of significance in understanding the value that the Muslim community identifies with Bo-Kaap, as it is home to several Mosques and cemeteries. INT-03 stated that residents have a strong relationship with Bo-Kaap because of their ancestral heritage. INT-03 further depicted the identities of Bo-Kaap to be so linked to the place that diversity has less value. INT-04 agreed that some of Bo-Kaap should remain the same but that there is also room for specific change. INT-04 disagreed with INT-03, who felt that although there is a need to retain Bo-Kaap for its heritage and history the idea of change is not something that should immediately be met with resistance.

Another dimension of social sustainability covered in the research is the relationships that exist within the framework of Bo-Kaap. Through the interviews a few barriers to communication with different role players were identified thereby stifling Bo-Kaap's ability to reach social sustainability. More significantly, that these barriers tend to be induced by the effects of gentrification. The first aspect of relationships looked at was identity. INT-03 emphasised that Bo-Kaap residents have identified with the physical space. INT-01 talked about how when the residents feel their culture is threatened, when their traditions and religious beliefs are threatened there is a tendency for any communications to be blocked. The fact that residents shut off when they feel threatened simply means that the approach taken needs to be done with a greater understanding of the positionality and consequent sensitivity of the residents. When done incorrectly this stifles any form of communication that could have existed.

The second aspect investigated included the government's role in Bo-Kaap. INT-04 stated there are high levels of mistrust between the residents and the government. In Bo-Kaap there are two political parties involved in the area: the Democratic Alliance (DA) and the African National Congress (ANC), who both have differing levels of involvement and associated trust. INT-03 believed that there is a lot less trust for the DA even though they have jurisdiction over Bo-Kaap, there has however been a stronger connection with the ANC. This tension leads to resistance from the residents to any form of change driven by either party.

The third dimension noted was the factions within the community. INT-02 stated that there are two main factions that exist within Bo-Kaap these are: "*Bo-Kaap Youth Development Forum, they are pro development, and they want to be involved in the development sector in the area. Then there is BOKCRA (Bo-Kaap Rate Payers and Civic Association) which are not anti-development but anti community participation in how things happen*". The main issue is the relational separations within Bo-Kaap are growing stronger and there isn't a movement forward.

The fourth dimension is about the commonality of understanding amongst the role players. Finding commonality in understanding amongst role players is clearly not a simple task. It is a slow progressive process, but it needs to be initiated by certain role players thereby allowing other role players to participate. INT-02 encouraged a consistent theme throughout the research, the idea of embracing another role players perception to understand both their positionality and sensitivity to development and/or gentrification.

The themes discussed above allowed an understanding of how different role players within the precinct of Bo-Kaap are affected by gentrification in different ways. This created the positionality that allows the role players to make more informed decisions about development in the area, which consequently impacts Bo-Kaap's social sustainability.

5 Conclusion

The research findings identified that the socially sustainable pursuit of urban precincts, when impacted by the effects of gentrification, will be influenced. The main dimensions of social sustainability in Bo-Kaap, which proved to be influenced by the effects of gentrification were diversity and social relations. This finding supports the literature in that socially sustainable urban precincts are influenced by the effects of gentrification. The case study undertaken portrayed the specific effects of gentrification that had a restrictive influence on social sustainability. The case study interviewees expressed the view that the pursuit of a more socially sustainable urban precinct is influenced by both the fear residents have for displacement and loss of identity; and the financially driven mindset of the developers and the government. This has resulted in a lack of commonality in understanding of what is expected and what is wanted from the different perspectives of opposing role players. It is clear from the findings that in mitigating the negative impacts of gentrification on historically marginalised communities, role players need both a level of understanding and a clear direction on how to manage this transition.

The intention of this research was to investigate the effects of gentrification in the creation of socially sustainable urban precincts. Through an in-depth analysis of the pertinent literature and conducting the single case study, an improved understanding of both gentrification and social sustainability was established. The research into the effects of gentrification, with relevance to the role players in Bo-Kaap was analysed. Bo-Kaap however is a unique precinct which makes the effects identified both independent and contextual and therefore difficult to apply outside of this singular case study. The research in this field is continually changing as urban landscapes transform and there is a need for continued research within the realm of both gentrification and social sustainability. This continued research will give the relevant role players direction on how to manage the effects of gentrification in the pursuit of socially sustainable urban precincts.

References

1. Atkinson R (2002) Does gentrification help or harm urban neighbourhoods? An assessment of the evidence-base in the context of new urban agenda. ESRC Centre for Neighbourhood Research Glasgow (2002)
2. Atkinson R (2004) The evidence on the impact of gentrification: new lessons for the urban renaissance? Euro J Housing Policy 4(1):107–131
3. Bahadure S, Kotharkar R (2012) Social sustainability and mixed land use, case study of neighbourhoods in Nagpur, India. Bonfring Int J Ind Eng Manage Sci 2(4):76–83
4. Bourne L (1993) The demise of gentrification? Commentary Prospect View Urban Geogr 14(1):95–107
5. Bourne LS (1993) The myth and reality of gentrification: a commentary on emerging urban forms. Urban Stud 30(1):183–189
6. Brown-Saracino J (ed) (2013) The gentrification debates: a reader. Routledge, Abingdon

7. Brundtland GH (1987) Report of the world commission on environment and development. Our Common Future General United Nations Assembly Resolution 42:1–36
8. Cape Argus (2019) Heritage protection for Bo-Kaap officially gazetted. Cape Argus Newspaper. https://www.iol.co.za/capeargus/news/heritage-protection-for-bo-kaap-officially-gazetted-21779117. Last accessed 2019/09/10
9. Capetownhistory.com (2019) Bo-Kaap|Cape Town history. http://capetownhistory.com/?page_id=236. Last accessed 2019/09/10
10. Cheshire PC (2006) Resurgent cities, urban myths and policy hubris: what we need to know. Urban Stud 43(8):1231–1246
11. Davidson M, Lees L (2005) New build 'gentrification' and London's riverside renaissance. Environ Plan 37(7):1165–1190
12. Davis R (2019) Bo-Kaap's complicated history and its many myths. https://ewn.co.za/2019/09/20/bo-kaap-s-complicated-history-and-its-many-myths. Last accessed 2019/10/08
13. Dempsey N, Bramley G, Power S, Brown C (2009) The social dimension of sustainable development: defining urban social sustainability. Sustain Dev 19(5):289–300
14. Emas R (2015) The concept of sustainable development: definition and defining principles. Brief for GSDR, pp 1–3 (2015)
15. Fainstein SS (2005) Cities and diversity: should we want it? Can we plan for it? Urban Affairs Rev 41(1):3–19
16. Glasson J, Wood G (2009) Urban regeneration and impact assessment for social sustainability. Impact Assess Project Appr 27(4):283–290
17. Grier G, Grier E (1978) Urban displacement: a reconnaissance. Office of the Secretary, Department of Housing and Urban Development, United States (1978)
18. Groebel A (2007) Sustainable urban development? Low-cost housing challenges in South Africa. Habitat Int 31(3–4):291–302
19. Hackworth J, Smith N (2001) The changing state of gentrification. Tijdschr Econ Soc Geogr 92(4):464–477
20. Hansmann R, Mieg HA, Frischknecht P (2012) Principal sustainability components empirical analysis of synergies between the three pillars of sustainability. Int J Sust Dev World 19(5):451–459
21. Klug N, Rubin M, Todes A (2013) Inclusionary housing policy: a tool for re-shaping South Africa's spatial legacy? J Housing Built Environ 28(4):667–678
22. Kotze NJ, Van der Merwe IJ (2000) Neighbourhood renewal in Cape Town's inner city: is it gentrification? J Consumer Sci 28(1)
23. Landman K (2012) Urban space diversity in South Africa: medium density mixed developments. Open House Int
24. Lemanski C (2007) Global cities in the south: deepening social and spatial polarisation in Cape Town. Cities 24(6):448–461
25. Lohnert B, Oldfield S, Parnell S (1998) Post-apartheid social polarisations: the creation of sub-urban identities in Cape Town. S Afr Geogr J 80(2):86–92
26. Ley D (1996) The new middle class and the remaking of the central city. Oxford University Press, New York
27. Ley D (2003) Artists, aestheticisation and the field of gentrification. Urban Stud 40(12):2527–2544
28. Magis K (2010) Community resilience: an indicator of social sustainability. Soc Nat Resour 23(5):401–416
29. Marra G, Barosio M, Eynard E, Marietta C, Tabasso M, Melis G (2016) From urban renewal to urban regeneration: classification criteria for urban interventions. Turin 1995–2015: evolution of planning tools and approaches. J Urban Regener Renewal 9(4):367–380
30. Office of the Deputy Prime Minister (ODPM) (2006) UK Presidency: EU ministerial informal on sustainable communities policy papers. ODPM, London
31. Porter L, Shaw K (eds) (2013) Whose urban renaissance? An international comparison of urban regeneration strategies. Routledge, Abingdon

32. Pieterse EA (2010) Counter-currents: experiments in sustainability in the Cape Town region. Jacana Media, Cape Town
33. Rohe WM (2009) From local to global: one hundred years of neighborhood planning. J Am Plann Assoc 75:209–230
34. Shaw WS (2000) Ways of whiteness: Harlemising Sydney's aboriginal Redfern. Aust Geogr Stud 38(3):291–305
35. Smith N (1979) Towards a theory of gentrification: a back to the city movement by capital not people. J Am Plann Associ:538–548
36. Smith N (2002) New globalism, new urbanism: gentrification as global urban strategy. Antipode 3:427–450
37. Spain D (1993) Been-Heres versus Come-Heres negotiating conflicting community identities. J Am Plann Assoc 59(2):156–171
38. Vigdor JL, Massey DS, Rivlin AM (2002) Does gentrification harm the poor? Brookings-Wharton papers on urban affairs, pp 133–182
39. Walsh S (2013) 'We won't move': the suburbs take back the center in urban Johannesburg. City 17(3):400–408
40. Yigitcanlar T, Kamruzzaman M, Teriman S (2015) Neighborhood sustainability assessment: evaluating residential development sustainability in a developing country context. Sustainability 7(3):2570–2602

Factors Influencing Successful Completion of Property Development Projects in Gqeberha, South African

Tandokazi Mkhwe, Brink Botha, and Sebata Mafereka

Abstract Property development plays a major role in the Gross Domestic Product (GDP) of global economies worldwide and, changes in the supply and demand values of property impacts on the performance of the property market. Therefore, property development and real estate can be considers as one of the largest suppliers of employment. However, most property development entrepreneurs lack the project management skills that are necessary for managing projects successfully. The aim of this study, therefore, was to investigated the factors influencing the successful completion of property development projects in Gqeberha, South Africa. This study intends to assist project managers/stakeholders with measures to identify and mitigate the effects of factors influencing the successful completion of property development projects. Moreover, this study also wishes to contribute the South African Property Development Body of Knowledge. In this study, the related literature was reviewed and the empirical study was conducted using a quantitative statistical approach. Research questionnaires were distributed to 100 individuals who were randomly selected due to their availability and willingness to participate in the study. A descriptive survey was conducted among professionals in the property industry that consisted of project managers and property development practitioners as well as clients and contractors respectively. The sample population was limited to the Built Environment Professionals registered with the South African Council for Project and Construction Management Professions (SACPCMP) and the Association of South African Quantity Surveyors (ASAQS) as well as Stakeholders in the property development industry in Gqeberha. The findings suggested that the variables researched in this study (namely technological, environmental regulation as well as project management factors) have a significant influence on how property development projects are ran by property managers. Therefore, this may cause delays and/or disruptions that have an influence on how successful the property development project gets completed.

T. Mkhwe (✉) · B. Botha · S. Mafereka
Faculty of Engineering, The Built Environment and Information Technology, Nelson Mandela University, Port Elizabeth 6031, South Africa
e-mail: Tandokazi.Mafereka@mandela.ac.za

T. C. Haupt et al. (eds.), *Construction in 5D: Deconstruction, Digitalization, Disruption, Disaster, Development*, Lecture Notes in Civil Engineering 245,
https://doi.org/10.1007/978-3-030-97748-1_24

Keywords Project management · Property development · Project property developer · Body of knowledge

1 Introduction

Project management can be described as being concerned with the achievement of complex goals, by integrating multifunctional inputs into a team relationship, under the guidance of a singular responsibility, authority and leadership [1]. Therefore, this is why project management (PM) has been recognised as a strategic delivery capability that can assist organisations to achieve their strategic objectives [2]. A project manager needs to apply general management principles in the project environment as the roles of project managers are diverse, and may include planning activities, tasks, schedules and budgets; organising, selecting and placing the project team; leading the project team; controlling activities and schedules; having the ability to select, develop and motivation a team; communicating with the project team and all stakeholders; problem solving, decision making or knowledge about where to find help; negotiating and persuading; and understanding the environment [3]. Therefore, it is a duty of project managers to known which factors have a negative influence on how property development project is successfully completed. The objective of this study was to unearth the issues that affect success of a property development projects in Gqeberha and finding out how these issues can be mitigated.

Furthermore, the paper presents the review of the literature, followed by the research material and methods that were used to undertake the study/collect data; presentation and discussion of the results, and how they link to the literature. Finally, the study presents a conclusion and areas for further research.

2 Literature Review

The project management body of knowledge (PMBoK)

The Project Management Institute developed a Project Management Body of Knowledge Guide ([4] to identify general project management knowledge, processes, techniques, tools, and skills that can asset project managers in a process of managing a construction project. If every project delivers a unique product, service or result, as outlined in the PMBoK [5], according to Tshiki [6], it is the responsibility of a project manager to define the desired product, service or end result that the project seeks to produce, and to lead the team to the path that will ensure the attainment of that end result, product or service. In essence when applying project management techniques, the project managers and project teams can boost their chances to reach project success. That is why PMBoK Guide is one of the most used project management tool in the property development industry [7]. While it is not clear whether PMBoK has successfully established a universal classification system, it is evident that the key goal of the document is to establish the foundation of universal standards. Hodgson and Cicmil [7] states that in the 1996 PMBoK, "Poject Management Institute (PMI) was founded in 1969 on the premise that there were many management practices that were common to projects in application areas as diverse as construction and pharmaceuticals". By the time of the Montreal Seminar/Symposium in 1976, the idea that such common practices might be documented as 'standards' began to be widely discussed [8]. The PMI Project Management Standards Program was later established to advance this standardisation, with a clear vision and a mission reflecting the perceived importance of standardisation [7]. The American National Standards Institute (ANSI) approved PMBoK as a standard on 21 September 1999 as it has vital implication on how projects are implemented and judged throughout work organisations [7].

Why are projects delayed?

Fuelled partly by the affluence and the availability of attractive government housing subsidies and loan programs, the magnitude of property development projects has overshadowed the development of other infrastructural facilities in the past years [9]. However, the growing boom in the property development activities has, in-turn, attracted the attention of a rapidly-growing number of small both qualified and non-qualified property development entrepreneurs [10]. South Africa has a serious problem of inadequate residential property resulting from many years of neglect, undeveloped property finance systems, a limited supply of long term funds, low household income levels, high unemployment, high inflation rate, high interest rate on mortgages, high cost of land and building materials, poor planning and poor implementation of housing policies and programmes, the existence of administrative bottlenecks that make the processing and securing of approvals for building plans, certificate of occupancy as well as other necessary government permits that a very difficult to acquire [11]. Additionally, the lack and/or insufficiency in government rules, regulations and specifications on the qualifications of property entrepreneurs such as financial, technical and experience, has also paved the way and encouraged growth in the establishment of small property development firms, most of which are unqualified and unfit for the job [12]. Therefore, such attributes can negatively influence completion duration of a project, as well as the quality of the service being provided.

Project time management

Project time management is about the processes and the activities that allow a project manager to complete the project on time [13]. Therefore, project time management is the process of comparing actual schedule performance to the baseline schedule. It is very crucial for a project manager to know the characteristics and important features of project time management [13]. Based on the definition of project time management, it is quite evident that Project time management is the process of comparing actual schedule performance to the baseline schedule [14]. Although the project duration and the milestones are decided in the initiation phase, the time management process happens mainly in the planning phase. Nevertheless, it is still the project manager's responsibility to plan the project activities and to meet the set project duration within the planned budget.

Risk identification

Several techniques can be applied by managers for elimination, transformation, retention and reduction of the risks. In the ISO 31000 it is mentioned that, risk is the "effect of uncertainty on objectives" and an effect is a positive or negative deviation from what is expected [15]. Risk management starts with identifying the risks followed by analysing, evaluating and eventually treating the problem [16]. In addition, the existing risks can be categorised into two major categories: unique and general risks. Unique risk such as political, social and cultural risk and general risks are regulatory, design, natural and management risks.

Risk management

If the intention is to become involved in property development, one must realize that there are potential risks involved [17]. Risk management is more than merely identifying, evaluating and treating of risks and, thus, it has become an important part of the overall project management process [18]. Neglecting or avoiding it would be like driving a car without checking whether all the wheel nuts are fastened. The same can be said for opportunities that need to be exploited which would otherwise be lost in attaining improved project performance. Nevertheless, as risk is also a function of the project duration and lengthy projects being inherently riskier due to legislative, environmental and organisational changes [19], it is evident a successful property development is a matter of understanding and managing the associated risks [17, 20]. Therefore, risk management is needed in all projects and system life cycle phases, if the delivery of the project is a technical system.

3 Materials and Methods

For the purpose of this quantitative study, structured questionnaires were presented to hundred (100) property development practitioners, namely: property project managers, real estate companies and consultants within the property development sector, staff as well as clients and contractors working for property developers, who were randomly selected due to their availability and willingness to answer the questionnaire. The questionnaires were presented neatly, politely, and concisely, to ensure that the data was not sensitive and that the respondents can also relate to the intentions of the research, thereby enhancing the value of the collected data. The research instruments were not unnecessarily lengthy and assured privacy towards respondents. However, this was a quantitative study and, therefore, the sampling survey engage a simple random sampling technique. This sampling method was used to generate appropriate sample size for the survey. The respondents were required to indicate the extent to which they agreed with a statement on a 5-point Likert scale. Out of the 100 questionnaires circulated, only 62 valid responses were received, representing a

62% response rate. To describe the findings in more detail, the following weighted Mean Scores (MS) terms were used:

$\geq$ 1.00 to $\leq$ 1.80—ineffective to near ineffective.
$>$ 1.80 to $\leq$ 2.60—ineffective to near ineffective/near ineffective.
$>$ 2.60 to $\leq$ 3.40—near ineffective to effective/effective.
$>$ 3.40 to $\leq$ 4.20—effective to near highly effective/near highly effective.
$>$ 4.20 to $\leq$ 5.00—near highly effective to highly effective/highly effective.

4 Results and Discussion

1. *The influence of technological factors on the effectiveness on the success of property development project in Gqeberha.*

Figure 1 illustrates the results from the respondents based on whether technological factors poses a significant effect on property development projects delivery. The respondents believed that technological factors have a significance difference Mean Scores on how property developers and/or project managers lead the property development projects towards a success. The highest factor found to have the most significance influences was social media with a mean score (MS) of 3.98, followed by creating a business website and anticipated prevention methods in place (preventing delays) with MS of 3.90 and 3.85, respectively, sharing the same confidence interval of 95%.

Table 1 has made it possible to distinguish the performance of the MS in the sense that average MS for the technological variables is 3.86 meaning that it is $>$ 3.40 but is $\leq$ 4.20. This goes to show that these variables have an effective to near highly effective significant difference on the success of property development projects with ranking of social media at 1 and the least being preventing delays at 2.

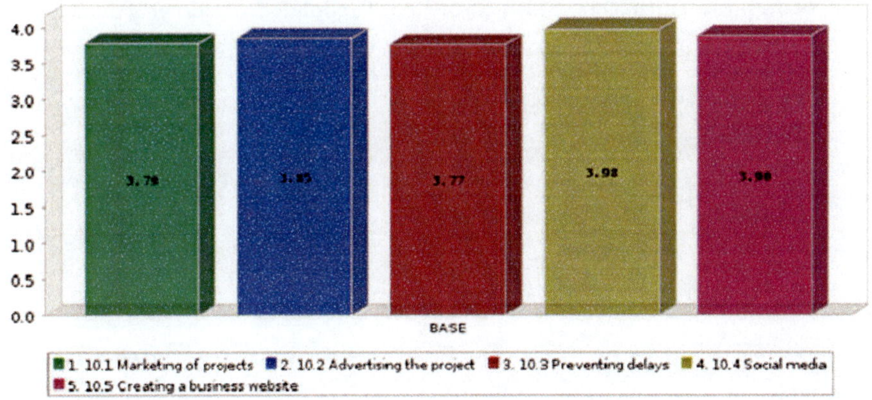

Fig. 1 The influence of technological factors on the effectiveness of property development success

Table 1 Ranking of the influence of the following technological factors on the effectiveness on the success of property development project in Gqeberha

S. No.	Variables	Score	Ranking
1	Social media	3.984	1
2	Creating a business website	3.902	2
3	Advertising the project	3.852	3
4	Marketing of projects	3.787	4
5	Preventing delays	3.770	5
		3.859	

2. *The influence of environmental regulations on the effectiveness on the success of property development project in Gqeberha.*

Figure 2 demonstrates that environmental regulations in property development have a significant influence on the success of property development project in Gqeberha. The highest factor that was found to be near highly effective in this regard was SANS 10400 & 204 with a mean of 3.82 and a standard deviation of 1.16 at a confidence interval of ninety-five (95)%. Although Green building Technology was found to be effective with mean score of 3.41, its effect may be minimal as its MS closely associated with mean score 3.40 having only 0.01 mean difference.

Table 2 presents the performance of the MS of environmental regulations in the sense that the average MS was 3.62 which means that it is > 3.40 but also ≤ 4.20. This means that respondents saw that these variables had a significant effective to near highly effective when ranking the influences of environmental factors that have a negative impact on delivery of property development projects.

3. *The influence of project management factors on the effectiveness of property development project success.*

Figure 3 illustrates that project management factors in property development have a strong influence on the success of property development, with the least significant factor being project managers with mean of 3.85 at a confidence interval of ninety-five

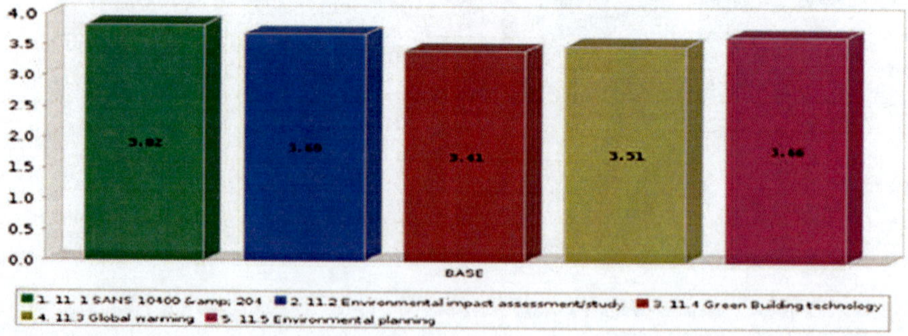

Fig. 2 The influence of environmental regulations on the effectiveness of property development project success

Table 2 Ranking of the influence of the following environmental regulations on the effectiveness of property development success in Gqeberha

S. No.	Variables	Score	Ranking
1	SANS 10400 & 204	3.820	1
2	Environmental impact assessment/study	3.689	2
3	Environmental planning	3.656	3
4	Global warming	3.508	4
5	Green building technology	3.410	5
Average		3.616	

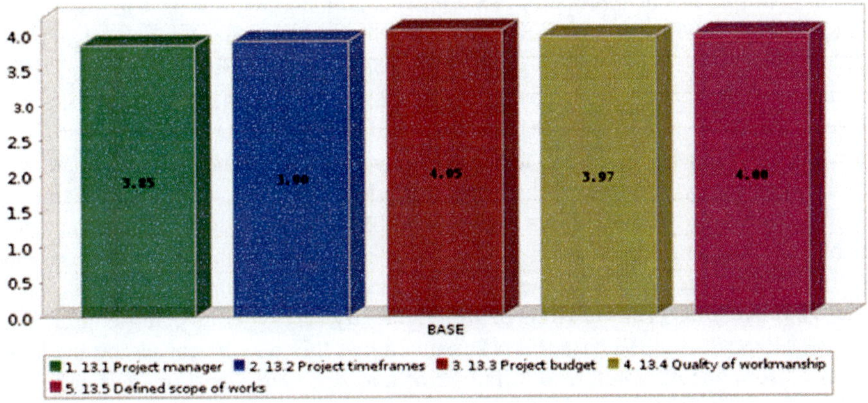

Fig. 3 The influence of project management factors on the effectiveness of property development project success

(95)%. The most significant variable that was found to be influential in the success of a project in property development in Gqeberha was project budget with a mean of 4.05 a followed by defined scope of works (4.00) and Quality of workmanship (MS = 3.98) at a confidence interval at ninety-five (95)%.

Table 3 demonstrating the significance difference in the performance of the MS in the sense that the average MS for the project management variables is 3.95, which

Table 3 Ranking of the influence of the following project management factors on the effectiveness of property development projects success in Gqeberha

S. No.	Variables	Score	Ranking
1	Project budget	4.049	1
2	Defined scope of works	4.000	2
3	Quality of workmanship	3.967	3
4	Project timeframes	3.902	4
5	Project manager	3.852	5
Average		3.954	

is > 3.40 but is ≤ 4.20. This suggestive that the variables shows results of effective to near highly effective when ranking the influences of project management factors that have a positive or a negative impact on the delivery of property development projects, the highest in this regard is project budget appear to be at one (1) and the lowest is project manager at five (5).

5 Discussion

From the results, it has been noted that all variable have a significant difference on the success completion of a property development projects, with project management factors (MS = 3.95) being the most influential followed by technological budget (MS = 3.86) and then environmental regulations (MS = 3.62). With such findings, it is evident that project management variables were near highly effective to contributing to the success of property projects because without any project management tools in place, the projects are bound to fail. Moreover, if there is no briefing, planning, timeframes or any one to lead the projects could cause chaos and hostility among the employees as well as serious financial problems. That is why project management factors are important for a project to be completed with a high success. On the other hand, one would assume that environmental regulations would play a vital role by preventing projects from being reluctantly left without being completed, however this was not the case. It was assumed that this may be caused by the lack of law enforcement, which lead to failure of many property development projects. For example, Environmental Planning and Assessment Act 1979 was established "to promote the proper construction of buildings, as well as to promote the social and economic welfare of the community and, better environment via the proper management, development and conservation of the State's natural and other resources". However, overlooking environmental regulations and allowing construction of buildings to take place in protected areas can result in project failure, even thought the construction was proper and the purpose of the building was to serve the nation. This is because building on a protected arears or areas that are declared as heritage sites, is prohibited under the National Environmental Management Protected Areas Act 57 of 2004. It is therefore important for a project manager or property developer to familiarise him/herself with such regulations from the early stages of project. Finally, technology factors were also illustrated to have as near highly effective as project management tools. It is assume this is because with the growing technology, more work friendly tools and innovative ideas are being established while knowledge of how to run a project successfully is shared throughout the world via literature. Moreover, using social media to look for project managers with more the know how and/or managers who are acquainted with the new technology, can results in high project success. Out of the 15 factors that were investigated, all project management factors (Table 3) was found to be the most influential of variables. Project budget and defined scope of works were found to be the most significant variables than all 15 variables with a mean scores of 4.05 and 4.00, respectively. These findings could either mean that if there is a poor definition of scope even though it was well budgeted for, property development projects can not succeed. This is because without well defined scope, how will the budget be distributed when the project resume? This can lead to over/under spending of funds on material that is required for the project to be successfully completed, resulting into project failure.

Green Building technology was found to be the least factors to have an influence on projects success. We assume this is because green building technology has just been discovered and many property developers may not be aware of such technology. Or, it may be because most contractors or quality of workmanship is outdated or not are acquainted with the type new technology. If this may be the case, most property development project will not succeed. Monetary policies have a direct effect on income patterns, which in return affects the availability of credit. Furthermore, it is clear from these findings that all the variable are interrelated, as a change in one variable can make a change in all variables, resulting in either a failure or success of property development projects. These findings are, therefore, in coherence with the findings of the studies mentioned in the literature. Therefore, project managers or property developers must, at all times, implement proper technics and adhere to the building regulation; define the scope of work, recruit high quality employees who are willing to take risks and reach for the stars and prepare project time fames that are attainable for projects to succeed.

6 Conclusions

The literature reviewed in this study indicated that there are problems experienced in property development projects and that they impact severely on completing projects with high success. Moreover, this study has made it possible for us to identify and highlight some of the most influential factors that may either have negative or positive influence towards property development projects success. We therefore recommend that property managers or developers must encourage every individual involved with property development projects to continuously develop their own skills through education, to ensure successful delivery of projects. Attention to detail during the planning of property development projects must be largely strengthened with the help of experienced and reliable project managers. Project delays experienced serve as a major inconvenience in all aspects of projects because they affect, scope, quality, cost and the stakeholders at large. Therefore, proper project control measures in the execution and post project cycle are quite critical for the success of property development projects.

7 Recommendations

Property managers should manage the relationship with project stakeholders even at planning stages, due to the fact that all property projects have people expecting their delivery and the feasibility studies should be communicated, should there be a need for project discontinuity due to non-compliance with legal requirements and environmental standards. All proceedings during the implementation need to be conveyed to stakeholders to strengthen their trust in the project managers. Furthermore, property developers should apply suitable planning techniques and be accountable for the project's duration as they are the ones who check the execution of plans and agree with the contractors or they may hire experienced projects managers. These programmes should be followed at all times and regular check-up and reviews need to be done after each milestone is achieved and if no milestone is achieved, a proper and sound explanation should be provided on the monthly or weekly reports and a plan of action to remedy the situation should be provided too. Project managers, however, should avoid such programme slip-offs as they can be costly and may delay completion of a project with high success.

References

1. Marnewick C, Labuschagne L (2010) A conceptual framework to improve the delivery capability of an organization. Professional Account 10(1):249–263
2. Jaafer YA, Behrang S (2019) The relationship between critical success factors and success criteria in construction projects in the United Arab Emirates. Int J Adv Appl Sci 6(7):43–53; J Eng Technol Manage 34:46–62; J Project Manage 37:820–838; J Project Organ Manage 10(3):191–221
3. Zulch B (2014) Leadership communication in project management. Proc Soc Behav Sci 119:172–181
4. Project Management Institute (2013) A guide to the project management body of knowledge: PMBOK© guide, 5th edn. Project Management Institute, Pennsylvania
5. Project Management Institute (2017) A guide to the project management body of knowledge: PMBOK guide, 6th edn. Project Management Institute, Washington, DC; Newtown Square, PA
6. Tshiki M (2015) Critical success factors for infrastructure construction projects in South Africa: project management and construction. South Afr Inst Civ Eng (SAICE) 23(6):19–24
7. Hodgson D, Cicmil S (2006) Makin projects critical: an introduction. Palgrave MacMillan, New York
8. Nguyen PT, Nguyen PC (2020) Risk management in engineering and construction: a case study in design-build projects in Vietnam. Eng Technol Appl Sci Res 10(1):5237–5241. https://doi.org/10.48084/etasr.3286
9. Kouski PA, Al-Rashid K, Kartam N (2005) Delays and cost increases in the construction of private residential projects in Kuwait. Constr Manage Econ 23:285–294
10. Ajibola MO, Oluwunmi AO, Eguh O (2012) Examining the factors contributing to affordable housing in Kosofe local government council area, Lagos, Nigeria. Journal of Asian Business Strategy 2(10):206–218
11. Briassoulis H (2001) Sustainable development and its indicators: through a (planner's) Glass Darkly. J Environ Planning Manage 44(3):409–427

12. Alvarez-Dionisi LE, Turner R, Mittra M (2016) Global project management trends. Int J Inf Technol
13. Burke R (2013) Project management: planning and control techniques. New Jersey, USA
14. Brook JW, Pagnanelli F (2014) Integrating sustainability into innovation project portfolio management—a strategic perspective
15. Shahbaz MS, Kazi AG, Othman B, Javaid M, Hussain K, Rasi RZRM (2019) Identification, assessment and mitigation of environment side risks for malaysian manufacturing. Eng Technol Appl Sci Res 9(1):3852–3858. https://doi.org/10.48084/etasr.2529
16. Serpella AF, Ferrada X, Howard S, Rubio L (2014) 2014 Risk management in construction projects: a knowledge-based approach. Proc Soc Behav Sci 119:653–662. https://doi.org/10.1016/j.sbspro.2014.03.073
17. Yardney M (2010) How to get started in property development. Metropole
18. Poon C, Silvius AJG (2019) Factors that stimulate project managers to consider sustainability; exploring the stimulus patterns of Canadian project managers. J Manage Sustain 9(2):90–114
19. Sabini L, Muzio D, Alderman N (2019) 25 years of 'sustainable projects'. What we know and what the literature says. International
20. Schipper R, Silvius AJG (2018) Towards a conceptual framework for sustainable project portfolio management. International

Lean Thinking as an Approach for Improving Client Communication During the Design Process

Ghada N. Mossad, Ayman A. E. Othman, and Nishani Harinarain

Abstract Lean thinking is perceived as a successful strategy for saving time and cost, while improving at the same time the position in the market. Lean thinking is about eliminating waste and increasing the quality. When applied correctly, lean thinking is a well understood and well examined platform upon which to build firm practices. In the architectural profession miscommunication has a lot of consequences including time and cost overruns, conflict and ultimately project failure. The aim of this research was to investigate the role of lean thinking in improving the communication between the client and the architect during the design process. To achieve the aim of this research, an indepth firstly, a literature review was used to build a comprehensive background about the research topic including the design process, causes and impacts of poor communication during the design process, and lean thinking. Secondly, an analysis of case studies was used to investigate the role of lean thinking in enhancing the client communication during the design process. It was found that comprehending, understanding and incorporating lean thinking principles in the design stages can assist in reducing the causes of poor communication between the architect and the client.

Keywords Lean thinking · Communication · Design process · Architecture design

G. N. Mossad · A. A. E. Othman · N. Harinarain (✉)
Architectural Engineering Department, Faculty of Engineering, The British University in Egypt (BUE), El Shorouk City—Cairo Suez Desert Road, Cairo, Egypt
e-mail: harinarain@ukzn.ac.za

G. N. Mossad
e-mail: ghada156481@bue.edu.eg

N. Harinarain
The School of Engineering, Construction Studies Discipline, The University of KwaZulu-Natal, Howard College, Durban, South Africa

T. C. Haupt et al. (eds.), *Construction in 5D: Deconstruction, Digitalization, Disruption, Disaster, Development*, Lecture Notes in Civil Engineering 245,
https://doi.org/10.1007/978-3-030-97748-1_25

1 Introduction

In the field of architecture, the communication between the client and the architect is based on using all forms of communication whether it is verbal, non-verbal or computer-generated architectural drawings to communicate [1]. Communication is defined as a process of exchanging ideas, opinions, and goals [1]. Communication is more than just gathering information. The information must be trustworthy, heard by the right people, and result in appropriate outcome. Communication has become more problematic in the architectural projects because of complicated design projects, the advancement in technology and design information management systems, as well as the diversity in the stakeholders [1].

Communication issues between the client and the architect can be classified into technical and social. The most complaints from the clients were the misunderstandings and dissatisfaction of the architectural services [2]. Though the client is the centre of the design process, they are not aware of the design processes and what information needs to be passed on to the architect [1]. Traditional methods created problems as they were not well coordinated and intelligent and the knowledge was not well managed [1]. Miscommunication has a lot of consequences including time and cost overruns, conflict and ultimately project failure [3]. To achieve the aim of this research, the researchers investigated how lean principles can improve communication between the architect and client utilising case studies.

2 Literature Review

2.1 The Design Process

The design process according to the architect is characterised by doing the same in a different way each time. Accordingly, every design project is unique, but the process and the methods used in creating it are somewhat similar each time. Several requirements are needed to develop architecture: consisting of legal provisions and regulations, brief by local authorities, economy, and most important clients brief. This assembly of requirements covers the creative process. The creative process attends to each of these specifications along with aesthetic and technological aspects and take constant considerations [4].

The design process does not signify the process of creating, however it is based on the principle of deduction of the final structure of the basic primary introductions, and the importance of those introductions and the final structure depends, however, on three basic methods:

1. Collecting information: which include gathering information, organising, and analysing them to ensure that they contribute to the design role, and how such information is incorporated into the picture.

2. Testing: which include the design choices through the process of showing the architect's analytical ability, intuitive, and personal experience.
3. Evaluation process: which requires a judgment on the suitability of these choices regarding the design problem [5].

2.1.1 Stages of the Design Process

The RIBA plan of work [2] identified eight stages that need to be conducted to assist anyone associated with building a project, from the professional architect through to a client going through their first project. This research discusses the first five stages of the RIBA plan of work. These five stages discuss the design process that the architect and the client go through.

0. Strategic Definition

The primary objective of Stage 0 is strategic confirmation of a construction project. It is the best way to achieve the client requirements. Stage 0 is not about the functional specifics or the design. It focuses on making and capturing the necessary strategic decisions in a business case. The stage involves stating the advantages and disadvantages, project risks, project budget, and execute site surveys, and recommending the best solution for carrying out the client requirements. In this stage, only the client team is involved. The client may ask for advice from a wide range of professional advisors to help them establish the client requirements and the business case [2].

1. Preparation and Briefing

The outcome of this stage is developing the project brief approved by the client and confirming that it could be established on the site. The client requirements are discussed in more detail, in connection with a certain site, and the result is documented in the project brief. The project outcomes, sustainability outcomes, and quality aspirations are all included in the project brief. This stage is mainly about collecting all the information needed for the design team to start stage 2. In this stage, the client team only is involved [2].

2. Concept Design

Stage 2 establishes the architectural concept for a project. Proposals that match with the site information, project brief, and spatial requirements are prepared during this stage. The proposals for the architectural concept must also be reconfirmed to satisfy input from the design team and specialist consultants, including the criteria for strategic engineering. Proposals will need to be synchronised with the project strategies, as well as with everything found in the stage report along with the cost plan. In this stage, the client team and the design team are core players, along with specialist consultants. In some cases, the construction team may also be included in this stage [2].

3. Spatial Coordination

The outcome of this stage is architectural and engineering information spatially coordinated. To confirm the assumptions made during stage 2 and to layer more detail on the concept, detailed design studies and engineering analysis are conducted. Stage 3 is not about changing the architectural concept, which should remain substantially unchanged, although adaptations to ensure that the building is spatially coordinated may be needed for detailed tasks or engineering tasks. In this stage the key players are the lead designer and design team. The client team are involved where decisions are required in stage 3. The construction team may also be needed if the chosen procurement route needs early inputs by a contractor or a subcontractor [2].

4. Technical Design

This stage involves preparing all details needed for a building to be manufactured and constructed. The basic documents needed at the start of stage 4 are the information requirements, responsibility matrix, and stage 4 design programme, which the procurement strategy has a great influence on. Before work starts on site, a building regulations application should be developed during stage 4. In this stage, the design team and the specialist contractors employed by the contractor complete the design. A client monitoring team may be assigned under certain types of procurement to review the information that is generated [2].

2.1.2 Client Issues Occurring During the Design Process

Several studies have highlighted different variables that rationalise the communication between the architect and the client and their participation in the design process [6]. These are some issues that occur during the design process:

- At the beginning of each design brief, there is a significant lack of information.
- Clients frequently have too many expectations.
- The final decision in the design is not decided by the architect.
- As the design progresses, the scope of the design frequently expands.
- The terms of definition differ according to the architect and the client.
- Architects are not making enough effort to keep clients connected to the progress. • Architects do not look up the client's background history in proper depth.
- Project solutions are based essentially on the subjective considerations arising from the clients need [6].

2.2 *Communication During the Design Process*

The initial phase of a project implementation is the design process, which is based on the requirements of the client, and greatly influence the value of the project [7]. Communication is a certain necessity during the design process, and it is important to communicate extensively between the architect and the client in order to prevent conflict by deciding the messages and information to be sent by the right sender to the right recipient in the correct manner by the right media [8]. In order to solve problems at the initial design phase of the project, it is important to develop a convenient communication management system [9].

2.2.1 Communication

In its broadest sense, communication is the transfer of information from one individual to another [10]. Communication is the method of information exchange between the sender and the recipient to equalise data on both. Communication can influence the architect-client interactions. The concept of communication was expanded by Gabriel and Maher [11] into the field of architecture when he said, "architecture is mainly about communication". The foundation of such claim lies in the fact that architects' theories are the product of close communication between the architect and the client, which is ultimately reflected in the constructed environment [6]. Problems can be overcome with respect to collaborative design by using different forms of communication instead of relying on additional communication [12]. Communication means more than receiving information, it also implies that the information must be reliable, heard by the right audience, and lead to a suitable response. The communication process is complicated but it is this process that eventually leads to project-relevant knowledge and ensures success [6].

2.2.2 Causes for Poor Communication During the Design Process

Miscommunication is defined by insufficient and incorrect communication; furthermore, frustration and misunderstandings are all sources of miscommunication. The architectural design process is harmed by ineffective communication between the architect and the client [13]. Table 1 depicts the causes for poor communication during the design process.

2.2.3 Impacts of Poor Communication During the Design Process

An architect is a person who plans, designs, and models a construction project. An architect uses a broad range of knowledge from a variety of fields, including arts, science, technology, to bring a building design to life. Previously, traditional approaches caused issues because they were not well-coordinated and intelligent, and clients were not given the useful details. As a result, the building design process would be hampered, and bad choices would be made. The client is the most essential person in the architect-client partnership. They are the only one who can have a close relationship with the architect, and they need detailed information about the building design [23]. Despite technological advancements, architects are faced with

Table 1 Causes for poor communication during the design process

Number	Causes for poor communication	Explanation	References
1	Physical barrier (face to face meetings)	It is the environmental and natural conditions that act as a barrier in communication when transmitting messages from sender to receiver. It is the most significant impediment to successful workplace communication, as well as the overall environment	[6, 13]
2	Linguistic barrier	Architectural language is confusing by nature, and its words are symbols, so if the client understood it in their own way, it may lead to misinterpretation	[6, 14]
3	Diversity of culture	Age, thinking behaviour, social position, economic status, ethics, motivations, and priorities are all examples of cultural barriers that occur when two people belong to different cultures	[6, 14]
4	Lack of honesty from the architect	Architect's honesty is essential for the success of the project as the architect should be honest and candid while avoiding deception. If the architect is not sure about his/her ability to carry out certain task, they should inform the client as soon as possible	[15, 16]
5	Poor feedback received from the client	Can delay the project and cause confusion to the architect	[17]
6	Work pressure on architectural firm	When people are under pressure, they are more likely to make mistakes. Personnel errors may manifest themselves in the form of incorrect or inaccurate outcomes. People may be affected psychologically and physically by pressure, which would have an impact on their output	[18]
7	Poor communication management	Failing this multidimensional process means inefficient information exchange, which will result is major project issues	[18]

(continued)

Table 1 (continued)

Number	Causes for poor communication	Explanation	References
8	Ineffective communication between the client and the architect	Means that it was completed but the results or the process were insufficient. Two major key contributors to ineffective communication are time and quality	[14, 18, 19]
9	Lack of collaboration and representation between stakeholders	Instead of leaving their knowledge base locked in the perceptual domain, stakeholders must take the time to express it in a truthful and clear manner	[16]
10	Unclear communication channels	Can cause many problems for both the architect and the client, causing miscommunication. So, the absence of this medium will result in message transmission and arrival failure, the architect and the client must establish an agreed-upon route	[18]
11	Unclear requirements from the client	Sometimes the client may only have a hazy notion of the problem that needs to be solved, maybe he cannot describe it precisely	[20]
12	Different level of education	A well-educated client will have more insight and will be able to be better understand the architect's conceptual solutions and make valid suggestions as a result contrary to an uneducated client	[1]
13	Lack of architect's experience	Many architecture schools place a greater emphasis on math, theory, and other technical subjects instead of on effective communication [20]	[20]
14	Unavailability of information in time of need	Will halt the team's readiness to communicate	[18]
15	Technology malfunctions	Have bad impression on the architect, as it implies poor work quality and unsatisfied client	[21, 22]

ever-increasing amounts of data to handle. In addition, the time it takes to complete projects has shrunk. For example, the use of ICT allows for fast communication between stakeholders, which speeds up the project by reducing wasted time [24].

There are numerous impacts of poor communication between the architect and the client have on the project during the design process which are listed below:

(1) **Project failure** is the most serious consequence of poor communication. This is typically the result of several causes that were not resolved over a long period of time, resulting in a project being labelled as a failure [25].

(2) **Disputes** are a common effect that can be seen throughout the project's life cycle. Dispute may occur because of inaccuracies in information or poor communication between organisations and individuals. This may also be the result of bad or dishonest feedback, causing the client to apply aggressively, resulting in a disagreement [26].

(3) **Over budget** is one of the most important risk factors of any construction project. This dangerous impact can be attributed to a lack of project management skills, poor coordination, and ineffective communication [25, 27].

(4) **Time overrun** is a common occurrence that can result in charges and losses for the client. To ensure that the project is delivered on time, the schedule must be continuously updated and communicated to the client [25, 28].

(5) **Poor quality of design** is a significant source of inefficiency in the design process, resulting in delays, rework, and variations as well as increase in project cost and time for both the client and the architect [26].

(6) **Reworks of project design** are both financially and time-consuming. The tiniest error or misinterpretation of data can result in days of redesign in the office [27].

(7) **Poor collaboration** between the client and the architect is caused by a variety of communication flaws. Collaboration is particularly important between the architect and the client, where goals are accomplished by combing the efforts of many different professions [28].

(8) **Unmotivated client** can cause the project's demise if they reach a point where the continuation or success of a project is insignificant to their interests [26].

(9) **Untimely reactions** in important situations can cause problems during the design process. The clarity with which a client with responsibilities provides feedback and reacts can either solve or cause a problem [28].

(10) **Low productivity rates** can be caused by a variety of causes, but the fact remains that bad work is performed in terms of quality and quantity, which has a direct impact on the design's outcome [27].

(11) **Poor understanding of data** means that the communication mechanism that carried this data from the sender to the receiver failed [26].

(12) **Misinterpretation** means that knowledge was misinterpreted, which can result in unsuccessful actions that waste time and money [26].

(13) **Mistakes** have different consequences depending on the severity of the harm they inflict. Effective communication can mitigate the quantity and consequences of mistakes during the design process [25, 27].

(14) **Unsatisfied client** is the product of several negative process that persisted unabated during the project lifecycle. Client communication is critical, and client satisfaction must be a goal for all parties involved in the project [25].

(15) **Client's constant design changes** during the design process in terms of additional resources, expense, and project length have a negative effect the design process [25].

2.3 *Lean Thinking*

Lean has its origins in the automobile industry, specifically the Toyota Production System. The Japanese company was able to build a long-term work environment in which they could cut costs, improve process quality, and sell their goods at a fair price. In reality, when Toyota first entered the US market, they were able to sell their vehicles for a fraction of the cost of American manufacturers. Since they were able to find a way to speed up the production process without losing efficiency, they were able to do this. They were also able to recognise and eliminate unnecessary practices in their methods as lean thinkers [29]. Lean thinking, is a method to specify value, line up value-creating behaviours in the appropriate sequence, execute these activities without interruption if anyone requests them, and execute them more and more efficiently [30].

2.3.1 Brief About Lean Construction

Lean Construction involve owners, designers, suppliers, architects, engineers, builders, and end users. Lean construction is a method of designing development systems to minimise waste of materials, effort, and time in order to produce the most value possible [31]. Lean construction employs the same principles as lean manufacturing to minimise the waste while improving productivity and efficiency in construction work. Workflow accuracy and labour flow are presumed to be the major factors of construction, but lean construction changed the conventional view of the project as transition and promotes the principle of flow and value generation. It also has the same goals as lean production, such as reducing cycle time, elimination waste, and reducing variability. The implementation of lean construction has been guided by continuous improvement, pull production control, and continuous flow [32].

2.3.2 Lean Architecture

The continuous process of rethinking and improving architectural methodology is known as lean architecture. It is the quest of better work through the application of lean principles to all areas of practice. It is about having smarter information flow and understanding how the architect realises and handles information in an attempt to be better communicators with themselves and with their clients. It is the process of determining what adds value and reducing or eliminating what does not [33]. Since architects are not making identical projects, architectural production practices must not be confused with lean manufacturing. Because architects work with people and organisations, design phase usually includes managing a great deal of subjectivity. Nevertheless, analysing the thinking behind good manufacturing processes and applying the theories and values to the more repeated aspects of the architecture practice can have a great impact [33].

2.3.3 Lean Thinking Principles

Lean thinking has five main principles, they were implemented to address the numerous challenges that exist as a result of variations in business culture and management thought process. To optimise the benefits of lean performance, there are five basic principles for lean thinking that must be followed.

1. The first lean thinking principle is to identify the value from the perspective of the client. The principle requires the architect to consider and rethink who their real clients are, as well as what the client consider valuable. The clients eventually determine the value of the design, so the principle emphasises determining the value from the client's perspective [26]. The goals for each of the activities surrounding the outcome from design to delivery to the client [34].
2. The second lean thinking principle is 'identify the value stream'. The definition of value stream differs from the conventional supply or value chain concepts. The pervious is a narrow view of the value adding process, referring to the activities that go into adding value to the development of the design, while the latter encompasses all the activities that are required in the design process [35]. The 'identify the value stream' principle encourages companies to:

   - examine and identify all activities involved in the development of the project,
   - identify activities that add value,
   - remove activities perceived as waste in a value stream [36].

   As a lean approach in architecture design these can include:

   - using only reliable, thoroughly tested technology during the design process and while communicating with the client, it can be like using BIM, 3D models [37].
   - Align the design to the quality requested by the client. Quality is described as absolute accuracy on what everyone agreed upon, since failure is viewed as inconvenience by the consumer, and providing something unique, which is difficult to measure but makes the client satisfied and connects them to the firm [38].
   - Respect the preferred modes of communication according to the client and the type of information delivered. Every mode of communication has its own set of advantages and disadvantages whether it is verbal, written, or visual communication [39].

3. The third principle of lean thinking is to implement flow in the value-added processes after removing the obvious wastes in the value stream step. The basic principle of flow, according to Lian and Landeghem [26] is to make parts preferably one piece at a time from production to delivery and to transfer them one

by one to the next workspace with no waiting in between. In case of communication, it is the flow of information to the client with minimum delay. This can be applied by the following:

- Levelling out the workload between the architects [40].
- Developing communication platform to facilitate communication and smoothly transfer information between the client and the architect [41].
- The encoding of information into proper communication like architectural terms and other important information.
- The information that is communicated must be transparent, descriptive, visible, and simple to understand [42].
- Repeat the information that was received to ensure correct understanding [34].

4. The fourth principle of lean thinking is pull production; it is the next essential part of lean thinking in ensuring clients receive their desired design when they want it [43]. When applying the pull principle to the flow of information, an architect should be able to control when and what information is delivered to him [44]. So, the architect should send the information to the client at the appropriate time when requested.
5. The final principle of lean thinking is to always strive for excellence. This can be implemented by training. There are two distinct fields of expertise and work in lean education and training. One region could be labelled lean teaching while the other could be called teaching lean. The teaching lean field can be described as the process of effectively training students and professionals in lean principles, concepts, and tools so that they can apply lean in real-world situations such as industries and other companies. The other field, lean thinking is characterised as the application of lean principles and concepts to learning and training practices [44].

2.4 Relationship Between Design Stages, Communication, and Lean Thinking

A relationship was created in Table 2 between the design stages and communication causes and impacts and how lean thinking principles and tools can be used to solve the causes of poor communication during the design process.

Table 2 Relationship between causes and impacts of poor communication during the design process and lean thinking methods

Number	Causes of poor communication during the design Process	Phases of architectural desig phases					Impacts of Poor communication during the design Process															Lean Thinking Principles									
																						Define Value from the client	Identify Value Stream			Flow				Pull	Perfection
		Startegic Definition	Preparation and Briefing	Concept Design	Spatial Coordination	Techincal Design	Project Failure	Dispute between the client and the archiect	Over Budget	Time overrun for the Design process	Poor quality	Reworks of Project design	Poor collaboration between the client and the architect	Unmotivated Client	Untimely reactions	Low productivity rate	Poor understanding	Misinterpretation	Mistakes	Unsatisfied client	Client's Constant design changes	Defining value from the client	Use only reliable, thoroughly tested technology	Align design to quality	Respect Preferred Modes of communication	Level out the workload	Develop a communication platform	Encoding of knowledge like certain architectural terms	Repeat the information that was received to ensure correct understanding	Deliever information as soon as it was requested	Training
1	Physical Barriers (face to face communication)	✓	✓														✓	✓									✓				
2	Linguistic Barriers between the client and the archiect			✓	✓	✓						✓						✓	✓										✓		
3	Cultural Barriers (difference between cultures, age difference, thinking behaviour, etc).	✓	✓					✓					✓															✓			
4	Lack of honesty from the archiect		✓		✓		✓	✓	✓			✓								✓				✓							
5	Poor Feedback Recevied from the client			✓	✓					✓		✓																		✓	
6	Work Pressure on Architectural team				✓	✓					✓					✓			✓							✓					
7	Poor communication management		✓	✓	✓	✓	✓	✓		✓																	✓				
8	Ineffective communication between the client and the architect		✓	✓				✓					✓				✓										✓				
9	Lack of collaboration and representation between stakeholders		✓		✓								✓								✓				✓						
10	Unclear communication channels			✓	✓	✓					✓		✓			✓											✓				
11	Unclear requirements from the client	✓	✓	✓						✓							✓				✓	✓									
12	Different level of education			✓	✓	✓		✓									✓			✓								✓			
13	Lack of architect's experience		✓	✓				✓						✓																	✓
14	Unavailability of information in time of need			✓	✓	✓						✓			✓															✓	
15	Technology Malfunctions			✓	✓						✓									✓			✓								

(Developed by authors)

3 Case Studies

3.1 *Kuwait University and Gulf Consult.*

Gulf Consult (GC) was assigned by Kuwait University as their local consultants in their College of Business package in collaboration with Cambridge Seven Associates, Inc. (C7A). During the design process of this project, it was clear that there was poor communication between the project partners. However, Gulf Consult was quick to identify and solve the problem by conducting unstructured interviews to find the root cause of the problem, which were identified as follows:

- Lack of knowledge was observed from some stakeholders;
- Lack of experience from the architects during the project;
- Unclear language circulated which caused miscommunication;
- Issues regarding the representation of stakeholders when the communication was needed.

These causes were met with proper action and solutions to solve the problem to prevent further communication problems that will affect the project greatly. The following where suggested to solve these problems which follows the lean thinking theories.

- Establishing a clear communication system;
- Level out workload between the staff by assigning tasks to capable staff members;
- Monitor and control the communication process.

In this case study, Gulf Consult used several lean thinking principles to solve the causes of poor communication. The lean thinking principles used were:

1. Identify the value stream which was reflected in focusing on the value adding processes, like the regular meetings and assigning tasks to competent staff members.
2. Flow which was used in making sure that the correct information is delivered to all staff members to ensure the flow of information.
3. Strive for perfection, which was reflected in continuous monitoring and controlling the communication process.

3.2 *Istanbul Grand Airport IGA*

Istanbul grand airport was built to operate Istanbul airport for 25 years. It has a total area of 76.5 million m^2. The method of designing, incorporating, and recording design and development details by creating an interactive BIM model developed during the design and construction stages in IGA for BIM-based project execution. The use of an optimised framework for automated data analysis during the project phases further represents lean architecture and construction practices. BIM implementation in the IGA project during design and engineering provided a shared simulated world for all parties involved in the process. This is made possible by data sharing procedures through cloud-based data management software, as well as the creation of BIM models with required engineering decisions by incorporating various types of design knowledge. On a regular and weekly basis, the Istanbul Grand Airport BIM team holds meetings to realise and define each important assignment, workflow, and implementation plan schedules of the IGA construction. The strategic benefits of these meetings are defined as coordination and workflow between all disciplines, as well as regular, weekly, and monthly checks and quality controls. The quality of the design product is improved by applying these implementations. During the design process for collaborative clash detection phase, this workflow ensures that all participants are synchronised. As a result, produced construction documents (shop plans, BIM models, etc.) are updated in terms of potential clashes on site to prevent any on-site rework.

The following results were the outcomes of applying lean in the IGA project via BIM:

- Total costs as predicted;
- Correct and high predictability;
- High rate of production;
- High level of profit; and
- Easy information flow.

4 Conclusion and Recommendations

The design process is the start of any construction project. A lot of problems between the client and the architect as a result of poor communication could be easily prevented if managed early. The architect should know how to manage the communication between the client to have a better understanding of the project and the client's requirements. Lean thinking can solve this problem, with its five main principles; identifying the value, value stream, implement flow, pull production, and ensure perfection.

The aim of this research was to investigate the role of lean thinking in improving the communication between the client and the architect during the design process. It was achieved by understanding the design stages and the causes of poor communication as well as comprehending the lean thinking principles. Two case studies were discussed to show how lean thinking was implemented during the design process and the improvement that it had on the project and the team. Based on the above, the research comes to the following recommendations:

- Architectural design firms (ADFs) in Egypt need to adopt lean thinking strategies to manage the miscommunication between architects and clients, thus obtaining communication and better project quality.

- ADFs should provide training to architects about lean thinking principles. In addition, raising awareness toward the importance of applying lean thinking through means of education and the introduction of philosophies into architectural education.
- Finally, local authorities and governments should promote more research towards adopting lean thinking to manage the communication between architects and clients.

References

1. Norouzi N, Shabak M, Bin Embi MR, Khan TH (2015) The architect, the client and effective communication in architectural design practice, pp 635–642
2. RIBA plan of work (2020). https://www.architecture.com/knowledge-and-resources/resources-landingpage/riba-plan-of-work. Retrieved on 13 Dec 2020
3. Gamil Y, Rahman IA (2017) Identification of causes and effects of poor communication in the construction industry: a theoretical review
4. SINTEF (2007) SINTEF building and infrastructure design process challenges—simple obstacles or complex building defects? Report from the R&D-programme "Climate 2000" CECILIE FLYEN ØYEN
5. Jones JC (1980) Design methods: seeds of human future. Op.Cit
6. Norouzi N, Shabak M, Embi MRB, Khan TH (2014) Participation problems and communication difficulties in architectural design practice. Life Sci J 11(9):984–990. ISSN:1097-8135. http://www.lifesciencesite.com
7. Senescu RR, Haymaker JR, Meža S, Fischer MA (2013) Design process communication methodology: Improving the effectiveness and efficiency of collaboration, sharing, and understanding. J Archit Eng 20(1):05013001
8. Kliem RL (2007) Effective communications for project management. CRC Press, 11; Taleb H, Ismail S, Wahab MH, Rani WNMWM (2017) Communication management between architects and clients. https://doi.org/10.1063/1.5005469
9. Oann Keyton PS-Z (2006) Case studies for organizational communication: understanding communication processes. Roxbury Publishing Company, p 421
10. Den Otter AF, Prins M (2002) Architectural design management within the digital design team. Eng Constr Archit Manage 9(3):162–173. https://doi.org/10.1046/j.1365232X.2002.00252.x
11. Smulders F, Lousberg L, Dorst K (2008) Towards different communication in collaborative design. Int J Manag Projects Bus 1(3):352–367. https://doi.org/10.1108/17538370810883819; Cheng EWL, Li H, Love PED, Irani Z (2001) Network communication in the construction industry. Corpor Commun Int J 6(2):61–70. https://doi.org/10.1108/13563280110390314
12. Gamil Y, Abdul Rahman I (2018) Identification of causes and effects of poor communication in construction industry: a theoretical review. Emerg Sci J 1(4). https://doi.org/10.28991/ijse-01121
13. HPH118 (2021) How to find a good architect—with Adrian Dobson from RIBA. (2016, Mar 2). Retrieved March 21, 2021, from house planning help website. https://www.houseplanninghelp.com/hph118-howto-find-a-good-architect-with-adrian-dobson-from-riba/
14. Standard 1: Honesty and Integrity (n.d.) Retrieved March 21, 2021, from architects registration board website. https://arb.org.uk/architect-information/architects-code-standards-of-conduct-andpractice/advisory-notes/honesty-integrity/
15. Rensselaer V, Griswold M (2013) Client and architect. Places J. Retrieved from https://placesjournal.org/article/client-and-architect/?cn-reloaded=1

16. Ahmed A, Othman E, Hussien M (2018) Causes and impacts of poor communication in the construction industry sustainable buildings and project waste management during the design and construction phase view project (bachelor dissertation) talent management: a novel lean-integrated approach to maximize creative heritage communities development in Egypt. View project
17. Hameed Mem A, Rahman IA, Mem AH, Karim AT (2013) Significant factors causing cost overruns in large construction projects in Malaysia. J Appl Sci 13(2):286–293.https://doi.org/10.3923/jas.2013.286.293
18. The challenges of communication in architectural design. ibds, P.C. (n.d.) Retrieved March 21, 2021, from https://ibdspc.com/2019/02/28/the-challenges-of-communication-in-architectural-design/
19. Dainty A, Moore D, Murray M (2006) Communication in construction: theory and practice. Taylor & Francis, London
20. Tai S, Wang Y, Anumba CJ (2009) A survey on communications in large-scale construction projects in China. Eng Constr Archit Manage 16(2):136–149
21. Abdul Rahman I, Memon AH, Karim A, Tarmizi A (2013) Significant factors causing cost overruns in large construction projects in Malaysia. J Appl Sci 13(2):286–293
22. Gómez-Ferrer AP (2017) Communication problems between actors in construction projects
23. Ponomareff D (n.d.) Lean thinking: its origins, pillars and principles. Retrieved from Kanban zone website. https://kanbanzone.com/resources/lean/
24. Womack JP, Jones DT (1996) Lean thinking: Banish waste and create wealth in your corporation. Simon and Schuster, New York
25. Enshassi A, Al-Najjar J, Kumaraswamy M (2009) Delays and cost overruns in the construction projects in the Gaza strip. J Fin Manage Proper Constr 14(2):126–151
26. Guevara JM, Boyer LT (1981) Communication problems within construction. J Constr Eng ASCE 107(CO4):552–557
27. Tessema Y (2008) BIM for improved building design communication between architects and clients in the schematic design
28. Penttilä H (2006) Managing the changes within the architectural practice—the effects of information and communication technology (ICT)
29. Emiliani ML (1998) Lean behaviors. Manage Decis 36: 615–631; Howell G, Ballard G (1998) Implementing lean construction: understanding and action. In: Proceedings of the international group of lean construction 6th annual conference (IGLC-6), Guaruja, Brazil
30. Koskela L, Huovila P, Leinonen J (2002) Design management in building construction: from theory to practice. J Constr Res 3(1):1–16
31. Lean Architecture (n.d.) Retrieved March 24, 2021, from lean architecture website. https://www.leanarchitecture.com/#:~:text=What%20is%20Lean%20Architecture%3F%20Lean%20Architecture%20is%20the
32. Hines P, Holweg M, Rich N (2004) Learning to evolve: a review of contemporary lean thinking. Int J Oper Prod Manage 24:994–1011
33. Duggan KJ (2012) Creating mixed model value streams: practical lean techniques for building to demand, 2nd edn. CRC Press. New York
34. Aziz RF, Hafez SM (2013) Applying lean thinking in construction and performance improvement. Alex Eng J 52(4):679–695. https://doi.org/10.1016/j.aej.2013.04.008
35. Suerth J (2017) How technology enables a lean approach to construction. Retrieved April 1, 2021, from pepper construction website. https://www.pepperconstruction.com/blog/how-technologyenables-lean-approach-construction
36. Balle M (2015) What is the lean approach to quality—is that what six sigma is all about? Retrieved from www.lean.org website https://www.lean.org/balle/DisplayObject.cfm?o=3106
37. Willkomm A (2018) Five types of communication—goodwin college of professional studies. Retrieved from Goodwin College of professional studies website.
38. https://drexel.edu/goodwin/professional-studies-blog/overview/2018/July/Five-types-of-communication/, 47

39. Liker DJK (2004) Principle 4: level out the workload (Heijunka). In: www.accessengineeringlibrary.com. Retrieved from https://www.accessengineeringlibrary.com/content/book/9780071392310/chapter/chapter10#:~:text=Heijunka%20is%20the%20leveling%20of
40. Varaksina S (2020) Building internal communication platforms for enterprises: the what and how—-mind studios. Retrieved April 2, 2021, from blog—mind studios website. https://themindstudios.com/blog/build-internal-communication-platforms/
41. Gîfu D, Teodorescu M (2014) Communication process in a lean concept. Int Lett Soc Human Sci 28:119–127. https://doi.org/10.18052/www.scipress.com/ilshs.28.119
42. Smith ATY (2015) Lean thinking: an overview. Ind Eng Manage 04(02). https://doi.org/10.4172/2169-0316.1000159
43. The Pull Principe—In Production, the Office & in Life|MudaMasters (n.d.) Retrieved April 2, 2021, from www.mudamasters.com, website https://www.mudamasters.com/en/lean-production-theory/pullprincipe-production-officelife#:~:text=When%20the%20pull%20principle%20is
44. Dinis-Carvalho J (2020) The role of lean training in lean implementation. Prod Plann Control:1–2.https://doi.org/10.1080/09537287.2020.1742376

A Bespoke Approach for Relating Building and Infrastructure Poor Performance to Inadequate Maintenance Management Policy Systems

Abdul Razak Musah, Derick Vosloo, and Wayne Draai

Abstract Assets maintenance management in accordance with the maintenance management policy systems remains a challenge not only nationally but globally. The research revealed that institutions are not complying with the practice of maintenance management policy systems. This is the gap the research aim to establish an improved maintenance management system for public high school buildings and other infrastructure based on standard practice as stipulated in the Department policy document. Literature shown that building and other infrastructure asset deterioration owned to inadequate maintenance. Quantitative methodology was employed in this research. The uThungulu District of KZN in South Africa was the selected geographical scope of this research. Thirty public high schools in the district were purposely selected for the study. The reasons include the proximity of the schools relative to where the researcher currently resides; resource constraints regarding the ability to fund transport costs to the various schools. The questions that were related to maintenance management policy systems, functional requirements of public high school buildings and other infrastructure were correlated with the data that were generated from the maintenance management stakeholders at the various institutions. The research revealed that most of the institutions do not comply with the existing National Education Infrastructure Management System and Provincial Maintenance Policy System and concluded that the majority (82.4%) of the respondents strongly agreed that maintenance conducted in accordance with policy documents enhances the functioning of the assets that satisfies users' needs and expectations.

Keywords Maintenance management policy systems · Functional requirements · Buildings and infrastructure · Maintenance

---

A. R. Musah · W. Draai (✉)
Faculty of Engineering, the Built Environment and Information Technology, Nelson Mandela University, Port Elizabeth 6031, South Africa
e-mail: Wayne.draai@mandela.ac.za

D. Vosloo
Department of Building and Human Settlement Development, Nelson Mandela University, Port Elizabeth 6031, South Africa

T. C. Haupt et al. (eds.), *Construction in 5D: Deconstruction, Digitalization, Disruption, Disaster, Development*, Lecture Notes in Civil Engineering 245,
https://doi.org/10.1007/978-3-030-97748-1_26

1 Introduction

According to [1], "the value of buildings depends on the quality of the maintenance invested in them". Moreover, [1] also stated that "maintenance management involves obtaining maximum benefit from the investment made in the maintenance activities."

The Department of Public Works tabled a Government Immovable Assets Management Act (GIAMA) in parliament to develop and improve asset management guidelines that will provide a government-wide policy framework for the management of assets, including:

- planning for new infrastructure and maintaining existing assets more efficiently so that they can continue to be used;
- compiling and updating asset registers, including the information required to be kept in asset registers, (for example: description of the asset; condition of the asset; maintenance history; value; utilisation etc.) and
- funding of infrastructure being changed to a life-cycle funding approach [7].

"Maintenance of buildings is a combination of all the technical and administrative work undertaken on a building and its elements to full fill under service conditions all the functions established at the design stage and therefore, satisfying the users" [10].

According to [7], project delivery needs to be universally understood as embracing not only the construction of the infrastructure but also the appropriate subsequent operation and maintenance thereof for the entire design life of the asset.

2 Literature Review

2.1 Practices of Building and Infrastructure Maintenance Management

A maintenance policy is a written document that provides a framework for maintaining a facility in accordance with the strategic objectives of the organisation [13]. Furthermore, [13] explain that policy enables management to clearly define a maintenance strategy and coordinate maintenance activities efficiently and cost-effectively. All objectives of maintenance management are aimed at "total asset life cycle optimisation" [15].

The National Education Infrastructure Management System (NEIMS) has documented a significant deterioration in the condition of schools owing to inadequate maintenance. The document states that the maintenance/preservation policy of each Provincial Education Department (PED) must consistently address the planning, budgeting and implementation of preventative and maintenance policies [7]. In emphasising the need for maintenance management, [17] hold that the "concept of the operability and maintainability integrate the function and experience of operation and maintenance into project planning and design. According to [17] maintainability review sessions need to be an integral part of the project plan, as are constructability programs.

2.2 Functional Requirements of the Assets

Owing to resource constraints, a decline in the condition of the building assets of educational institutions has been observed since the early 1980s in the United Kingdom [18].

> Infrastructure delivery does not in fact end with the commission of the infrastructure asset. 'Delivery' needs to be universally understood as embracing not just constructing the infrastructure, but the appropriate operation and maintenance thereafter, for the whole design life of the asset. [8]

As stated by [6], teacher satisfaction, (and by implication, performance), is influenced by the condition of the school building. Furthermore, [6] posits that the cost to taxpayers of allowing schools to deteriorate increases when communities defer the maintenance of building systems.

2.3 Significance of Maintenance

> Buildings exist (and are created) largely for the benefits of their occupants and for what goes on in the building. For instance, an office building is for administrative functions, blocks of flats for people to live in and enjoy social interaction between residents. Building elements (floors, walls, roofs, etc.) separate residents from one another and keep at bay the external elements of wind, rain, snow, weather changes, etc. However, building maintenance is carried out to allow those functions to continue to carry out in a way and to the standard of that original envisaged designed and at least satisfactorily. [21]

Along with [17], maintenance increases the useful life of infrastructure assets, reduces costs and increases the benefits of the final delivered project. The maintainability concept is concerned with the ease of maintenance of the projects by bringing knowledge and experience of the possible maintenance concern and issues during the planning stage [17]. In addition, [18] said, it should be considered that certain buildings and infrastructure in public high schools, such as science laboratories, administration buildings etc., cannot easily be closed completely for major maintenance exercises, whereas classrooms are not used during holiday periods. Maintenance should be scheduled to accommodate operational seasons and during school holidays to avoid inconvenience and the disruption of services. In addition, Building maintenance is done to prolong the useful life of the property; retention of the property value; ensuring that facilities are in the best functional condition to promote the use or owner- organisation activities as well as to maintain the aesthetic values of the property and its environment [2].

2.4 *Classification of Maintenance*

Maintenance can be classified into Strategic repairs and maintenance; and Tactical repairs and maintenance [18].

2.4.1 Strategic Repairs and Maintenance

This represents work required for the long-term preservation of an asset and includes planned maintenance of the building fabric (decoration and routine replacement), the maintenance of engineering services installations and major repair items such as roofing. These are normally items that can be planned for because, to some extent, the need for repair can be foreseen and included in the budget.

2.4.2 Tactical Repairs and Maintenance

These items relate to day-to-day work of a minor nature in response to immediate needs. The commission indicated that 'tactical maintenance' is not necessarily the same as responsive maintenance, as some immediate response items are clearly strategic, for example, flat roof failure.

However, [18] indicated that the maintenance workload would consist of a mix of the following list of items.

- Planned maintenance
- Unplanned maintenance
- Preventive maintenance
- Corrective maintenance
- Emergency maintenance
- Condition-based maintenance
- Scheduled maintenance.

2.5 *Maintenance Management Policy*

According to [4], maintenance policy is "a set of administrative, technical and managerial actions to apply during the life cycle of a machine used to guide maintenance management decision making towards retaining certain operating conditions of a machine or dedicated to restoring the machine to said conditions". The definition is affirmed by [11] that a positive policy approach for implementing rational facilities maintenance can be promoted based on accumulated data, especially a plan or maintenance management strategy about the facilities' characteristics. In addition, [11] further argued that optimal maintenance management as an attempt to minimize life-cycle costs by carrying out an appropriate maintenance policy promotes facilities' longevity. Asset life-cycle is the cycle of activities that an asset goes through, including planning, design, initial acquisition and/or construction, cycles of operation and maintenance, capital renewal and finally, disposal. The following items discussed below are included in the maintenance management policies structure [12].

Policy statement

Maintenance policy—the overall management policy in the organisation. Departmental maintenance policy should identify how, and by whom, the maintenance of the department's building assets is to be managed. The policy should be structured to include the following components [17]:

- statement of the policy's intents and objectives
- the scope of the policy
- the details of the policy (i.e. the policy requirements)
- allocation of responsibility for implementing the various policy requirements
- continuous improvement arrangements for the policy, including policy review procedures.

The department policy statement [17] put it that, the maintenance department policy is required to make it clear their approach to addressing the following:

- establishment and periodic review of condition standard ratings for all building assets.
- preparation of a departmental maintenance strategy incorporating a balance of planned (i.e. preventative, statutory and condition-based) maintenance and unplanned maintenance).
- formulation of a Strategic Maintenance Plan (SMP) that reflects the department's maintenance needs over the immediate, medium and long term.
- development and implementation of a program of regular condition assessments.
- an assessment of maintenance demand across the department's building portfolio.
- formulation of a budget based on a realistic calculation of the level of funding needed to maintain the department's portfolio to specified condition standard ratings.
- production of annual maintenance works program based on condition assessments, existing programs, historical data and the agency asset plan.
- procurement of maintenance services in accordance with government policy.
- ongoing monitoring and periodic review of maintenance performance, including the development of performance measures and their integration with other asset management performance measures.
- establishment of processes for the collection and utilisation of maintenance information to facilitate maintenance management and meet minimum reporting requirements.
- establishment of processes for the retention of technical and asset information from handover and commissioning • arrangements for the establishment of feedback loops between maintenance service providers and building planners and

designers (effective feedback loops can facilitate improvements in maintainability, thereby minimising the maintenance needs of future buildings).

- Legal consideration and responsibilities—lease terms, repair covenants and restrictions to access that are applicable to statutory standards. According to [3]The South African Schools Act states that "the governing body of a public school must administer and control the school's property, and buildings and grounds occupied by the school, including school hostels, if applicable" (Section 20). In addition, the Department of Basic Education has developed guidance documents for use by School Principals and School Governing Bodies for maintenance implementation. These comprise:

  i. The National maintenance manual "Management of First Line Maintenance at Teaching and Learning Environment which provides detailed guidance on facilities maintenance and management. This manual is often referred to as the Schools Infrastructure Support Programme or SISP.
  ii. The standard templates (SISP) for the Maintenance plan, budget, and performance report are to be completed by each school.
  iii. Quotation templates to request quotations for maintenance work.
  iv. Contact details of the provincial Department of Public Works cost centres within the province.
  v. Prescripts for the use of the schools' fund allocation" defining the top priority maintenance work to be done using the norms and standards funding, the amount ring-fenced for facilities repairs and maintenance, and deadlines for submission of the annual maintenance plan, budget, and performance report.

- Environmental policies—on waste, noise, dust generation, sourcing of materials and energy management. construction sustainability is a way forward to balance the need to continue development without ignoring the responsibility to care for the natural environment whilst creating healthy, comfortable and economically prosperous spaces for people to live, work and play [22]. The policy as it relates to building users.
- Standard of maintenance should retain asset functioning status.
- Health and safety compliance must be clear.
- Security and access should be specified.
- Work in occupied premises.

In addition, the Security of facilities has so far improved the appearance of buildings and reduced the operating budget—due to the reduced replacement of vandalised items [12]. The subject of security in buildings includes the means and equipment to stop unauthorised entry onto land or into buildings, whether with the intention of committing a further felony or not. Such items as fences and barriers, as well as door or window locking systems and security glazing materials, fall into this category. Entry systems, which provide selective access to various parts of the building to some people—and the exclusion of others—are clearly included, as is the equipment needed to raise the alarm if an intrusion takes place. Kenneth further indicated that improving security in the ordinary running of buildings does not necessarily add to the cost of the building. The avoidance of undesirable features may actually reduce costs. Above all, the full costs-in-use must be considered: replacement costs of articles stolen, repair costs, time spent on employees and police in preventing and apprehending culprits, as well as the psychological effect on the victims. As indicated, maintenance policy should always consider school asset security.

Organisation

The organisational structure of the school maintenance programme should clearly define duties and responsibilities. Representatives from parent/teacher associations, SMT, SGB, students, and any other school organisations should be responsible for establishing the school maintenance programme [12].

According to [3] Section 2 of the South African Schools Act (SASA), No. 84 of 1999 states that the School Governing Body must maintain and improve the school property and buildings and grounds occupied by the schools, including school hostels. Basic Education Department [3] further indicates that School Principals and School Governing Bodies are responsible for school operation under Section 20, and most are responsible for school maintenance under Section 21. The policy further clarified that preventive maintenance activities such as replacing doors, windows and fittings, repairing water systems, plumbing and toilets, repairing the electrical system, emptying toilets, one building to be repaired and painted every year as necessary, repair and paint the roof of one building each year to ensure that every roof is painted every 5–15 years, Termite treatment every 5–10 years as necessary, repairing roof leaks, cleaning gutters and stormwater drains every year. 'The policy specified that major maintenance (roof trusses are rotten or broken, cracks in a wall more than 5 mm wide, rehabilitation or replacement of severely damaged buildings, additional classrooms and administration blocks, major rehabilitation of buildings, Major civil works for the stormwater management system should be carried by Provincial department'.

Moreover, [12] believe that maintenance organisational structure should include:

- Management staff.
- Preventive maintenance and planning cycle proportion and type of preventive maintenance and timeframes for inspections/service.
- Identifying maintenance needs—reports of defects, inspection types and cycles, prioritizing maintenance needs.
- Routine servicing requirements and cycles.

Procurement

A procurement strategy determines the methods used to procure maintenance services [17]. The procurement policy should reflect the Department of Basic Education Immovable Asset procurement policy. According to [3] school assets maintenance procurement is done under three levels of authority, School Principals and School Governing Bodies, Circuits and/or District Management and The provincial infrastructure directorate. Then a technician is appointed to submit the assessed condition of the facilities report to the authorities. Based on the report including other maintenance information available to the relevant authorities, the appointment of the contractor is done. Usually, the Department of Basic Education uses outsource for execution of the maintenance [12] and said the procurement policy should include the following:

- Use of contractors and direct labour.
- Approve contractor list policy—contractor selection and vetting.
- Contracts—timeframes, costs, levels of service.
- Emergency and disaster routines.

Monitoring and Feedback

The condition of a building asset—i.e. a building's physical state of repair—influences its physical and functional performance [17] and this can be achieved through physical monitoring and inspection. A preliminary school building inspection needs to be conducted in order to prepare a school maintenance plan. The information gathered during the preliminary school building inspection is the basis for the maintenance programme. A school building report should be prepared in order to summarise the inspection. The general coordinator and the person responsible for each team should have a meeting to discuss the purpose of the inspection and the instruments used for data collection. The inspection should be conducted by technical personnel capable of identifying major deficiencies that may affect the normal activities in the school. The preliminary inspection is intended to assist in the evaluation of the overall condition of the school building [12] and said maintenance monitoring and feedback policy should include:

- Reporting to the management board.
- Quality management routines.
- Benchmarking.

2.5.1 Maintenance Schedule/Manual

The effects of a maintenance schedule in relation to the maintenance context has been described as follows [5]:

- A maintenance activity schedule is used to inform the maintenance organisational structure to perform the critical work that is to be executed.
- A maintenance activity schedule is also used as the basis for determining the tools and other equipment required to perform the necessary maintenance.
- Outsourcing or the use of alternative delivery mechanisms to perform tasks, or groups of maintenance tasks, must be considered.

2.5.2 Establishment of an Asset Maintenance System

Maintenance system has been described as a process of monitoring, evaluating and modifying existing information systems to make required or desirable improvements [20]. It was further stated that the results obtained from the evaluation process assist the organisation to determine whether or not its information systems are effective and efficient. This indicates that to achieve the maintenance objective, it is important to effectively establish an asset maintenance system for the institution.

- The tasks listed below can be utilised to establish an asset maintenance system.
- Maintenance activities may be scheduled and controlled using an appropriate system(s), such as a computerised maintenance management system (CMMS), or an integrated maintenance management system (IMMS).
- The maintenance system(s) include(s) the functions listed below:
- Recording the progress of activities. Have they been completed or reprogrammed?
- Recording the maintenance costs, time and other resources consumed against assets and facilities.
- Linking the activities to the financial management system so that reconciliation of maintenance budgets can be performed.
- Employing built-in maintenance analysis tools or export information to other applications to enable maintenance analyses to be undertaken.
- Analysing asset performance to use it as an input for planning future maintenance.
- Establishing a link between the maintenance management system and the customer complaints system, as this is one of the main points of origin of unplanned maintenance activities [5].

2.5.3 Maintenance Performance Indicators

According to [21], performance indicators, or key performance indicators (KPIs), are types of performance measurements that are used to evaluate the success of an organisation or of an activity, such as projects, programs, products and other initiatives in which it engages. In addition, [15] said, the level of performance of maintenance operations carried out is gauged by set criteria that act as indicators for achieving maintenance because they describe the condition that exists when maintenance tasks are completed. The criteria, or indicators, act as instruments for measuring maintenance standards.

In the context of this research, the following were used to determine maintenance performance indicators.

- Appropriate key performance indicators (KPIs) were identified and used to monitor maintenance performance.
- The maintenance management KPIs were obtained from the service delivery and budget implementation plan and performance management system, where possible [5].

2.6 Maintenance Generators

According to [9] defects in buildings refer to the degree of physical deterioration of building elements and services, including a decline in operational effectiveness, aesthetic appearance and decayed materials that present a hazard to occupants/users. Improper use, inappropriate maintenance strategies and ignorance of the appropriate use of building facilities are the major factors responsible for the acceleration and intensity of deterioration [9]. Factors responsible for the process are ageing, wear and tear. Maintenance generators fall into five main categories [18]:

- Technical
- Managerial
- Political and Legal
- Environmental and Biological
- Social and Cultural.

The maintenance generators may be narrowed into two sections relative to the problems under investigation.

2.6.1 Maintenance Generators

Technical defects/maintenance related defects: According to [17] Building characteristics and design defects mostly arise the need for extension of initial three principles of integration, construction knowledge and team skills, whilst construction related issues occurs because construction knowledge and methodologies extremely need to be operability and maintainability oriented [17].

- Faulty design or construction or both can result in defects in the building
- Poor quality of material and workmanship during initial construction phase
- Building facilities are subjected to wear and tear when in use ageing is inevitable, whether the building is in use or not, therefore, there are defects or deterioration that are related to the age of the building
- Poor quality of material and workmanship when undertaking maintenance works.

2.6.2 Maintenance Generators

Social and cultural defects/user related defects: this has to do with the Safety and security of end-users. Social and cultural factors have direct influence on operability and maintainability implementation process. Social and cultural generators include:

- Misuse of the building—i.e. exceeding the functional design of the asset.
- Exceeding the designed performance capability and capacity of the building.
- Vandalism.
- Delay of maintenance work.
- High expectations and needs of the users.

- Problems with reporting delays.
- Faulty use [17].

2.7 *Building and Other Infrastructure Maintenance Standards*

The quality of maintenance services invested in a building enables it to attain certain acceptable physical, functional and economic standards set by the organization as part of its policy [2]. According to [9] technological, economic, sociological and political environments in which an organisation finds itself affect the standard it sets for maintaining its built environment, as these standards are based on, but not limited to, the following factors:

- Financial: standard determined by financial capability;
- Functional: use of the building and;
- Environmental: availability and quality of the infrastructure.

> Maintenance standards may differ or change based on the utility and value. For instance, the maintenance standard of a hotel will obviously be different from that of a student hostel, hospital, although all facilities are categorised as hospitality structures. Thus, utility (function of the building, its public image, and even national prestige) and commercial value of a residential building have a great influence on the maintenance standard. [14]

3 Research Methodology

The research methodology utilized a quantitative approach which began with formulating a problem statement and identifying the objectives of the study. The first step of conducting this research was formed based on reviews of the literature to ascertain why public high school buildings and infrastructure are not meeting their functional requirements. Then, the operationalization of the established maintenance generator and maintenance performance indicators into a questionnaire were carried out. Subsequently, pilot testing of the questionnaire was carried out and the developed format of the questionnaire was formed.

3.1 *Data Collection Methods*

Data collection method comprised the questionnaire design, pre-testing of the questionnaire, questionnaire administration and research population and sample. Data achieved using questionnaires from respondents was gathered and quantitatively analysed. A total number of thirty (30) questionnaires (hard and soft copies) was distributed to all parties involved in the maintenance department of the Department of

Education in South Africa, Kwazulu Natal Province in uThungulu District, including Principals, Deputy Principals, Head of Departments, Director of Department, Staff, Senior Staff, Supervisor and Trainee. The respondents' working experience ranged from less than three years to more than thirty (30) years and they had different levels of education from Diploma to B Tech degrees. However, only Seventeen (17) questionnaires were returned, which constitutes a sum of a 52% percent response rate.

4 Results

4.1 Relationship Between Performance of Maintenance Management and Asset Functional Requirements

See Tables 1, 2, 3, 4, 5, 6, 7 and 8.

Table 1 Analysis: building maintenance conducted in accordance with policy documents

Maintenance conducted in accordance with policy					
1 Strongly disagree	2 Disagree	3 Agree	4 Strongly agree	5 Unsure	Total
0	0	3	14	0	17
0%	0%	17.6%	82.4%	0%	100%

	Mean	Mode	Range	Minimum	Maximum
Maintenance is conducted in accordance with policy	3.82	4	1	3	4

Table 2 Analysis: delay in maintenance work results in deterioration of facilities

Delay in maintenance work results in deterioration of such facilities					
1 Strongly disagree	2 Disagree	3 Agree	4 Strongly agree	5 Unsure	Total
0	0	3	13	0	16
0%	0%	18.8%	81.2%	0%	100%
No response					1

	Mean	Mode	Range	Minimum	Maximum
Delay in maintenance work	3.81	4	1	3	4

Table 3 Analysis: damage as a result of vandalism must be repaired without delay

Damage to buildings as a result of vandalism is to be repaired without delay					
1 Strongly disagree	2 Disagree	3 Agree	4 Strongly agree	5 Unsure	Total
0	0	4	13	0	17
0%	0%	23.5%	76.5%	0%	100%

	Mean	Mode	Range	Minimum	Maximum
Damage to the buildings as a result of vandalism to be repaired without delay	3.76	4	1	3	4

Table 4 Analysis: impact of the school buildings' condition on teachers' work satisfaction

Impact of the school buildings' condition on teachers' work satisfaction					
1 Strongly disagree	2 Disagree	3 Agree	4 Strongly agree	5 Unsure	Total
0	0	3	13	0	16
0%	0%	18.8%	81.2%	0%	100%
No response					1

	Mean	Mode	Range	Minimum	Maximum
Teachers' work satisfaction is influenced by the condition of the school buildings	3.8	4	1	3	4

Table 5 Analysis: maintenance management's objective is to achieve asset life cycle optimisation

Maintenance management's objective is to achieve total asset life cycle optimisation					
1 Strongly disagree	2 Disagree	3 Agree	4 Strongly agree	5 Unsure	Total
0	0	0	17	0	17
0%	0%	0%	100%	0%	100%

	Mean	Mode	Range	Minimum	Maximum
Maintenance management's objective is to achieve total asset life cycle optimisation	4.00	4	0	4	4

Table 6 Analysis: maintenance manual/schedule

Buildings			Infrastructure		
	Number	Percentage (%)		Number	Percentage (%)
No	12	70.6	No	11	64.7
Yes	5	29.4	Yes	6	35.3
Total responses	17	100	Total responses	17	100

Table 7 Analysis: maintenance manual/schedule assistance level

Assistance level of maintenance manual/schedule						
1 Very good assistance	2 Good assistance	3 Average assistance	4 Poor assistance	5 Very poor assistance	6 Unsure	Total
3	7	0	5	0	0	15
20.0%	46.7%	0%	33.3%	0%	0%	100%
No response						2

	Mean	Mode	Range	Minimum	Maximum
Maintenance manual/schedule assistance level feedback	2.47	2	3	1	4

Table 8 Analysis: indication of properly implemented maintenance management policy benefitting the quality of teaching and learning

Benefits of maintenance policy to teaching and learning	Number	Percentage (%)
Create a sound learning environment for learning and teaching	6	40.0
Safer learning environment	2	13.3
Improve teachers' performance	7	46.7
Total responses	15	100
No response	2	

4.2 Discussions and Interpretation of Functional Requirements of the Assets

The interpretation of the feedback pertaining to the functional requirements of public high school buildings and infrastructure in relation to the impact of maintenance policies, the delay in maintenance work, vandalism, users' needs and teachers' work satisfaction led to the conclusions discussed below:

- The information pertaining to buildings' and infrastructure's functional requirements provides a good indication of the assets' shortcomings in relation to their attainment of the users' needs and satisfaction, as well as policy direction pertaining to the ways in which maintenance should be conducted.
- As per Table 1, the majority (82.4%) of the respondents strongly agreed that maintenance conducted in accordance with policy documents satisfies users' needs and expectations. This implies that policy-directed maintenance will enhance the functioning of the assets and satisfy the users' needs.
- As per Tables 2 and 3, an overwhelming proportion (81.2% and 76.5% respectively) of the respondents, (with regard to general maintenance and vandalism), indicated that delayed maintenance results in the deterioration of buildings and infrastructure. The interpretation of the findings led to the conclusion that a

delay in maintenance activity could be attributed to constrained resources (monetary, human or equipment). Management should attempt to resolve such issues timeously and avoid such delays.

- According to Table 4, the majority (81.2%) of the respondents indicated that in their opinion the condition of the buildings and infrastructure has a direct impact on teachers' work satisfaction. This implies that teachers' and learners' performance could improve if the buildings in which they function are in a good condition.
- In Table 5 all the respondents (100%) reflected that the core objective of maintenance management is to achieve total asset life cycle optimisation. This implies that to achieve total asset life cycle optimisation, effective maintenance management activities should be sustained.
- As per Table 6, the majority (70.6%) of the respondents stated that they do not have a maintenance manual/schedule for buildings and 64.7% do not have a maintenance schedule for infrastructure. This could lead to maintenance being delayed, as access to detailed specifications and specific parts of the facility were lacking and this could result in total failure of the asset and excessive costs.
- A significant portion (46.7%) of the respondents (Table 7) noted that the level of assistance provided by a maintenance manual/schedule is good. This result indicates that the respondents understood the importance of a maintenance manual/schedule, as it provides easy access to specifications and the requirement for parts.
- According to Table 8, 46.7% of the respondents mentioned "Improve teachers' performance" as the key benefit derived from a properly implemented maintenance management policy and 40.0% stated, "Create a sound learning environment for learning and teaching". These results indicate that significant benefits could be gained if maintenance management policy is applied effectively.

5 Findings

The research revealed that currently, there is a need to improve the understanding and practices pertaining to the existing maintenance management policy system. The salient findings of the research revealed that most of the institutions do not comply with the existing maintenance management policy system. The finding further revealed that the maintenance management policy currently in place is reactive and lacks proper planning for maintenance activities.

6 Summary

In summary, the research was aimed at determining an effective maintenance management policy system for public high school buildings and other infrastructure within the Department of Education in South Africa, based on standard criteria practice with the key objective of *"To implement effective maintenance management policy systems to enable public high school buildings and infrastructure to perform their required functions*".

The research found that it is accepted as a fact that learning and teaching can be improved by an asset environment that is in a sound condition. It is required that maintenance be conducted in accordance with the maintenance policy guidelines to achieve maximum performance of the asset and to attain total life cycle optimisation.

6.1 Conclusion

The prime objective of providing buildings and other infrastructure for public high schools is to promote academic excellence. Academic performance is enhanced by a sound learning environment provided by the proper maintenance and repair of the asset. To keep assets in sound condition, strategic planning should be practised, as a planning gap leads to top management paying less attention to maintenance activities.

The research concluded that to avoid maintenance delays and to meet users' needs with improved asset performance and user performance (teachers and learners), and to maintain assets in an acceptable condition, maintenance activities should be guided by maintenance policy documents. This will lead to the achievement of the core maintenance objective of total asset life cycle optimisation.

6.2 Recommendations

Maintenance managers should make management and the occupants/users aware of the importance of the work performed by the assets maintenance department and the role they have in assisting the assets maintenance department by reporting defective items.

Management should engage competent employees for the maintenance department and provide sufficient tools and equipment for maintenance activities.

There should be a clear schedule of duties for all workers that are involved in managing the buildings and other infrastructure with an adequate understanding of the span of the organisational structure.

The research further encourages that maintenance manual should be prepared for all the existing buildings and other infrastructure that has no maintenance manual and should be updated regularly if there are any changes to components or the rate of services when maintenance work is carried out.

Policy best practice standards should be set to determine the minimum acceptable state or condition of the buildings and other infrastructure to eliminate both minor and major breakdowns of the buildings resulting in academic performance being affected negatively.

References

1. Abdul Lateef O (2009) Building maintenance management in Malaysia. J Build Appraisal 1
2. Abdul Lateef O, Khamidi MF, Idrus A. Sustainability in the context of maintenance: building defects in the Malaysian University Campuses. Kuala Lumpur, International
3. Conference 2010 on sustainable building and infrastructure (2010)
4. Basic Education Department (2012) Facilities maintenance guidelines for public schools. Basic Education Department, South Africa
5. Business Dictionary (2017) Encyclopedia of business analytic and optimization 2016 [online]. Available through www.igi-global.Com/dictionary/maintenance-policy/39946. Accessed 7 Aug 2017
6. Cape Winelands Municipality (2015) Maintenance management policy [online]. Available at http://www.capewinelands.gov.za/DocumentsDownload/Reports/Budget%20Related. Accessed 10 Aug 2017
7. Council of the Great City Schools (2016) Reversing the cycle of deterioration in the Nation's Public-School Buildings 2014. Pennsylvania Ave N.W. Available through www.Cgc.org/cmc/lib/centricity/Domain/87/facilitiesReport. Accessed 12 Feb 2016
8. Department of Basic Education (DBE) (2015) National school infrastructure maintenance guide. DBE online. Available through http://www.kzneducation.gov.za/portals/0/Infrastructureplanninganddelivery/Maintenance. Accessed 13 Mar 2015
9. Department of Public Works (DPW) (2010) National infrastructure maintenance strategy in support of ASGISA and government growth objectives 2010. Department of Public Works, CIDB, Council for Scientific and Industrial Research: South Africa
10. Dzikiwi A (2015) Maintenance management systems of on-campus student hostels at Nigerian Universities. Nelson Mandela Metropolitan University. Unpublished Ph.D. thesis in Construction Management
11. Flores-Colen I, de Brito J, Freitas V (2010) Discussion of criteria for prioritisation of predictive maintenance of building façade. J Performance Constr Facilities Los Angeles 1
12. Hamza S, Kobayashi K (2012) Utilizing mid-long-term maintenance management policy for sustainable maintenance of infrastructure facilities. In: The 3rd international conference 2012 on sustainable future for human security. Elsevier B.V., Indonesia
13. Kenneth B (2011) Maintenance of university facilities in developing country. Nelson Mandela Metropolitan University. Unpublished M.Sc. treatise in construction management
14. Lee H, Scott D (2009) Overview of maintenance strategy, acceptable maintenance from a building maintenance operation perspective. J Retail Leisure Property:29–30
15. Overstrand, Municipality (2017) Overstrand municipality maintenance management policy. Council of Overstrand municipality. Available at http://www.Overstrand.gov.za/en/documents/policies/168-maintenance-management/file. Accessed 07 Aug 2017
16. Overveen (2011) Effective maintenance management practice [online]. Available through http://www.trimitra.com/services/effective-maintenance-management-practices. Accessed 23 Nov 2015
17. Queensland Department of Housing and Public Works (2017) Maintenance management framework guideline: building maintenance policy, standards and strategy development. The State of Queensland. Department of Housing and Public Works. Australia
18. Saghatforoush E, Trigunarsyah B, Too E (2012) Assessment of operability and maintainability success factors in provision of extended constructability principles. In: Isfatan, 9th international congress on civil engineering
19. Swallow B, Chanter P (2007) Building maintenance management, 1st edn. John Wiley and Sons, UK
20. Swallow B, Chanter P (2010) Building maintenance management, 2nd edn. Blackwell, Oxford
21. Wikipedia (2017) Performance indicator. Wikipedia [online]. Available at http://en.wikipedia.org/wiki/performance-indicator. Accessed 9 Aug 2017

22. Wood B (2009) Building maintenance. Blackwell, United Kingdom
23. Zainul Abidin N (2010) Investigating the awareness and application of sustainable construction concept by Malaysian developers. Habitat Int 34(4):421–426

Influencing Building and Infrastructure Poor Performance to Inadequate Maintenance Budget and Funding

Abdul Razak Musah, Derick Vosloo, and Wayne Draai

Abstract This research aimed to establish the financial resource requirements for the maintenance management of public high school buildings and infrastructure. Quantitative methodology was employed in this research. The research included a study of relevant literature and the distribution of questionnaires to the target sample to gather the required data that seek to establish an improved maintenance management policy system for public high schools based on standard practice. The uThungulu District of KZN in South Africa was the selected geographical scope of this research. Thirty public high schools in the district were purposely selected for the study. The reasons include the proximity of the schools relative to where the researcher currently resides; resource constraints regarding the ability to fund transport costs to the various schools. The questions that were related to a maintenance funding and budget, functional requirements of public high school buildings and other infrastructure were correlated with the data that were generated from the maintenance management stakeholders at the various institutions. The research finding was that resource constraints continue to pose a challenge to public high schools' maintenance departments. The research revealed that most (37.5%) of the respondents perceive that the method used for determining maintenance costs was to use the previous year's budget and add a percentage amount and concluded that effective estimation techniques should be employed to satisfy the financial requirements of ten (10) to fifteen (15)% of the annual budget for maintenance activities of the institutions.

A. R. Musah (✉) · W. Draai
Faculty of Engineering, the Built Environment and Information Technology, Nelson Mandela University, Port Elizabeth 6031, South Africa
e-mail: s214231976@mandela.ac.za

W. Draai
e-mail: Wayne.draai@mandela.ac.za

D. Vosloo
Department of Building and Human Settlement, Nelson Mandela University, Port Elizabeth 6031, South Africa
e-mail: s214231976@mandela.ac.za

T. C. Haupt et al. (eds.), *Construction in 5D: Deconstruction, Digitalization, Disruption, Disaster, Development*, Lecture Notes in Civil Engineering 245,
https://doi.org/10.1007/978-3-030-97748-1_27

Keywords Maintenance funding and budget · Functional requirements · Buildings and infrastructure

1 Introduction

Department of Basic Education [7] holds that "Until a more specific guideline can be made available, an amount at least 4–5% of the Provincial Department of Education's (PED's) budget figure will need to be ring-fenced by the PED for sole application in maintenance activities."

According to [1], "the value of buildings depends on the quality of the maintenance invested in them". [1] also stated that "maintenance management involves obtaining maximum benefit from the investment made on the maintenance activities."

The Department of Public Works tabled a Government Immovable Assets Management Act (GIAMA) in parliament to develop and improve asset management guidelines that will provide a government-wide policy framework for the management of assets, including:

- Planning for new infrastructure and maintaining existing assets more efficiently so that they can continue to be used;
- Compiling and updating asset registers, including the information required to be kept in asset registers, (for example description of the asset; condition of the asset; maintenance history; value; utilisation etc.) and Funding of infrastructure is being changed to a life-cycle funding approach [8].

"Maintenance of buildings is a combination of all the technical and administrative work undertaken on a building and its elements to full fill under service conditions all the functions established at the design stage and therefore, satisfying the users" [10].

According to [8], project delivery needs to be universally understood as embracing not only the construction of the infrastructure but also the appropriate subsequent operation and maintenance thereof for the entire design life of the asset.

2 Literature Review

2.1 Practices of Building and Infrastructure Maintenance Management

A maintenance policy is a written document that provides a framework for maintaining a facility in accordance with the strategic objectives of the organisation [13, 14]. Lee and Scott [13, 14] explain that policy enables management to clearly define a maintenance strategy and coordinate maintenance activities efficiently and cost-effectively. All objectives of maintenance management are aimed at "total asset life cycle optimisation" [16, 17].

The National Education Infrastructure Management System (NEIMS) has documented a significant deterioration in the condition of schools owing to inadequate maintenance. The document states that the maintenance/preservation policy of each Provincial Education Department (PED) must consistently address the planning, budgeting and implementation of preventative and maintenance policies [7]. In emphasising the need for maintenance management, [19] hold that the "concept of the operability and maintainability integrate the function and experience of operation and maintenance into project planning and design. [19] further claim that maintainability review sessions need to be an integral part of the project plan, as are constructability programs.

2.2 *Functional Requirements of the Assets*

Owing to resource constraints, a decline in the condition of the building assets of educational institutions has been observed since the early 1980s in the United Kingdom [20, 21]

> Infrastructure delivery does not in fact end with the commission of the infrastructure asset. 'Delivery' needs to be universally understood as embracing not just constructing the infrastructure, but the appropriate operation and maintenance thereafter, for the whole design life of the asset. [8]

According to the [6], teacher satisfaction, (and by implication, performance), is influenced by the condition of the school building. The Council [6] posits that the cost to taxpayers of allowing schools to deteriorate increases when communities defer the maintenance of building systems.

2.3 *Significance of Maintenance*

> Buildings exist (and are created) largely for the benefits of their occupants and for what goes on in the building. For instance, an office building is for administrative functions, blocks of flats for people to live in and enjoy social interaction between residents. Building elements (floors, walls, roofs, etc.) separate residents from one another and keep at bay the external elements of wind, rain, snow, weather changes, etc. However, building maintenance is carried out to allow those functions to continue to carry out in a way and to the standard of that original envisaged designed and at least satisfactorily. [24]

According to [19], maintenance increases the useful life of infrastructure assets, reduces costs and increases the benefits of the final delivered project. The maintainability concept is concerned with the ease of maintenance of the projects by bringing knowledge and experience of the possible maintenance concern and issues during the planning stage [19]. Swallow and Chanter [20, 21] said, it should be considered that certain buildings and infrastructure in public high schools, such as science laboratories, administration buildings etc., cannot easily be closed completely for major maintenance exercises, whereas classrooms are not used during holiday periods. Maintenance should be scheduled to accommodate operational seasons and during school holidays to avoid inconvenience and the disruption of services. In addition, Building maintenance is done to prolong the useful life of the property; retention of the property value; ensuring that facilities are in the best functional condition to promote the use or owner- organisation activities as well as to maintain the aesthetic values of the property and its environment [2].

2.4 *Classification of Maintenance*

[20] believe that maintenance can be classified into Strategic repairs and maintenance; and Tactical repairs and maintenance.

Strategic repairs and maintenance

This represents work required for the long-term preservation of an asset and includes planned maintenance of the building fabric (decoration and routine replacement), the maintenance of engineering services installations and major repair items such as roofing. These are normally items that can be planned for because, to some extent, the need for repair can be foreseen and included in the budget.

Tactical repairs and maintenance

These items relate to day-to-day work of a minor nature in response to immediate needs. The commission indicated that 'tactical maintenance' is not necessarily the same as responsive maintenance, as some immediate response items are clearly of a strategic nature, for example, flat roof failure.

Swallow and Chanter [20] indicated that the maintenance workload would consist of a mix of the following list of items.

- Planned maintenance
- Unplanned maintenance
- Preventive maintenance
- Corrective maintenance
- Emergency maintenance
- Condition-based maintenance
- Scheduled maintenance.

2.5 Maintenance Management Policy

According to [4], maintenance policy is "a set of administrative, technical and managerial actions to apply during the life cycle of a machine used to guide maintenance management decision making towards retaining certain operating conditions of a machine or dedicated to restoring the machine to said conditions". The definition is affirmed by [11] that a positive policy approach for implementing rational facilities maintenance can be promoted based on accumulated data, especially a plan or maintenance management strategy about the facilities' characteristics. Hamza and Kobayashi [11] further argued that optimal maintenance management as an attempt to minimize life-cycle costs by carrying out an appropriate maintenance policy promotes facilities' longevity. Asset life-cycle is the cycle of activities that an asset goes through, including planning, design, initial acquisition and/or construction, cycles of operation and maintenance, capital renewal and finally, disposal. Kenneth [12] posits that the following items discussed below are included in the maintenance management policies structure.

Policy statement

Maintenance policy—the overall management policy in the organisation. Departmental maintenance policy should identify how, and by whom, the maintenance of the department's building assets is to be managed. The policy should be structured to include the following components [18]:

- statement of the policy's intents and objectives
- the scope of the policy
- the details of the policy (i.e. the policy requirements)
- allocation of responsibility for implementing the various policy requirements
- continuous improvement arrangements for the policy, including policy review procedures.

The department policy statement [18] put it that, the maintenance department policy is required to make it clear their approach to addressing the following:

- establishment and periodic review of condition standard ratings for all building assets.
- preparation of a departmental maintenance strategy incorporating a balance of planned (i.e. preventative, statutory and condition-based) maintenance and unplanned maintenance).
- formulation of a Strategic Maintenance Plan (SMP) that reflects the department's maintenance needs over the immediate, medium and long term.
- development and implementation of a program of regular condition assessments.
- an assessment of maintenance demand across the department's building portfolio.
- formulation of a budget based on a realistic calculation of the level of funding needed to maintain the department's portfolio to specified condition standard ratings.

- production of annual maintenance works program based on condition assessments, existing programs, historical data and the agency asset plan.
- procurement of maintenance services in accordance with government policy.
- ongoing monitoring and periodic review of maintenance performance, including the development of performance measures and their integration with other asset management performance measures.
- establishment of processes for the collection and utilisation of maintenance information to facilitate maintenance management and meet minimum reporting requirements.
- establishment of processes for the retention of technical and asset information from handover and commissioning • arrangements for the establishment of feedback loops between maintenance service providers and building planners and designers (effective feedback loops can facilitate improvements in maintainability, thereby minimising the maintenance needs of future buildings).
- Legal consideration and responsibilities—lease terms, repair covenants and restrictions to access that are applicable to statutory standards. According to [3] The South African Schools Act states that "the governing body of a public school must administer and control the school's property, and buildings and grounds occupied by the school, including school hostels, if applicable" (Section 20). In addition, the Department of Basic Education has developed guidance documents for use by School Principals and School Governing Bodies for maintenance implementation. These comprise:

  i. The National maintenance manual "Management of First Line Maintenance at Teaching and Learning Environment which provides detailed guidance on facilities maintenance and management. This manual is often referred to as the Schools Infrastructure Support Programme or SISP.
  ii. The standard templates (SISP) for the Maintenance plan, budget, and performance report are to be completed by each school.
  iii. Quotation templates to request quotations for maintenance work.
  iv. Contact details of the provincial Department of Public Works cost centres within the province.
  v. Prescripts for the use of the schools' fund allocation" defining the top priority maintenance work to be done using the norms and standards funding, the amount ring-fenced for facilities repairs and maintenance, and deadlines for submission of the annual maintenance plan, budget, and performance report.

- Environmental policies—on waste, noise, dust generation, sourcing of materials and energy management. construction sustainability is a way forward to balance the need to continue development without ignoring the responsibility to care for the natural environment whilst creating healthy, comfortable and economically prosperous spaces for people to live, work and play [25]. The policy as it relates to building users.
- Standard of maintenance should retain asset functioning status.
- Health and safety compliance must be clear.

- Security and access should be specified.
- Work in occupied premises.

In addition, the Security of facilities has so far improved the appearance of buildings and reduced the operating budget—due to the reduced replacement of vandalised items [12]. The subject of security in buildings includes the means and equipment to stop unauthorised entry onto land or into buildings, whether with the intention of committing a further felony or not. Such items as fences and barriers, as well as door or window locking systems and security glazing materials, fall into this category. Entry systems, which provide selective access to various parts of the building to some people—and the exclusion of others—are clearly included, as is the equipment needed to raise the alarm if an intrusion takes place. Kenneth further indicated that improving security in the ordinary running of buildings does not necessarily add to the cost of the building. The avoidance of undesirable features may actually reduce costs. Above all, the full costs-in-use must be considered: replacement costs of articles stolen, repair costs, time spent on employees and police in preventing and apprehending culprits, as well as the psychological effect on the victims. As indicated, maintenance policy should always consider school asset security.

Organisation

The organisational structure of the school maintenance programme should clearly define duties and responsibilities. Representatives from parent/teacher associations, SMT, SGB, students, and any other school organisations should be responsible for establishing the school maintenance programme [12]. According to [3] Section 21 of the South African Schools Act (SASA), No. 84 of 1999 states that the School Governing Body must maintain and improve the school property and buildings and grounds occupied by the schools, including school hostels. Basic Education Department [3] further indicates that School Principals and School Governing Bodies are responsible for school operation under Section 20, and most are responsible for school maintenance under Section 21. The policy further clarified that preventive maintenance activities such as replacing doors, windows and fittings, repairing water systems, plumbing and toilets, repairing the electrical system, emptying toilets, one building to be repaired and painted every year as necessary, repair and paint the roof of one building each year to ensure that every roof is painted every 5–15 years, Termite treatment every 5–10 years as necessary, repairing roof leaks, cleaning gutters and stormwater drains every year. 'The policy specified that major maintenance (roof trusses are rotten or broken, cracks in a wall more than 5 mm wide, rehabilitation or replacement of severely damaged buildings, additional classrooms and administration blocks, major rehabilitation of buildings, Major civil works for the stormwater management system should be carried by Provincial department'.

Moreover, [12] believe that maintenance organisational structure should include:

- Management staff.
- Preventive maintenance and planning cycle proportion and type of preventive maintenance and timeframes for inspections/service.

- Identifying maintenance needs—reports of defects, inspection types and cycles, prioritizing maintenance needs.
- Routine servicing requirements and cycles.

Procurement

A procurement strategy determines the methods used to procure maintenance services [18]. The procurement policy should reflect the Department of Basic Education Immovable Asset procurement policy. According to [3] school assets maintenance procurement is done under three levels of authority, School Principals and School Governing Bodies, Circuits and/or District Management and The provincial infrastructure directorate. Then a technician is appointed to submit the assessed condition of the facilities report to the authorities. Based on the report including other maintenance information available to the relevant authorities, the appointment of the contractor is done. Usually, the Department of Basic Education uses outsource for execution of the maintenance [12] and said the procurement policy should include the following:

- Use of contractors and direct labour.
- Approve contractor list policy—contractor selection and vetting.
- Contracts—timeframes, costs, levels of service.
- Emergency and disaster routines.

Monitoring and Feedback

The condition of a building asset—i.e. a building's physical state of repair—influences its physical and functional performance [18] and this can be achieved through physical monitoring and inspection. A preliminary school building inspection needs to be conducted in order to prepare a school maintenance plan. The information gathered during the preliminary school building inspection is the basis for the maintenance programme. A school building report should be prepared in order to summarise the inspection. The general coordinator and the person responsible for each team should have a meeting to discuss the purpose of the inspection and the instruments used for data collection. The inspection should be conducted by technical personnel capable of identifying major deficiencies that may affect the normal activities in the school. The preliminary inspection is intended to assist in the evaluation of the overall condition of the school building [12] and said maintenance monitoring and feedback policy should include:

- Reporting to the management board.
- Quality management routines.
- Benchmarking.

2.5.1 Maintenance Schedule/Manual

The [5] highlighted the effects of a maintenance schedule in relation to the maintenance context as described as follows:

- A maintenance activity schedule is used to inform the maintenance organisational structure to perform the critical work that is to be executed.
- A maintenance activity schedule is also used as the basis for determining the tools and other equipment required to perform the necessary maintenance.
- Outsourcing or the use of alternative delivery mechanisms to perform tasks, or groups of maintenance tasks, must be considered.

2.5.2 Establishment of an Asset Maintenance System

Thakur [22] described a maintenance system as a process of monitoring, evaluating and modifying existing information systems to make required or desirable improvements. It was further stated that the results obtained from the evaluation process assist the organisation to determine whether or not its information systems are effective and efficient. This indicates that to achieve the maintenance objective, it is important to effectively establish an asset maintenance system for the institution.

The tasks listed below can be utilised to establish an asset maintenance system.

- Maintenance activities may be scheduled and controlled using an appropriate system(s), such as a computerised maintenance management system (CMMS), or an integrated maintenance management system (IMMS).

  The maintenance system(s) include(s) the functions listed hereunder.

- Recording the progress of activities. Have they been completed or reprogrammed?
- Recording the maintenance costs, time and other resources consumed against assets and facilities.
- Linking the activities to the financial management system so that reconciliation of maintenance budgets can be performed.
- Employing built-in maintenance analysis tools or export information to other applications to enable maintenance analyses to be undertaken.
- Analysing asset performance to use it as an input for planning future maintenance.
- Establishing a link between the maintenance management system and the customer complaints system, as this is one of the main points of origin of unplanned maintenance activities [5].

2.5.3 Maintenance Performance Indicators

According to [23], performance indicators, or key performance indicators (KPIs), are types of performance measurements that are used to evaluate the success of an organisation or of an activity, such as projects, programs, products and other initiatives in which it engages. [16] said, the level of performance of maintenance operations carried out is gauged by set criteria that act as indicators for achieving maintenance because they describe the condition that exists when maintenance tasks are completed. The criteria, or indicators, act as instruments for measuring maintenance standards.

In the context of this research, the following were used to determine maintenance performance indicators.

- Appropriate key performance indicators (KPIs) were identified and used to monitor maintenance performance.
- The maintenance management KPIs were obtained from the service delivery and budget implementation plan and performance management system, where possible.

2.6 Maintenance Generators

According to [19], maintenance generators fall into five main categories.

- Technical
- Managerial
- Political and Legal
- Environmental and Biological
- Social and Cultural.

The maintenance generators may be narrowed into two sections relative to the problems under investigation.

2.6.1 Maintenance Generators

Managerial defects/financial defects

- Inadequate budgeting for maintenance work.
- Insufficient resources for maintenance, creating maintenance backlog.
- Poor information and communication technology in the organisation (lack of information logistic resources).
- Low maintenance cost estimates and cost implications of delayed repairs.

2.6.2 Maintenance Generators

Social and cultural defects/user related defects

- Misuse of the building—i.e. exceeding the functional design of the asset.
- Exceeding the designed performance capability and capacity of the building.
- Vandalism.
- Delay of maintenance work.

- High expectations and needs of the users.
- Problems with reporting delays.
- Faulty use [19].

2.7 Building and Other Infrastructure Maintenance Standards

The quality of maintenance services invested in a building enables it to attain certain acceptable physical, functional and economic standards set by the organization as part of its policy [2]. Dzikiwi [9] is of the opinion that the technological, economic, sociological and political environments in which an organisation finds itself affect the standard it sets for maintaining its built environment, as these standards are based on, but not limited to, the following factors:

- financial: standard determined by financial capability;
- functional: use of the building and
- environmental: availability and quality of the infrastructure.

> Maintenance standards may differ or change based on the utility and value. For instance, the maintenance standard of a hotel will obviously be different from that of a student hostel, hospital, although all facilities are categorised as hospitality structures. Thus, utility (function of the building, its public image, and even national prestige) and commercial value of a residential building have a great influence on the maintenance standard [14, 15]

3 Research Methodology

The research methodology utilized a quantitative approach which began with formulating a problem statement and identifying the objectives of the study. The first step of conducting this research was formed based on reviews of the literature to ascertain why public high school buildings and infrastructure are not meeting their functional requirements. Then, the operationalization of the established maintenance generator and maintenance performance indicators into a questionnaire were carried out. Subsequently, pilot testing of the questionnaire was carried out and the developed format of the questionnaire was formed.

3.1 Data Collection Methods

Data collection method comprised the questionnaire design, pre-testing of the questionnaire, questionnaire administration and research population and sample. Data achieved using questionnaires from respondents was gathered and quantitatively analysed. A total number of thirty (30) questionnaires (hard and soft copies) was distributed to all parties involved in the maintenance department of the Department of Education in South Africa, Kwazulu Natal Province in uThungulu District, including Principals, Deputy Principals, Head of Departments, Director of Department, Staff, Senior Staff, Supervisor and Trainee. The respondents' working experience ranged from less than three years to more than thirty (30) years and they had different levels of education from Diploma to B.Tech degrees. However, only Seventeen (17) questionnaires were returned, which constitutes a sum of a 52% response rate.

4 Results Presentation and Analysis

4.1 Relationship Between Maintenance Budget and Funding, and Asset Functional Requirements

See Tables 1, 2, 3, 4, 5, 6, 7 and 8.

Table 1 Analysis: methods of budgeting and financing of maintenance work

Methods of maintenance budgeting and financing	Number	Percentage (%)
Use the budget of the previous year and add a percentage amount	6	37.5
Use historical figures to estimate maintenance costs	0	0
Estimate the actual cost of maintenance	5	31.2
Use the institution's maintenance plan to determine the cost of maintenance work	3	18.8
Appoint outside consultants to estimate the cost of maintenance work	2	12.5
Other	0	0
Total responses	16	100
No response	1	

Table 2 Analysis: maintenance fund availability to repair the damage caused by vandalism

Maintenance fund availability to repair the damage caused by vandalism					
1 Very limited funds	2 Limited funds	3 Reasonable available funds	4 Very available funds	5 Unsure	Total
5	6	5	0	0	16
31.3%	37.4%	31.3%	0%	0%	100%
No response					1

	Mean	Mode	Range	Minimum	Maximum
Maintenance fund availability to repair the damage done by vandalism	2.00	2	2	1	3

Table 3 Analysis: delay in maintenance work results in deterioration of facilities

Delay in maintenance work results in deterioration of such facilities					
1 Strongly disagree	2 Disagree	3 Agree	4 Strongly agree	5 Unsure	Total
0	0	3	13	0	16
0%	0%	18.8%	81.2%	0%	100%
No response					1

	Mean	Mode	Range	Minimum	Maximum
Delay in maintenance work	3.81	4	1	3	4

Table 4 Analysis: damage as a result of vandalism must be repaired without delay

Damage to buildings as a result of vandalism is to be repaired without delay					
1 Strongly disagree	2 Disagree	3 Agree	4 Strongly agree	5 Unsure	Total
0	0	4	13	0	17
0%	0%	23.5%	76.5%	0%	100%

	Mean	Mode	Range	Minimum	Maximum
Damage to the buildings as a result of vandalism to be repaired without delay	3.76	4	1	3	4

Table 5 Analysis: vandalism to buildings influences users and their needs with regard to their functional requirements

Level of influence of vandalised buildings over users and their needs					
1 Extremely positive influence	2 Positive influence	3 No influence	4 Negative influence	5 Extremely negative influence	Total
0	0	0	16	0	16
0%	0%	0%	100%	0%	100%
No response					1

	Mean	Mode	Range	Minimum	Maximum
Level of influence of vandalised buildings over users and their needs	4.00	4	0	4	4

4.2 Discussions and Interpretation of Maintenance Budget and Financing in Relation to Assets Performance

The interpretation of the feedback pertaining to the maintenance budget and financing of public high school buildings and infrastructure and their performance in relation to the methods of budgeting and financing of maintenance work, maintenance fund

Table 6 Analysis: impact of the school buildings' condition on teachers' work satisfaction

Impact of the school buildings' condition on teachers' work satisfaction					
1 Strongly disagree	2 Disagree	3 Agree	4 Strongly agree	5 Unsure	Total
0	0	3	13	0	16
0%	0%	18.8%	81.2%	0%	100%
No response					1

	Mean	Mode	Range	Minimum	Maximum
Teachers' work satisfaction is influenced by the condition of the school buildings	3.8	4	1	3	4

Table 7 Analysis: maintenance management's objective is to achieve asset life cycle optimisation

Maintenance management's objective is to achieve total asset life cycle optimisation					
1 Strongly disagree	2 Disagree	3 Agree	4 Strongly agree	5 Unsure	Total
0	0	0	17	0	17
0%	0%	0%	100%	0%	100%

	Mean	Mode	Range	Minimum	Maximum
Maintenance management's objective is to achieve total asset life cycle optimisation	4.00	4	0	4	4

Table 8 Analysis: maintenance manual/schedule

Buildings			Infrastructure		
	Number	Percentage (%)		Number	Percentage (%)
No	12	70.6	No	11	64.7
Yes	5	9.4	Yes	6	35.3
Total responses	17	100	Total responses	17	100

	Mean	Mode	Range	Minimum	Maximum
Maintenance manual/schedule assistance level feedback	2.47	2	3	1	4

availability to repair the damage caused by vandalism, the delay in maintenance work, users' needs and teachers' work satisfaction, led to the conclusions discussed hereunder.

- The respondents' feedback regarding their management of, and participation in providing funds in the budget for the maintenance of buildings and infrastructure, provided a good indication of the shortcomings related to insufficient funding of maintenance work for buildings and infrastructure.

- Table 1 indicates that most (37.5%) of the respondents perceive that the method used for determining maintenance costs was to use the previous year's budget and add a percentage amount. This implies that the actual cost of maintenance cannot be ascertained and this could lead to inadequate funding. In addition, [13] research found that maintenance cost estimation method is vital in determine the actual expenditure of asset maintenance.
- With regard to the availability of funds for maintenance (as per Table 2), the majority of the respondents (37.4%) said that funds remain limited. This is an indication that funds allocated for maintenance work continue to be insufficient and can lead to the deterioration of facilities. On the Contrary, [13] further found that, quality of maintenance exercise is significantly influenced by the way the maintenance is conducted couple with amount of budget allocated.
- As per Tables 3 and 4, an overwhelming proportion (81.2% and 76.5% respectively) of the respondents, (with regard to general maintenance and vandalism), indicated that delayed maintenance results in the deterioration of buildings and infrastructure. The interpretation of the findings led to the conclusion that a delay in maintenance activity could be attributed to constrained resources (monetary, human or equipment). Management should attempt to resolve such issues timeously and avoid such delays.
- All of the respondents (100%) in Table 5 indicated that a damaged building has a negative influence on users' needs. This implies that a building damaged by vandalism should be repaired without delay.
- According to Table 6, the majority (81.2%) of the respondents indicated that in their opinion the condition of the buildings and infrastructure has a direct impact on teachers' work satisfaction. This implies that teachers' and learners' performance could improve if the buildings in which they function are in a good condition.
- In Table 7 all the respondents (100%) reflected that the core objective of maintenance management is to achieve total asset life cycle optimisation. This implies that to achieve total asset life cycle optimisation, effective maintenance management activities should be sustained.
- A significant portion (46.7%) of the respondents (Table 8) noted that the level of assistance provided by a maintenance manual/schedule is good. This result indicates that the respondents understood the importance of a maintenance manual/schedule, as it provides easy access to specifications and the requirement for parts.

5 Summary

In summary, the research was aimed at determining an effective maintenance management policy system for public high school buildings and other infrastructure within the Department of Education in South Africa, based on standard criteria practice stipulated in the Department asset maintenance policy document with the key objective of "*To establish the financial resources requirements for the maintenance management of public high school buildings and infrastructure*". The research found that it is accepted as a fact that learning and teaching can be improved by an asset environment that is in a sound condition. And in addition, Effective estimation techniques should be employed to satisfy the financial requirements for maintenance activities. Maintenance should be included in the institutions' annual budgets and must realistically reflect the actual amounts needed to maintain the buildings and infrastructure in an optimal operational condition to achieve maximum performance of the asset and to attain total life cycle optimisation.

5.1 Conclusion

The prime objective of providing buildings and other infrastructure for public high schools is to promote academic excellence. Academic performance is enhanced by a sound learning environment provided by the proper maintenance and repair of the asset. To keep assets in sound condition, strategic planning should be practiced, as a planning gap leads to top management paying less attention to maintenance activities.

The research concluded that to avoid maintenance delays and to meet users' needs with improved asset performance and user performance (teachers and learners), and to maintain assets in an acceptable condition, maintenance activities should be guided by maintenance policy documents.

Financial constraints intensify as maintenance activities are carried out on an ad hoc basis instead of according to a maintenance schedule and the maintenance management policy guidelines. Management ought to employ a formal system of estimation costs for maintenance work instead of adjusting historical figures.

5.2 Recommendations

Maintenance managers should make management and the occupants/users aware of the importance of the work performed by the assets maintenance department and the role they have in assisting the assets maintenance department by reporting defective items.

Management should engage competent employees for the maintenance department and provide sufficient resources (money, tools and equipment, technical personnel) for maintenance activities.

There should be a clear schedule of duties for all workers that are involved in managing the buildings and other infrastructure with an adequate understanding of the span of the organisational structure.

The research further encourages that maintenance manual should be prepared for all the existing buildings and other infrastructure that has no maintenance manual and should be updated regularly if there are any changes to components or the rate of services when maintenance work is carried out.

Policy best practice standards should be set to determine the minimum acceptable state or condition of the buildings and other infrastructure to eliminate both minor and major breakdowns of the buildings resulting in academic performance being affected negatively.

References

1. Abdul Lateef O (2009) Building maintenance management in Malaysia. J Build Appraisal 1
2. Abdul Lateef O, Khamidi MF, Idrus A (2010) Sustainability in the context of maintenance: building defects in the Malaysian University Campuses. In: Kuala Lumpur, international conference 2010 on sustainable building and infrastructure
3. Basic Education Department (2012) Facilities maintenance guidelines for public schools. Basic Education Department, South Africa
4. Business Dictionary (2017) Encyclopaedia of business analytic and optimization 2016 [online]. Available through http://www.igi-global.Com/dictionary/maintenance-policy/39946. Accessed 7 Aug 2017
5. Cape Winelands Municipality (2015) Maintenance management policy [online]. Available at http://www.capewinelands.gov.za/DocumentsDownload/Reports/Budget%20Related. Accessed 10 Aug 2017
6. Council of the Great City Schools (2016) Reversing the cycle of deterioration in the Nation's Public-School Buildings 2014. Pennsylvania Ave N.W. Available through www.Cgc.org/cmc/lib/centricity/Domain/87/facilitiesReport. Accessed 12 Feb 2016
7. Department of Basic Education (DBE) (2015) National school infrastructure maintenance guide. DBE online. Available through http://www.kzneducation.gov.za/portals/0/Infrastructureplanninganddelivery/Maintenance. Accessed 13 Mar 2015
8. Department of Public Works (DPW) (2010) National infrastructure maintenance strategy in support of ASGISA and government growth objectives 2010. Department of Public Works, CIDB, Council for Scientific and Industrial Research: South Africa
9. Dzikiwi A (2015) Maintenance management systems of on-campus student hostels at Nigerian Universities. Nelson Mandela Metropolitan University. Unpublished Ph.D. thesis in Construction Management
10. Flores-Colen I, de Brito J, Freitas V (2010) Discussion of criteria for prioritisation of predictive maintenance of building façade. J Perform Constr Facilities Los Angeles 1
11. Hamza S, Kobayashi K (2012) Utilizing mid-long-term maintenance management policy for sustainable maintenance of infrastructure facilities. In: The 3rd international conference 2012 on sustainable future for human security. Elsevier B.V., Indonesia
12. Kenneth B (2011) Maintenance of university facilities in developing country. Nelson Mandela Metropolitan University. Unpublished M.Sc. treatise in construction management
13. Lee H, Scott D (2009) Overview of maintenance strategy, acceptable maintenance from a building maintenance operation perspective. J Retail Leisure Property:29–30
14. Olagunju RE (2011) Development of mathematical model for the maintenance of residential building in Niger State, Federal University of Technology, Mina, Nigeria. Unpublished Ph.D. thesis
15. Overstrand, Municipality (2017) Overstrand municipality maintenance management policy. Council of Overstrand municipality. Available at http://www.Overstrand.gov.za/en/documents/policies/168-maintenance-management/file. Accessed 07 Aug 2017
16. Overveen (2015) Effective maintenance management practice [online]. Available through http://www.trimitra.com/services/effective-maintenance-management-practices. Accessed 23 Nov 2015

17. Pintelon L, Parodi-Herz A (2008) Maintenance: an evolution perspective. complex system maintenance handbook. Spring Series in Reliability Engineering, London
18. Queensland Department of Housing and Public Works (2017) Maintenance management framework guideline: building maintenance policy, standards and strategy development. The State of Queensland. Department of Housing and Public Works. Australia
19. Saghatforoush E, Trigunarsyah B, Too E (2012) Assessment of operability and maintainability success factors in provision of extended constructability principles. In: Isfatan, 9th international congress on civil engineering
20. Swallow B, Chanter P (2007) Building maintenance management, 1st edn. John Wiley and Sons. UK
21. Swallow B, Chanter P (2010) Building maintenance management, 2nd edn. Blackwell, Oxford
22. Thakur D (2017) System maintenance [online]. Available at http://ecomputernotes.com/mis/implementation-andevaluation/what-is-system
23. Wikipedia (2017) Performance indicator. Wikipedia [online] Available at http://en.wikipedia.org/wiki/performance-indicator. Accessed 9 Aug 2017
24. Wood B (2009) Building maintenance. Blackwell, United Kingdom
25. Zainul Abidin N (2010) Investigating the awareness and application of sustainable construction concept by Malaysian developers. Habitat Int 34(4):421–426

Stakeholder Participation Trends in the South African Green Building Industry: 2016–2018 Perspectives

T S. Ngobeni, Tebogo H. Ngele, and Danie J. Hoffman

Abstract The participation of stakeholders involved in the construction of buildings is crucial to the successful implementation of green building practice in South Africa. This paper investigated stakeholders' participation trends in the new South African office buildings for 2016–2018 with an As-built or Design rating. The study expanded on a 2016 study covering the period 2009 to 2016. The study described participation trends and the market share of major stakeholders. Seven stakeholder disciplines were analysed: owners, architects, quantity surveyors, green consultants, main contractors, and structural and electrical engineers. Data was made available by the Green Building Council of South Africa. A total of 156 new office buildings received a green building certification between 2009 and 2018. The extent of participation between firms of the different disciplines varied significantly: Owners (75), architects (63), quantity surveyors (55), green consultants (27), main contractors (34), structural engineers (42) and electrical engineers (48). The market share of the top three firms also differed significantly: Owners (19.9%), architects (29.3%), quantity surveyors (26.4%), green consultants (57.7%), main contractors (50.3%), structural engineers (41.4%) and electrical engineers (31.9%). The study indicated an acceptance of green building as a practice by different stakeholders, which was validated by a growing stakeholder participation trend and by the number of projects firms had undertaken.

Keywords Green buildings · Participation trends · Market share · Stakeholders · South Africa

1 Introduction

Climate change has become a reality in the twenty-first century. The shortage of food and water, rising sea levels, and an increased likelihood of severe weather events are global concerns [1, 2]. The Economist reported that the realisation that

T. S. Ngobeni · T. H. Ngele · D. J. Hoffman (✉)
Department of Construction Economics, University of Pretoria, Gauteng, South Africa
e-mail: danie.hoffman@up.ac.za

T. C. Haupt et al. (eds.), *Construction in 5D: Deconstruction, Digitalization, Disruption, Disaster, Development*, Lecture Notes in Civil Engineering 245,
https://doi.org/10.1007/978-3-030-97748-1_28

the earth cannot sustain the current rate of natural resource exploitation leads to the consideration of green building principles [3].

The construction industry contributes extensively to the problem. The worldwide impact of the built environment accounted for 40% of end-use energy consumption, 12% of freshwater consumption, 40% of solid waste generation, and 33.3% of greenhouse gas emissions worldwide [4, 5]. Green building practices can be the panacea to this problem by offering energy saving, land saving, stormwater runoff-reducing, material conservation, and pollution reduction [6]. There is, therefore, a need for efficient stakeholder participation in green building construction and upgrades in South Africa. For South Africa to benefit from new green developments, stakeholders must fully commit and participate in green building practices [7].

2 Literature Review

2.1 *Climate and Climate Change*

Climate change and rising temperatures are unavoidable realities in the twenty-first century. NASA [1] projected that the rise in temperatures would continue into the foreseeable future. Ozone radiation from the sun, oceanic changes, volcanic eruptions, human activities and construction activities contribute to global warming [8]. The IPCC reported that between 1970 and 2004, global greenhouse gas emissions due to human activities increased by as much as 70% [9]. The built environment's share of pollution is estimated to be between 30 and 40% of solid waste, 30–40% global greenhouse gas (GHG) emissions and roughly 20% water effluents [10, 11].

2.2 *Stakeholder Participation*

Built environment stakeholders may have differing reasons for undertaken green building projects [12]. Architects may focus on the conservation of resources; engineers are concerned about structural designs and energy efficiency aspects, while property owners are anxious about penalties due to greenhouse gas emissions [13]. Nelson [14] indicated the role of governments to formulate regulations and policies supporting green building. Macaluso and Walker [15] stated that to ensure cost efficiency, there is a need to appoint a Quantity Surveyor at the earliest possible stage.

2.3 Green Buildings

According to the GBCSA, green building involves energy and resource-efficient construction and operating methods exercised with environmental responsibility [4]. The WGBC [16] defines green buildings in terms of their efficiency on the utilisation of energy, water and other materials while at the same time reducing the impact of the building on people's health and environment using enhanced design, construction, operation, maintenance and removal.

Benefits of Green Buildings. Green buildings may offer many benefits and rewards compared to conventional buildings.

Energy savings. Zuo and Zhao [17] claimed that green buildings use about 30% less energy than conventional buildings. Dodge Data and Analytics [18] claimed that in 2018 green technologies reduced energy consumption in buildings by 66%.

Environmental benefits. Green building defends biodiversity, preserves natural resources, decreases water wastage and enhances air quality. This will ultimately positively contribute to saving the environment and reduce the rate at which climate change [19].

Productivity and health. According to the USGBC most conventional commercial buildings in the USA offer poor indoor environmental quality, costing companies billions of dollars annually from lost time to sick leave, medical claims and ultimately poor performance [20]. Green buildings significantly reduced illness symptoms, absenteeism and increased productivity compared to conventional buildings [21].

Financial benefits. Green buildings offer lower energy use and reduce the cost of waste, water, environmental and emissions, operations and maintenance while increasing [22]. In addition, the WGBC [23] also identified higher property value, higher occupancy levels, job creation, marketability, higher return on investment, insurance discounts and reduced liability and risk as some of the financial benefits associated with green building [24].

Barriers to Green Buildings. South Africa remains the leading African country in adopting the green building concept [18]. However, there are still potential barriers to the green building sector's growth.

Cost. Yudelson [25] argues that green building projects are often more expensive when compared to conventional buildings. However, a recent study on South African green building costs and trends found an average green building cost premium of only 3.9% for Green Star SA certified office buildings [26]. To minimise the green building cost premium requires sustainability to be part of conceptual designs [27, 28].

Attitude and Market. Häkkinen and Belloni [29] argued that property owner buy-in on the green building concept is often difficult to achieve. Hwang and Ng [30] found resistance to change to be a challenge is for green building.

Management and Government support. Olubunmi, Xia and Skitmore [31] stated that governments need to promote the adoption of green technologies and should formulate financial and non-financial incentives and regulations to promote green buildings.

False green claims. According to Hwang and Tan [32], the underperformance of green products leads to mistrust from clients and the public and tarnish the green concept and hinders the development of the green building industry.

3 Methodology

This study expanded on a previous study that described stakeholder participation trends in the South African green building industry from 2010 to 2016 [7]. The study data was secured from the GBCSA. Microsoft Excel was used as the primary software tool to capture data, design graphs and present findings. The study included test–retest reliability testing. The findings' reliability and validity were verified through the available raw data sheet received from GBCSA. The results are for the period 1 January 2016 to 31 December 2018.

4 Research Findings/Results

The available data revealed a total number of 159 new offices which were certified from 2009 to 2018. Of the 159 certified projects, 69 were certified between 2016 and 2018. The GBCSA data confirmed green building is growing in South Africa from only one project certified in 2009, 12 certified in 2012 and 26 projects certified in 2018.

The data revealed that on occasion, two consultants were appointed such as for example as architects on one project. In such instances, that project was counted for both firms. The rationale was that when two firms were involved in one project, that implied a split market share of 50% each, resulting in each stakeholder achieving some green building experience. There is also an implicit assumption that where firms are working together, they would be treated as equal partners.

4.1 Dispersion of Participation in Green Building Projects Per Discipline

Table 1 displayed the overall participation of all the stakeholders for the period 2009–2018. The data revealed that the 159 green office buildings are owned by 75 legal entities, the highest number of participants per discipline. The same number of green buildings was undertaken by only 27 green building consultants and was built by only 34 main contractors. This finding confirmed that the participation amongst different stakeholder disciplines in the South African green building industry differs significantly.

Table 1 Dispersion of participation in green building projects per discipline

	Owners	Architects	Quantity Surveyors	Green Consultants	Main Contr.	Struc. Eng.	Elec. Eng.
Total number of firms with green building projects	75	63	55	27	34	42	48
Number of projects (expanded total)	171	181	174	163	163	162	163
Number of firms with single project	37	34	29	15	16	21	23
% of firms with single projects	49,3%	54,0%	52,7%	55,6%	47,1%	50,0%	47,9%
Number of firms with multiple projects	38	29	26	12	18	21	25
% of firms with multiple projects	50,7%	46,0%	47,3%	44,4%	52,9%	50,0%	52,1%
Number of projects by firms with a single project	37	34	29	15	16	21	23
Market share (%) of firms with single projects	21,6%	18,8%	16,7%	9,2%	9,8%	13,0%	14,1%
Number of projects by firms with multiple projects	134	147	145	148	147	141	140
Market share (%) of firms with multiple projects	78,4%	81,2%	83,3%	90,8%	90,2%	87,0%	85,9%
Number of projects held by top 3 firms	34	53	46	94	82	67	52
Market share of top 3 firms	19,9%	29,3%	26,4%	57,7%	50,3%	41,4%	31,9%
Number of projects held by the largest firm	16	25	20	45	48	24	25
Market share of top firm	9,4%	13,8%	11,5%	27,6%	29,4%	14,8%	15,3%

Owners. Figure 1 detailed the cumulative number of projects certified by the GBCSA until 2018. The large majority of 115 green buildings belonged to 45 owners from the commercial real estate sector. Other well-represented sectors were the financial services sector, with 15 owners who owned 27 buildings and nine public sector owners with 20 buildings. Of the 75 owners, a total of 37 (49, 3%) companies owned a single green building project.

Figure 2 indicated the inclining accumulative number of green building owners and the growth in the number of new owners entering the market between 2009 and 2018. The majority of 54.2% of the projects were owned by private owners, followed by listed companies (33.7%) and owners from the public sector with 12.1% of projects.

The number of firms participating in green building projects per discipline grew over time (see Fig. 3). The percentages of firms per discipline who participated in

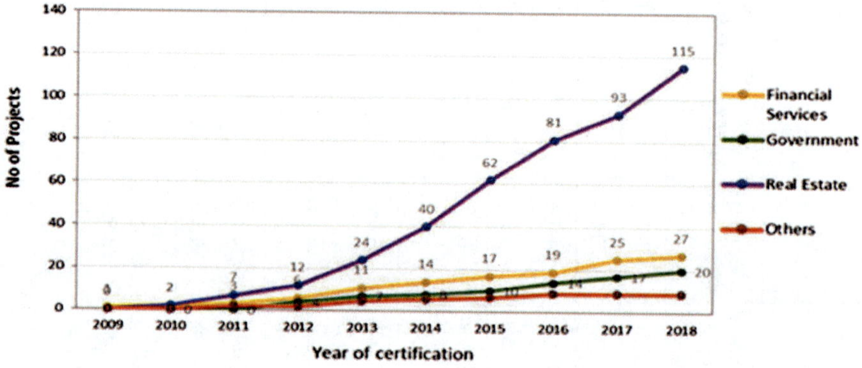

Fig. 1 Cumulative project by sector

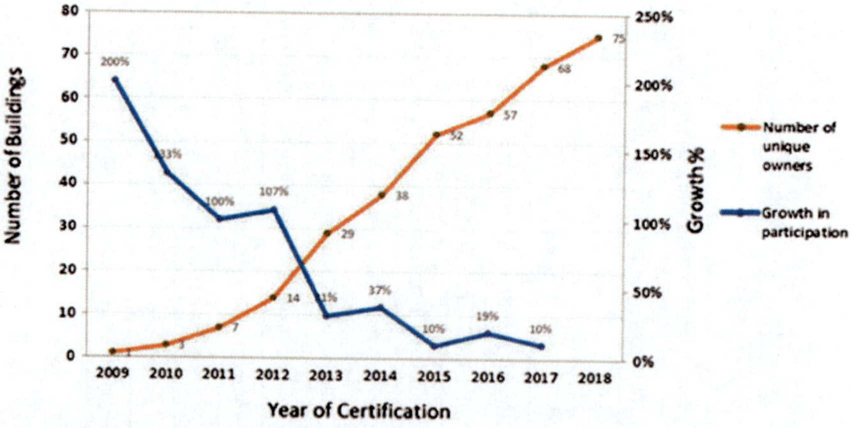

Fig. 2 Owners participation over time

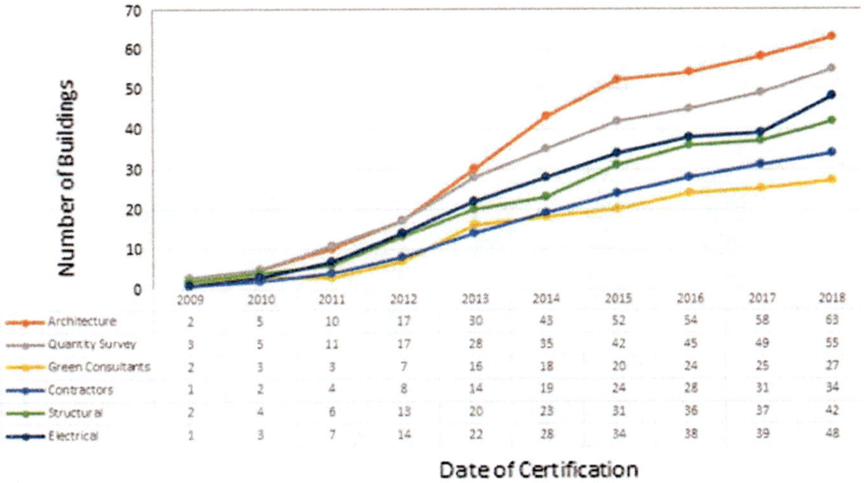

	2009	2010	2011	2012	2013	2014	2015	2016	2017	2018
Architecture	2	5	10	17	30	43	52	54	58	63
Quantity Survey	3	5	11	17	28	35	42	45	49	55
Green Consultants	2	3	3	7	16	18	20	24	25	27
Contractors	1	2	4	8	14	19	24	28	31	34
Structural	2	4	6	13	20	23	31	36	37	42
Electrical	1	3	7	14	22	28	34	38	39	48

Fig. 3 Cumulative number of firms participating in green building projects over time

only one green building were very similar—Architects (54%), Quantity Surveyors (52.7%), Green Consultants (55.6%), Main Contractors (47.1%), Structural Engineers (50%), and Electrical Engineers (47.9%). However, the number of architect firms with green building experience was more than double that of green buildings consultancy firms (63 firms vs. 27 firms).

The relatively small number of 27 green building consultancy firms and 34 building contractors with green building experience indicated that some of the firms in these disciplines may have achieved significant market share. Table 1 confirmed that specific disciplines did confirm dominant stakeholders. The combined market share of the three most experienced firms in each discipline differs significantly. The disciplines of green building consultancy firms (57.7%) and building contractors (50.3%) has the highest market share between the three most experienced firms. For architects (29.3%) and quantity surveyors (26.4%), these market share percentages were much less dominant.

4.2 2016–2018 Results

Figures 4, 5 and 6 summarised the number of firms for discipline that achieved experience with green building projects from 2016 to 2018.

Green building consultants with multiple project experience again achieved the highest market share percentage.

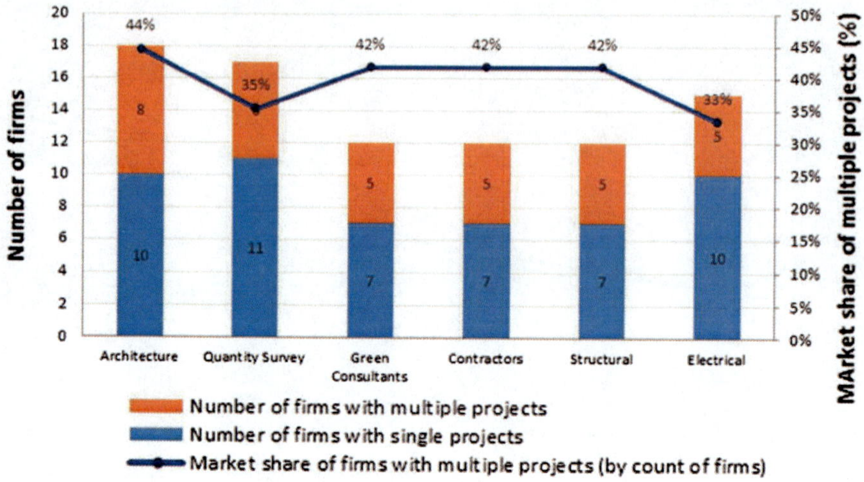

Fig. 4 2016 Number of firms with different experience in green building projects

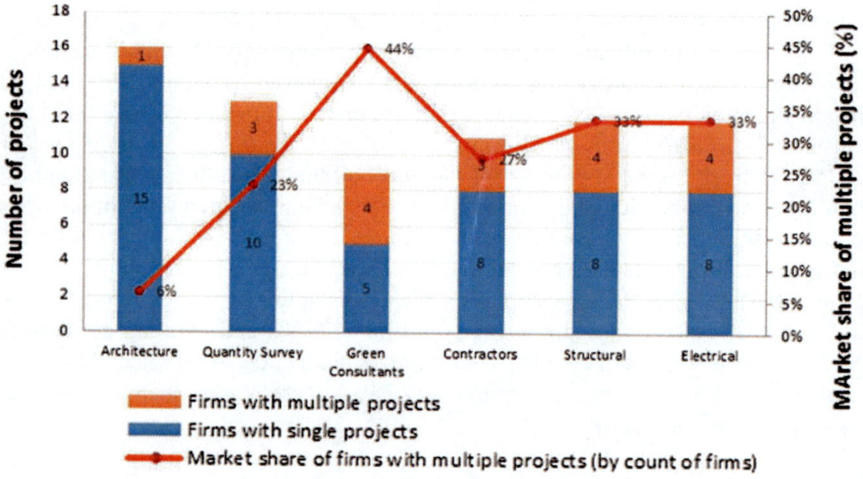

Fig. 5 2017 number of firms with different experience in green building projects

5 Conclusion and Recommendations

The study indicated a widely distributed acceptance of the green building practice amongst South African stakeholder firms. The study also found a wider spread of participation in green building amongst architects than in the other six disciplines included in the study from 2016 to 2018. Many architect firms indicated a willingness to accept the risk and new challenges presented by green building. However, the study

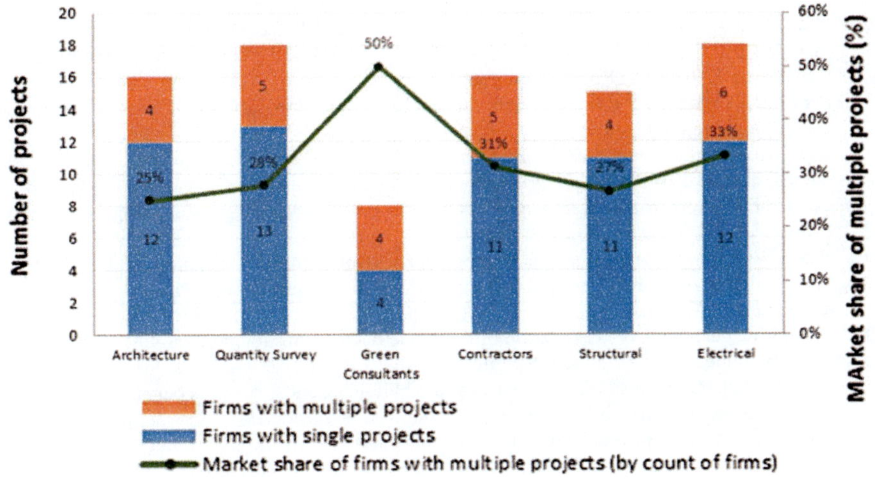

Fig. 6 2018 Number of firms with different experience in green building projects

has revealed that many architectural firms lacked continued experience. Compared to other disciplines, more architect firms participate in only a single green building project. In contrast, the green building consultant firms were the discipline with the lowest number of participating firms, but also with the firms that achieved the highest market share for participating in green buildings.

References

1. NASA (2017) Evidence of global warming [online]. Available from http://climate.nasa.gov/evidence. Last accessed 2019/09/29
2. United States Environmental Protection Agency (2016) Climate change indicators in the United States [online]. Available from https://www.epa.gov/climate-indicators. Last accessed 2019/09/29
3. The Economist (2004) The rise of the green building, 4th edn. Technology quarterly, December [online]. Available from https://www.economist.com/technology-quarterly/2004/12/04/the-rise-of-the-green-building. Last accessed 2019/09/27
4. GBCSA (2016) GBCSA integrated annual report 2016 [Online]. Available from https://gbcsa.org.za/wp-content/uploads/2018/01/GBCSA-Integrated_Report_2016_FA_low-res.pdf. Last accessed 2019/11/05
5. WGBC (2017) WGBC about us: our story [online]. Available from https://www.worldgbc.org/our-story. Last accessed 2019/09/15
6. Yellamraju (2010) LEED-new construction project management. McGraw-Hill Inc [online]. Available from https://www.greenbiz.com/article/leed-and-future-green-building. Last accessed 2019/09/15

7. Hoffman DJ, Cloete J, van der Schijff L, Wagner L (2017) Stakeholder participation trends in green buildings during 2009–2016. In: The 21st PAQS2017 conference, Vancouver, BC Canada, pp 139–149. ISBN: 978-1-896606-33-0
8. Persson J, Gronkvist S (2015) Drivers for and barriers to low-energy buildings in Sweden. J Clean Prod:296–304
9. Intergovernmental Panel on Climate Change (2014) Using a modified legend, published in: climate change 2014: the physical science basis. Contribution of working group I to the fourth assessment report of the intergovernmental panel on climate change. Cambridge University Press
10. World Wide Fund (2014) Living planet report 2004, World wide fund for nature, Gland, Switzerland [online]. Available from http://www.wwf.be/eco-footprint/docs/LPR2004_en.pdf. Last accessed 2020/09/15
11. Suzuki M, Oka T (2014) Estimation of life cycle energy consumption and CO_2 emission of office buildings in Japan. J Energy Build 28:33–41
12. Fitch E, Laquidara-Carr D (2013) World green building trends: business benefits driving new and retrofit market opportunities in over 60 countries. McGraw-Hill Construction
13. Ma T, Luu HT (2013) The changing role of quantity surveyors in the green building development in South Australia. In: 38th AUBEA international conference, pp 1–10
14. Nelson A (2010) Green buildings, niche becomes mainstream. Deutsche Bank Research, Frankfurt, Germany
15. Macaluso J, Walker A (2011) Green building project planning and cost estimating, 3rd edn. Wiley, New Jersey
16. WGBC (2018) World green building council [online]. Available from http://www.worldgbc.org/index.php?cID=220. Last accessed 2019/04/13
17. Zuo J, Zhao Z (2014) Green building research–current status and future agenda: a review. Renew Sustain Energy Rev 30(2):271–281
18. Dodge Data and Analytics (2018) World green building trends 2018. SmartMarket Report
19. WGBC (2016) The business case for green building: a review of the costs and benefits for developers, investors and occupants [online]. Available from https://www.worldgbc.org/news-media/business-case-green-building-review-costs-and-benefits-developers-investors-and-occupants. Last accessed 2020/09/15
20. United States Green Building Council (USGBC) (2016) Strategic plan [online]. Available from http://www.usgbc.org/sites/default/files/usgbc-strategic-plan. Last accessed 2020/09/11
21. Singh A, Syal M, Grady SC, Korkmaz S (2015) Effects of green buildings on employee health and productivity. Am J Public Health 100(9):1665–1668
22. Yudelson J (2014) The business case for green buildings [online]. Available from http://www.greenbuildconsult.com/pdfs/PPA-Yudelson.pdf. Last accessed 2020/06/23
23. WGBC (2019) The benefits of green buildings [online]. Available from https://www.worldgbc.org/benefits-green-buildings. Last accessed 2021/01/08
24. Smit AM, Du Toit F (2015) Investigating the financial benefits of green buildings. Environ Econ 6(3)
25. Yudelson J (2012) The green building revolution. Island Press, Washington, DC
26. GBCSA, ASAQS and UP (2019) Green building in South Africa: guide to cost and trends. 2019 edition [online]. Available from https://gbcsa.org.za/wp-content/uploads/2019/08/2019GreenBuildingBooklet-Final-1.pdf. Last accessed 2019/09/28
27. Milne N (2012) The rands and sense of green building: building the business case for green commercial buildings in South Africa. Green Building Council, Cape Town, South Africa
28. Hoffman DJ, Cowie D (2014) Property developer's perspective on the current status of green building in South Africa. Durban. ASOCSA 2014. In: The 8th built environment conference—reflection on direction in construction. Durban, South Africa, 27–29 July 2014
29. Häkkinen T, Belloni K (2014) Barriers and drivers for sustainable building. Build Res Inf 39(3):239–255

30. Hwang BG, Ng WJ (2013) Project management knowledge and skills for green construction: overcoming challenges. Int J Project Manage 31(2):272–284
31. Olubunmi OA, Xia PB, Skitmore M (2016) Green building incentives: a review. Renew Sustain Energy Rev 59:1611–1621
32. Hwang BG, Tan SJ (2015) Green building project management: obstacles and solutions for sustainable development. J Project Manage 1(10):21–26

Construction Sustainability: Empirical Evidence on SME Development and Delivery Challenges of Road Infrastructure in South Africa

Kgashane Stephen Nyakala, Andre Vermeulen, and Jan-Harm Pretorius

Abstract This research aim is to investigate the factors influencing small and medium enterprises (SMEs) development and delivery challenges of road infrastructure in the Limpopo province of South Africa. Sustainable construction, transformation and technology are characteristically dynamic and complex. The analysis of qualitative data gathered through in-depth interviews with contractors, architects and project/construction managers in all five cases resulted in the emergence of variety of factors influencing delivery challenges of road infrastructure in construction SMEs. Primary data were collected through a number of depth interviews with construction SMEs experts and project managers. Secondary data were collected through construction SME organisations' documents written procedures, transformation and technological issues. The main findings of this study showed there was a huge delivery challenges related to road infrastructure in the construction industry, i.e. poor construction design, lack of technical and engineering abilities, inadequate management involvement and support, unclear targets to sustainability in building construction, lack of technological skills, improper training of the team members and inadequate use of information and communication technologies.

Keywords Small and medium enterprises · Construction sustainability · Challenges · Road infrastructure

---

K. S. Nyakala (✉)
Department of Operations Management, Tshwane University of Technology, Pretoria, South Africa
e-mail: nyakalaks@tut.ac.za

A. Vermeulen · J.-H. Pretorius
School of Engineering Management, Faculty of Engineering and the Built Environment, University of Johannesburg, Johannesburg, South Africa

T. C. Haupt et al. (eds.), *Construction in 5D: Deconstruction, Digitalization, Disruption, Disaster, Development*, Lecture Notes in Civil Engineering 245,
https://doi.org/10.1007/978-3-030-97748-1_29

1 Introduction

The study was conducted in order to identify key causes of failure and quantify sustainable development process among small-and –medium enterprises (SMEs) contractors operating in the road infrastructure and service delivery in South Africa. The SMEs construction industry are recognized as an important instrument for stimulating economic growth [27]. SMEs are key to the economy and growth and development of the South African economy. Economic development and growth in South Africa can be measured in relation to physical infrastructural development such as roads, buildings and bridges [3]. In South Africa, although construction SMEs are currently at the forefront of local economic growth and are supposed to resolve socio-economic complications, this sector faces a varied range of limitations, which constrains them from sustainable construction and maintaining a competitive advantage in their respective industries [1].

Sustainable construction is defined as the creation and responsible maintenance of a healthy built environment based on resourceful and ecological values [21]. The SME construction project is made up of a mixture of very dynamic processes which rarely go in hand with the kind of implementation plan adopted. The key problems to sustained development and growth among emerging contractors in developing countries are lack of technology and transformation, shortage of technical skills in road construction projects and scarce network with well-established construction SMEs [14]. Construction industry in developing countries are often characterised by lack of accountability and transparency, inadequate fundamental principles of good governance. Poor construction of road infrastructure is related to performance problems [7]. There are various factors behind the occurrence of such lack of construction sustainability in rural infrastructure and delivery services. The difficulties experienced in the performance of the road construction industry in developing countries can be grouped into three classes namely; difficulties related to scarcities in the construction industry, difficulties due to consultants and clients and lastly difficulties rising as a result of ineffectiveness of the construction SMEs [27].

One of the most effective road infrastructure is to develop sustainable construction of a built environment during its lifetime and definitive deconstruction, and recycling of resources in order to reduce the waste stream. Skill, transport and economic development contribute to a more reliable resources costs as well as improving its capability of reporting deviation in a rural area [6]. Literature on sustainable construction and delivery challenges related to SME construction organisations in South Africa is fairly limited [17]. Much of the existing research concentrated on mentorship programmes within the SMEs contractor development programme and elements of project management performance. It is possible to evaluate the contribution of construction professionals in budgeting for infrastructure development 16. Regardless of the issues investigated, road deterioration, poor infrastructure, and ineffective management of the construction process [3].

2 Literature Review

2.1 Infrastructure Development

Effective infrastructure development process begins by recognising that a distinctive development process comprises of a fairly expectable set of phases that can be successfully achieved by establishing a series of reviews 7. In a construction industry, infrastructure development refers to the provision of fundamental infrastructure facilities such as the construction of roads, availability of transportation, bridges, ports, and telecommunication systems [6, 9]. Many construction project professionals recognise the difficulty of correlating cost with the benefits of an excellence program [25]. They maintain that placing a cost figure on quality is difficult and that accounting is incapable to capture the true costs of quality. Monitoring and controlling of road construction projects are required to conform to the specifications. The closer the observance to the construction specifications, the standards, and the costs associated with conforming to these standards can be calculated [11].

New technologies have reduced the inherent failure rates of materials and limited the direct labour content of most infrastructure facilities. Good communications are vital for the effective running a project and should be considered at a very early stage. These include the designer, engineer, assessors, architects and main contractor. In order to function they must monitor their environment, take decisions, communicate their intentions and ensure that what they intended to happen does happen [19]. In construction SMEs, these information flows generate and control flows of materials as well. Construction sustainability is essentially an organisational innovation –the identification of a team responsible for ensuring the effective delivery of the project mission for the client [13]. Conversely, it has become associated with a particular set of tools and techniques –most notably critical path analysis. Construction process analysis has become increasingly influential in a number of industries- both in the re-engineering of business processes to maximise the benefits of ICT systems and in the diffusion of lean thinking [8].

Effective infrastructure development necessitates that infrastructures be clear. Poor planning leading to inefficient practice and accidents. Use of construction process provisions which influences innovation and enhance the scope for integration of sustainability. It is about the relationship of the client organisation to its economic and social environment2. The construction industry is wholly reliant upon effective communication between individuals, teams and construction SMEs braces meanings such as knowledge, processed data, skills and technology [1]. Within construction, information is exceptionally diverse given the huge number of parties involved with construction sustainability.

2.2 Innovation and Construction Sustainability

Amer et al. [4] investigate how technological capital such as innovation influences sustainable development on road infrastructure in construction SMEs. The result of the literature studied shows that when there is a lack of innovation, such as the use of an unknowledgeable and unskilled cheap workforce as well as the inaccessibility of networks, high costs are incurred and ultimately the implementation of innovation by construction SMEs is affected [5, 10]. Competitiveness is critical to the efficiency of construction SMEs. Construction SMEs, by contributing in international markets via export, can access potential sources of information [4, 27]. As a result, this type of innovation helps these SMEs to acquire knowledge that is more external; which empowers their businesses to sustain their delivery of essential infrastructures such as fine roads [15, 17]. These construction SMEs become innovative by benchmarking with other innovative construction SMEs, employing workers with experience and training coming from innovative businesses, and collaboratively with academies on research and development and/ or with other research representatives and industries [8].

In terms of unnoticeable assets such as knowledge and experience to innovation, costs are reduced by construction SMEs and contribute to the delivery of important infrastructure development in rural construction projects. So, the absence of knowledge hinders implementation of innovation among construction SMEs [24]. The role of different levels of local societies in construction stands a function of the technical requirements of the task. Construction SMEs initiatives have a critical role in infrastructural development because construction activities rely on local materials and familiar technologies [1]. There is also more incentive for communities to take responsibility for construction phase if they had significant involvement in the design phase [23].

2.3 Technological Skills

Nyakala et al. [18] identified capital factors such as information technology and communication application on management of road projects. Findings from the literature studied show that employees exposed to Information Technology (IT) skills make provision for the productivity of construction SMEs [22, 26]. As a result, the working environments of construction SMEs continues to improve. According to Construction Industry Development Board [7], technological skills in construction SMEs influence the productivity of their businesses. When technical knowledge or managerial skills are available, construction SMEs add value to their productivity. Training in technical know-how and skills in infrastructure efficiency is essential to employees of construction SMEs (Mofokeng and Thwala 2014). Conversely, construction SMEs that have acquired specialised knowledge in built environment through constructing a communiqué board of brochures, employee phone directories, and other brochures such as catalogues and price lists. This is a simple way that construction SMEs can simplify the flow of information to clients. Technological skills play an important role in the construction SMEs [24]. These skills are embedded by training employees to prepare them to be engaged in achieving results of construction SMEs by contributing to productivity [2].

2.4 *Networking*

As pointed out by Barret and Sexton [6] networks such electronic data interchange, internet, extranet, websites, barcoding, and facsimile are used to ensure productivity enhancement in construction SMEs. The importance of teamwork through networking for decision making in construction industry [25]. Several studies shows that working as a team in SMEs from top to bottom, improves the degree of integration within the business as well as that between the buyer and contractor, encourages innovation through suggestions, and cuts down construction costs by implementing process of project control systems and performance of rural roads construction projects [20–22]. One of the components of innovation influencing growth in construction SMEs is networking. When knowledge is shared among the members of these SMEs manage to sustain their construction. Knowledge transfer plays a pivotal role in the construction process of construction industry. The transfer of knowledge through networking among the members of the construction SMEs becomes critical to improve the construction industry [13].

2.5 *Technological Tools and Techniques*

A tool such as information and communications technology (ICT) ensures that the information of new product design and development, which influences the productivity of construction SMEs [25]. Findings from the literature show that the absence of ICT in most SMEs hinders their productivity [16]. The use of ICT ensures monitoring and sustainable development of road infrastructure that support operational management in making informed decisions on the use of per product. This monitoring and identification of energy consumption patterns though the use of sufficient information reduces costs and contributes to the performance, development and growth of construction SMEs [17, 19]. Use of information and communication technologies by SMEs in rural road construction projects helps to integrate infrastructure development facilities. An increasing amount of literature advocates that design decisions and choice of technology for rural roads are more suitable when made at the poorer levels [12].

3 Research Methodology

The analysis of qualitative data gathered through in-depth interviews with contractors, designers and construction managers in all five cases resulted in the emergence of numerous aspects influencing delivery challenges of road infrastructure in construction SMEs. Primary data were collected through a number of depth interviews with construction SMEs experts and project managers. Secondary data were collected through construction SME organisations' documents written techniques, transformation and technological skills issues.

4 Findings and Discussion

The aims of this study were to identify critical factors that influenced a construction SMEs effectiveness and determine delivery challenges of road infrastructure in the Limpopo province of South Africa. For the study, concepts of infrastructure development in building construction, sustainable construction practices, innovation and construction sustainability, networking, technological skills and transformation, interview questions and NVivo software were used to analyse the data of 15 participants across the construction industry in Mopani District Municipality, Limpopo Province of South Africa. The results indicated that project participant perceptions of innovation, effectiveness of top management and infrastructure development were positive from their perceptions of construction sustainability and service delivery challenges. Furthermore, their perceived management support and commitment was shaped by their perception of employee empowerment. The results also indicated that road construction project participants' judgement in relation to sustainable construction practices and total control of construction work was more likely to skew towards how they had been involved and consulted about sustainable construction and development. SME contractors will need to share knowledge, skills and information with employees in order to make significant modifications during the construction process and include active participants in all construction levels. However, when asked about financial situations, the project participants were highly- sensitive to management commitment and networking of their counterparts. Collectively, the results implied that project participants were unconsciously rational toward quality assurance if process implementation processes were effective or not [8, 19, 21]. As such, in road building and keeping good relationship with participants, construction managers should be respectful and communicate regularly to ensure workers remain updated of any decisions made and attend to their feedback.

Lastly, the results revealed that lack of technological skills and inadequate construction methods in construction SMEs could be viewed as a consequences of society's relationship exchanges between participants. The findings showed that when project participants believed that their socio-emotional wishes were fulfilled and that they had been well taken care of, they might feel more grateful to return the humanity by engaging in beneficial of the overall construction team. On the microcomputer level of sustainable communications, the findings also pointed to the prominence of road-building managers not only having good people skills but also including computer-based information technology and professionalism as well as the ability to improve road infrastructure and soft skills of construction SMEs. It may well be applicable and significant to identify critical success factors influencing sustainable development that can lead to higher profitability in the built environment, looking at the interview results, which raised some questions about respective road construction projects. Some literature cautions against modestly replicating existing project-based approaches in the construction industry, suggesting that road construction SMEs need to move away from building's restrictive rationale. This position, and the significance of this study in terms of the sustainable development of a construction SMEs as well as its approaches to sustainability in building construction and the need to compare it against a comparable built environment equivalent, if one existed (Table 1).

Table 2 provides a summary of infrastructure development during construction process in road construction industry against which performance can be benchmarked. Given that better performing construction SMEs are also likely to be innovative.

Table 3 shows the responses from the respondents regarding effective management and leadership on road building projects in their area of work. The results of the survey demonstrated that effective management and leadership was critical for SMEs development and delivery challenges of road infrastructure in the construction industry. The leadership aspect could also extensively be observed in all of interviewee responses. Indicative of a committed leadership was the fact that the majority of construction functions were facilitated by top managerial or supervisory officers in all cases explored. In addition, management and leadership training programmes (see Table 3) seemed to coordinate the process of overwhelming barriers for successful operation, which was aligned with the recommendation that effective management training required satisfactory planning, operations and review at all levels of an organisation [15]. Table 3 illustrates a summary of evidence of effective management and leadership.

From Table 3, it was evident that commitment was widely-recognised as an important factor for the sustainable development in building construction. It reflected that all project members and affected people were strongly concerned with management commitment and support [19, 21]. Commitment was also found to be an important factor in the study conducted by Banihashemi et al. [5]. Sustainable communications and networking are both known as concentrating on management and leadership, commitment and participation, expectation and finding of faults and deficiencies, customer satisfaction, and effective transformation [20]. Thus, combining the positive aspects of the two approaches could be beneficial, provided it was possible to strengthen the techniques and developments of both approaches and convert the same into an integrated system [4]. The leader's vision, in turn, should reach the team members so that they trust in it and become delighted by it [17]. As a result, the management of shared resources is one of the key factors for an efficient and fluent

Table 1 Critical success factors influencing on SMEs development and delivery challenges of road infrastructure

Critical success factors (CSFs)	Greater Tzaneen Municipality	Ba-Phalaborwa Municipality	Greater Letaba Municipality	Greater Giyani Municipality	Maruleng Municipality
Infrastructure development	V	V	V	V	V
Innovation and construction sustainability	V	V	V	V	V
Technological skills	V	V	V	V	V
Sustainable communications and feedback systems	V	V	V	V	V
Enhancing planning, monitoring and controlling of road projects	V	V	V	V	V
Use of information and communication technologies	V	V	V	V	V
Development of new skills in various roles	V	V	V	V	V
Networking	V	V	V	V	V
Defective work is reworked or improved	na	V	na	V	V
Transformation and service delivery challenges	V	na	na	V	V
Effective management to better quality of life	V	V	V	V	V
Sustainable building construction	V	V	V	na	na
Sustainable development	V	V	V	na	na

Note The indicator "V" documents the source of the associated Critical Success Factor (CSFs) as emerged from the interviews

Table 2 Summary of infrastructure development evidence

CSFs	Indicative statements	Source of evidence
Infrastructure development	All employees are accountable for infrastructure development actions to a construction project	Community Liaison Officer-Greater Tzaneen Municipality
	Local community leaders are consulted before the road project starts	Ward Councillor-Ba-Phalaborwa Municipality
	Construction managers did provide accurate and report to the Local Organising Committee (LOC)	Community Liaison Officer-Greater Giyani Municipality
	Appointed consultants monitor and assess quality technique of construction project	Ward Councillor- Maruleng Municipality
	Construction project function related SME needs to establish good relationship with other functions in the business	Greater Letaba Municipality Manager

construction process [27]. Table 4 illustrates a summary of evidence of sustainable communications and feedback systems.

From Table 4, it is evident that people should be informed to know the sustainable communication and feedback systems, expected sustainable building construction and particularly their roles [16, 24]. The building technology and construction is a building technical knowledge and skills that presents a sequence envisioned to grow business proficiency and customer satisfaction [22, 26]. Table 5 illustrates a summary of information and communication technology evidence.

From Table 5, it is evident that there is a global increase in demand for the use of information and communication technologies on construction SMEs across the globe. The current global demand should include utilisation of up-to-date technology, appropriate emphasis on past skill and knowledge, capable project team and awarding tenders to the right designer/SMEs [4, 10]. Networking are also among the current approaches used in the delivery of construction projects. This could be due to the complex and critical nature of construction SMEs. Ideally, the purpose of a Networking is to enable the companies involved to achieve the common goal of the project, with all having shared ownership and control, while using the strengths of one another. A serious challenge to service delivery in developing countries is their inability to adopt or adapt established practices already working in other countries [1, 9, 14].

Table 3 Summary of evidence of effective management and leadership

CSAs	Indicative statements	Source of evidence
Management commitment and support	During the service delivery of road construction project the great majority of top project managers needs to active participants in all phases	Ward councilor-Greater Tzaneen Municipality
	Top management commitment is very critical the road construction and service delivery in the construction projects	Community Liaison Officer-Greater Letaba Municipality
	I have observed that commitment of management and skills development are contributing factors to successful service delivery in the construction industry	Ward Councillor-aruleng Municipality
	Top management provide accurate technological skills and networking activities required by all employees	Community Liaison Officer-Greater Ba-Phalaborwa Municipality
	I have learned that issues of management attitude and lack of support to empower all employees can be seen as obstacles that may affect effective transformation and technological issues	Ward Councillor-Greater Giyani Municipality

5 Conclusions and Recommendations

This study has contributed to the critical success factors influencing on small and medium enterprises development and delivery challenges of road infrastructure in the construction industry and promoting local government growth as well as economic development as indicated in the findings and recommendations. In a sustainable construction practices studies, there is a need to identify and determine the importance that different stakeholders assign to various success factors. Since SMEs sustainable development involve sustainability in building design and construction, the management need to allow each person to get assistance required to be successful individually, and collectively as a team. This study provides critical success factors of the construction sustainability and innovation, which contributes to the theory of SMEs success and transformation. The review of literature presented numerous innovation and construction sustainability studies within the field of SME infrastructure development, technological skills, networking, effective management commitment and

Table 4 Summary of sustainable communication and feedback systems evidence

CSFs	Indicative statements	Source of evidence
Sustainable communication and feedback systems	SME contractors need to improve communication as well as the feedback systems to the local communities	Ward Councillor-Greater Giyani Municipality
	Most of the departments of these construction SMEs do not bother to provide a good built environment	Ward Councillor-Ba-Phalaborwa Municipality
	SME contactors needs to increase employee awareness and knowledge about the feedback mechanism used	Greater Tzaneen Municipality Manager
	There was a lack of strategic planning channels, witnessed that some SME contractors do not communicate the opportunities within their business environment as well as clear plans for training	Project/construction manager-Greater Letaba Municipality
	Rural communities road projects are often seen to be low quality standards, this unfortunate situation has a negative effect because sometimes completion of projects results in poor cost estimation, time overrun, and disputes	Ward Councillor- Greater Giyani Municipality
	Customer satisfaction is of primary focus, but most SME-led projects never completed on time as planned	Ward Councillor-Ba-Phalaborwa Municipality
	Financial difficulties often encountered in most rural road projects due abandoned of important projects in this area	Greater Tzaneen Municipality Manager
	There is bad reputation of the financial difficulties encountered SME contractors due to their lack of financial skills	Project/construction manager-Greater Letaba Municipality

support, use of information and communication technologies, technical and engineering abilities, sustainable communication and feedback systems as well as transformation in building environment [3, 16, 27]. The main findings of this research clearly showed that the construction industry is faced with a disconcerted design and threatening economic conditions. Concerns have surfaced that the road deterioration impact using building technical knowledge and skills. Poor building design

Table 5 Summary of information and communication technology evidence

CSAs	Indicative statements	Source of evidence
Information and communication technology (ICT)	The use of information and communication technology helps to develop sustainable communication programs for construction SMEs	**Ward Councillor-Greater Giyani Municipality**
	Web-based sustainable communications need to be promoted for effective information sharing purposes	**Ward Councillor-Ba-Phalaborwa Municipality**
	Microsoft project software usually help project leaders to integrate scheduling, can be helpful in saving time especially when recording the field capacities and test results needed in the checklist	**Greater Tzaneen Municipality Manager**
	Very helpful in scheduling, designing and integrating sustainable developments in the construction industry	**Project/construction manager-Greater Letaba Municipality**

may cause high accident rates and economic instabilities, in addition to damaging vehicles and roadway. This has had an impact on both business performance and project. The study also found a strong similarity in the most commonly used project management techniques by both groups for time/cost control and planning. A recent survey of South African municipalities and countries identified some interesting strategies for dealing with improper construction design. Much of the data from the study covers practices and includes issues such as identifying the project construction specification in the construction industry. It is evident from the data that there is gap between the construction sustainability, technological skills and developments of road construction projects. The use of very high quality basic resources, the establishment of classy construction construction, analysis and project techniques using up-to-date scientific achievements are important in the construction project. Communicating effectively internally and externally can have positive impacts on the development of distinct procedures for enhancing road quality building even in extreme conditions, decreasing the whole life cycle costs of road re-engineering, and consequently, minimal traffic disruption. The study also found a strong similarity in the most commonly used information and communication technologies for time/cost control and planning. These factors were the main difficulties facing road construction SMEs to ensure successful operation of tasks effectively. A good understanding of the impact of road construction SME-led projects in identifying and transforming technological issues and aligning service delivery with construction sustainability in the local government as well as difficulties faced by the SME road construction projects or municipal managers in South Africa was clearly-identified and discussed in detail. Suggestions or recommendations of these impacts and difficulties were also discussed.

The results clearly indicated that quality of implementation leads to business growth, customer satisfaction and decrease of poor quality. In identifying and determining CSFs and effectiveness of road construction, the relationship between top management, consultants, employees and customers was imperative to the success of the business. The results clearly indicated that knowledge sharing, technical and sustainable communications were defined by the top project management to ensure that resources were used effectively. Sustainable construction and development also had a positive impact on technological skills. The construction SMEs should be able to provide necessary training which would lead to the improvement of employees in terms of sustainable construction methods and technological skills. The SMEs in construction should take the opportunity to develop skills in different roles/areas so as to meet the expectations of all employees. Top management should also strive to facilitate employees learning new skills.

References

1. Aigbavboa CO, Thwala WD (2014) Challenges faced blacked owned small and medium construction companies: a case study of Nelspruit Mbombela Municipality, South Africa. J Econ Behav Stud 6(10):771–778
2. Aigbavboa C, Aghimien D, Oke A, Mabasa K (2018) A preliminary study of critical factors impeding the growth of SMMEs in the construction industry in Zambia. In: Proceedings of the international conference on industrial engineering and operations management, Washington DC, USA, Sept 27–29
3. Abbott C, Jeong K, Allen S (2006) The economic motivation for innovation in small construction companies. Constr Innov 6(3):187–196
4. Amer N, Okasha A, Arafa A (2019) Enhancing planning, monitoring and controlling of road construction projects in Egypt. In: CSGE conference, 12–15 June 2019
5. Banihashemi S, Hosseini MR, Golizadeh H, Sankaran S (2017) Critical success factors (CSFs) for integration of sustainability in construction project practices in developing countries. Int J Project Manage 35:1103–1119
6. Barret P, Sexton M (2006) Innovation in small, project-based construction firms. Br J Manage 17(4):331–346
7. Construction Industry Development Board (CIDB) (2012) The construction industry as a vehicle for contractor development and transformation. CIDB, Pretoria, South Africa
8. Hardie M, Newell G (2011) Factors influencing technical innovation in construction SMEs: an Australian perspective. Eng Constr Archit Manage 18(6):618–636
9. Hong J, Chu Z, Wang Q (2011) Transport infrastructure and regional economic growth: evidence from China. Transportation 38(5):737–752
10. Gachungi RW, Mugambi M (2017) Influence of information and communication technology application on management of road projects in Kenya: a case of National Highway Authority. Int J Project Manage 1(6):98–118
11. George H (2016) Eliciting the financial challenges facing emerging contractors in developing countries using the critical incident technique: a case of South African construction industry. Bus Manage Horizons 4(2):23–33
12. Manggat T, Zain R, Jamaluddin Z (2018) The impact of infrastructure development on road communities: a literature review. Int J Acad Res Bus Soc Sci 8(1):647–658

13. Manley K (2008) Implementation of Innovation by manufacturers subcontracting to construction projects. Eng Constr Archit Manage 15(3):230–245
14. Mbande C (2010) Overcoming construction constraints through infrastructure deliver. In: Proceedings: the Association of Schools of Construction of Southern Africa (ASOCSA) fifth built environment conference. Durban, South Africa, 18–20 July
15. Mofokeng G, Thwala WD (2012) Mentorship programmes within the small and medium sized contractor development programme: a case study of the Free State province, South Africa. J Econ Behav Stud 4(12):712–722
16. Munyai TT, Nyakala KS, Mbohwa C (2017) Knowledge transfer model for improving productivity rate of the cable manufacturing industry: a South African perspective. Afr J Sci Technol Innov Dev 9(6):749–759
17. Nyakala KS, Munyai T, Vermeulen A, Pretorius JHC (2018) Applications and assessment of a quality assurance process measurement model in construction projects. J Constr 11(2):25–37
18. Nyakala KS, Pretorius JHC, Vermeulen A (2019) Factor analysis of quality assurance practices in small and medium-sized road-construction projects: a South African perspective. Acta Structilia 26(1):1–41
19. Obare JO, Kyalo DN, Mbugua J (2016) Implementation process of project control systems and performance of rural roads construction projects in Kenya: role of project team experience diversity. Eur Sci J 12(29):408–426
20. Ofori G, Hindle R, Hugo F (1996) Improving the construction industry of South Africa: a strategy. Habitat Int 20(2):203–220
21. Opawole A, Jagboo, Bababola O, Babatunde SO (2012) Evaluation of the contribution of construction professionals in budgeting for infrastructure development in Nigeria. Int J Sustain Constr Eng Technol 3(2):83–95
22. Owen D, Terence H, Green A (2012) Skill, transport and economic development: evidence from a rural area in England. J Transp Geogr 21:80–92
23. Rumane AR (2011) Quality management in construction projects. CRC Press, Boca Raton, FL
24. Statistics South Africa. (2015) Producer price index (PPI), Apr 2015. Statistical releaseP0142.1
25. Wilkinson S (2013) An analysis of the use of information technology for project management in the New Zealand construction industry. Austr J Constr Econ Build 23:47–56
26. Windapo AO, Cattell K (2013) The South African construction industry: perceptions of key challenges facing its performance, development and growth. J Constr Develop Countries 18(2):65–79
27. Van Wyk L (2003) A review of the South African construction industry: economic, regulatory and public sector capacity influences on the construction industry. CSIR, Pretoria

Critical Factors Influencing Energy Use Behaviour of Building Occupants: A Literature Review

Comfort Olubukola Iyiola and Modupe Cecilia Mewomo

Abstract The wasteful and inefficient energy consumption behaviour of energy users has caused negative effects on the environment and the economy at large. It has also been found to lead to an increase in various environmental effects like ozone layer depletion, scarcity in electricity supply, increase in energy bills, global warming, and climate change. Consequently, the factors influencing this energy use behaviour becomes a crucial area of study which needs urgent attention. This paper, therefore, discusses the critical factors influencing the energy use behaviour of occupants in buildings. The methodology involves a thorough review and analysis of selected published journals. The search included journal articles, books, and conference proceedings on the factors influencing energy use behaviour of building occupants from different databases including Science Direct, Google Scholar, Research gate, Web of Science. Five (5) critical factors influencing occupants' energy use behaviour were identified from the 71 reviewed literature. These factors are; awareness factors, personal factors, socio-demographic factors, management factors and motivational factors. The challenges faced by the environment and the ecosystem call for urgent attention to the factors that influence occupants' energy use behaviour. This paper provides an understanding of the critical factors that can influence occupants' energy use behaviour in buildings. The understanding of these critical factors will not only assist policymaking but will also provide knowledge of their prevalence in ascertaining their severity or benign nature.

Keywords Energy · Energy consumption · Energy use behaviour · Energy conservation · Energy efficiency

C. O. Iyiola (✉) · M. C. Mewomo
Department of Construction Management and Quantity Surveying, Durban University of Technology, Durban 4001, South Africa
e-mail: 22176415@dut4life.ac.za

M. C. Mewomo
e-mail: Modupem@dut.ac.za

T. C. Haupt et al. (eds.), *Construction in 5D: Deconstruction, Digitalization, Disruption, Disaster, Development*, Lecture Notes in Civil Engineering 245,
https://doi.org/10.1007/978-3-030-97748-1_30

1 Introduction

The relevance of energy consumption in the growth of an economy cannot be underestimated [1]. The significance derived from its impact on infrastructure, a range of economical processes, and, as a result, the nation's standard of wealth is improved. Despite its importance and universality, questions have been expressed concerning its use [2]. Concerns have been expressed because of the growing global usage of energy in buildings, which is promoting various environmental effects (ozone depletion, environmental degradation, global warming, climate change, etc.) [3–5]. For this reason, energy management in buildings is now a key objective at regional, national, and international levels [3]. Understanding energy use behaviour is considered one of the strategies to achieving sustained energy management in buildings [6].

Recent studies on energy use behaviour in developing countries such as Nigeria revealed that there is a major retrogression in the efficient use of energy owing to a lack of understanding of occupants' behaviour [7, 8]. This lack of understanding of occupants' energy use behaviour has led to high energy bills [9]. It has also been found to lead to an increase in maintenance problems, wastage, and inefficiency in energy use [10, 11]. Several behavioural studies carried out investigations on causes and way forward to achieve sustained efficient energy use in buildings [7, 9, 11–18]. Although the studies acknowledged the value of architectural and technological solutions to the challenges of high energy demand in buildings, there seems to be a consensus that the behaviour of energy users equally affects energy use. These views and reports are consistent with the findings of [19, 20] who further opined that it is essential to study the position of individuals rather than concentrating exclusively on energy-efficient technologies because energy use in buildings has been considered a social problem (attitudes, beliefs, lifestyles, among others) rather than a technological issue. Studying users' behaviour could help in developing proper energy awareness devices and huge savings can be done in the energy sector [17]. Frederiks et al. [17] also noted that understanding users' behaviour could help energy experts to comprehend the cause of the problem and develop an appropriate method that could mitigate the negative impacts of users' attitudes on energy saving. Kate [9] mentioned that behavioural studies could open ways to more energy-saving sources. Jeff [21] mentioned that consumers' behaviour towards energy use can be affected by individuals seeing energy as 'invisible; that is, the invisibility of energy which are influenced by knowledge, attitudes, values and beliefs of energy consumers. Ambrose [22] further mentioned that the invisibility of energy does not only make it hard to comprehend the environmental harm resulting from inefficient energy use behaviour, it could also make it hard for consumers to obtain efficient and comparative feedback on their level of consumption. References [9, 19, 23, 24] also opined that users' energy use behaviour is influenced by their attitudes, norms, lifestyle, and beliefs towards energy use. From the foregoing, it is apparent to understand the factors affecting the energy use behaviour of building occupants. This study aims at discussing the critical factors influencing the energy use behaviour of building occupants. This is a literature-based paper and reports the findings of the preliminary literature review which form the first part of the proposed research.

2 Literature Review

Energy use behaviour refers to human actions that affect the way energy is being consumed. It can also be referred to the way in which energy-related technologies are being used and the mental processes that relate to these actions whether positive or negative [9]. Various types of energy use behaviours have been found in literature.

Some of these behaviours are linked with personal comfort, luxury, preference and so on [25]. Energy use behaviour such as leaving a light on when not needed, unplugging electrical appliances when not in use, among others are evidently related to the everyday habitual element of individuals' lifestyles as they undertook daily activities [17]. Therefore, even though the world seems to be highly focused on improving designs and technology for energy sustainability, wastage of energy due to poor occupants' behaviour towards the use of household appliances seems to warrant more serious attention. Angel Ancha and Gladman [26] carried out a study on action for increasing energy-saving behaviour in student residences at Rhodes University, South Africa. The result reviewed that a substantial proportion of the respondents reported pro-environmental actions when it comes to turning off lights when they left their rooms or common rooms, unplugging chargers and turning off electronic devices when not in use, use of task lighting and laundry lines for drying clothes. The respondents attributed non-engagement in pro-environmental energy use actions to security, laziness, convenience, and lack of control of the situation. In the same vein, [27] measured the electricity consumption of 72 households in the UK over a two-year monitoring period. They observed that the large variation in the annual energy consumption resulted from variations in the number of occupants, the number and type of appliances and the behaviour pattern of the occupants. The result noted that electricity consumption was measured at 5-min intervals, and they observed that it significantly fluctuated according to changes in the behaviour of occupants. Anju and Saswat [28] revealed that residence students in Indian higher education institutions exhibit behaviour such as leaving their light on when not in use, hibernating the laptops when they are not in use, leaving computers on when not needed. Selin et al. [29] carried out a study on occupants' behaviour and household electrical appliances. In the study, sixteen appliances behaviour metrics were proposed to describe the behaviour of household electrical appliances use which were the number of switch-on events that occurred over a specified period (for instance, a day or a month), the time of the day when each switch-on events occurred and the duration of the appliance use (the length of time that the appliances were switched on). The study noted that the way in which appliances are being used in buildings was directly driven by occupants' behaviour (for instance, switching on a television). Al-Ammar [30] reported that the energy use behaviour of students in King Saud University was due to students not switching off their lighting systems when not in use and also, failure to turn off air-conditioning systems when not in use. Otegbulu and Egbenta [31] investigated the level of compliance to electricity energy efficiency practice by residential households in Lagos Metropolis. The result found that the type of appliances used by consumers in households and their compliance to energy efficiency influence energy conservation in the area. Also, the result revealed that consumers in households have poor attitudes towards energy efficiency practices both behaviorally and technologically. For this literature review to thoroughly reflect energy management and its conservation, key factors that make up these energy use behaviours must be identified and analysed.

3 Research Methodology

This study critically reviews and analyses the factors affecting occupants' energy use behaviour in buildings. To achieve the aim and focus of this study, the methodology adopted in this research is purely literature review. The review was scoped to present arguments on the issues on factors that influence energy use behaviour published in academic (peer-reviewed) journals and conference papers. To retrieve relevant articles for this study, a systematic literature search was conducted with the help of search databases like Google Scholar, Web of Science, Research gate, Science Direct. Suitable search keywords used were; "barriers", "factors", "causes", "energy use behaviour in buildings". The initial search identified a total number of 114 articles. However, not all the identified presented studies on factors influencing energy use behaviour were used. Since the aim of the study was to review literature studies on factors influencing energy use behaviour, it was necessitated to filter out unrelated articles. During the filtering process, articles that were not relevant to the study were removed. After filtering, 71 articles were found to be relevant and valid for further analysis. The study, therefore, adopted a systematic review technique to review the 71 relevant articles.

After retrieval of the selected articles which were critically reviewed and analyzed, the result and factors affecting energy use behaviour were itemized. The study adopted the flow indicated by Fig. 1. The final stage involved discussions, conclusions, and recommendations based on the findings from the literature.

4 Findings

From the reviewed literature, several researchers were able to identify major factors influencing energy use behaviour of building occupants as; awareness, personal, socio-demographic, management, motivational and management factors.

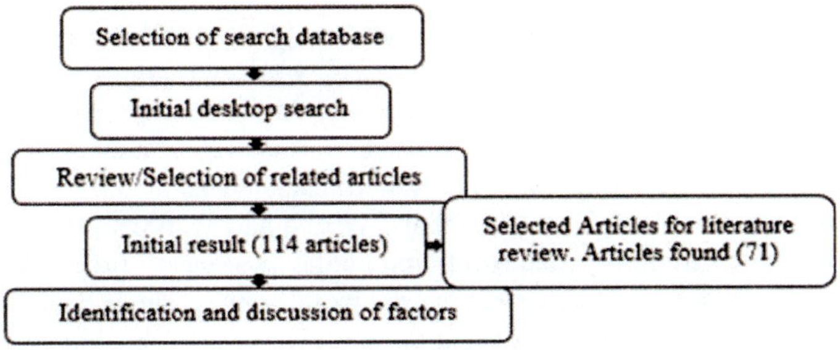

Fig. 1 Flow diagram of the searching process and paper retrieval

4.1 Awareness Factors

Brounen et al. [32] studied the role of awareness and energy literacy on Dutch household behaviour. They found that households that were more aware of their consumption of energy were also more efficient because their consumption of energy was better preserved and structured. Gyberg and Palm [33] argued that details of dwellings could help achieve sustained energy technologies and control electrical energy demand in people's daily lives. Al-Ammar [30] measured and evaluated the degree of awareness and behaviour of electricity conservation at King Saud University (KSU). The study found a low level of awareness among students towards the negative impact of energy consumption in the institution. Choong et al. [34] mentioned that energy wastage action was due to the absence of knowledge among students in public and private universities in Malaysia and no precise pattern was accessible to guide students to enhance the condition. In line with this, [35] also propagated that the issue for high energy consumption behaviour in University of Sheffield buildings was that the occupants cited a lack of awareness about the energy consumption of the building and a lack of personal control and responsibility for energy conservation. In addition, [36] noted that in Kenya Universities, energy wastage occurred in the institutions due to a lack of awareness among students in the institution on the negative impact of energy consumption in buildings and they have also not been trained on energy-saving techniques. Syed Hussain et al. [37] surveyed on energy-saving awareness among students in Universiti Utara Malaysia. The result found that the level of awareness about energy saving among UM students was low and this has caused a great burden on the university management in terms of the cost of electricity bills. Wai et al. [38] stressed that to change the ongoing character of individuals', information, education and awareness-raising measures can be used to make the public aware of their behaviour and consumption patterns. Energy awareness can be developed through the transmission of a message of knowledge and understanding that is suitable. Barr et al. [39] argued that awareness of action is an important requirement for proper behaviour and would constitute an important obstacle to action if the amount of awareness is limited.

People cannot be concerned about environmental issues unless they are aware of them and cannot act in an environmentally friendly manner unless they understand the consequences of their specific behaviour [40]. On the contrary, [41] pointed that education has played a role in raising awareness of energy efficiency and environmental problems but did not necessarily result in sustained behavioural modifications across university campuses among students. Wadha and Watkins [42] investigated domestic energy consumption behaviour and public awareness of renewable energy in Qatar. The study found that efficient energy monitoring, usage and environmental impact was dependent on user's education and awareness. This suggested that a greater awareness of the effect of excessive energy use was required through sustainable education to motivate behavioural change. The result of the study was analysed using self-determination and hierarchal needs theories indicating that education and awareness were the best options for domestic energy conservation. Alkhayyal et al. [43] analyzed sustainability awareness among higher education faculty members in Saudi Arabia. The study noted that there was no definitive understanding among faculty members about the concept of sustainability in higher education in the university. In addition, the study opined that the role of faculty members was crucial to strengthen the awareness and knowledge about sustainability in higher education among the new generation. Msengi et al. [44] assessed the knowledge and awareness of sustainability initiatives among college students. The findings indicated that only a minority of the students knew what sustainability was, but 95.8% indicated it was important, while most of the students were not aware of it. The study concluded that most of the students were not conversant with sustainability issues and were largely unaware of campus sustainability initiatives.

4.2 *Socio-Demographic Factors*

Lhendup et al. [45] divided various factors that influence pro-environmental behaviour into socio-demographic factors (age, gender, tribe, education (level of study), nationality, political ideology, marital status) external factors (social, cultural, economic, institutional) and individual factors (motivation, understanding of the environment, consciousness, values, attitudes, emotions, locus of control, duties, priorities). Gender appears to have an inconsistent, minimal or statistically insignificant impact on the use of household energy. Research suggested that females had more environmentally friendly attitudes and behaviours than males [45], while others had no important connection [46, 47]. Similar research by [48] examined the position of socio-demographic variables in the pro-environmental behaviour of users, the outcome showed that men exhibited greater pro-environmental behaviour than females, married consumers exhibited more pro-environmental behaviour than single individuals, and mid-aged users (36–50 years) also exhibited more pro-environmental behaviour than young people and old-aged consumers. Aytekin and Buyukahraz [49] researched to examine the behaviour of consumers in buying eco-friendly products, the study showed that there was a positive correlation among the demographic factors (gender, age, education level, marital status, and income level). Gender was found to have a significant difference on green product purchasing behaviour as women showed greener product purchasing behaviour than men.

Consumers' age group was also found to exhibit differences on green product purchasing behaviour as aged consumers purchased greener products than the younger age group. Married consumers were also found to exhibit green product purchasing behaviour than unmarried consumers. References [50, 51] in their findings noted that females were more environmentally aware than males and females have shown higher involvement in environment-friendly behaviour than males. Another study by [52] analyzed the gender differences of adolescents in Hong Kong and concluded that female adolescents were more concerned with issues related to the environment than male adolescents. Abrahamse and Steg [47] conducted a study to determine socio-demographic and psychological factors relating to households direct and indirect energy use and saving, the findings showed that energy consumption was determined by socio-demographic variables, whereas energy savings were mainly determined by psychological factors; that is, a contextual variable such as income was found to shape household's opportunities for energy consumption while a reduction in energy use required conscious effort to change behaviour or adopt energy-saving measures. Saba et al. [53] investigated the influence of socio-demographic variables on electrical energy management practices among residents in Niger State. The findings of the study showed that low-income consumers wasted more electricity, young aged from 18 to 39 years wasted more electricity compared to other aged groups and consumers with low education contributed significantly to energy wastage. Gram-Hanssen [20] discovered beneficial connections between income and residential energy consumption in research, indicating that households with higher income tended to consume more energy than households with lower income. Moreso, there was also proof that higher-income homes might be more energy conscious because they could afford energy-saving investment expenses such as the acquisition of innovative sustainable technology. McLoughlin et al. [54] categorized the influencing variables into several groups; socio-demographic factors (income, age, population, time of stay at home); physical parameters (family size, number of rooms, construction age, indoor temperature); user behaviour variables (attitude, understanding, motivation); electrical equipment (number of appliances, age of appliances) and climatic aspects (location, weather conditions, among others). Frederiks et al. [17] in their study found that socio-demographic factors such as income, employment status, dwelling type/size, home ownership, household size, stage of family life cycle) and psychological factors such as beliefs and attitudes, motives, and intentions, perceived behavioural control, cost–benefit appraisals, personal and social norms) were major factors affecting the energy use behaviour of various individuals. Abrahamse and Steg [47] proposed that both personal factors (preferences, attitudes, values, abilities, opportunities) and economic factors (availability of new technology, economic and population growth, government regulations and policies, socio-cultural change) could influence household energy consumption. Abrahamse and Steg [47] further asserted that household size and earnings had a major impact on the direct and indirect electricity use of households.

4.3 Personal Factors

Jeff [21] attributed factors influencing the energy use behaviour of students to 'invisibility factor'. The study noted that energy invisibility was caused by a lack of awareness and knowledge about energy issues. The study pointed that not only did the invisibility of electricity make it hard to comprehend the environmental harm resulting from the behaviour of energy consumption, it also made it hard for students to obtain efficient and comparative feedback on their level of consumption. Anju and Saswat [28] assessed energy consumption behaviour among students of Indian higher education institutions. The study found that personality traits of the students towards building regulation, environmental concern, social aspiration and comfort along with education, activities, social marketing and personality variables of openness, conscientiousness, extraversion, agreeableness and emotional stability were influencing factors affecting energy consumption pattern of students in the institution. Therefore, understanding its implication would reduce or prevent the impact that would be caused to the economy, environment and the university bills itself. Shaharon and Jalaludin [55] in their research carried out in Malaysia categorized the factors influencing energy consumption behaviour into two namely internal factors and external factors noting that external factors were factors that were uncontrollable by the individual; that is, building regulations, availability of technology, education, social marketing while internal factors; that is, device and setting, environmental concerns, activity, social aspiration and comfort were controllable by the individual.

4.4 Management Factors

Anthony et al. [56] seeks to understand manager's perspective of residential students' energy use behaviour in higher education institutions in Ghana. The result of the study showed that residential students' energy use behaviour was influenced by institutional factors which consisted of regulation-related issues and monitoring of students' energy activities; awareness factors in the form of lack of awareness and knowledge of energy issues and energy-saving strategies from authorities of the institutions; attitudes, habits and social factors. References [57, 58] noted that improper management and implementation could also influence student's energy use behaviour in students' housing. This study attributed the influencing factor to ineffective rules and regulations guiding the use of energy in students' housing and poor imposition of penalty for inefficient energy use [58].

4.5 Motivational Factors

Steg and Vlek [59] noted that the effectiveness of behavioural interventions generally increased when they were aimed at important antecedents of the relevant behaviour and at removing barriers for change. According to [60] in research in Sonora University in Mexico carried out a long-term implementation of an energy conservation and efficiency programme yielded a substantial energy reduction with a reasonable amount of savings in three years. In the same vein, [61] found that electricity use was reduced by 32% in students' residences at a US university after a combination of feedback on energy performance and incentives for energy saving. The study also conforms with the findings of [62]. Similarly, [63] found that providing regular feedback and incentives to students resulted in energy savings ranging from 10.7 to 16.2% in students' residences at Otago University. A study by [41] at the University of Kent, UK found energy savings of 5–15% via energy feedback meters and real-time electricity display.

Angel Ancha and Gladman [26] investigated action for increasing energy-saving behaviour in students' residences at Rhodes University. The findings showed that participant residences that received a mix of different interventions in the form of pamphlets, face to face discussions, incentives and feedback recorded more energy reductions up to 9% than residences that received a single or no intervention. The result concluded that information provision of energy-saving tips combined with regular feedback and incentives could result in energy use reductions in students housing in tertiary institutions. Ucci [64] suggested that energy-saving behaviour in office buildings was more likely encouraged by the concern for the environmental issues rather than knowing the amount of energy cost reduced since the energy costs were not paid by the occupants. In addition, [65] compared the impact on energy consumption of generic energy advice through postcards sent directly to individuals, with the impact of combining this energy advice alongside e-mailed feedback on energy use at the building level, peer education involving employees acting as points of contact and information for colleagues. The study found that feedback and peer education generated energy savings of 7% and 4% respectively while the provision of information alone led to an increase in energy use by 4%. On the contrary, [66] found that participants considered e-mail a better means of communicating information than either posters or leaflets. The study draws attention to the frequency and duration over which feedback information was provided suggesting emails sent weekly would be effective in providing information. The study reported sustained energy use behaviour over a 3-month period post-intervention, indicating the potential for savings over a longer-term. References [67, 68] also mentioned that the energy use behaviour of students can be improved by raising awareness and education of energy saving. Direct information provision (through signs and stickers) and peer education were also compared by [69], the study noted that there was a reduction in the percentage of light that was being left on.

5 Conclusion and Recommendation

This study was conducted as a systematic review of literature on the factors influencing energy use behaviour, with the help of databases which were used to collect relevant academic (peered-reviewed) journal and conference papers. From the comprehensive literature review, it was revealed that there are many factors affecting energy use behaviour, but most reported barriers are awareness factors, personal factors, socio-demographic factors, management factors and motivational factors. The study thereby presents the five most reported factors as the major factors that affect the energy use behaviour of building occupants. Regarding recommendations to overcome the factors influencing energy use behaviour of occupants, the study recommends that a strong collaboration between policymakers, industry associations and companies, management of institutions should be established regulations, training, feedback, and strategic plans to create awareness on how wasteful energy consumption behaviour can be mitigated. According to [70, 71], the effort put into action and adoption towards the concept of anything depends on the awareness, knowledge and understanding of individuals. In fact, awareness and knowledge are the first stumbling blocks that must be conquered in achieving sustained energy use. These awareness-raising measures will help create a society and environment for sustained energy use and provide opportunity for collaboration and knowledge sharing. Emeakaroha et al. [41] also mentioned that energy users can be stimulated to support an energy management program if they are informed of the amount of energy they are using, the cost involved, the seriousness of the energy problem and its potential effects upon a nation's economy in the future. This study has therefore contributed to the knowledge of factors influencing energy use behaviour by identifying the most reported factors (major barriers) in literature. From a global perspective, the findings are liable to provide better understanding of how the behaviour, attitudes and lifestyles of consumers affect energy consumption.

References

1. Stern D (2011) The role of energy in economic growth. Ann N Y Acad Sci 1219:26–51
2. Aliyu SA, Ramli TA, Saleh AM (2013) Nigeria electricity crisis. Power Gener Capacity Expansion Environ Ramification 61:354–367
3. Perez-Lombard L, Ortiz J, Pout C (2008) A review on buildings energy consumption information. Energy Build 40:394–398
4. Oyedepo SO, Adeleke T, Leramo RO, Kilanko O, Babalola OP, Balogun AO, Akhibi MO (2015) A study on energy demand and consumption in Covenant University, Ota, Nigeria. In: International conference on African development issues (CU-ICADI): renewable energy track, pp 203–211
5. Olanipekun EA, Nunayon SS (2017) Electricity end use characteristics of public universities in Southwestern Nigeria. Br J Environ Sci 5(3):30–46
6. Khan K, Shah A, Jaffar K (2016) Electricity consumption patterns: comparative evidence from Pakistan's public and private sectors. Lahore J Econ 1:99–122

7. Bernard J (2016) Energy behaviour in Dutch student houses. An analysis of energy use and energy time of use in a field lab setting
8. Lefkeli S, Manolos E, Ioannou K, Georgios T (2018) Socio-cultural impact of energy saving: studying the behaviour of elementary school students in Greece. J Sustain 10:737
9. Kate C (2014) Learning energy systems: a holistic approach to low energy behaviour in schools
10. Gadenne D, Sharma B, Kerr DV, Smith T (2011) The influence of consumers' environmental beliefs and attitudes on energy saving behaviour. J Energy Policy 39(12):7684–7694
11. Chiang T, Gokhan M, Sukumar N, Julian P, Walker I (2014) Inducing subconscious energy behaviour through visually displayed energy information. A case study in university accommodation. Department of Computer Science, University of Bath, Bath, UK
12. Brandon G (2009) Energy use in Penn's student housing and techniques to achieve energy reduction. Senior Seminar Univ Pennsylvania Eng J 11
13. Black R, Davidson P, Retra K (2010) Intrinsic changes: energy saving behaviour among resident university students. Aust J Environ Educ 26:85–99
14. Stephenson J, Barton B, Carrington G, Gnoth D, Lawson R, Thorsnes P (2010) Energy cultures: a framework for understanding energy behaviours. Energy Policy 38:6120–6129
15. Sutterlin B, Brunner TA, Siegrist M (2011) Who puts the most energy into energy conservation? A segmentation of energy consumers based on energy related behavioural characteristics. Energy Policy 39:8137–8152
16. Galis V, Gyberg P (2011) Energy behaviour as a collectif. The case of Colonia students' dormitories at a Swedish University. J Energy Efficiency 4(2):303–319
17. Frederiks ER, Stenner K, Hobman EV (2015) The socio-demographic and psychological predictors of residential energy consumption: a comprehensive review. Energies 8:583–609
18. Kaile Z, Shanlin Y (2016) Understanding household energy consumption behaviour: the contribution of energy big data analytics. Renew Sustain Energy Rev 56:810–819
19. Janda KB (2009) Buildings don't use energy, people do. In: Conference on passive and low energy architecture
20. Gram-Hanssen K (2014) New needs for better understanding of household's energy consumption behaviour, lifestyle or practices. Archit Eng Design Manage 10(1–2):91–107
21. Jeff V (2016) The effectiveness of the Colby college electricity competition on promoting electricity conservation to students. J Environ Resour Econ 3
22. Ambrose A (2019) Walking with energy: increasing energy visibility through research participation. ECEEE Summer Study Proc:83–92
23. Hori S, Kondo K, Nogata D, Ben H (2013) The determinants of household energy saving behaviour. Survey and comparison in five major Asian cities. Energy Policy 52:354–362
24. Nair G, Gustavsson L, Mahapatra K (2010) Factors influencing energy efficiency investments in existing Swedish residential buildings. Energy Policy 6(38):2956–2963
25. Ghasemi M (2015) Behaviour related energy use in single family homes: a study of residential houses in Sweden. Unpublished thesis submitted to KTH School of Industrial Engineering and Management Energy Technology EGI-2014
26. Angel Ancha LB, Gladman T (2018) Action for increasing energy saving behaviour in students residences at Rhoda University. Int J Sustain Higher Educ 19(4)
27. Firth SK, Lomas AW, Wall R (2008) Identifying trends in the use of domestic appliances from household electricity consumption measurements. J Energy Build 40(5):926–936. https://doi.org/10.1016/j.enbuild.2007.07.005
28. Anju K, Saswat B (2017) Factors influencing the energy consumption behaviour pattern among the Indian higher educational institution students. In: International conference on technological advancements in power and energy, pp 1–6
29. Selin Y, Steven KF, David A (2017) Occupant behaviour modelling in domestic buildings. The case of household electrical appliances. J Build Perform Simul 10(5–6):582–600
30. Al-Ammar E (2010) Meaning and evaluating degree of awareness and behaviours of electricity conservation at King Saud University. In: International conference on energy planning, energy saving, environmental education, pp 1790–5096

31. Otegbulu AC, Egbenta I (2016) Energy efficient practice among residential households in Lagos, Nigeria. J Trop Environ 13(1):8–18
32. Brounen D, Kok N, Quigley JM (2013) Energy literacy, awareness and conservation behaviour of residential households. J Energy Econ 38:42–50
33. Gyberg P, Palm J (2009) Influencing households' energy behaviour-how is this done and on what premises? J Energy Policy 37(7):2807–2813
34. Choong WW, Abdul HM, Buang A (2012) A conceptual framework of energy awareness development process. Malaysian J Real Estate 1(1):58–67
35. Whittle C, Jones C (2013) User perceptions of energy consumption in university buildings: a university of Sheffield case study. J Sustain Educ
36. Richard KR (2018) Level of awareness of energy measures of electricity users at universities in Kenya. Int J Eng Comput Sci 7(4):23875–23882
37. Syed Hussain TPR, Ismail MK, Noh MK (2013) PERKEM Proc 2
38. Wai C, Mohammed A, Alias B (2006) Energy consumption—a conceptual framework of energy awareness development process, pp 58–67
39. Barr S, Gilg AW, Ford N (2005) The household energy gap: examining the divide between habitual and purchase-related conservation behaviours. Energy Policy 33(11):1425–1444
40. Stavroula K, Marina L, Mat S (2014) Models of behaviour change and residential energy use: a review of research directions and findings for behaviour based energy efficiency. Adv Build Energy Res 8(2):137–147
41. Emeakaroha A, Ang CS, Yan Y (2012) Challenges in improving energy efficiency in a university campus through the application of persuasive technology and smart sensors. Challenges 3(2):290–318
42. Wadha A, Watkins M (2018) An investigation into domestic energy consumption behaviour and public awareness of renewable energy in Qatar. J Sustain Cities Soc 41:639–646
43. Alkhayyal B, Labib W, Alsulaiman T, Abdelhadi A (2019) Analyzing sustainability awareness among higher education faculty members. A case study in Saudi Arabia. J Sustain 11:6837
44. Msengi I, Doe R, Wilson T, Fowler D, Wigginton C, Olorunyomi S, Banks I, Morel R (2019) Assessment of knowledge and awareness of sustainability initiatives among college students. J Renew Energy Environ Sustain 4(6):1–11
45. Lhendup T, Lhundup S, Wangchuk T (2010) Domestic energy consumption patterns in urban Bhutan. J Energy Sustain Dev 14(2):134–142
46. Sardianou E (2007) Estimating energy conservation patterns of Greek households. Energy Policy 7(35):3778–3791
47. Abrahamse W, Steg L (2009) How do socio-demographic and psychological factors relate to house-holds' direct and indirect energy use and savings? J Econ Psychol 5(30):711–720
48. Jayesh DP, Ashwin M, Justin P (2017) Pro-environmental behaviour and socio-demographic factors in an emerging market. Asian J Bus Ethics 6(2)
49. Aytekin M, Buyukahraz G (2014) Demographic characteristics of consumer buying behaviour effects of environmentally friendly products and an application in Gaziantep. Bus Manage Rev 5(1):72–81
50. Oerke B, Bogner FX (2010) Gender, age and subject matter: impact on teachers' ecological values. Environmentalist 30:111–122
51. Xiao C, Hong D (2010) Gender differences in environmental behaviours in China. Popul Environ 32:88–104
52. Lee K (2009) Gender differences in Hong Kong adolescent consumers' green purchasing behaviour. J Consum Mark 26:87–96
53. Saba TM, Tsado J, Bukar B, Raymond E (2016) Influence of socio-demographic variables on electrical energy management practices among residents of Niger State, Nigeria. Br J Econ Manage Trade 13:1–10
54. McLoughlin F, Duffy A, Conlon M (2012) Characterizing domestic electricity consumption patterns by dwelling and occupant socio-economic variables: an irish case study. Energy Build 48:240–248

55. Shaharon MN, Jalaludin J (2012) Thermal comfort assessment—a study towards workers' satisfaction in a low energy office building. Am J Appl Sci 9(7):1037–1045
56. Anthony A, Maimunah S, Ting S, Eugene OK (2018) Factors affecting higher education residential students energy use behaviour, managers' perspective. J Comput Theor Nanosci 24(6):4089–4093
57. Arbuthnott K (2014) Education for sustainable development beyond attitude change. Int J Sustain High Educ 10(2):152–163
58. Dzulkefli MH (2017) The energy audit process for universities accommodation in Malaysia: a preliminary study. In: 7th international conference on environment and industrial innovation, vol 67, pp 1–8
59. Steg L, Vlek C (2009) Encouraging pro-environmental behaviour; an integrative review and research agenda. J Environ Psychol 29:309–317
60. Manguia N, Zavala A, Velazquez L (2011) Energy end use efficiency issues. World Renew Energy Congr 8(13):907–912
61. Peterson JE, Vladislav S, Kathryn J, Gavin P, Kate W (2007) Dormitory residents reduce electricity consumption when exposed to real-time visual feedback and incentives. Int J Sustain High Educ 1(8):16–33
62. Dixon GN, Deline MB, McComas K, Chambliss L, Hoffman M (2015) Using comparative feedback to influence workplace energy conservation. A case study of a university campaign. J Environ Behav 45:667–693
63. Bekker M, Cummin T, Osborne K, Bruining A, Leland L (2010) Encouraging electricity savings in a university residential hall through a combination of feedback, visual prompts and incentives. J Appl Behav Anal 43(2):257–278
64. Ucci M (2014) Behaviour change potential for energy saving in non-domestic buildings: development and pilot-testing of a benchmarking tool. Build Serv Eng Res Technol 35(1):36–52
65. Carrico AR, Riemer M (2011) Motivating energy conservation in the workplace: an evaluation of the use of group-level feedback and peer education. J Environ Psychol 31:1–13
66. Kamilaris A, Neovino J, Kondepudi S, Kalluri B (2015) A case study on the individual energy use of personal computers in an office setting and assessment of various feedback types towards energy savings. J Energy Build 104(5991):73–86
67. Hilas A (2011) Determining energy saving behaviour and energy awareness of secondary school students according to socio-demographic characteristics. J Educ Res Rev 6(3):243–250
68. Malan M (2014) An investigation of the impact of information on energy consumption at a Tertiary Institution. Unpublished master's thesis, Stellenbosch University
69. Werner CM, Cook S, Colby J, Lim HJ (2012) Light out in university classrooms: brief group discussion can change behaviour. J Environ Psychol 32(4):418–426
70. Spagnolli A (2011) Eco-feedback on the go. Motivating Energy Awareness 44(5):38–45
71. Craig CA, Allen MW (2015) The impact of curriculum-based learning on environmental literacy and energy consumption with implications for policy. J Util Policy 35:41–49

The Challenges with Developing and Applying Knowledge, Skills and Competences (KSC) in the Nigerian Engineering and Construction Industry (NECI)

Uzor Onyia, Charles Egbu, and Obas John Ebohon

Abstract In Nigeria, the construction industry contributes over 3% of the annual gross domestic product and is the fourth highest employer of labour in Nigeria. Despite its contributions and the huge potential of the Nigerian Engineering and construction industry (NECI), inadequate attention is given to its significance in driving the Nigerian economy. Knowledge, skills, and competences are the main critical success factor of the construction industry. Several challenges are responsible for sloth of the NECI. This study explored these challenges evident in developing and applying knowledge, skills, and competences (KSC) in management of the NECI. The NECI employs an extremely diverse range of workers from extensive and various backgrounds, who are deployed to be managed and supervised. Ineffective management caused by lack of updated KSC, has been identified in academic literatures as a major setback in the NECI. This study identified and extensively discussed these challenges through a review of literature and data from the NECI. The data, on which this study was based, was sourced from 155 completed and usable survey questionnaires and 30 semi-structured interviews with registered engineering and construction professionals in Nigeria. Participants were drawn from membership database of the Nigerian Society of Engineers, practicing in both private and public sectors.

U. Onyia
Department of Civil Engineering, University of Nigeria, Nsukka, Nigeria

C. Egbu
Leeds Trinity University, Leeds, England

O. J. Ebohon (✉)
Sustainability and Environmental Law, London South Bank University, London, England
e-mail: ebohono@lsbu.ac.uk

School of the Built Environment and Architecture, London South Bank University, London, UK

T. C. Haupt et al. (eds.), *Construction in 5D: Deconstruction, Digitalization, Disruption, Disaster, Development*, Lecture Notes in Civil Engineering 245,
https://doi.org/10.1007/978-3-030-97748-1_31

1 Introduction

Modern and organised construction arguably began in the early 1940s in Nigeria, since then, the industry through the 60s and 80s, has largely been dominated by foreign companies, creating jobs, and generating revenue for the people and government of Nigeria respectively, though with lots of imported resources as against the use of local alternatives [78]. Poor quality of material and human resources was identified by Owolabi et al. [75] as the main challenge in the NECI. This resource includes knowledge, skills, and competences of practitioners in the NECI. Wang et al. [99] established that poor development and application of KSC is one of the principal causes of moderate performance in quality and safety issues. KSC contribute to the performance of the construction industry, and it's expected success [4]. Developing and applying KSC can be argued as the main cause of the underperformance of the NECI. There exist several approaches to developing and applying KSC but Zhu et al. [102] revealed that traditional development of KSC is through "learning by doing" and some by "apprenticeship". To support Zhu et al.'s [102] solution and policy for the Chinese construction industry, Sun et al. [91] revealed that the Chinese government regulated main contractors to provide onsite KSC training to their workers before the start of construction projects in China.

Construction is a labour-intensive industry, is particularly attractive to many due to its easy entry and relatively high pay [99]. Hence, the continuous need for KSC development and application. Several challenges resist the development and application of KSC in the construction industry. These challenges are: Measuring KSC Attainment and Reward System; Commitment, Values and Unguaranteed loyalty of Employees; General Education and Knowledge Sharing Culture; KSC Development is Expensive and Lacks Sponsorship; Under-Priced Project Budget, Corruption and Ethical Issues; others are Training Infrastructures and Established Standard; Understanding, Agreement and Conflicts; Lack of Effective Communication and Teamwork; Multicultural Challenges; Lack of Proper Planning, Scheduling and Directing; Lack of Leading Creativity; Time, Technologies and Environment; Political Interference, Regulations and Policies; Research, Strategy and External Challenges [27]. Construction organisations face challenges when developing and applying KSC for effective construction management among practitioners. Understanding these challenges and their peculiarities to their industries is the first step to responding to them with effective solutions.

2 Literature Review: The Challenges of Knowledge, Skills and Competences (KSC)

Previous studies emphasised that the challenges faced in today's world by construction management are more complex and ill-defined than is discussed in the common bodies of knowledge [23]. There exists a gap between school education and construction KSC [99], with education perceived as being unable to reveal the challenges of projects appropriately [23]. Researchers are continuously trying to understand KSC development, and the challenges associated with it. The issue of detailed understanding posed a major challenge in any development. This was elaborated by Fellows and Liu [33] when they argued that understanding is essential for effective delivery or realisation of any meaningful KSC development.

The main established issue is the fact that the three most widely accepted project management institutions: the International Project Management Association (IPMA), Association for Project Management (APM) and the Project Management Institute (PMI) who virtually run project management by training, assessing, and certifying member practitioners, as well as set standards, which influence academic and industrial thinkers, and also accredit academic programmes, show that the institutional frameworks pay greater attention to functional and cognitive rather than social competences [23]. Some researchers [9, 47] suggested that much greater salience ought to be paid to social KSC in construction management. Camilleri [21] and Nixon et al. [69] report that technical KSC is good but in today's industry, managerial and behavioural KSC are most paramount to project success. Similarly, several studies rank human KSC higher than technical KSC in project management [23]. This research was designed to investigate and understand the precise challenges inherent in KSC practiced in Nigeria through a combination of literature review and evidence from the NECI.

2.1 Attainment and Reward

A significant challenge has been how to measure and demonstrate workers' Knowledge, skills, and competences (KSC) outcomes before engaging or assigning tasks to them [70]. Most KSC frameworks lack guidance on expected and basic standards of performance for these skills and behaviours in workers as they progress through different stages of their trainings to the workplace [34, 81]. As employability KSC typically involve a significant proportion of worker selection criteria, workers themselves must appreciate, not only the KSC they have acquired, but can also communicate and demonstrate to employers how they meet expected levels of development required of them [81]. It will be very productive if employers can assess the level of workers' KSC without going through the workers. Further research is needed to enable organisations determine the exact level of workers' KSC. It is productive to measure and monitor KSC of management members through their commitment, the values they add and the attitude they bring on the job [70].

Professionalism and best practices are rarely encouraged nor rewarded in the NCI, and this discourages KSC development. Untrained workers (quacks) are hired instead of experts on the altar of favouritism, cutting cost and avoiding responsibility of training [73]. According to an interview participant: "Best practice is not encouraged; well trained personnel are not assigned responsibilities in areas of their expertise in the industry. Many seek and encourage the services of unskilled and untrained workers in the industry, so why would anyone bother to develop or improve their KSC in a market that does not patronise skilled and trained professionals." Another participant stated that "the reward system is poor; most construction managers are underpaid and under motivated. Many organisations subscribe to the use of contract staff, which is cheaper but then they deliver cheap and substandard projects as they lack trainings, motivation, and rewards. This orientation forces most of the young managers to focus on financial benefits from a project instead of experience." It is evident that the current reward system in the NECI does not motivate professionals to develop their KSC, corruption and shunning of ethics in the construction profession is the current practice and it has direct negative impact on project outcomes. Separation of professional disciplines is not the popular practice in the NECI. Nigerian construction professionals take up responsibilities outside their field of specialisation, making them quacks in those areas. There is strong advocacy from participants of this study for the discontinuation of quackery in the NECI, for the industry to meet its fully potential of driving the Nigerian economy and hence be reckoned among the best in the world.

2.2 *Commitment and Loyalty*

Martin and Nicholls (1987) define an individual's commitment as the willingness to contribute much more to the organisation than their prescribed contractual obligation. Also, Merriam–Webster defines commitment as the act of committing to a charge or trust. Tagaza and Wilson (2004) found out that once the enthusiasm of managers and supervisors, or any other construction professional is low on any aspect of work, their results and performance is directly low. Tagaza and Wilson (2004) also pointed out that problems such as the managers and supervisors' attitudes and values, which were demonstrated in their lack of commitment to common goals and objectives and in neglecting to develop their own professional KSC showed a resistance to change. Resistance to change is best expressed in the managers' reluctance in sticking to routines and active resistance. Staff's resources were also found to be a problem because some of the managers suffered from work exhaustion [35]. Managers' exhaustion by age and mental fatigue and managers' activity levels and motivations vary significantly at different times, which complicates systematic development of managerial and supervisory KSC [35]. Interest amongst managers and supervisors in any form of training they are to undergo is very important [22]. For instance, workers' resistance to changing their traditional practices or upgrade has been the most critical challenge faced by trainers and project managers before and during trainings [44].

Another major challenge is the concern of losing a staff to a competitor after training them. After the development, due to the investment made by organisations,

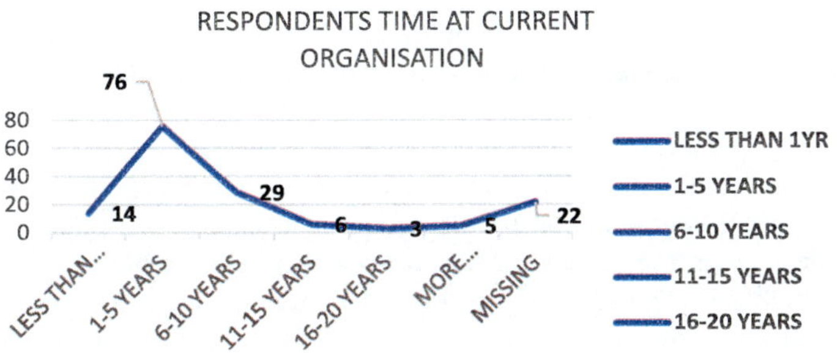

Fig. 1 Respondents time at current organization

there is a serious concern that their newly developed worker could be poked and snatched by a competitor with more financial power. This discourages the employer from funding any major development for employees. An executive director of a private firm, who participated in the interview asked, "what guarantee do we have that after investing this huge amount on training, that they (employees) will not move to our competitors when offered higher pay"? This opinion was corroborated by the fact that 67.6% of the survey respondents have spent less than five years in their respective organisations as revealed in Fig. 1.

2.3 *Education and Knowledge Sharing Culture*

According to result of this study shown in Fig. 2, 93% of the registered Nigerian engineering and construction professionals possesses degrees and advanced degrees, other 7% have acquired other educational qualifications. This result supports previous studies conducted by Abidoye and Chan [2] and Inuwa et al. [45]. This is a significant progression over almost two decades, different from the situation Koivula and Paunonen-ilmonen [56] found in their research which revealed that managers and supervisors had a low level of education, which hinders the development of KSC. Though this progression in general education does not apply for all supervisors,

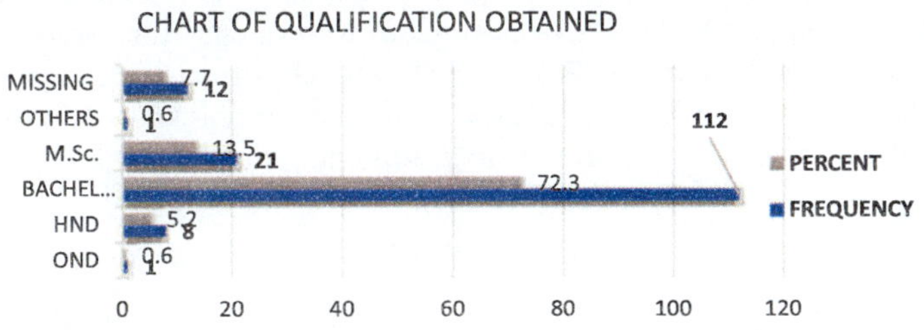

Fig. 2 Qualifications obtained by respondents

foremen and craftsmen. Managers with low level of education feel less valued than their well-educated counterparts; a situation, which if not well addressed, may affect their performance during any skill training and KSC development they undertake [44]. Secondly, low level of general education mainly causes inadequacies in learning and knowledge development [98]. Another major educational challenge is the time it takes to finish a degree in construction or project management; the shortest one is the masters programme, which takes at least two years in Nigerian academies. Secondary to this, is the high cost of quality education in Nigeria as there is no functional student loan structure [3]. Even the Council of Registered Builders of Nigeria (CORBON, 2011) and construction site management for builders as cited by Jimoh et al. [48] approved that a small construction site can be managed and supervised by a trade foreman who does not have professional education. As indicated by Jimoh et al. [48], small building construction projects consist of mainly residential buildings. It is widely believed in the NECI that what these managers and supervisors lack in proper education, they make up for in their long experiences in the industry.

As a support strategy, a lot of companies also expect and enrol their managers and supervisors in regular short-term courses annually (Dave et al. 2017). Take China, for instance, where similar challenges exist in their wind power sector. By the end of 2011, over 200,000 managers were hired directly or indirectly in the various sectors of Chinese wind energy industry. Almost all of them have no previous wind energy education background and are taking short-term or long-term wind power managerial training courses. It is a huge market for training centres in China and funding of such trainings is also a challenge. Obviously, the number of training centres is inadequate, and the Chinese government is aiding further expansion and strengthening of the training centres (Salami et al. 2017). The Nigerian government can borrow a leaf from this practice.

Secondly, the lack of knowledge sharing orientation and visionary leadership at the top of the industry is one of the huge challenges associated with the development of construction [6]. This is one view shared by all the 30 participants in the interview. They shared the opinion that; there is no orientation for proper and structured channel for transfer of knowledge from more experienced and skilled senior managers to less experienced and skilled fresh managers. According to one of the interviewees, "*the knowledge, skills and experiences of the senior managers retire with them while the next generation start from the avoidable struggling stages to build their own experiences*". This is unacceptable as knowledge is a continuous development and should be built on as recommended by CKT (Continuous Knowledge Transfer) of European Patent Office (Schombacher et al. 2016). The lost for construction managers, organisations, HR managers and the entire NECI is immeasurable, as knowledge that could have been transferred or shared are lost to deaths and retirements.

2.4 Funding and Sponsorship

This study was designed to gather the perspectives of construction professionals in Nigeria on the current challenges associated with developing and applying some of the listed KSC in the industry. Hong et al. [41] identified cost of KSC development as one of the challenges. From the results of the survey conducted, Fig. 3 shows that over 66.2% of respondents are of the view that the training is expensive. Investigation during interviews revealed that most organisations lack the interest to develop KSC of their employees and that the employees cannot afford the training themselves because they have marked them as "expensive".

According to the result published in Fig. 3, the perspective of managers in the NECI are that the cost of KSC development is expensive. The implication of this result is that leaving managers to bear the responsibility of funding their KSC development will yield no desired result as 66% of respondents indicated that KSC development is expensive. Their position was further emphasized by exonerating themselves from sponsoring their KSC development and instead transferred the responsibility to their employers with, perhaps support from government as shown in Fig. 4.

What this means is that organisations, HR managers, training providers will need to focus on how to secure sponsorship of KSC development from employers and not employees. During the interviews, a participant stated that most private and public companies do not comply or commit to training needs of their managers and supervisors. Hence, training lacks the required sponsorship. The companies do not have the sense of commitment to invest and improve the competences of their workers. They do not consider it a worthy investment as they are more interested in maximizing profits on every project.

Lack of comprehensive strategy on how to achieve employers' sponsorship of KSC development which is marked as "expensive" by managers and supervisors, will mean low numbers of qualified professionals with updated KSC managing Nigerian construction projects. 100% of the participants of this research study indicated that KSC development directly influences project outcomes in the NECI. Lack of KSC development as a result of cost of KSC development programmes, will damage the image of any construction industry as is the current case with the NECI, where recent newsflashes of the incessant building collapses are regular across the country.

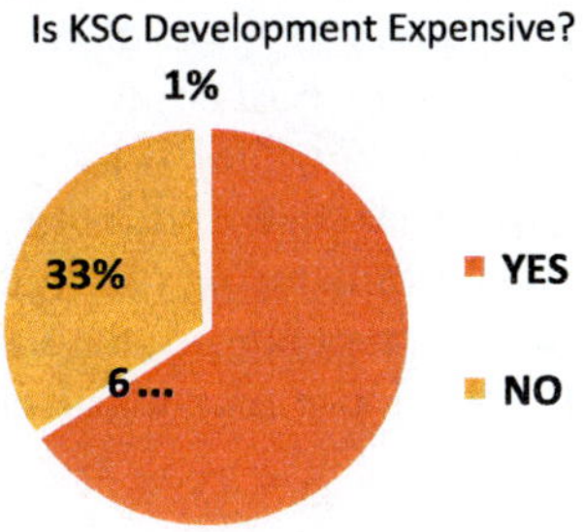

Fig. 3 KSC development is expensive

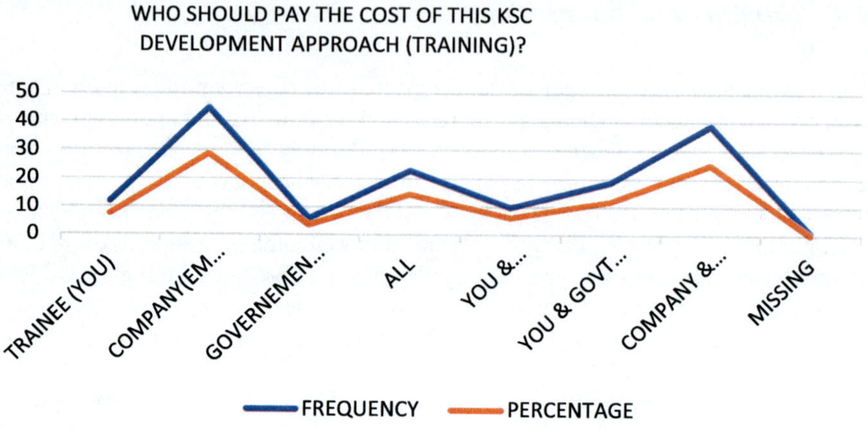

Fig. 4 KSC development responsibility

2.5 *Corruption and Ethics*

The research participants helped us understand KSC development challenges occasioned by budget, corruption, and ethics. An interviewee averred that "*under-priced projects and unethical contractors mean no room for training of engineers, managers and supervisors working on the project.*" Another interviewee noted that: "*due to stiff competitions because of the state of the economy, some projects are under-priced, which puts the contractors in a financial fix and greedy contractors do not want to spend the approved percentage on training for their workers engaged on the project.*" This means that under-priced construction projects, unethical practice and corruption have direct implications and directly undercut training budgets of large construction projects in Nigeria.

Another interviewee, who is an assistant director in the federal ministry responsible for construction related developments in Nigeria, said this about corruption: "*corruption is one of the major challenges that hinders the development of KSC. The right (qualified) people are not given the right opportunities; rather, there are constant cases of tribalism, nepotism, and unethical practices. Budgets on trainings are carted away for personal aggrandisement. Recruitment and nomination for jobs and training are based on strong family and social ties not on professional qualifications. In most cases, your state, local council of origin, religious and political affiliation play far more important roles than your professional qualification.*" The participants went further to reveal how the code of conduct for public and private employees have been shoved aside for more selfish financial rewards. This implies that selection process for KSC development in the NCI is not based on merit, and this practice impact negatively on the outcomes of construction projects in Nigeria.

2.6 Infrastructures and Standards

Shalleya and Gilson [87] reported that the failure to modernize recruitment and training can result in skill shortages, higher prices, and poorer quality output. There is a long-term trend in industrial change which relates closely to KSC and training. Fluctuations in output, smaller size and specialisation of projects make it difficult for individual firms to plan labour requirements [28]. Moreover, frequently short-term competitive pressures lead to poaching of skilled labour and act as a further disincentive to training [87]. For these reasons, few construction firms have KSC development plan; and most take insufficient interest in integrating workers KSC development into their corporate strategies, where these exist at all. Hou et al. [42] stated that onsite trainings offered by some construction organisations and associations is very limited, moreover training facilities that have been established are far from sufficient to the growing standard of construction professionals for industry expansion. So, there is need for creativeness in leading KSC application in construction management.

In expressing the lack of a generally known and accepted standards for developing managers' KSC, a participant in the interview of this study stated that: "*There is also a lack of framework, guidelines, specialised professionals and resourceful trainers and conducive training facilities. Where there are a few, they are usually very expensive*". The interviewee's expression confirmed the lack of established framework, guidelines, resourceful trainers and facility for development of KSC in the NECI. The implication of this is lack of acceptable standards and models for the development of managers KSC. See how one of the interviewees puts it: "*There are no specialized and recognised professional trainers who are ready to research or work with researchers to gain current and future local insight and understanding of the current and possible future challenges of the industry.*" By "specialised" and "recognised", the interviewee meant an expert recognised and licensed by reputable professional or educational institution. This lack of standards and guidelines in the NECI robs it of the credibility and value of training. Because of this, most organisations and institutions sort and develop their own internal training policies, guidelines which are rarely tailored to the needs of their construction managers and supervisors, and the uniqueness of the NECI. This lack of a comprehensive and generally accepted framework and guidelines for KSC development in the NECI means no standard gauge for KSC in Nigeria.

2.7 Agreement and Understanding

It is believed that managers possess a good level of written and oral communication skills, and often these may not have been formally taught communication skills, instead they would have been experiences acquired on-the-job. Most managers in the NECI, possess further and higher education qualifications, and have formal basic education (Fig. 2). About two decades ago, in the United Kingdom, only three of the 20 executives who were interviewed by Briscoe et al. [19] had degrees or HND qualifications. Although these individuals become directors, senior managers, and owners of smaller organisations in the construction supply chain, some still lack key oral and written skills that are essential for successful partnering and collaboration required in managing and delivering a successful construction project. A good instance of such deficiencies is the inability to understand, interpret and effectively respond to contracts and other technical documents. Although the use of external expert advice in this area is being practiced, an inability of managers to comprehend the finer points of a binding legal contract is a significant handicap to supervision and management (Briscoe et al. 2001).

According to Hardison et al. [38], unprofessional, disrespectful or attempts by a manager or supervisor to change worker's performance or behaviour have the potential to undermine a positive work atmosphere. Nevertheless, it is important for managers and supervisors to understand that, when corrective action is imminent, there are ethical and effective ways to reprimand workers on best practices. As Lu and Wang [61, 62] revealed, agreement and conflict must be managed effectively to ensure a positive outcome: and that failure to control conflict between two sides deteriorates the quality of existing relationship and lead to poor performance. Conchie et al. [25] suggested that managers must facilitate relationships between all workers and must possess the authority and knowledge of analysing and understanding disruptive behaviours to be able to resolve conflicts and discipline when necessary. Managers without these leadership skills continuously fan crisis and escalate conflicts by their actions or inaction on the projects.

2.8 Communication and Teamwork

To be successful, managers need to manage information and communication with many suppliers, subcontractors, and project team members [53, 86]. Managers recounted the problems they had with information acquisition, either between operative units or at their own management level [70]. Communication is particularly critical for any construction KSC development, as this is the only way to convey the sustainable practices expected between trainers, managers, supervisors, and team members [7]. Ineffective communication creates lack of understanding of the real interests of team members before KSC development [44]. Communication is especially critical in the training and development of supervisors and managers to convey the updated and sustainable practices expected of the trained team members after the training [44]. In the construction industry, conflicts mostly seem inevitable due to high differences in interests amongst the project stakeholders. According to Mitkusa and Mitkusa [66], the relationship between the client and the contractor is defined by communication.

Unsuccessful communication is known to be the main cause of conflicts between the project, client, and the contractor in the construction industry [36]. Secondly, the true and major cause of construction related conflicts is unsuccessful communication between and among the participants in a construction project [74]. Similarly, in a partnering and collaborative agreement, managers need to possess strong presentation and negotiating skills to guarantee they get a fair share of the profits adequate for the risks incurred [49]. Most managers, especially in Nigeria, lack these skills and cannot afford them, whereas each of these skills needs to be acquired, developed, and improved [72]. To ensure genuine satisfaction at all levels of the project and delivery of quality services and products, there is a growing need for improved communication skills in managing relationships with clients and workers [38, 74]. Managers must understand the need for them to strive to establish positive exchange relations among employees in efforts to improve job performance and job satisfaction (Michael et al. 2006). This is necessary as they relate and interact with employers than any other level of management [53].

Several contractors usually work as part of the main team and sub teams in different units on most construction projects [29]. And irrespective of the size and scope of the project, it requires more than one person to successfully deliver construction projects [61, 62, 79]. The teamwork required of construction projects, often brings into contact other professionals such as the design professionals, engineers, builders, surveyors, tradesmen all of whom are grouped into sub-teams based on the services they provide or the materials they supply. The inability of managers be able to communicate effectively with these sub-teams of professionals, who normally will possess KSC developed through experiences in their respective areas of specialisation, is a major challenge in construction management [74]. Increasingly, such professionals like structural engineers, architects, electrical or mechanical engineers will seek to transfer plans and drawings down to the managers for onward transfer to the clients, using computer technology. As the client's representative, the managers need to contribute fully to the design and planning process of every task; some ICT skills will thus be needed for speedy inter-team communications. In addition, managers need to be in touch with advanced Information and communication technology [43]. It is important that managers and supervisors build positive atmosphere for their employees. Studies [92] believe that team building skills will have positive effect on building a pleasant atmosphere needed for employees to do their work, this will help create an enabling environment and genuine willingness to consider new ideas that will improve overall job performance.

2.9 *Multicultural Challenges*

Abankwa et al. [1] assert that modern project teams are more likely to be multicultural and active. Therefore, attention should be paid to these, lest the active energy be abused by poorly managed multiple cultures [23]. According to Kim [54] and Thompson [95], management varies from one culture to another, therefore application of managerial KSC should be adjusted to the new situation. For examples, the natives of over 250 ethnic groups in Nigeria form the NECI work force. There are Yoruba, Igbo, Hausa, Efik, Tiv, Borgu, Igala, Ijaw, Urhobo, Ishekini, Fulani, Gwarri,

Nupe, Kanuri, Idoma, Annang, Benin and other tribes. This composition indicates differences in language, values, attitudes, beliefs, religion, and education.

From his research, which was carried out over three decades ago in 40 countries around the world, Hofstede [40] found that leadership style varies from one country to another irrespective of the economic situation. But Cullen and Parboteeah [26] support the economic school which thinks that as nations of the world converge and become equally industrialized, managers all around the world are forced to use the same management tools and methods to execute their managerial duties, this eradicates international and cultural differences in management. One fact against this thought is that all nations of the world cannot at the same time be equally industrialized [10]. Surely, these KSC are still deployed within the cultural linings of respective communities and countries. The economic school is like the universal school where scholars like Blake and Mouton [14] long contributed to. They reported that organisations gradually became similar across cultures. They suggested that it would be more appropriate to establish universal theories for practices in the management of multicultural organisations [20]. Many scholars [18, 39, 52, 82] believe there are psychological and sociological Schools, but Barrett and Bass [8] observed that there is a great deal of overlapping and interaction between psychological, sociological and cultural dimensions, therefore, they categorized the cross-cultural studies into only three comprehensive groups: universal, economic and cultural cluster.

The main theory of the cultural school is that culture is an independent variable which influences managerial attitudes and behaviours. Nowotny et al. [71] established that tradition and socio-cultural values, and environmental variables, are the core cause of management philosophy differences between Europe and North America. Over half a century ago, Megginson and McCann [65] conducted a study on the application of management principles in developing countries and conclude that the functions and principles of management are global, but that methods and management application are vastly dependent on cultural variables. Although various factors are influential today, Megginson and McCann [65] findings are still relevant [55]. Therefore, they resolved that culture is a crucial determinant of management effectiveness. Hence, ill managed cultural differences among the workforce could be disastrous [79].

2.9.1 Planning and Scheduling

Construction work is a sequential activity; it therefore relies heavily on detailed planning and programming to be successful [101]. Normally, a given activity or task cannot be undertaken until a preceding dependent activity has been completed [5]. Planning skills are very important and such skills are once again associated with managers' successful performance [19]. From Iyer and Banerjee's [46] report, managers have exhibited high levels of deficiency in project planning and execution, and it could be argued that often, these managers are not given the opportunity to be involved in project scheduling. Hardison et al. [38], found that improved planning by the construction managers and supervisors leads to improved productivity and vice versa. Organisations and clients exclusively plan and schedule the projects and force it down on the frontline managers and operatives. This contradicts Hardison et al. [38] findings and creates planning, development, and operational crisis on construction projects, especially, in Nigeria. Lingard et al. [60] emphasised that line managers, supervisors and foremen are more likely to have a significant impact on daily project planning, scheduling and management, compared to organisations or owners. According to Manuele [64], a manager's leadership in the implementation of pre-job or pre-training planning meetings and job analyses with workers is key to preventing serious issues and confusion that always characterises the absence of leadership. This planning process must be completed before the task or development commences [13]. This is also not the practice in most organisations and therefore issues abound.

This practice and being close to the workers will help managers and supervisors monitor and respond to worker's stress level and issues like task/development-related pressures, conflicting job demands, extreme time pressures, incentives, cutting corners and taking un-assessed risk. This helps managers maintain job satisfaction and improve employee performance and safety behaviours. It is exclusively important for managers and supervisors to be competent in the procedures for directing workers' tasks, responsibilities, and development in a way that the instructions and operating lines are clear, precise and detailed [53]. Properly delegating responsibilities and tasks/developments is a key component to increasing the general performance of the line level work force [53]. According to Choy et al. [24] the reports of these researchers, the inability of managers and supervisors to effectively delegate and assign responsibilities and tasks or workers development is a huge challenge to successful construction management.

2.9.2 Leadership and Creativity

Most managers and supervisors on construction sites do not creatively inspire or encourage their workers [85]. Most prefer the traditional ways and refute most proposed new practice, this impedes discovery and learning. Kozlo and Shemshurina [57] disagreed with this view, and argued that creativity has already existed for years, though more need to be done, to encourage it. In fostering creativity, organisations need to take active role in encouraging and supporting it [94]. Hence, the role of leaders is to ensure that in their structures of the work environment, the climate and culture, and the human resource practices such as, for example, rewards, resources, goals, and expected evaluations enables creative outcomes. Managers forget they are leaders of day-to-day construction operations, and inspiring and managing creativity is an important responsibility for managers [85]. For decades, researchers have submitted that some level of creativity is vital in almost any job [87, 97]. However, there is need to understand that there is a range of what would be considered, as creative outcome which is essential for managers and supervisors who can evaluate and lead creativity. Mumford and Gustafson [68] described creativity as outcomes that can range from minor adaptations in products or workflow to major breakthroughs and the development of new processes or products.

In every project and organisation, it is important to identify the roles that leadership plays in encouraging workers creativity [67]. Managers want creative workforce, but the complete responsibilities of managers in leading creative performance is unclear [96]. Therefore, if creativity is desired, organisations should try to hire and train individuals that are more predisposed to be creative for the roles of managers and supervisors [80]. Additionally, they can use a person's predisposition for creativity as a factor in placing them in managerial roles where creativity may be more desirable or critical. Lack of full application of Shalleya and Gilson's [87] proposal means many managers still lack the creative appetite [80]. It still goes to show that most managers are risk averse: it is much easier for them to keep on performing in more routine ways instead of taking a chance with a new, and possibly better, approach [80].

Motivation and central leadership of the construction workforce are often critical skills for managers [17]. When managers, irrespective of the scope of projects or size of organisation, lack the ability to lead their workforce through the life span of the project, issues are developed. Some managers on sites do not have any higher-level management qualifications; it is very unlikely that they possess any formal leadership skills—although, in construction, learning by experience will produce significant leadership abilities [88]. But the question is whether these abilities go towards providing the essential motivational influences for problems of satisfactory construction management. It is also a leadership responsibility to be able to train others in an organisation in new working methods appropriate for managers and supervisors [16]. For continuous improvement of performance, managers and supervisors need to be able to identify and resolve problems. This is still the sole responsibility of the project managers or senior personnel within construction organisations [38]. Other management problems in organisations include working conditions, working times and access to information [70]. In some organisations as well as projects, work is based on shift, which according to the managers makes it difficult to assemble the staff and disseminate information. Unhealthy or unconducive working environment also impede KSC development within the organisation [50].

2.9.3 Time and Environment

Koivula and Paunonen-ilmonen [56] in their findings referred to lack of time as one of the problems in KSC development. They elaborated that the working time of managers is not sufficient for work and development and that they are compelled to work under excessive pressure. Over a decade later, nothing seems to have changed as Sanghi [84] shares the same view, elaborating that because of shift work and time pressure, workers fail to find time for KSC development. Explaining further that in some situations, the number of management or supervisory staff is too small in relation to the volume and quality of the projects being executed, therefore, management does not always have time and resources to spare on KSC development. In addition, Tagaza and Wilson [93] stated that time is spent on random checks, cross-checks, and on-site practices of the new KSC acquired, to ensure the applicability and the sustainability of the KSC acquired. This practice is essential but still a challenge as managers and workers may tend to forego time-consuming sustainable practices when they are under time pressure to complete a project [44]. Also, Sanghi [84] detected that the inability to manage time pressure and effectively monitor and measure KSC attainment could be traced to the level of general education acquired.

Most construction managers and supervisors, apart from being equipped with a rich range of technological and organisational methods and frameworks, do not understand the specifics of the organisation and work environment in which they work [31]. They do not know how to cope with emerging technologies and the existing traditional methods [31]. Recorded in Weisinger [100] study for Formtek, a global technology ranking organisation, the construction industry lags other industries in terms of embracing and taking advantage of innovative practices and new technologies. Thus, these compromises awareness in development of KSC of managers and supervisors, training cost effectiveness, quality of life and training offered, competitiveness with other sectors and productivity among others. The general infrastructure of their operational base also poses a great challenge [27]. Recently, Madanayake and Egbu [63] reveal that there is a huge demand for data management capabilities among professionals, to help embed the use of new technology across the NECI and generate value creation. These key challenges and environment for KSC and economic development must be addressed for any system to achieve good result in development [76]. All proposed training curricula needs highlight the need to identify and determine unsafe working conditions, acts associated to an activity, and recognise areas where lack of prevention efforts could lead to accidents in KSC development [42].

2.9.4 Politics and Policies

According to Bhattacharyya's [12] report, most of the policies and regulation are not updated on time and obtaining permissions from relevant authorities and institutions often take a long time and this is a deterrent for any advance update of knowledge, skills, and competences project. This is not a Nigerian problem alone. India [12] and the United Kingdom [11] face similar challenges. Understanding government policies and institutionalised guidelines related to construction management is also vital and was ranked highly. These policies and regulations are designed to foster best practices in KSC acquisition, protect human health and ensure environmental issues; failure to comply may cause delays, termination of KSC development or fines. Even though professional trainers are involved during most of the skills development programmes, it is vital that project managers have a good understanding of the guideline, policies [44]. This will help in planning and managing these challenges.

There is a lack of strong government support to implement and enforce some of the KSC regulations that require government backing and enforcement [51]. According to Kaufmann [61, 62], government political interference kills best practices. Government, through its agencies, interfere in the professional standing of the industry through the influence of regulatory agencies. Kaufmann [61, 62] emphasized that there is no enforcement of strong legislative backings for reasonable policies that will strengthen the development of KSC in the industry. Areas of enforcement of the existent laws and policies are yet to receive strong government support. This view was shared by thirty of the interviewees. They emphatically stated that government interference through regulations is not good for the industry. What the industry needs from the government is support and backing. The government should allow the professional bodies to lead in formation, shaping and enforcing policies and standards that affect the construction industry in Nigeria.

2.9.5 Research and Strategy

For the leading business and management schools, most if not all, note that their unique focus area is "strategy" and "leadership" [30]. While this does not sound like a unique focus, it demonstrates the importance given to strategy in organisations [89]. Improving efficiency in KSC development has been an area of great interest for both practitioners and researchers for some decades now and is sought to be accomplished through various measures [77], these include implementing information technology solutions and data analysis that enhance the awareness of updates on KSC [32, 63], as well as restructuring the organisational structure [37] among others. Generally, few project practitioners and researchers have given some attention to the actual challenges experienced by practitioners in managing projects [90].

As Hwanga and Ng [44] opined, most challenges faced by organisations during KSC development programmes are planning-related. Planning is a long-time requirement for the pre-development process; and it has proven to be one with the most frequently encountered challenges during skills development. These challenges change and get more challenging over times because of the complicated nature of construction projects.

According to Kubba [58], over time, building projects progressively incorporate more advanced and intricate systems. A failure to consider the integration of construction technologies and its impact on KSC development elements results in confusions and conflicts, which in turn leads to delay to address such problems before KSC development. The reason is due to the need for a more detailed and comprehensive KSC development plan, which incorporates all features required by the current construction practices [53]. This means more involvement and interaction with different stakeholders, which is also likely to slow down the pre-skills development process. Managers must have the ability to assess the factors at hand and come to the best solutions for the project [53].

External challenges can impact KSC development in many ways. Mostly when faced with unforeseen circumstances, according to Bolden [15], project managers or organisers of KSC development may not be equipped with any prior knowledge or experience to handle the situation and steer the development to a successful outcome. This unforeseen change poses considerable challenges to the project manager and organisation in rescheduling the KSC development process according to any new development [44]. Another big challenge is that the number of KSC development centres are still small compared to sectors like finance, oil and gas, power. Findings from this study revealed lack of interest in many organisations to sponsor research in the NCI. Research helps in formation of strategy. Lidelöw and Simu [59] cautioned construction professionals against not researching and recommended that they should plan their operations strategies with a focus on organisation, quality, human resources development and planning.

3 Conclusion

This study presented the complexity of challenges associated with practice and development of knowledge, skills, and competences (KSC) in the Nigerian engineering and construction industry (NECI). It, through a combination of a thorough review of literature, survey and interviews recognised: Attainment and Reward; Values and Loyalty; Education and Knowledge Sharing; Funding and Sponsorship; Corruption and Ethics; Infrastructures and Standard; Understanding, Agreement and Conflicts; Lack of Effective Communication and Teamwork; Multicultural Challenges; Planning and Scheduling; Leadership and Creativity; Time and Environment; Politics and Policies; Research and Strategy.

Not enough has been written towards dealing with these current challenges. Stakeholders in the NECI and government have a role to play. With the current challenges, evidence from this study suggests that government should lead right behind the professional organisations in policy making for regulation and enforcement in the industry. For their part, professional construction organisations should ensure research are encouraged, this is useful in tackling these challenges and the development of an effective approach. Employers should be responsible for the funding of their employees KSC development.

References

1. Abankwa DA, Rowlinson S, Adinyira E (2019) Conceptualizing team adaptability and project complexity: a literature review. Int J Innov Manage Technol 10(1)
2. Abidoye RB, Chan AP (2016) Critical determinants of residential property value: professionals' perspective. J Facil Manage 14(3):283–300
3. Ahmed S (2015) Public and private higher education financing in Nigeria. Euro Sci J ESJ 11(7)
4. Almatrooshi B, Singh SK, Farouk S (2016) Determinants of organizational performance: a proposed framework. Int J Product Perform Manage 65(6):844–859
5. Argyris C (2017) Integrating the individual and the organization Routledge. Eng Constr Arch Manage 17(6):536–548
6. Arif M, Al Zubi M, Gupta AD, Egbu C, Walton RO, Islam R (2017) Knowledge sharing maturity model for Jordanian construction sector. Eng Constr Archit Manage 24(1):170–188

7. Banihashemi S, Hosseini MR, Golizadeh H, Sankaran S (2017) Critical success factors (CSFs) for integration of sustainability into construction project management practices in developing countries. Int J Project Manage 35(6):1103–1119
8. Barrett GV, Bass BM (1972) Cross-cultural issues in industrial and organizational psychology (No. TR-45). Rochester University NY Management Research Center
9. Bashkarev A (2015) Social aspects of specialist training in the construction industry. Proc Eng 117:60–65
10. Bendix R (2017) Nation-building and citizenship: studies of our changing social order. Routledge
11. Berr (2007) Department for Business Enterprise and Regulatory Reform
12. Bhattacharyya S (2010) Shaping a sustainable energy future for India: management challenges. Energy Policy 38(8):4173–4185
13. Bielefeld B (2017) Basics construction scheduling. Birkhäuser
14. Blake RR, Mouton JS (1985) The managerial grid III: the key to leadership excellence
15. Bolden R (2016) Leadership, management and organisational development. In: Gower handbook of leadership and management development, pp 143–158. Routledge
16. Borkovskaya V, Bardenwerper W, Roe R (2018) Interactive teaching of risk management in the Russian construction industry. IOP Conf Ser Mater Sci Eng 365(6)
17. Bratton J, Gold J (2017) Human resource management: theory and practice. Palgrave
18. Bridges KMB (2017) The social and emotional development of the pre-school child. Routledge
19. Briscoe G, Dainty A, Millett S (2001) Construction supply chain partnerships: skills, knowledge and attitudinal requirements. Euro J Purchasing Supply Manage 7:243–255
20. Bromley P, Meyer JW (2017) "They are all organizations": the cultural roots of blurring between the non-profit, business, and government sectors. Admin Soc 49(7):939–966
21. Camilleri E (2016) Project success: critical factors and behaviours. Routledge
22. Cheche SG, Muathe S, Maina SM (2019) Employee engagement, demographic characteristics and performance of state research and training corporations in Kenya. IUP J Organ Behav 18(1)
23. Chipulu M, Neoh JG, Ojiako U, Williams T (2013) A multidimensional analysis of project manager competences. IEEE Trans Eng Manage 60(3):506–517
24. Choy J, McCormack D, Djurkovic N (2016) Leader-member exchange and job performance: the mediating roles of delegation and participation. J Manage Dev 35(1):104–119
25. Conchie SM, Taylor PJ, Charlton A (2011) Trust and distrust in safety leadership: mirror reflections? Saf Sci 49(8):1208–1214
26. Cullen JB, Parboteeah KP (2013) Multinational management. Cengage Learning
27. DeBerry-Spencea B, Elliotb E (2012) African micro entrepreneurship: the reality of everyday challenges. J Bus Res 65(12):1665–1673
28. DeCenzo DA, Robbins SP, Verhulst SL (2016) Fundamentals of human resource management, binder ready version. Wiley
29. Dhurup M, Surujlal J, Kabongo DM (2016) Finding synergic relationships in teamwork, organizational commitment and job satisfaction: a case study of a construction organization in a developing country. Proc Econ Fin 35:485–492
30. Doh JP, Stumpf SA (2007) Executive education: a view from the top. Acad Manage Learn Educ 6(3):388–400
31. Doppelt B (2017) Leading change toward sustainability: a change-management guide for business, government and civil society. Routledge
32. Durmuşoğlu SS, Barczak G (2011) The use of information technology tools in new product development phases: analysis of effects on new product innovativeness, quality, and market performance. Ind Mark Manage 40(2):321–330
33. Fellows R, Liu A (2015) Sense-making in the cross-cultural context of projects. Int J Project Manage 34:246–257
34. Hampson I, Junor A, Barnes A (2009) Articulation work skills and the recognition of call centre competences in Australia. J Ind Relat 51(1):45–58

35. Freedy J, Hobfoll SE (2017) Conservation of resources: a general stress theory applied to burnout. In: Professional burnout, pp 115–129
36. Gamil Y, Rahman IA (2018) Identification of causes and effects of poor communication in construction industry: a theoretical review. Emerg Sci J 1(4):239–247
37. Gokpinar B, Hopp W, Iravani S (2010) The impact of misalignment of organizational structure and product architecture on quality in complex product development. Manage Sci 56(3):468–484
38. Hardison D, Behm M, Hallowell MR, Fonooni H (2014) Identifying construction supervisor competencies for effective site safety. Saf Sci 65:45–53
39. Hewstone M, Stroebe W, Jonas K (2016) An introduction to social psychology. Wiley
40. Hofstede GH (1984) Culture's consequences—international differences in work-related values. In: Abridg (ed) Sage Publications, Newbury Park, Calif, London
41. Hong J, Shen GQ, Li Z, Zhang B, Zhang W (2018) Barriers to promoting prefabricated construction in China: a cost–benefit analysis. J Clean Prod 172:649–660
42. Hou et al (2017) A framework of innovative learning for skill development in complex operational tasks. Autom Constr 83:29–40
43. Hu Q, Kapucu N (2016) Information communication technology utilization for effective emergency management networks. Public Manage Rev 18(3):323–348
44. Hwang B, Ng W (2013) Project management knowledge and skills for green construction: overcoming challenges. Int J Project Manage 31(2):272–284
45. Inuwa II, Napoleon Daniel Usman ND, Dantong JS (2015) The effects of unethical professional practice on construction projects performance in Nigeria. Afr J Appl Res (AJAR) 1(1)
46. Iyer KC, Banerjee PS (2016) Measuring and benchmarking managerial efficiency of project execution schedule performance. Int J Project Manage 34(2):219–236
47. Izatul LJ, Faridah I, Nur MA, Nurul AJ (2013) Construction manager's competency in managing the construction process of IBS projects. Soc Behav Sci 105:85–93
48. Jimoh R, Oyewobi L, Suleiman S, Isa R (2017) Influence of supervision on labour productivity on construction sites in Abuja-Nigeria. Independent J Manage Prod 8(1)
49. Johnston MW, Marshall GW (2016) Sales force management: leadership, innovation, technology. Routledge
50. Jumah L, Sulo T (2018) Influence of job assignment on employee turnover in selected medium sized hotels in Kisumu city
51. Kaufmann W (2017) Going by the book: the problem of regulatory unreasonableness. Routledge
52. Kelly JG (2017) Adolescent boys in high school: a psychological study of coping and adaptation. Routledge
53. Kerzner H, Kerzner HR (2017) Project management: a systems approach to planning, scheduling, and controlling. Wiley
54. Kim MS (2017) Culture-based conversational constraints theory. In: The international encyclopedia of intercultural communication, pp 1–10
55. Kimani JW (2014) Principles and practice of effective accounts receivable management in Kenya: a case of selected manufacturing firms in Thika municipality (Doctoral dissertation)
56. Koivula M, Paunonen-ilmonen M (2001) Ward sisters' objectives in developing nursing and problems with development. J Nurs Manage 9(5):287–294
57. Kozlo AV, Shemshurina SA (2018) Fostering creativity in engineering universities: research activity and curriculum policy. Int J Instr 11(4):93–106
58. Kubba S (2010) Green construction project management and cost oversight. Butterworth-Heinemann
59. Lidelöw H, Simu K (2015) Understanding construction contractors and their operations strategies. Proc Econ Fin 21:48–56
60. Lingard L, Vanstone M, Durrant M, Fleming-Carroll B, Lowe M, Rashotte J, Tallett S (2012) Conflicting messages: examining the dynamics of leadership on interprofessional teams. Acad Med 87(12):1762–1767

61. Lu W, Wang J (2017) The influence of conflict management styles on relationship quality: the moderating effect of the level of task conflict. Int J Project Manage 35(8):1483–1494
62. Macomber M, Eller K, Dolan A (2018) Teamwork-engineering, operations, and procurement working together for reliable treatment facilities. Proc Water Environ Fed 9:4466–4479
63. Madanayake U, Egbu C (2019) Critical analysis for big data studies in construction: significant gaps in knowledge. Built Environ Project Asset Manage 9(4):530–547
64. Manuele FA (2008) Advanced safety management focusing on Z10 and serious injury prevention. Wiley
65. Megginson LC, McCann EC (1965) Applicability of management principles in underdeveloped economies. International handbook of management. McGraw-Hill, New York
66. Mitkusa S, Mitkusa T (2014) Causes of conflicts in a construction industry: a communicational approach. Soc Behav Sci 110:777–786
67. Mittal S, Dhar RL (2015) Transformational leadership and employee creativity: mediating role of creative self-efficacy and moderating role of knowledge sharing. Manage Decis 53(5):894–910
68. Mumford MD, Gustafson SB (1988) Creativity syndrome: integration, application, and innovation. Psychol Bull 103(1):27
69. Nixon P, Harrington M, Parker D (2012) Leadership performance is significant to project success or failure: a critical analysis. Int J Product Perform Manage 61(2):204–216
70. Noe RA, Hollenbeck JR, Gerhart B, Wright PM (2017) Human resource management: gaining a competitive advantage. McGraw-Hill Education, New York, NY
71. Nowotny H, Scott PB, Gibbons MT (2013) Re-thinking science: knowledge and the public in an age of uncertainty. Wiley
72. Nwosu JC, Makinde OG (2014) Conflict management in an organization. Kuwait Chap Arabian J Bus Manage Rev 3(6):28
73. Ogunyomi P, Bruning NS (2016) Human resource management and organisational performance of small and medium enterprises (SMEs) in Nigeria. Int J Human Resour Manage 27(6):612–634
74. Osei-Kyei R, Chan AP, Yu Y, Chen C, Dansoh A (2019) Root causes of conflict and conflict resolution mechanisms in public-private partnerships: comparative study between Ghana and China. Cities 87:185–195
75. Owolabi JD, Amusan LM, Oloke CO, Olusanya O, Tunji-Olayeni PF, Dele O, Peter NJ, Omuh IO (2014) Causes and effect of delay on project construction delivery time. Int J Educ Res 2(4):197–208
76. Palmer R (2009) Skills development, employment and sustained growth in Ghana: sustainability challenges. Int J Educ Dev 29(2):133–139
77. Piedras S, Pouey J, Moraes P, Cardoso D (2006) Lethal concentration (CL50) of un-ionized ammonia for pejerrey larvae in acute exposure. Sci Agricola 63(2):184–186
78. Nigerian Bureau of Statistics (2018) Key Statistics for Nigerian Gross Domestic Product Report (Expenditure and Income Approach) Arrears of 2018. http://www.nigerianstat.gov.ng/
79. Reason J (2016) Managing the risks of organizational accidents. Routledge
80. Renault BY, Agumba JN, Balogun OA (2016) Drivers for and obstacles to enterprise risk management in construction firms: a literature review. Proc Eng 164:402–408
81. Riebe L, Jackson D (2014) The use of rubrics in benchmarking and assessing employability skills. J Manage Educ 38(3):319–344
82. Riza Yosia S, Bonaventura H, Ogunlana S (2006) Emotional intelligence and leadership styles in construction project management. J Manage Eng 23(4)
83. Rogers C (2017) A social psychology of schooling: the expectancy process. Routledge
84. Sanghi S (2016) The handbook of competency mapping: understanding, designing and implementing competency models in organizations. SAGE Publications India
85. Schuh SC, Zhang XA, Morgeson FP, Tian P, van Dick R (2018) Are you really doing good things in your boss's eyes? Interactive effects of employee innovative work behavior and leader–member exchange on supervisory performance ratings. Hum Resour Manage 57(1):397–409

86. Senaratne S, Ruwanpura M (2016) Communication in construction: a management perspective through case studies in Sri Lanka. Arch Eng Des Manage 12(1):3–18
87. Shalleya C, Gilson L (2004) What leaders need to know: a review of social and contextual factors that can foster or hinder creativity. Leadersh Quart 15(1):33–53
88. Singh R, Kumar N, Puri S (2017) Thought self-leadership strategies and sales performance: integrating selling skills and adaptive selling behaviour as missing links. J Bus Ind Market 32(5):652–663
89. Söderlunda J, Maylor H (2012) Project management scholarship: relevance, impact and five integrative challenges for business and management schools. Int J Project Manage 30(6):686–696
90. Sommer A, Dukovska-Popovska I, Steger-Jensen K (2014) Barriers towards integrated product development—challenges from a holistic project management perspective. Int J Project Manage 32:970–998
91. Sun JD, Wang XC, Shen LF (2017) Chinese construction workers' behaviour towards attending vocational skills trainings: evolutionary game theory with government participation. J Differ Equations Appl 23(1/2):468–485
92. Swuste P, Frijters A, Guldenmund F (2012) Is it possible to influence safety in the building sector? A literature review extending from 1980 until the present. Saf Sci 50(5):1333–1343
93. Tagaza E, Wilson JL (2004) Green buildings: drivers and barriers e lessons learned from five Melbourne developments. Report prepared for building commission by University of Melbourne and Business Outlook and Evaluation
94. Tang H, Hammack C, Ogden SC, Wen Z, Qian X, Li Y, Yao B, Shin J, Zhang F, Lee EM, Christian KM (2016) Zika virus infects human cortical neural progenitors and attenuates their growth. Cell Stem Cell 18(5):587–590
95. Thompson M (2018) Cultural theory. Routledge
96. Thundiyil TG, Chiaburu DS, Li N, Wagner DT (2016) Joint effects of creative self-efficacy, positive and negative affect on creative performance. Chin Manage Stud 10(4):726–745
97. Unsworth CA (2001) The clinical reasoning of novice and expert occupational therapists. Scandinavian J Occup Therapy 8(4):163–173
98. Viswanathan M, Rosa JA, Ruth JA (2010) Exchanges in marketing systems: the case of subsistence consumer–merchants in Chennai, India. J Market 74(3):1–17
99. Wang X, Wang X, Huang Y (2019) Chinese construction worker reluctance toward vocational skill training. J Eng Des Technol 17(1):155–171
100. Weisinger D (2016) Construction IT: industry lags all others in technology spend. Eng Data Manage Form
101. Zhang S, Zhao C, Wang S, Wang F (2017) Pseudo time-slice construction using a variable moving window k nearest neighbour rule for sequential uneven phase division and batch process monitoring. Ind Eng Chem Res 56(3):728–740
102. Zhu H, Feng J, Wang M, Xu F (2017) Sustaining regional advantages in manufacturing: skill accumulation of rural–urban migrant workers in the coastal area of China. Sustainability 9(1):1–23

A Study on Knowledge, Attitudes and Commitment of Managers Within Construction Firms Towards the Recent Construction Health and Safety Legislation Changes

Mohlomi Terah Raliile and Theo C. Haupt

Abstract This study seeks to evaluate the knowledge, attitudes and commitment of managers within construction firms towards the recent construction H&S legislation changes. This study further aims to identify the differences in responses between management personnel and construction workers within construction firms. An empirical research approach was adopted using a quantitative method of data collection and data were analysed using IBM Statistical Package for Social Sciences (SPSS) version 25. Furthermore, descriptive statistics was used to analyse data and data were interpreted using inferential statistics. Cronbach's Alpha reliability test was conducted to determine the internal consistency of the constructs. Furthermore, independent t-test was conducted to determine the statistical significance of the means between the groups of respondents. The findings from management and construction workers within construction firm revealed that although there was a small correlation between management commitment and knowledge of H&S legislation, there was no statistical significance between the two constructs. Also, management was not fully committed in applying all the aspects of the construction regulations to improve construction workers wellbeing. The findings revealed that there is a need for more knowledge of H&S legislation and management commitment. The paper proposes measures for managers to engender full compliance and to apply all aspects of the construction regulations such as to; improve knowledge; improve working conditions, consider the health and wellbeing of workers and treat health and safety as a value and not just a priority.

Keywords Health and safety · Legislation · Workers wellbeing · Management commitment

T. C. Haupt
Department of Engineering, Mangosuthu University of Technology, 511 Mangosuthu Highway, Umlazi, Durban 4031, South Africa

M. T. Raliile (✉)
Department of Quantity Surveying and Construction Management, University of the Free State, 205 Nelson Mandela Drive Park West, Bloemfontein, Free State, South Africa

T. C. Haupt et al. (eds.), *Construction in 5D: Deconstruction, Digitalization, Disruption, Disaster, Development*, Lecture Notes in Civil Engineering 245,
https://doi.org/10.1007/978-3-030-97748-1_32

1 Introduction

The standards of health and safety set out to protect workers are often overlooked by management [1]. Reference [2] state that "A fundamental human right of every worker is to be able to return home at the end of each working day; alive and healthy in the same physical condition that he/she commenced that working day. According to the CIDB [3], the South African construction industry has less than 50% rate of compliance with health and safety requirements with unacceptably high rates of incidents as a result of poor workmanship and lack of proper supervision on site. Some accidents are attributed to senior managements' lack of commitment, lack of communication and lack of training [4, 5]. ILO [6] reports that, despite government and trade union efforts, accidents will continue to happen on-site due to lack of safety adherence by contractors.

The increase in construction accidents has led to the revision of Construction Regulations 2014 in South Africa. However, instead of contractors focusing on improving the quality of life and sustainability of the construction workers, the primary focus is on compliance with the regulations in order to avoid fines and having construction activities stopped as a result of non-compliance [7]. The other reasons for compliance are profit maximisation and satisfying the requirements of the Department of Labour or the Department of Public Works [7]. Improving construction workers' quality of life goes beyond just providing PPE, ticking checkboxes and securing projects [7].

There is a seeming knowledge gap in the previous research about the attitudes and commitment of managers towards the recent construction H&S regulations. Further, there is an apparent knowledge and population gap in relation to the construction workers' knowledge and participations in studies relating to health and safety legislation in SA. Prior research in SA mainly focused on the impacts of the regulations, constructors' attitudes and the knowledge of construction professionals [8–18]. Therefore, it is important to investigate in the context of both construction workers and managers (on-site) to gainer an in-depth understanding of the underlying factors that constitute to non-improvement of safety on-site between the implementation of the construction regulations 2003 (old regulations) and the construction regulations 2014 (recent). This study seeks to explore the knowledge, attitudes and commitment of management towards the construction regulations 2014 in order to improve workers wellbeing. The study further compares the responses given by management to those reported by construction workers about management commitment as well as identifying the attitudes and level of knowledge both respondents possessed.

2 Literature

The construction industry is a heavy industry where workers perform repetitive manual tasks [19]. The tasks require heavy lifting, twisting, turning in awkward and cramped positions, working in unpleasant conditions with noise, dust and sometimes with little light, low ventilation and high rates of staff turnover [19]. Moreover, the industry is still facing further challenges dealing with the increasing complexity in mega projects. In comparison to other industries, construction is believed to have the 3rd highest death rate [20]. The low price culture in competitive tendering does not encourage health and safety requirements and promotes low wages in the industry [1]. Consequently, there is a lack of provision for PPE, workers' insurance and the influx of informal workers for whom no tax or social security are paid [1]. Although procurement methods contribute significantly towards the improvement of workers' health and safety, the most common method for awarding tenders in developing countries is through competitive tendering by the government [21]. This method of procurement is usually evaluated based on the lowest price, meaning that contractors must keep their prices low, and this affects labour as it forms part of the major cost items. In this regard, for contractors to secure lower rates, they compromise on welfare facilities, protective equipment and a safe working environment for the workforce [21].

2.1 The Impacts of CR 2014

The Construction Regulations are perceived to have had a positive impact, especially through the increase of health and safety awareness and increased consideration among project managers and general contractors [15]. Furthermore, although the impacts of Construction Regulations cannot be quantified, it may be inferred that they have had a positive impact on the reduction of accidents. The Construction Regulations have had a desired upstream, midstream and downstream impact [8]. However, despite the perceived contributions made by the Construction Regulations 2003, there is still need for further improvement of overall health and safety. It is argued that health and safety should not be driven solely by legislation, but instead must be seen as a value [11]. Management skills are essential to the overall realisation of health and safety in the workplace [11].

2.2 Application of CR 2014

The construction workers' quality of life is often overlooked, and usually, discussions about this topic place more emphasis on safety and prevention than health [13]. Additionally, it is astounding how rarely health and well-being issues are discussed despite the construction industries reliance on healthy, capable and physically fit workers [12]. If employees are mentally fit and in good health, they are less likely to injure themselves [12]. Starting health and wellness programs can sustain the workforce for many more years [13]. Contractors must take responsibility to ensure their workers' sustainability by improving the quality of life as some construction workers coin the belief "too tough to worry about it" when it comes to health and safety [13]. It is necessary to use legislation to ensure that the safety, health and wellness of workers are taken seriously [10].

Worth noting is the fact the wellness programs are prevalent in sedentary jobs that require much less physical exertion [12]. Sites of projects designed for manufacturing adhere fully to the requirements of health and safety legislation due to the involvement of heavy machinery and commitment by line manager as opposed to sites operated by uncommitted managers [11]. Compliance with health and safety legislation requires a combination of both building site type and site managers' attitude [11]. South Africa does not lack in terms of occupational health and safety legislation, however occupational health and safety should not be driven solely by legislation but must be regarded more as a value [15, 17]. The construction industry requires more than just legislation to make it safer; it requires an attitude change towards safety [9].

The intent of the Construction Regulations is useful in creating general requirements for health and safety in all involved in the construction work; however, various practical problems have been pointed out [9]. It is argued that occupational health and safety aspects are covered at large within the OH&S Act; however, there is a lack of legislation regarding employees' quality of life [10]. Furthermore, psychological stressors at work have often been overlooked in the past as were traditionally considered unimportant and even taboo; if considered, were often treated as an individual case [10].

2.3 *Methodology*

An extensive review of literature on the topic was conducted from online databases, books, articles, reports and other studies. Furthermore, a quantitative study was employed and data were analysed using IBM Statistical Package for Social Sciences (SPSS) version 25. Descriptive statistics was adopted for the data analysis and further interpreted using inferential statistics. The total sample size for the study was 120 participants (80 managers and 40 construction workers) where a total of 80 construction companies in the Kwa-Zulu Natal province were conveniently sampled based on proximity and familiarity. The response rate for the study was 78.3%. A total of 64 managers (as representatives of the companies) and 30 construction workers completed close-ended questionnaires. Cronbach's Alpha reliability test was used to determine the internal consistency of the constructs that had been used to identify the knowledge, attitudes and commitment of management towards the construction regulations 2014 in order to improve construction workers wellbeing and the internal consistency of the various scales was deemed acceptable for interpretation.

Independent t-test was done to determine the statistical significance of the means between management and construction workers' responses with a further Spearman's correlation to identify the relationship between the constructs.

2.4 Profile of the Respondents

The median age for managers was 35 years ranging from 22 years minimum to 55 years' maximum. The median years of experience were 6.5 years ranging from a minimum of 2 years to a maximum of 20 years. The median years of experience suggested that managers were knowledgeable in most aspects of the construction industry. Most respondents were Health and Safety Managers/Officer (29.70%) followed by Civil Engineers (23.40%) and Quantity Surveyors (21.90%). The respondents had either obtained a technical/vocational or university education. Furthermore, 32.80% of the respondents had obtained their qualifications from technical/vocational schools, and the other respondents obtained their qualifications from universities, 67.20%.

The median age of the construction workers was 33.50 years ranging from 26.00 years to 43.00 years and the median years of experience were 6.00 years with a minimum of 3.00 years and a maximum of 21.00 years. Most workers (60%) were employed as artisans with different trades skills. Moreover, 33.30% of the construction workers had obtained junior high school certificate while 30.00% obtained their Matric/OLevel and 36.70% had technical/vocational training qualifications.

3 Findings and Discussions

3.1 Reliability

The Cronbach's Alpha coefficients between $0.70 \leq \alpha \geq 0.80$ were considered 'acceptable' while between $0.80 \leq \alpha \geq 0.90$ were considered 'good' and coefficients $0.9 \leq \alpha$ were 'excellent'. Therefore, the internal consistency of the various scales was deemed acceptable for further interpretation:

Reliability was acceptable across all the scales used, and further analysis of the results was conducted.

3.2 Knowledge of Construction Health and Safety Legislation

Table 1 presents 5-knowledge based statements where respondents had to rate their knowledge of existing health and safety legislation affecting the construction industry

Table 1 Summary of Reliability Statistics

Construct	Management		Construction workers	
	Cronbach's Alpha	Reliability	Cronbach's Alpha	Reliability
Knowledge of construction health and safety legislation	0.838	Good	0.806	Good
Management commitment	0.813	Good	0.828	Good

using a 5-point Likert scale where 1 = Poor, 2 = Fair, 3 = Average, 4 = Good and 5 = Excellent.

Management: The results in Table 2 indicated that the managers were most knowledgeable about the Occupational Health and Safety Act 85 of 1993. Knowledge of the Construction Regulations 2014 and 2003 was ranked second and third, respectively. It is also evident that the respondents had high knowledge of the three legislations (means range between 3.77 and 3.38) and a medium knowledge of two legislations (means range between 3.27 and 3.05). Although management seemed knowledgeable about the relevant health and safety legislation, it is expected that management must become more knowledgeable on the subject than what the findings suggested. Therefore, there is a need for managers to gain more knowledge of health and safety legislation.

Construction Workers: In Table 2, the construction workers' knowledge of the Construction Regulations 2014 and knowledge of OH&S Act 85 of 1993 were ranked

Table 2 Knowledge of H&S legislation

Statement	Management				Construction workers			
	Mean	SD	Intr.	Rank	Mean	SD	Intr.	Rank
Knowledge of OH&S Act 85 of 1993	3.77	1.16	H	1	3.35	1.43	H	2
Knowledge of Construction Regulations 2014	3.50	1.17	H	2	3.48	0.96	H	1
Knowledge of Construction Regulations 2003	3.38	1.21	H	3	3.16	1.27	M	4
Knowledge of COID Act 130 of 1993	3.27	1.17	M	4	3.23	1.18	M	3
Knowledge of the SA Constitution	3.05	1.10	M	5	3.16	1.27	M	4

high and considered good (mean values were 3.48 and 3.35). However, knowledge was medium for the three statements COID Act 130 of 1993; the Construction Regulations 2003; and the SA Constitution (means range between 3.23 and 3.16). From the findings, it may be inferred that although construction workers were somewhat knowledgeable in some aspects of legislation, there is a need for more knowledge of construction health and safety legislation.

3.3 Management Commitment

The respondents were requested to indicate their level of agreement on management commitment and attitudes towards the Construction Regulations 2014, based on a 5-point Likert scale where 1 = Strongly Disagree, 2 = Disagree, 3 = Neutral, 4 = Agree and 5 = Strongly Agree. Under Table 5–11, the participants were presented with 11-statements on management commitment to health and safety legislation and their attitudes towards the Construction Regulations 2014.

Management: Table 3 indicated a high level of agreement that all workers had medical certificates of fitness; firms employed competent health and safety staff as stipulated in the construction regulations, and firms conducted health and safety inspections at least once daily (means ranged from 4.36 to 4.13). Although workers' possessed medical certificates, there was a reduction in the mean value for the validity of medical certificates even though the response rate was still high (mean = 3.98). There was also a high level of agreement that head office ensures commitment with the regulations; were intolerant of poor health and safety practices, and insisted on hazard elimination (means range between 3.92 to 3.86).

Although there was a high level of agreement that construction workers wellbeing was important to management, the statement was not the highest-ranking as would have been expected (mean = 3.72). This is suggestive that management merely complies because it is a requirement by law but do not fully commit to health and safety practices such as promoting workers wellbeing. It was also found that workers were penalised for poor health and safety practices (mean = 3.55). Although still high ranking, there was a lower level of agreement concerning the consultation of health and safety committees; representative unions or representative group of employees, on the monitoring and reviewing of the site-specific risk assessments (mean = 3.45). There was a medium level of response for construction workers being rewarded for good health and safety practices on-site (mean = 2.72) over and above the finding that they were penalised for poor health and safety practices. There is a need for more management commitment to improve workers wellbeing.

Construction Workers: The data in Table 3 suggested that the construction workers had a high level of agreement that firms employed trained health and safety staff and had medical certificates of fitness (means range between 4.47 and 4.23). There was also a high level of agreement that head office management ensured compliance with the regulations; insisted on hazards elimination and further agreed that they were penalised for poor health and safety practices (means ranged between 4.13

Table 3 Management commitment

Statements	Management				Construction workers			
	Mean	SD	Intr.	Rank	Mean	SD	Intr.	Rank
All workers possess medical certificates of fitness	4.36	1.10	H	1	4.23	1.22	H	2
The firm employs trained H&S staff on site	4.34	0.93	H	2	4.47	0.82	H	1
H&S inspections are done regularly and at least daily	4.13	0.89	H	3	3.47	1.28	H	6
All workers have medical certificates of fitness are valid	3.98	1.39	H	4	3.40	1.77	H	7
The head office management ensures compliance with CR 2014 for wellbeing	3.92	0.86	H	5	4.13	1.07	H	3
The head office management are intolerant of poor construction H&S	3.91	1.08	H	6	3.40	1.30	H	8
The head office management insists on the elimination of hazards by complying with CR	3.86	0.75	H	7	4.00	0.83	H	4
The construction workers' wellbeing is important to the head office management	3.72	1.19	H	8	3.00	1.58	M	10
The firm penalises workers for poor H&S practices on site	3.55	1.11	H	9	3.80	1.67	H	5
Management consults with the H&S committee, representative union or representative group of employees, on the monitoring and reviewing of the risk assessments of a specific site	3.45	1.05	H	10	3.33	1.18	M	9
Workers are rewarded for good H&S practices on site	2.72	1.00	M	11	2.57	1.48	L	11

Table 4 Summary of composite mean values

Management					Construction workers			
	Mean	SD	Inter.	Rank	Mean	SD	Inter.	Rank
Management commitment	3.81	0.62	H	1	3.62	0.80	H	1
Knowledge of legislation	3.39	0.91	H	2	3.26	0.90	M	2

and 3.80). Furthermore, there was a high level of agreement about daily health and safety inspections, even though inspections were not carried out sufficiently (mean = 3.47). Workers had valid medical certificates and head office management was intolerant of poor health and safety (both means were 3.40), as indicated by the high agreement level though still far less than satisfactory. There, however, was a medium response whether management consulted with the health and safety committees; representative unions or representative group of employees, on the monitoring and reviewing of the site-specific risk assessments (mean = 3.33). Workers also gave a medium response whether management cared for their wellbeing (mean = 3.00) and disagreed that they were rewarded for good health and safety practices on site (mean = 2.57). From these findings, it is suggestive that construction workers' health and safety are still not prioritised.

It could be argued from the findings in Table 3 that construction workers, who are physically on-site, did not view inspections as sufficient enough. Further similarities indicated that workers possessed medical certificates of fitness, although they were not all valid. Also, construction workers level of agreement was very low compared to that of management. There seemed to be a lack of consensus between what the managers reported and what the workers reported, which further suggested a lack of commitment by management. Workers were not consulted in reviewing site-specific risk assessments and of significance was the fact that workers were almost unsure whether head office management cared about their wellbeing. Construction workers were punished for poor health and safety practices and were not rewarded for good health and safety practices. This could be an indication that construction companies may still use a reactive approach rather than a proactive approach when dealing with health and safety issues.

The finding presented in Table 4 indicated the composite means of two constructs used in the study. A comparison of the composite means indicated that both constructs ranked high.

3.4 T-Testing of the Results

Furthermore, an independent t-test was conducted (Table 5) to determine the statistical significance of the composite means between managers and workers (Table 4). Moreover, Levene's test (f-test) was used to test the assumptions of the homogeneity for equal variances.

Table 5 Independent samples T-Test

		f-test		t-test for Equality of Means						
		F	Sig.	t	df	Sig. (2tailed)	Mean Difference	Std. E Dif	95% Con. Int. of the difference	
									Lower	Upper
Management commitment	Equal variances assumed	4.430	0.038	1.281	92	0.203	0.194	0.152	−0.107	0.495
	Equal variances not assumed			1.170	46.001	0.248	0.194	0.166	−0.140	0.529
Knowledge of legislation	Equal variances assumed	0.106	0.746	0.651	92	0.516	0.131	0.201	−0.268	0.529
	Equal variances not assumed			0.652	57.016	0.517	0.131	0.200	−0.270	0.532

Management Commitment: In comparison, the high composite means suggested that there was no statistically significant difference between the means for management (mean = 3.62, SD = 0.62) and construction workers (mean = 3.81, SD = 0.80); t (46) = 1.170, $p = 0.248$. The findings suggested that both respondents shared similar views regarding management commitment to improving workers wellbeing. However, based on the lower mean values, it is evident that full commitment is still not engendered.

Knowledge of Legislation: Comparatively, managements' knowledge was high (mean = 3.39, SD = 0.91), while the construction workers had medium knowledge (mean = 3.26, SD = 0.90) about the relevant construction health and safety legislation. There was no statistically significant difference between the means, t (92) = 0.651, p = 0.516; suggesting that both managers and workers had the same level of knowledge of construction health and safety legislation. From Table 4, of significance was the fact that knowledge of legislation was ranked the lowest for both management and construction workers. Ideally, both respondents were expected to be very knowledgeable (especially management) about legislation and not merely high or medium knowledge. It may be inferred that both management and construction workers' possessed knowledge about the relevant health and safety legislation even though this knowledge was less than satisfactory; therefore, there is a need for more education and training. A study conducted by Matete, Emuze & Smallwood (2016) on the Construction Regulations reported similar findings.

3.5 *Spearman's Correlation*

Spearman's rank-order correlation (rho/rs) was used to measure the strength and direction of association/relationship between the constructions: Knowledge of Legislation and Management Commitment (Table 6).

Management: There was no statistical significance between Knowledge of Construction Legislation and Management although there was a small positive correlation (r = 0.177, p = 0.162). The findings suggest that when the Knowledge of Construction Legislation increased, Management Commitment also improved although slightly.

Construction Workers: There was no statistically significant correlation between Knowledge of Construction Legislation and Management commitment (r = 152, p = 0.421). However, there was a small positive correlation between the constructs suggesting that an increase in knowledge of legislation increased management's commitment slightly.

Table 6 Spearman's correlations, management (N = 64)

	Knowledge of legislation	Management commitment
Knowledge of legislation	1.000	0.177
		0.162
Management commitment	0.177	1.000
	0.162	
Construction workers (N = 30)		
	Knowledge of legislation	Management commitment
Knowledge of legislation	1.000	0.152
		0.421
Management commitment	0.152	1.000
	0.421	

*. Correlation is significant at the 0.05 level (2-tailed)
**. Correlation is significant at the 0.01 level (2-tailed)

4 Conclusions and Recommendations

This study analysed the knowledge, attitudes and commitment of managers within construction firms in South Africa towards the recent construction H&S legislation changes. From the findings of the study, it may be inferred that there was an average knowledge of construction health and safety legislation. However, the expectation would have been that both management and construction workers within construction companies possessed more knowledge in order to improve their level of compliance and improve construction workers wellbeing. Furthermore, it was perceived that legislation has an impact on improving workers wellbeing, though the findings further suggested there could be other reasons that impact workers wellbeing based on the less than excellent agreement levels. There is a need for more education on health and safety legislation and integration among management and workers in order for contractors to realise the benefits of legislation and to engender full commitment towards implementation and compliance. This study was limited to the Kwa-Zulu Natal province. There is a need for further studies across all South African provinces. There is also a need for an in-depth analysis regarding the lack of management commitment as there was no statistical significance between knowledge of H&S legislation and management commitment from both management and construction workers. Therefore, future research could be conducted on behaviour, attitudes and perceptions of contractors towards construction health and safety legislation in South Africa.

Acknowledgements The authors much acknowledge the support of the National Research Fund (NRF).

References

1. Murie F (2007) Building safety—an international perspective. Int J Occup Environ Health 13(1):5–11
2. Occupational Health and Safety Act 85 of 1993, (OHSA) (1993)
3. CIDB (2016). ANNUAL REPORT. [online] Available at: http://www.cidb.org.za/publications/Documents/Annual%20Report%202016-2017.pdf
4. Laney JC (1982) Site safety. Construction Press
5. Zwetsloot G, Kines P, Ruotsala R, Drupsteen L, Merivirta M, Bezemer R (2017) The importance of commitment, communication, culture and learning for the implementation of the Zero Accident Vision in 27 companies in Europe. Saf Sci 96:22–32
6. International Labour Organisation (2011) Construction OS&H Safe & healthy working environment. ILO
7. Raliile MT, Haupt TC (2019) The construction health and safety regulation 2014 impact on construction workers well-being. Dissertation
8. Smallwood J, Haupt T (2007) Impact of the South African construction regulations on construction health and safety. J Eng Des Technol 5(1):23
9. Jacobs E (2010) Health and safety in South African construction: status Quo. ASOCSA 5:446–447
10. Sieberhage C (2008). Employee health and wellness practices in South Africa. Ph.D. North West University
11. Windapo A (2013) Relationship between degree of risk, cost and level of compliance to occupational health and safety regulations in construction. Aust J Const Econ Build 13(2):67–82
12. Orvitz K (2018) Increase employee wellness in construction. [online] Constructionexec.com. Available at: http://constructionexec.com/article/increase-employee-wellness-in-construction
13. Fry G (2017) Bring health and wellness to your construction business. [online] Viewpoint Surveyor. Available at: https://blog.viewpoint.com/health-and-wellness-in-construction-2/
14. Deacon C (2017) World day for safety and health at work: 28 April 2017: Construction H&S | Health and Safety and the South African Entrepreneur. [online] Cda.co.za. Available at: http://www.cd-a.co.za/News/entryid/298/world-day-for-safety-andhealth-at-work-28-april-2017-construction-h
15. Smallwood J, Haupt T, Shakantu W (2009) Construction health & safety in South Africa status & recommendations. [online] Available at: http://www.cidb.org.za/publications/Documents/Construction%20Health%20and%20Safety%20in%20South%20Africa.pdf
16. Matete R, Smallwood J, Emuze F (2016) Exploring perceived implementation issues of the permit-to work requirement of the construction regulations in South Africa. J Const: Assoc Schools Constr Southern Africa 9(2):14:23
17. Agumba J, Haupt T (2012) Identification of health and safety performance improvement indicators for small and medium construction enterprises: A Delphi consensus study. [online] Available at: https://www.researchgate.net/publication/287745042_Identification_of_health_and_safety_performance_improvement_indicators_for_small_and_medium_construction_enterprises_A_delphi_consensus_study
18. Blitz Report (2007)
19. Asanka W, Ranasinghe M (2015) Study on the impact of accidents on construction projects. [online] Available at: http://www.civil.mrt.ac.lk/conference/ICSECM_2015/volume_4/Extract/SECM-15-088.pdf
20. US Bureau of Labor and Statistics (2019) National Census of Fatal Occupational Injuries in 2019. United States Department of Labor
21. Wells J, Hawkins J (2010) Promoting construction health and safety through procurement: a briefing note for developing countries. Engineers against Poverty. Institute of Civil Engineers

Assessing the Resilience of UK Construction SMEs: A Cross-Sectional Analysis of the Effects of Covid-19 Pandemic

Rafiu Dimeji Seidu, Obas John Ebohon, Bert Ediale Young, and Daniel Fong

Abstract The Construction of small and medium size enterprises (CoSMEs) are the backbone of the UK economy, Covid-19 have affected many of this sectors. This has not only affected the construction industry negatively, but has caused many health and safety concerns on construction sites, resulting in increased unemployment, cash flow problems, and insolvencies. The research analyses the implications of coronavirus COVID-19 for UK Construction small and medium size enterprises (CoSMEs) in terms of their resilience using secondary data and information from articles and public data. Findings from the research reveal that small and medium size construction firms in the UK were adversely affected by the pandemic with many going into liquidation. The impact of government interventional measures varied across the various subsectors, overall, the measures were deemed inadequate and implemented rather late to stop many CoSMEs from insolvency. Additionally, most CoSMEs that operate largely on self-employed basis could not immediately take advantage of these measures. Mostly, it was the large construction firms that took advantage of government schemes and survived. The government concentrated on the huge and indivisible capital projects which offered a soft landing for many big construction firms, benefitting the upper end CoSMEs. The research concludes that understanding the effects of external shocks on CoSMEs will help in formulating robust strategies and policies that will enhance their resilience, and by extension, the UK economy given the pivotal role the construction industry plays in the UK economy.

Keywords Covid-19 · Construction SMEs · Government · Measures · Strategy · Construction resilience · Unemployment

R. D. Seidu (✉) · O. J. Ebohon · B. E. Young · D. Fong
School of Built Environment and Architecture, London South Bank University, 103 Borough Road, London SE1 0AA, UK
e-mail: seidur@lsbu.ac.uk

T. C. Haupt et al. (eds.), *Construction in 5D: Deconstruction, Digitalization, Disruption, Disaster, Development*, Lecture Notes in Civil Engineering 245,
https://doi.org/10.1007/978-3-030-97748-1_33

1 Introduction

Resilient CoSMEs will increase productivity where they are able to overcome external shocks, this will give the stakeholders confidence about the industry. It means the sector can withstand the vagaries of the market and reinvent itself to emerge stronger and more resilient. The outbreak of covid-19 revealed some structural weaknesses in the sector, this was exacerbated by some of the inherent characteristics of the industry notably, the low profitability margin and consecutive nature of the construction process made sector particularly vulnerable to external shocks such as the pandemic. The challenges and opportunities posed to UK CoSMEs require innovative and coordinated strategic management and leadership response owing to the multifaceted nature of the pandemic. The outbreak of coronavirus COVID-19 caused major disruption to the global economy and the manners in which business was conducted [1]. The dimension of the pandemic and attendant deaths and panics affected both private and public businesses, including the third sector of the economy [2]. The fall in construction activities in the UK, according [3] was the largest declined witnessed in eleven years.

Government lock down to contain the virus had a severe disruption on construction supply chain, particularly the restrictions on local, regional, and international travels restricted the movements of workers and freights. The government has adopted the strategy of mass vaccination to restore the economy and allow workers back to work is not having much effect on the construction. Although no research has been done on this, given the age structure of construction industry on the one hand, and on the other, that the corona virus killed more of the older generations, it is fair to say that its effects on construction industry will become more pronounced, as events unfold. The effects are already being felt in industry with shortages of construction skills and materials becoming pronounced. The combined effects of these challenges have challenged the resilience of UK CoSMEs, this study seeks to determine the resilience of UK CoSMEs.

2 Literature Review

2.1 COVID-19 and Construction Performance

Construction performance in the UK is expected to experience a sudden shock in the range of 25% reduced output due to the pandemic [4]. The recession created because of the pandemic will continue until the end of the year 2021 and start to pick up slowly as major organisations will begin to recover. The UK has seen its Gross domestic product GDP decreased by 5.8% in March since the COVID-19 outbreak, the worse after the previous recession and since 1997. The 2.0% reduction in first quarter of 2020 represent the largest reduction since the last quarter of 2008 [5].

UK construction sites was shut down due to the lockdown and the effect of social distancing rules on site productivities caused about 60% loss of productivities output. This varies depending on industry and country's strategy in place to combat the outbreak. Almost 85% of work in House building sector has disappeared with reduced 60% work in repair and maintenance sector whist commercial sector drop by 60% and commercial maintenance sector experience a minimal of 20% reduction because most programme in this sector/s were moved forward [6]. Francis [7] agrees, stating that for the fortunes of construction more generally, the near-term effect of covid-19 on the economy and employment are likely to be considerably greater than those faced during the financial crisis of 2008/09. Harris et al. [8], reaffirmed that in addition to these issues around the economy and construction demand, productivity on site has fallen significantly due to social distancing and other safety factors, which means that construction activity will take longer and cost more. Even in our most optimistic scenario, construction output could bounce back by 25.5% in 2021 but, with growth starting from a low base, output will still be 6% lower than it was in 2019 [9].

UK construction companies are continuing to experience a reduction in working period with 5.8% less ahead of the accommodation and food services in first position [10]. This is mainly due to government advising workers to work from home, wherever possible with only 25.6% of construction workers who have had such experience previously working from home. The government directives were affected by challenges of getting suitable locations and getting the required technology to aid such remote working [11]. It created a new way of working with the disruption with virtual working, space management and increasing the use of prefabricated unit [1]. Construction output has fallen by 2.6% during the same period whilst the construction products association [7], predicted a further reduction of 25% before the end of the year due to the pandemic. The number of positions advertised had fallen by 86.2% and construction vacancies declined by 24.9% when compared with previous year [11].

The construction industry is experiencing the biggest financial constraints due to the impact of COVID-19 on the industry, despite the support in place from the government, e. g supporting the self-employed and contributing about 80% towards staff wages. This is a global challenge, in US profit is expected to be squeezed whilst in the UK, the profit margin is anticipated to fall compared to the same period last quarter. Ellison [2] concurs that it will provide a ray of light amid the gloomier outlook for the sector, perhaps pointing to a way out of the downward spiral. The pandemic has resulted in many discussions and questions such as: if it constitutes force majeure event. It can therefore, safe guide the contractor from fulfilling its contractual obligation based on the provision of the contract terms. This will create a lot of disputes in the implementation or interpretation of this provision, deciding if COVID-19 outbreak could be claimed under force majeure event [12], such event could be external, unexpected, and unavoidable [13, 14] or unforeseeable, unavoidable, uncontrollable, impracticable and beyond a party's responsibility [12] with a consequence of termination or suspension of the contract [14, 15] while each party covers their risk and can entitle contractors in some situation to claim for extension of time [15] and prevent most contractors from being held responsible from paying liquidated and ascertain damages. Invariably this could lead many construction companies into liquidation, if not addressed promptly.

The outbreak of the pandemic and its impact on construction companies in terms of delay on various projects, could prevent contractors winning future projects or facing penalties for non-performance. James [16] stated that the virus will affect supply chains due to the lockdown, causing delays to delivery of supplies by more than two weeks. Such delays are unaffordable with many self-employed workers affected not covered by government support/provisions, while causing projects to be suspended for unknown periods. This largely explains the high failure rates of many CoSMEs.

2.2 Construction Company Failure

Construction companies play major contributory roles for economic development and the efficiency of construction industry depends on the contribution of the construction firms [17]. Apart from the industries fragmentation, the construction industry continuing to be confronted with other internal and external factors that contribute to its performance [17] the recent pandemic is another influencing factor that have affected the industry performance [3].

The construction company has been an employer of labour, with small scale company employing less than 250 operatives to large organisation engaging over 250 employees have experienced the second largest failure, compared with other companies, just below the restaurant in the hospitality industry [18]. The effect of failure goes beyond any direct causative to the owners, creditors, or the employees [19], but with a great influence on the economic and social performance of the economy [17]. Construction company failure can be described as liquidation or administration [20], when the company cannot perform its obligation [21] or realised rate of return on investment [22], incompetent or in debt [23] and discontinuation for other reasons [24].

Many studies had been carried out for investigating the causes of construction company failure. Arditi el al. [25] conclude that failure in USA construction company is because of budgetary and macroeconomic issues whilst Schaufelberger [26] study on subcontractor's company revealed insufficient capital/excessive debt, lack of managerial maturity, lack of early warning measures, increase in project scope, poor billing procedures, failure to evaluate project profitability, unfamiliarity with new geographical areas, and poor use of accounting systems. Also, Kivrak and Arslan [27] agrees that inexperience and economics conditions were the critical factors that causes company failure. The research conducted by Arslan et al. [28] on main contractor reveals that such failure is intertwined on cash flow and relationship problem, also some construction companies experience failure due to low percentage profit allowance and padded bid proposal. Mahamid [17] highlight the main factors of contracting failure as: fluctuation in cost of material, payment delay, inexperience in contract management, low profit margin and location or movement limitation. Lowe and Moroke [29] research reveal corporate insolvency and individual bankruptcy is due to low productivity, cash flow, supply chain issues and fluctuating demand and Langdon [30] state some of the challenges that are mostly been overlook as cash flow shortages, falling profits, failure to pay suppliers, delayed and /or reduced valuation certificates, progress of works slowing, insufficient resources deployed on the project, falling asset values, excessive borrowing, or even boardroom tensions while Whitehill and Ainsworth [31] conclude by summarised the main cause of business failure in the construction organisation to include management incompetency, insufficient capital, lack of business knowledge, fraud, industry weakness, poor technical and technological capacity, poor relations with clients and government. Construction company must strengthen the procurement and contractual process in place to avoid failure and Quantity Surveyor who are responsible for analysing tender must act professionally when dealing with the issues of front loading, speculative pricing and corrections of error in other to prevent contractor failure as a result of quest to win a bid to be in operation [32] while BSI suggest operation, supply chain and information as the three main organisation area that need to be well managed in other to build organisational resilience into organisations strategies and practices.

Construction organisation resilience is a requirement for increasing future growth, there is need for strategy in place to enhance the operation of organisation and attitude to resilience. It is a forward looking, strategic enabler that require government support to thrive [33]. Whitehill and Ainsworth [31] linked the merit of organisation resilience to long term viability, reputation, improved competitiveness while the challenges include urgent financial need, insufficient management focus, lack of budget, lack of skills and knowledge. Organisation resilience is the power of predicting, preparing and putting the necessary procedure in place to ameliorate or reducing any unforeseen disruption or shock and continuing in business while Whitehill and Ainsworth [31] concur and describe it as "the ability of an organisation to anticipate, prepare for, respond and adapt to incremental change and sudden disruptions in order to survive and prosper." There an urgent call for renew approach to the existing British standard BS-65000 to inculcate these challenges and provide guidance which will improve organisational resilience.

3 Methodology

Exploratory research is utilised to explore the research question, allowing the gathering of multiple sources of information, whilst allowing supplementary research around the research question to give a more holistic view and understanding of the research topic and does not aim to reach conclusive solution to the research question [34]. Furthermore, quantitative research in the form of analysis of secondary data and collating key industry views on the subject, backed up by data to support these views have jointly allowed for the analysis and discussions. The data involve latest development effect on construction organisation, these data include relevant media, company and government reports which was analysed to assess the effect of Covid-19 on construction company failure.

4 Analysis and Discussion

The industry have experienced an increased in the number of construction company going into administration, about 3100 firm were in administration 12 months before the Covid-19, a higher rate of liquidation than other industry [35], this became worst in April 2020, when the lockdown take its effect with many construction sites closing down due to the interruption of labour and materials supply indicating similar level of activities as witness during previous recession.

The predicament caused by the pandemic is a major contributing factor, even though the government have provided some palliative measure. This has not in any way stopped construction failures, rather it has only reduced the effect it might have had on employees. Construction performance have reduced sharply during this stage and is continuing to decline despite the efforts made by government in asking construction companies to continue with site operation while maintaining social distancing on site. This has not prevented failure, as more CoSMEs finds it difficult to continue in business due to inability to service their debt and sustain cash flow because of the lockdown. Also, this has caused another tendering cycle as projects affected by possible insolvency will have to undergo another tendering process, which will add to the contract sum, therefore increase the client's budget, hence affecting the continuity of such project. All this would not stop at the client, employees or contracting businesses table but goes beyond that, to affect the economy since the construction industry is a major contributor to any nation's GDP.

Table 1 shows 332 total number of CoSMEs contracting organisations who are involved in different construction activities at different levels of administration due to the effect of ongoing pandemic for the month of April 2021, it covers every area of construction activity except the fact that it affects one services than the other. Development of building projects is taking the lead among organisation in insolvencies at different level with 68 number of organisations follow by other specialised construction activities n.e.c which accounted for 44 numbers, third on the list is Plumbing, heat and air-conditioning installation with 31 numbers while Construction of domestic buildings is in the fourth place with 29 numbers and other building completion and finishing is taking the fifth position 26 numbers. This numbers has continued to increase despite government policy in place to reduce the effect, such as the bounce back loan scheme, suspensions of creditor recovery actions such as winding up petitions and the introduction of the Corporate Insolvency and Governance Act 2020 as not reduced the backlog of cases going through the court process. The lifting of the lockdown as not generate mayor effect due to low consumer confidence while consumers are dealing within their locality, with the end to most of the measure in place coming to an end the construction organisation need to seek appropriate advice

Table 1 Lists insolvencies and administrations of construction-related firms, filed in the UK and Ireland during April 2021

Number of companies	Status	Types of work/project	Total numbers
8	In liquidation	Other building completion and finishing	
12	Appointment of liquidator		
6	Meeting of creditors		26
13	Appointment of liquidator	Construction of domestic buildings	
9	Meeting of creditors		
7	In liquidation		29
2	Administrative receiver appointed	Construction of domestic buildings	
5	Appointment of liquidator	Construction of commercial buildings	
2	Meeting of creditors		
8	In liquidation		17
10	Appointment of liquidator	Electrical installation	
5	Meeting of creditors		
7	In liquidation		22
1	Appointment of liquidator	Steam and air conditioning supply	1
2	Appointment of liquidator	Plastering	
1	Meeting of creditors	Plastering	
3	In liquidation		6
2	Appointment of liquidator	Painting	
3	Meeting of creditors		
3	In liquidation		8
32	Appointment of liquidator	Development of building projects	
6	Meeting of creditors		
24	In liquidation		
2	Administrator appointed		
2	Administration order		
1	In administration		
1	Administrative receiver appointed		68

(continued)

Table 1 (continued)

Number of companies	Status	Types of work/project	Total numbers
2	Appointment of liquidator	Site preparation	
1	Meeting of creditors		
1	In liquidation		4
14	Appointment of liquidator	Plumbing, heat and air-conditioning installation	
6	Meeting of creditors		
11	In liquidation		31
1	Appointment of liquidator	Specialised design activities	
1	In liquidation		2
5	Appointment of liquidator	Joinery installation	
5	In liquidation		10
9	Appointment of liquidator	Other construction installation	
3	Meeting of creditors		
6	In liquidation		18
27	Appointment of liquidator	Other specialised construction activities n.e.c	
3	Meeting of creditors		
14	In liquidation		44
1	Appointment of liquidator	Scaffold erection	
1	Meeting of creditors		
1	In liquidation		3
1	Appointment of liquidator	Roofing activities	
1	Meeting of creditors		
1	In liquidation		3
1	Appointment of liquidator	Manufacture of other furniture	
1	In liquidation		2
2	Meeting of creditors	Glazing	2
1	Meeting of creditors	Construction of roads and motorways	1

(continued)

Table 1 (continued)

Number of companies	Status	Types of work/project	Total numbers
3	Appointment of liquidator	Floor and wall covering	
1	Meeting of creditors		
2	In liquidation		6
1	Appointment of liquidator	Construction of utility projects for fluids	1
1	Appointment of liquidator	Construction of railways and underground railways	
1	Meeting of creditors		2
5	Appointment of liquidator	Construction of other civil engineering projects n.e.c	
4	Meeting of creditors		
5	In liquidation		14
1	Appointment of liquidator	Manufacture of other builders' carpentry and joinery	1
1	Appointment of liquidator	Buying and selling of own real estate	
1	Administrative receiver appointed		2

Source Adopted from Construction News [36]

to developed a resilience strategy into their operation, supply chain and information management process.

Figure 1 reveals that the construction organisations are likely going to end in administration compare with other sectors of the economy, the three most affected industries in Q2 2021 are Construction industries with 1801 (16%) follow by Accommodation and food services 1474 (13%) and Wholesale, retail trade and repair of vehicle with 1,366 (12%) insolvencies. The UK economy has seen decreases in insolvencies across all industries in the last one year while the construction sectors as experience 36% lower Q2 2021 and Q2 2020 which is due to government intervention, policy, and investment to support the economy. The dynamic as change due to the various government scheme coming to end with cash flow becoming an hindrances to the company continuing in business, despite the government intervention and capitalisation of the economy with increasing investment in infrastructure facility the construction sectors is still susceptible compare to other sectors, this is seen in recent increase in June 2021 by 1.71% to 1,724 in July of the same year and 11.34% increase during the same period in July 2020 while the construction only contributing 271 representing 16% to the insolvencies total for July 2021 follow by hospitality industry with 187 and wholesale and retail trade with 175 companies during the same month.

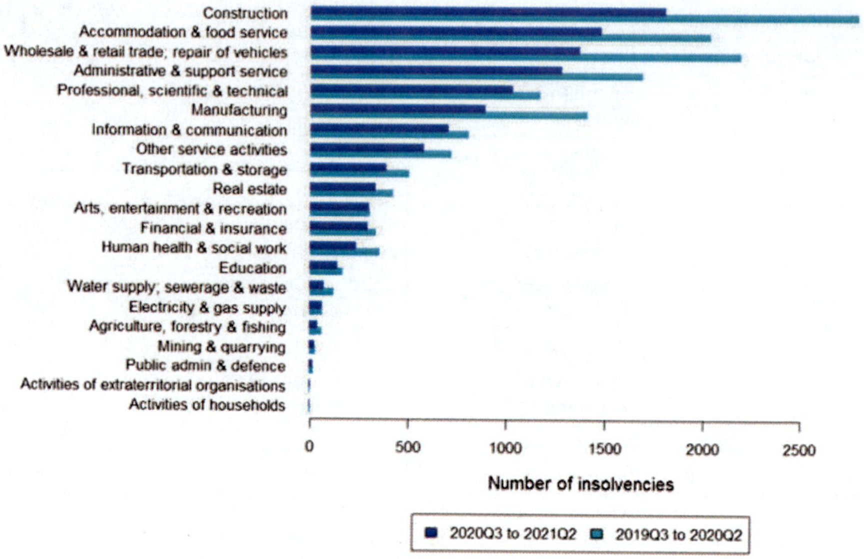

Fig. 1 Insolvencies in England and Wales by industrial sector year ending Q3 2019–Q2 2021. [10]
Source: Office of National Statistic

5 Government Responses During Covid-19

Government wages support scheme to the self-employed construction workers is in place to support the construction company at this crucial time which need to be extended to self-employed directors of small enterprises. The furlough scheme is also in place to help the industry cushion the effect of Covid-19 on workers. Majority of contracting organisations have started implementing the social distancing protocols on site, making the use of masks mandatory, and reducing the numbers of workers working on a location [35].

The government subsidy 'furlough scheme' was brought in to lessen redundancy rate and construction business failures is planned to end by October 2021. It is hoped that this will reduce the cash flow challenges that construction organisation would be experiencing. The scheme will be responsible for 7.5 million employees' wages working across 935,000 organisations, with 1 in 4 of these employees working in private sector. Construction organisation can defer their VAT payment [37]. Some construction organisation continued to work during the lockdown while maintaining social distancing and observing that the government have eased the restriction in May 2021 to boost the economy, as the construction organisation are trying to be back on track. Though, there has been an increase in disputes due to time and cost overruns [37], which is causing the present increase in company failure.

Many organisations have adopted different innovative idea to combat the effect on construction site with the use of technology for virtual reality to showcase design, organised site meeting, project exhibitions and planning application, including the use of artificial intelligence for site monitoring and valuation of work to date [32]. Despite all of this in place, many construction organisations have experienced reduced cash flow with many already insolvent.

6 Conclusion

The construction industry continues to provide employment and contribute to the economy and as a tool used by government to achieve its macroeconomic objectives, which has been greatly affected by pandemic. This, as said earlier, has led to redundancies of employees while others went into administration. Despite this, the overall effect cannot be ascertained, and claims are springing up due to non-performance from respective contractors. Strategic leadership training among construction leader in the industry must be in place to support organisational resilience and for construction building and services to build resilience into the organisation plan and construction procurement process. Any ambiguity in dealing with cases arising from pandemic should be made clear in the various form of contract in use to avoid any dispute or abandonment of project. Many of the affected construction organisation must seek advice on the legal intricacies and try to explore the best opportunity available to them rather than going into administration as it will not augur well for all the parties to the contract, the construction industry, and the economy at the receiving end.

References

1. Ogunnusi M, Hamma-Adama M, Salman H, Kouider T (2020). COVID-19 pandemic: the effects and prospects in the construction industry. Int J Real Estate Stud [online], 14(Special Issue 2):120–128. Available from: https://www.utm.my/intrest/files/2020/11/2_Final_MS_CRES-Covid-025.pdf
2. Ellison S (2020) How has global construction activity fared in the face of COVID-19? Global construction monitor, Available from https://www.rics.org/uk/news-insight
3. Raskino M (2020) 7 macro factors that will shape the 2020s. Smarter with Gartner, Available from https://www.gartner.com/smarterwithgartner/
4. Adian I, Doumbia D, Gregory N, Ragoussis A, Reddy A, Timmis J (2020) Small and medium enterprises in the pandemic: impact, responses and the role of development finance. The World Bank
5. RICS 2020 How has COVID-19 impacted construction and infrastructure globally?
6. Marshal J (2021) London office starts up 20% as commercial sector roars back into action, Building. Available from https://www.building.co.uk/Londonofficestartsup20%25ascommercialsectorroarsbackintoaction
7. Francis N (2020) Covid-19 still leaves uncertainty for construction industry, despite promises to 'build, build, build'. Construction Product Association (CPA) Available from https://www.constructionproducts.org.uk/news
8. Harris F, McCaffer R, Baldwin A, Edum-Fotwe F (2021) Modern construction management. Wiley

9. de Palma A, Vosough S (2021) Long, medium, and short-term effects of COVID-19 on mobility and lifestyle (No. 2021-06). THEMA (THéorie Economique, Modélisation et Applications), Université de Cergy-Pontoise
10. ONS (2020) Coronavirus and the impact on output in the UK economy, UK. Online at https://www.ons.gov.uk/economy/grossdomesticproductgdp/articles/coronavirusandtheimpactonoutputintheukeconomy/october 2020. Accessed on 1 Sept 2021
11. Garner-Purkis Z (2020) Construction working hours and vacancies nosedive, construction news online at www.constructionnews.co.uk
12. Hansen S (2020) Does the COVID-19 outbreak constitute a force majeure event? A pandemic impact on construction contracts. J Civ Eng Forum
13. Azfar F (2012) The force majeure "excuse." Arab Law Q 26(2):249–253
14. Ezeldin AS, Helw AA (2018) Proposed force majeure clause for construction contracts under civil and common laws. J Legal Affairs Dispute Resol Eng Constr 10(3)
15. Hagedoorn J, Hesen G (2007) Contract law and the governance of inter-firm technology partnerships: an analysis of different modes of partnering and their contractual implications. J Manage Stud 44(3):342–366
16. James M (2020) Covid-19 and the construction industry online at www.ukconstructionmedia.co.uk
17. Ibrahim M (2018) Factors affecting contractor's business failure: contractors' perspective. Eng Const Arch Manage 19(3):269
18. Clough RH, Sears GA (1994) Construction contracting, 6th edn. Wiley, New York, NY
19. Mason RJ, Harris FC (1979) Predicting company failure in the construction industry. Proc Inst Civil Eng 66:301–307
20. Dun and Bradstreet (1986) Dun's census of American business. Dun & Bradstreet Corporation, New York, NY
21. Frederikslust RAI (1978) Predictability of corporate failure. Martinus Nijhoff Social Sciences Division, Leiden
22. Altman EI (1971) Corporate bankruptcy in America. Health Lexington Books, Lexington, MA
23. Storey DJ (1994) Understanding the small business sector. International Thomson Business Press, London
24. Watson J, Everett J (1993) Defining small business failure. Int Small Bus J 11(3):35–48
25. Arditi D, Koksal A, Kale S (2000) Business failures in the construction industry. Engineering
26. Schaufelberger JE (2003) Causes of subcontractor business failure and strategies to prevent failure. Paper presented at the Construction Research Congress 2003, Hawaii
27. Kivrak S, Arslan G (2008) Factors causing construction company failure. Build Abroad October, pp 297–305
28. Arslan G, Tuncan M, Birgonul MT, Dikmen I (2006) E-bidding proposal preparation system for construction projects. Build Environ 41(10):1406–13; Construct Arch Manage 7(2):120–32
29. Lowe J, Moroke E (2010) Insolvency in the UK construction sector. In: Proceedings of the 26th annual ARCOM conference. ARCOM
30. Langdon D (2008) Construction business recovery news. Issue Number 1
31. Whitehill T, Ainsworth A (2018) The organisational resilience of the UK construction industry. A Report written by Project Five Consulting. Online at http://www.resilienceinconstruction.report. Accessed on 4 Sept 21
32. Seidu RD, Young BE, Kangwa J, Ebohon JO, Ige O (2021). Following the trail: a covariance plotting of the crippling effect of COVID-19 on small and medium sized—SMEs firms in the UK's construction industry. In: Proceedings of the 4th European international conference on industrial engineering and operations management, Rome, Italy, August 2–5, 2021
33. BSI (2015) Organizational Resilience: harnessing experience, embracing opportunity. Online at https://www.bsigroup.com/en-GB/standards/benefits-of-using-standards/reducing-risk-to-your-business-using-standards/howard-kerr-organizational-resilience-article/
34. Stebbins RA (2001) Exploratory research in the social sciences (vol 48). Sage
35. Dlamini S (2012) Relationship of construction sector to economic growth. In: International Congress on construction management research, Birmingham City University, Canada, pp 213–224

36. Construction News (2021) Insolvencies and administration. Online at https://www.constructionnews.co.uk/data/insolvencies-and-administrations. Accessed on 05 May 2021
37. Harwood B (2020) Can construction hold back the insolvency tide? Online at www.naismiths.com
38. Maindonald E (2020) How construction is responding to COVID-19 online at www.ukconstructionmedia.co.uk
39. ONS (2021) Commentary—company insolvency statistics April to June 2021. Online at https://www.gov.uk/government/statistics/company-insolvency-statistics-april-to-june-2021. Accessed on 5 Sept 2021

Critical Success Elements for Central Government Real Estate Asset Management in South Africa—A Literature Survey

Patricia Sekalo, F. Dowelani, and D. J. Hoffman

Abstract The need to manage public real estate assets effectively and efficiently is increasing globally. The widespread appreciation for assets management is due to ongoing government reforms and budget cuts that compel government departments to raise revenue and cut spending. The National Department of Public Works and Infrastructure (NDPWI) in South Africa owns diverse and capital-intensive assets which require systematic management practices, including following local and global trends. A comprehensive literature review covering Public Real Estate Asset Management in South Africa (PREAM), public sector reforms, Asset Life Cycle Management (ALCM), current and future public asset management trends was carried out. The review aimed to identify the elements necessary for a successful PREAM. By presenting this review, it is also intended to establish the extent to which the discovered elements could improve the usefulness of state buildings at the lowest possible cost. The results show that asset management information systems, ALCM, regular asset needs analysis, and performance management are essential elements for effective and efficient PREAM. Rapidly changing information technology, globalisation, increased focus on performance management, rising service delivery standards and changing socio-economic circumstances among the current trends driving PREAM changes. The study further found that through adherence to legislation and application of asset management best practices, the central government could enhance asset management effectiveness and efficiency and ultimately improve service delivery quality.

Keywords Asset management · Central government · National department of public works and infrastructure · Public real estate asset management

P. Sekalo · F. Dowelani · D. J. Hoffman (✉)
Department of Construction Economics, University of Pretoria, Gauteng, South Africa
e-mail: danie.hoffman@up.ac.za

T. C. Haupt et al. (eds.), *Construction in 5D: Deconstruction, Digitalization, Disruption, Disaster, Development*, Lecture Notes in Civil Engineering 245,
https://doi.org/10.1007/978-3-030-97748-1_34

1 Introduction

The public sector worldwide owns a substantial and diversified number of properties [1]. Due to the state-owned portfolio's significant importance and substantial value, the government is compelled to manage real estate more effectively and efficiently. As the biggest custodian of properties, the government is expected to drive asset management transformation and innovation [2]. However, public authorities in charge of state-owned properties manage these assets reactively, not as scarce resources with a value that needs to be nurtured and optimised [1].

According to Schuman and Brent [3], the increased focus on asset management was driven mainly by the New Public Management (NPM) and New Public Financial Management (NPFM) reforms introduced four decades ago. The NPM and the NPFM focus mainly on the increased productivity, performance management, accountability, transparency, efficient and effective asset management. As a result, PREAM is often considered the last option to generate revenue in developing countries and minimise government spending.

Government buildings in South Africa contribute significantly to supporting the government's constitutional mandate of service delivery. As per the Constitution of the Republic of South Africa Act No. 108 of 1996 and Government Immovable Asset Management Act (GIAMA), No. 19 of 2007, the NDPWI has the constitutional and legislative mandate to provide accommodation to other state departments [4]. The NDPWI further have a duty to manage state-owned immovable properties [5]. Boshoff and Chidi [4] also states that the Constitution of the Republic of South Africa and GIAMA further requires the NDPWI to provide infrastructure services and different types of accommodation such as:

- Specialised functional accommodation, for example, police stations, courts
- Land parcels
- Office accommodation
- Residential accommodation
- Infrastructure development
- Facilitate public–private partnership (PPP) in infrastructure development
- act as an agent between the client departments (user departments) and private landlords
- Construction and project management on behalf of other government departments that need to establish new buildings.

The department is further mandated to provide efficient and effective asset management and property-related services throughout the asset life cycle (GIAMA) [4, 6]. To achieve this overarching goal of managing all the state immovable assets, integrated, and coordinated management processes must be in place, which has been a challenge in the NDPWI. This research contributes to the consistent dialogue on the importance of PREAM.

2 Literature Review

2.1 Asset Management Information System

Most developing countries, including South Africa, still struggle to keep accurate records on assets portfolios. It has been found that the NDPWI asset register contains assets that the department does not own which is a cause for concern (NDPWI) [7]. Grubišić, Nušinović, & Roje stated: "The lack of reliable information on public assets hinders determination of the assets' value, budgeting for asset management activities, decision making and evaluating public asset portfolio performance" [8]. To successfully manage immovable assets, information relating to the asset portfolio must be always readily available and accessible to all stakeholders. Kaganova and Mckeller confirmed that advanced reformers such as New Zealand and Canada have vigorously maintained accurate asset registration for state-owned properties [9].

Hanis et al. agreed that an accurate and complete account of the immovable asset portfolio remains the foundation for any attempt to manage assets in any real estate institution [10]. A reliable asset register further remains a prerequisite for adequate maintenance, facilities management, property management, asset life cycle management, financial planning, and reporting. For big real estate institutions like the NDPWI, accurate data backed by a reliable and accessible asset management information system is essential for efficient and effective asset management.

An immovable asset register directly affects asset portfolio performance, lifecycle costing, financial planning, user satisfaction and decision making. Most institutions use an immovable asset register (IAR) to record critical information about immovable assets' status, value and condition. Buys and Mavasa [11] mentioned that real estate asset management efficiency and effectiveness rely upon the accuracy and completeness of the IAR. Hanis et al. [10] further indicates that immovable asset management challenges originate from poor inventory records. Therefore, the availability of data required to manage real estate is a cause for concern in many countries.

2.2 Asset Life Cycle Management (ALCM)

The second element identified as critical in enhancing government immovable asset management is asset life cycle management (ALCM). According to GIAMA, No 19 of 2007, immovable assets must be managed throughout their life cycle until disposed of strategically. One of the main concerns raised by GIAMA is life cycle asset management [6]. A well-designed building may be functionally obsolete if it is not adequately managed and maintained over time. Inadequate and reactive maintenance can lead to financial losses and unnecessary budget constraints. The NDPWI has a considerable responsibility to ensure that properties are functional throughout their entire life cycle. However, the empirical study conducted by Buys and Mavasa [11] found asset life cycle management in the NDPWI ineffective. Effective ALCM and its costing are still a challenge in the NDPWI. Although the business units recognised the importance of ALCM and its implications, the clients and user departments are still not engaged (NDPWI) [7].

Due to the nature and longevity of real estate assets, the management processes are divided into two main aspects: strategic asset management and operational asset management [2]. Strategic asset management is mainly focused on long-term strategic issues, including life cycle asset management. Simultaneously, the other facet focuses on operational activities, making the two aspects critical in achieving real estate asset management and asset life cycle goals. Abdullah et al. [2] stated that the strategic aspect of asset management focuses on the medium to long-term aspects of an asset, including planning, establishment, use, and disposal. In contrast, operational strategy, on the other hand, focuses on day-to-day property management activities.

Asset life cycle management recognises the inherent cost and risks associated with real estate long life span. According to Attwater et al. [12], for public real estate organisations to adequately manage real estate assets until disposed of, much attention must be placed on whole-life costing. Therefore, asset life cycle management is critical in maintaining the value and functionality of real estate assets. The considerable economic benefit accrues when infrastructure assets are adequately maintained. Similarly, a lack of proper maintenance regimes ultimately leads to additional costs on the user and lower production capacity. Figure 1 depicts the sequential phases of lifecycle asset management.

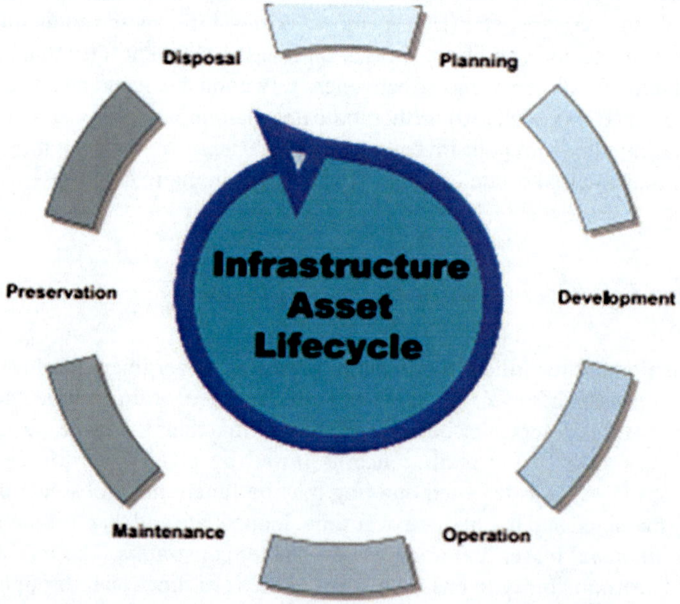

Fig. 1 Sequential phases of lifecycle asset management. *Source* Giglio et al. [13]

Sadly, public real estate assets, like any other assets, deteriorates over time. Many organisations fail to recognise asset life cycle management's value by focusing on the day-to-day operational activities. Due to limited budgets and poor resource management, the public sector fails to maintain assets and focus only on emergency maintenance, increasing assets operational failure, reduced revenue, and user dissatisfaction. According to Giglio et al. [13], "life cycle asset management is essential for cost-effective maintenance and long-term economic viability" while simultaneously reducing capital, operating and management cos.

2.3 *The Regular Asset Needs Analysis*

Another critical success element for efficient and effective PREAM, according to Kaganova and Nayyar-Stone [14], is regular asset needs analysis. In most developing countries, government ownership of land parcels and buildings results in illegal constructions, vacant properties exposed to illegal occupation and vandalism, negatively impacting revenue potential [14, 15]. As previously indicated, a fundamental requirement for any attempt to manage or reform the immovable asset portfolio is to have an accurate account of all assets.

An accurate account of assets is critical for asset needs analysis and decision-making regarding an organisation's current and future needs. For effective asset management, authorities responsible for the public sector must have a consistent immovable asset management strategy with clear and concise objectives. Government buildings must always be functional, and if they are no longer fit for purpose, alternative use must be found, or the buildings must be disposed of.

According to the IMF [16], there is a general understanding that public real estate organisations do not know assets they own or their value, who occupy their buildings and condition. Therefore, asset needs analysis is essential in determining if all organisation assets are still relevant to meet the organisations' current needs. Kaganova and Nayyar-Stone [14] emphasised that regular review of individual assets can help determine each asset's appropriateness to organisational goals. The regular review can also assist in carrying cost-benefit analysis and assessing each asset's long-term sustainability.

The NDPWI property portfolio consists of buildings, infrastructure assets and land parcels estimated to have a value of over 100 billion [5]. With such an extensive portfolio, a competent asset management team continuously need to ask questions about their portfolio to check if all assets are still appropriate for business. Hentschel and Kaganova [17] stated that regular review of assets is crucial to measure asset performance and determine which assets are still necessary for fulfilling organisational mandates.

Regular reviews on assets can also help determine if current use is appropriate and matches owners' long-term objectives [14]. Therefore, it is essential to conduct regular reviews and property inspections to determine essential or non-essential properties while evaluating each property's occupation, condition and appropriate maintenance required. Vacant properties or properties not needed for service delivery can alternatively be used to generate revenue. The public sector needs to disinherit no longer necessary assets to reduce operating costs and other ownership-related problems. Lyons [18] emphasised that good asset management will optimise the asset value, increase service delivery, and engender a budget surplus.

2.4 *Accountability and Performance Management*

Due to limited funding, regulatory organisations in the public sector have increased pressure on organisations to implement the principles of efficiency, effectiveness, transparency, and accountability. These principles further compel the government to put systems in place to measure the performance of real estate assets throughout the life cycle. Asset management requires funding; simultaneously, the funding must be just and equitable to ensure value for money. A massive government property portfolio requires an asset management system aligned with the overall business goals and strategies. The government must determine current asset management performance to know what to improve and how [19]. To effectively manage assets and measure performance, organisations need to have well-established systems and control measures. In addition, public authorities should balance efficient use today to accommodate society's present needs and not compromise the future's needs [20]. When it comes to real estate asset management, sustainability is fundamental.

Sustainable real estate asset management starts with aligning real estate strategy with the business units' goals. The real estate asset management framework must establish a feedback loop between managerial activities and actual property performance across the portfolio. An effective asset management performance system starts with preconditioned asset management performance indicators, which can be managed manually or systemically depending on the asset portfolio's size. According to Buys and Mavasa [11], there is a lack of monitoring and evaluation systems within the NDPWI. This results in assets not maintained at their optimal value and being stripped of the inherent value over time. Performance measurement as a control-ling element focuses on immovable asset management and strategic thinking of public assets ownership [10]. Public sector accounting reforms have increased transparency, efficiency, and accountability for the public sector financial management [21].

Abdullah et al. [2] found that the lack of a systematic public asset management strategy and performance management system, public assets fail to generate sufficient revenue to finance maintenance and other costs associated with their operations [22]. Private sector assets are financially self-sustaining. Private organisations view property ownership as an investment, while government properties are viewed as a service delivery tool. In emerging countries such as South Africa, public sector reforms remain a necessary and ongoing policy objective. The non-profitability element of government immovable asset management makes it more difficult for the government to measure performance.

In the public sector, efforts relating to public asset management are channelled towards providing service and achieving the government socio-economic objectives. The government can only achieve its economic growth and employment creation objective if it can move from current bureaucratic asset management strategies and processes [11]. NPM and NPFM reforms have distinct benefits related to client-driven service delivery, improved accountability, risk management, and financial management. Based on the literature surveyed, the elements discussed and their implementation is critical to achieving a real estate asset management breakthrough for any organisation.

The study further found rapidly changing information technology, globalisation, increased focus on performance management, rising service delivery standards and changing socio-economic circumstances among the current trends driving Public Real Estate Asset Management (PREAM) changes.

3 Conclusionss

Based on the literature reviewed, asset management generally has goals depicted in Fig. 2. The goals apply to most real estate organisations, regardless of whether the reason for existence is service delivery or profit-making. Both private and public institutions need to manage real estate assets throughout their whole life cycle effectively and efficiently. One of the prerequisites for PREAM is adequate resources, skills, relevant experience, and the necessary Knowledge [23].

Literature has revealed the gross importance of real estate assets and their management in society worldwide. In most countries, the government owns and manages diversified and extensive property portfolios, including land, infrastructure and building. Urgent implementation of effective and efficient PREAM practices is

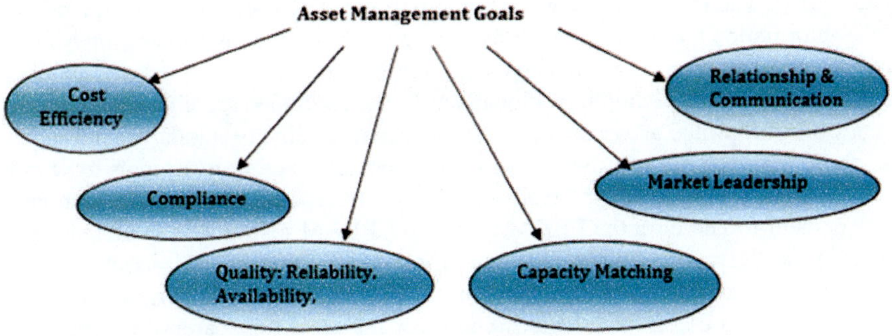

Fig. 2 Asset management goals in real estate organisations [24]

needed to fully utilise state-owned properties in South Africa in all three government spheres.

The public sector reforms were practically introduced in the 1980s, motivated by public sector management's inefficiency and ineffectiveness; however, they became popular in most countries in the 1990s [19, 20]. Real estate assets in public and private sectors were acknowledged as a strategic resource that can be exploited to meet strategic and organisational goals. Based on the literature, PREAM is a contagious issue worldwide and needs urgent attention theoretically and practically.

According to the reviewed literature, Australia, Canada, New Zealand, the U.S., and the U.K. are the most advanced reformers in asset management practices. However, overall asset management practices in most countries are still lacking [25]. Therefore, enhancing and transforming asset management practices in South Africa using international experiences and strategies might require some adjustments.

The New Public Management and the Public New Financial Management reforms have motivated governments worldwide to look at their assets as a portfolio, limited resource with the value that needs to be optimised, and a tool to generate revenue. Similar reforms also drive the NDPWI to find alternative ways of managing the immovable assets portfolio. The unprecedented global financial crisis in 2008 has further presented the public sector with a massive challenge of budget deficits across the world, which the South African government is not immune to the global crisis.

To a certain extent, these initiatives also demonstrate the NDPWI willingness to reform and rebuild asset management. However, even if these reforms occur, local and national governments continue to be unprepared to deal with property owners and managers [10]. Effective immovable asset management practice is needed to ensure that state assets are fully functional and of optimal value. Asset management can be an essential vehicle in improving financial performance and adding value to a 'company's operations.

The financial crisis has caused the public sector to develop effective and efficient financial management strategies, accounting and performance management systems. Scarcity of public funds, maintenance legacy, service delivery standards and socio-economic changes have also contributed significantly to asset management transformation [14]. The increased use of technology is currently becoming the new norm. "The information technology revolution has had a profound and multidimensional impact on public sector asset management" [26].

Rapidly changing technology, globalisation, the continued search for greater efficiency of state-owned assets utilisation, rising service delivery standards, and socio-economic changes have further pushed asset managers and professionals to search for alternative solutions for more advanced asset management. Changing economic environment forced both the PREAM and the CREAM institutions to increasingly focus on productivity, effectiveness and efficiency in fulfilling their asset management roles. Government departments, state-owned enterprises, and government entities worldwide are continuously searching for the most transparent, effective and efficient ways to deliver good quality service to stakeholders and further use state property for economic development.

The NDPWI, too, is not immune to the challenges faced by other public sector institutions globally. After the global financial crisis of 2008, South Africa, like many countries, found itself in unpleasant financial strain. Following international trends, the NDPWI started to engage in transforming its immovable asset management practices. To follow the trends, the NDPWI, with the approval of Parliament and the National Treasury, established an entity that focuses specifically on professionalising immovable assets management (NDPWI) [7]. As much as the main objective of acquiring the public asset is non-profitmaking, state assets' revenue potential cannot be overlooked.

The establishment of the Property Management Trading Entity under the control of the National Department of Public Works and Infrastructure proves that there is growing appreciation from the South African government to professionalise PREAM. The need to manage central government real estate assets efficiently and effectively in South Africa is critical to achieving its developmental goal, job creation, quality service delivery and inclusive growth. Worldwide asset management has been acknowledged as a vital tool in establishing and defining more efficient, accountable and effective organisations.

The central government could reduce asset management spending by applying best practices and processes, adhering and compliance to asset management legal frameworks, and carefully following trends. Implementing asset management best practices will help eradicate poor maintenance legacy, increase revenue, increase user satisfaction, and ultimately improve service delivery. Based on all the initiatives mentioned in the study, the NDPWI seems to be moving in the right direction regarding New Public Management and New Public Financial Management reforms, even though challenges still exist.

The media briefing held by the former Minister and subsequent NDPWI published annual reports indicates clear intentions in the department to improve PREAM, reduce public spending, and increase service delivery quality. The below list of policy documents and acts demonstrates the initiatives taken by the South African government to enhance public asset management legal framework:

- The development of 1997 Public Works White Papers,
- The development of 1999 Public Works White Papers,
- Public Finance Management Act, No. 1 of 1999, (amended with Act, No. 29 of 1999)
- Asset Management framework in 2004,
- Government-wide Immovable Asset Management policy of 2005,
- Government Immovable Asset Management Act, No 19 of 2007,
- The seven-year turnaround strategy in 2012, and
- Operationalisation of the PMTE in 2014.

The NDPWI took significant steps, including implementing the seven-year turnaround strategy and accrual accounting system and operationalising the PMTE to improve its asset management practices. The overall aim behind these changes is to reform the department's asset management processes and reclaim its role as the biggest custodian of immovable assets in the country and deliver high service standard user departments.

Investing in sustainable asset management practice is critical in any organisation, especially public institutions that utilise property assets to fulfil their mandates. Due to limited finance, governments worldwide are under severe pressure to improve service delivery quality while containing costs. Government real estate assets, if adequately managed, have a tremendous potential to generate revenue and finance their operations. Therefore, these assets serve as a critical backbone for the country's economy.

If adequately implemented and monitored, the identified key elements can improve real estate management's effectiveness, efficiency, and confidence level in the NDPWI. The research further indicates that although general recommendations can be helpful to all institutions responsible for asset management, there is no standard procedure that fit all organisations [27]. The study suggests that the critical component of managing immovable assets efficiently and effectively is accurate and reliable data. The sooner steps are taken to repair, maintain, and rehabilitate existing government immovable assets, the lower the asset's lifecycle cost.

There is no one-size-fits-all asset management strategy. Each real estate organisation needs to develop customised standard operating procedures that suit its organisational objectives, legal framework, and internal policies. Successful asset management depends entirely upon the asset management team's ability to alter the recommended strategies to fit their organisational structure.

4 Recommendations

The NDPWI should clearly define its real estate asset management objectives and link its strategic policy to business goals. The NDPWI current asset management strategies, policies, and procedure manuals must be revised to cater for the newly established trading entity placed under the control of the NDPWI ministry.

Secondly, the NDPWI, as the custodian of state-owned buildings, needs to encourage collaboration and regular engagement with occupiers of the state-owned buildings to increase productivity and efficiency. Based on the literature, an efficiently designed building may perform poorly if its systems are poorly managed and maintained. If the tenants do not use the building system as intended, the set objectives and expected results will not be achieved; hence, the NDPWI must encourage open communication with tenants to avoid disputes and inefficiencies.

Furthermore, the NDPWI must regularly conduct an asset needs analysis to identify non-productive assets and non-essential assets. The unutilised and non-essential assets must be disposed of, reallocated, or leased out to the private sector, simultaneously considering the financial implications of each approach.

Lastly, the world is changing from traditional bureaucratic asset management to vigorous digital asset management practices. Therefore, the NDPWI needs to take a proactive approach by introducing sustainable ways of managing state-owned properties through advanced information technology, including implementing an asset management information system that caters for life cycle asset management from the conceptual phase until disposal.

References

1. Wojewnik-Filipkowska A, Rymarzak M, Lausberg (2015) Current managerial topics in public real estate asset, management. World Real Estate J 4(94):5–9
2. Abdullah S, Razak AR, Hanafi MH, Pakir AHK (2012) Towards implementing property asset management: assessing current government practices in property management. Int J Soc Sci 4(1):1–8
3. Schuman CA, Brent AC (2005) Asset life cycle management: towards improving physical asset performance in the process industry. Int J Oper Prod Manage 25(6):566–579
4. Boshoff D, Chidi M (2013) The investigation into the process of providing office accommodation for national state departments in South Africa. Afr J Bus Manage 7(33):3233–3247
5. South Africa (2017) National Department of Public Works and Infrastructure. Annual report for period 1 April 2016 to 31 March 2017. Government Printers, Pretoria
6. South Africa (2007) Government Immovable Asset Management Act (No. 19 of 2007). Government Printers, Pretoria
7. South Africa (2019) National Department of Public Works and Infrastructure. Annual report for period 1 April 2018 to 31 March 2019. Government Printer, Pretoria
8. Grubišić M, Nušinović M, Roje G (2009) Towards efficient public sector asset management. Fin Theory Pract 33(3):329–362
9. Kaganova O, McKellar J (eds) (2006) Managing government property assets: international experience. The urban institute press, Washington, DC
10. Hanis M, Trigunarsyah B, Susilawati C (2010) Elements of public asset framework for local governments in developing countries. In: Wang Y, Yang J, Shen G, Wong J (eds). Proceedings of the 2010 international conference on construction and real estate management 1, 1–3 December 2010. Royal on the Park hotel, pp 418–423
11. Buys F, Mavasa T (2007) The management of government immovable assets. Acta Structilia 14(1):81–92
12. Attwater A, Wang JQ, Parlikad A, Russell P (2014) Measuring the performance of asset management systems. In: Asset management conference
13. Giglio JM, Friar JH, Crittenden WF (2018) Integrating lifecycle asset management in the public sector. Bus Horiz 61(4):511–519
14. Kaganova O, Nayyar-Stone R (2000) Municipal real property asset management: an overview of world experience, trends and financial implications. J Real Estate Portfolio Manage 6(4):307–326
15. South Africa (2015) National department of public works and infrastructure. Annual report for period 1 April 2014 to 31 March 2015. Government Printers, Pretoria
16. Tanzi V, Prakash T (2002) International Monetary Fund (IMF) The cost of government and the misuse of public assets, Working paper, pp 1–26
17. Hentschel JJ, Kaganova O (2007) Real estate 101 for government. The Urban Institute, © 2012 http://www.urban.org
18. Lyons M (2004) Towards better management of public sector assets: a report to the Chancellor of the Exchequer. See www.Hm-treasury.Gov.UK Sourced, 24, pp 03–09
19. Kaganova O, Telgarsky J (2018) Management of capital assets by local governments: an assessment and benchmarking survey. Int J Strateg Prop Manag 22(2):143–156
20. Wojewnik-Filipkowska A, Rymarzak M, Lausberg C (2015) Current managerial topics in public real estate asset, management. World Real Estate J 4(94):5–9

21. South Africa (2004) National treasury: asset management framework, Version 3.3. Government Printers, Pretoria
22. Abdullah S, Nah MNM, Razak AA (2015) Real Estate management in Malaysian public schools: determination of success factors. Real Estate Manage Valuat 23(2):52–67
23. Rymarzak M, Trojanowski D (2013) Asset management of the public sector in poland. Real Estate Manage Valuation 21(1):5–13
24. Too EG (2020) A framework for strategic infrastructure asset management. In: Definitions, concepts and scope of engineering asset management. Springer, London, pp 31–62
25. Phelps A (2011) Municipal property asset management–a comparative study of U.K. and Russia. Int J Strategic Property Manage 15(4):416–437
26. Kaganova O, Amoils JM (2020) Central government property asset management: a review of international changes. J Corporate Real Estate 22(3):239–260
27. Kaganova O, Rutledge CRE, Dragana Markovic KB (2012) Guidebook on real property asset management for local governments. Urban Institute Centre on International Development and Governance, IDG Asset Management Toolkit (1)

The Effective Implementation of the Framework for Infrastructure Delivery and Procurement Management

Neil Sirbadhoo and Nishani Harinarain

Abstract This study investigated government's ability to apply infrastructure delivery management tools through the implementation of the FIDPM towards improving public sector service delivery from the construction project manager's perspective. This quantitative study utilised online questionnaires. The questionnaire was distributed to 1185 professionally registered construction project managers with the South African Council for Project and Construction Management Profession (SACPCMP) through a stratified random sampling approach. Four hundred and ten responded representing 34.59% response rate and the data was analysed using IBM SPSS v27. Key policies and frameworks developed by the South African government for addressing service delivery and its challenges since its democracy in 1994 have not been successful in their implementation. Government still struggles with the implementation of the current policy, namely the FIDPM. Government is not effectively managing public procurement and needs to train, develop, and build capacity to achieve service delivery targets through robust policy implementation. Project managers through their knowledge, awareness and application of the infrastructure delivery management system (IDMS) and FIDPM are in a key position to assist government improve policy implementation towards improving service delivery.

Keywords Construction project managers · FIDPM · Policies and frameworks · Public service delivery

N. Sirbadhoo
University of KwaZulu-Natal, Durban, South Africa

N. Harinarain (✉)
The School of Engineering, Construction Studies Discipline, The University of KwaZulu-Natal, Howard College, Durban, South Africa
e-mail: harinarain@ukzn.ac.za

T. C. Haupt et al. (eds.), *Construction in 5D: Deconstruction, Digitalization, Disruption, Disaster, Development*, Lecture Notes in Civil Engineering 245,
https://doi.org/10.1007/978-3-030-97748-1_35

1 Introduction

The most important issue in South Africa remains improving the delivery of public services, as any democratic society needs to exhibit efficient, equitable and accountable public service [1]. The South African government like many other governments has placed infrastructure spending at the centre of its economic growth strategy. In the global context and with its many past disparities of oppressive history, South Africa has had to play catch up in dealing with unemployment, poor economic growth, underdevelopment, poverty and inequality. In the third decade of the countries post-apartheid constitutional democracy, serious concern has been raised about the government and its ability to deliver public services that its citizens painfully yearn for and are entitled to [2].

The South African government had a vision to create opportunities for social and economic growth through infrastructure investment. In 2002, the National Treasury conducted a review of provincial service delivery systems with the intention to enhance infrastructure delivery. A key recommendation from the review highlighted that a framework was to be developed to guide and structure the management of infrastructure delivery. To address the issues and gaps identified in the review, the Infrastructure Delivery Improvement Programme (IDIP) was established. Within this programme, the concept of the Infrastructure Delivery Management System (IDMS) was developed and later adopted as the chosen government wide system for infrastructure delivery [3].

This study highlights the role of the public sector and its efforts towards addressing service delivery in South Africa. However, through numerous policies and frameworks developed by government over the recent decades, service delivery challenges are still present. Various aspects hindering service delivery include incompetent public servants, a lack of accountability, poor human resources practices, inadequate procurement practices and a lack of leadership [4]. In addition, key problem areas include service delivery difficulties, poor management of finances, high level of unemployment, nepotism, and corruption [5].

As a result, government through National Treasury has released the Framework for Infrastructure Delivery and Procurement Management (FIDPM) in October 2019 as an initiative towards ensuring a systematic and structured approach for infrastructure procurement and delivery management for the successful delivery of construction projects [2, 3]. There is minimal literature available on its current performance with the framework only in its second year of implementation. Therefore, the research investigates the South African government's ability to apply the infrastructure delivery management tools the IDMS offers through the implementation of the FIDPM, towards improving public sector service delivery from the project manager's perspective.

2 Government Policies and Frameworks to Drive Service and Infrastructure Delivery

2.1 Public Sector Policies in South Africa

Public service policies since the new government came to power in 1994 have attempted to deal with the complex needs of the service delivery institutions including service delivery itself and had to be done against the background of the unfortunate legacies of the past [6]. Public service and service delivery policies were designed and promulgated to deal with the simultaneous transformation and quest for improved, service delivery orientated public service in South Africa [7]. The most important public service delivery policies and initiatives implemented in South Africa were:

- The Constitution and Public Administration [8].
- The White Paper on the Transformation of the Public Service [9].
- Project Batho Pele [9].
- The Code of Conduct for Public Servants [10].

2.2 Implementing Service Delivery Through Public Procurement

Public procurement refers to government activity of purchasing the goods and services needed to perform its functions through contractual means [11]. Public procurement operates in an increasingly intense environment which is scrutiny driven by technology, programme reviews including public and political expectations for service improvements [12]. In South Africa, procurement is of particular importance in the public sector and has been used as a policy tool due to the inequitable and unfair practices during apartheid [13]. Public procurement has since been granted constitutional status and is a recognised means of addressing past discriminatory policies and practices [13].

The South African Constitution of the Republic of South Africa (Act 108 of 1996) allows public procurement policy to provide for categories of preference in the allocation of contracts and the protection or advancement of persons or categories of persons, disadvantaged by unfair discrimination [8]. South Africa has subsequently become known for its use of procurement preferences as affirmative action to advance the economic status of groups who had suffered discrimination under apartheid [14]. Recent policy developments have been aimed at placing greater support on public procurement as a tool for achieving economic transformation and addressing socio-economic inequalities deriving from South Africa's pre-democratic past. In addition, the policies aim to use public procurement as a lever to promote socio-economic transformation, empower small enterprises, rural and township enterprises including designated groups and to promote local industrial development [15]. However, South African public procurement implementation too, has its challenges.

2.3 Public Procurement Challenges in South Africa

An important tool for managing public sector procurement in South Africa is Supply Chain Management also known as SCM. SCM is an important part of careful financial management in public sector management. It operates within a regulatory framework set by the national government and extended by provinces and local government bodies to specific policies, legislations and regulations and aims to add value at each stage of the procurement process [16, 17].

SCM has been identified a fundamental part of procurement in the public sector and is used as a tool for the management of public procurement practices. However, despite being used as a strategic tool, there are still challenges in public procurement in South Africa, namely:

- Lack of proper knowledge, skills and capacity,
- Non-compliance with SCM policy and regulations,
- Inadequate planning and linking of demand to the budget,
- Accountability, fraud and corruption,
- Inadequate monitoring and evaluation of SCM,
- Unethical behaviour,
- Too much decentralisation of the procurement system, and
- Ineffectiveness of the black economic empowerment (BEE) policy [18].

2.4 Infrastructure Delivery Management System for Improved Service Delivery

In 2004 the Infrastructure Delivery Improvement Programme (IDIP) was established as a partnership between National Treasury, the Construction Industry Development Board (CIDB), the Department of Public Works (DPW) and the Development Bank of South Africa (DBSA) to create a capacity building programme dealing with service delivery failures across provincial departments. Within this programme the concept of the Infrastructure Delivery Management System (IDMS) was created and informed [19].

The IDMS is the South African government's policy for implementing its strategy to enhance socio-economic growth and development through infrastructure delivery [20]. The IDMS includes three core systems, namely a planning and budgeting system, a supply chain management system and an asset management system, all of which have forward and backward linkages. These core systems are located within portfolio, programme and project management, and operation and maintenance processes. Collectively these processes and systems, together with a performance management system, establish the institutional system for infrastructure delivery [20].

2.5 *Service and Infrastructure Delivery Challenges Still Present*

Public service delivery challenges still experienced in 2020 were the same raised in previous years regarding capacity and skills shortages with public servants. According to the National Development Plan (NDP) the main challenge which affects infrastructure management and delivery has been the unevenness in capacity that leads to uneven performance in local, provincial and national government. This was caused by a complex set of factors, including tensions in political administrative interface, instability of administrative leadership, a skills deficit, no accountability and authority, poor organisational design and low staff morale. The weakness in capacity and performance are most serious in historically disadvantaged areas, where state intervention is most needed to improve people's quality of life [21].

Although the IDMS has made some good progress when implemented through provincial treasuries and provincial departments, it has not been sufficiently institutionalised in most departments. The implementation and sustainability of the IDMS stills remains a challenge.

The National Treasury Instruction No. 3 of 2019/2020 issued and effective from the 01st October 2019 gave legal effect to the Framework for Infrastructure Delivery and Procurement Management (FIDPM) which is applicable to:

- All departments, constitutional institutions and public entities listed in Schedules 2 and 3 of the Public Finance Management Act (PFMA) and
- Any organ of state, as identified in Section 239 of the Constitution, which implements infrastructure projects on behalf of institutions referred to in (a) [21].

3 Framework for Infrastructure Delivery and Procurement Management (FIDPM)

The National Treasury FIDPM establishes governance for infrastructure delivery and infrastructure procurement within the ambit of Clause 217 of the Constitution of the Republic of SA, 1996. It is primarily focused on governance decision making points as well as alignment and functions to support management of infrastructure delivery and procurement processes. It prescribes the minimum requirements for the implementation of the IDMS through the infrastructure delivery management processes portfolio, programme, projects, operations and maintenance of infrastructure, and the infrastructure procurement gates. In addition, the framework promotes value for money by organs of state throughout all phases of infrastructure delivery and management as well as promote optimal use of resources to achieve the intended outcomes. Furthermore, it facilitates the allocation of clear responsibilities for performing activities and making decisions at control points, stages, and procurement gates [20, 21].

This study focused primarily on the project processes section of the FIDPM referred to as the project management process in the IMDS training module (2019).

The importance of the public sector's capability to successfully provide infrastructure service delivery to the public underpins the platform upon which society depends and yet in many cases the causes of infrastructure failure can be traced back to a lack of good governance, poor procurement, and inadequate delivery management practices, all of which are under the control of the client (in this case, government departments). Successful infrastructure delivery does not rely solely on effective built environment professionals. The role of government departments is crucial in terms of setting the direction and providing effective governance [22].

Since the implementation of the FIDPM on 01 October 2019, real time performance of its application is developing. As a result, it is critical for government to ensure all necessary precautions have been taken (based on decades of consistent service delivery challenges) to ensure the FIDPM's successful implementation. This study seeks to investigate government's ability to apply infrastructure delivery management tools the IDMS offers through the implementation of the FIDPM during the project management processes towards improving public sector service delivery from the project manager's perspective. This allows a rich and valuable contribution to the academic body of knowledge as outcomes of performance are captured and reported on as they occurred.

4 The Construction Project Management Profession in South Africa

The SACPCMP defines Construction Project Management as, *"the management of projects within the built environment from conception to completion, including management of related professional services. The Construction Project Manager is one point of responsibility in this regard"* [21: 3]. The SACPCMP is a statutory body established by Section 2 of the Project and Construction Management Act, 2000. The SACPCMP was established to provide for statutory professional certification, registration and regulation of Project and Construction Management Professions to protect public interest and advance construction and project management education [23].

4.1 Construction Project Management and Its Role Towards Addressing Service Delivery Challenges

The significance of project management is regarded as a cornerstone for any institutions responsible for rendering services to the public or customers. A project is a vehicle for service delivery and project management is a strategic tool available at any organisations disposal to enhance productivity or service delivery [24]. In addition, the FIDPM recognises that project management plays a critical role in the delivery of public services in South Africa as all spheres of government structures implement and deliver services through policy, portfolio, programmes and projects [3].

While it is argued, the public sector is less innovative than the private sector, construction project management plays an important role towards innovative delivery approaches in public service [25]. Arguably the main driver behind the application of construction project management in government is to improve state institution's ability to deliver efficient, effective and high-quality services. Construction project management can support the achievement of institutional goals and provide greater assurance to stake holders that the resources are effectively managed as construction project managers vest responsibility, authority, and accountability [26].

5 Research Methodology

The study employed a quantitative study in which online questionnaires were distributed through the SACPCMP. This allowed for a large of amount of data to be collected from a sizeable population in very economical way [27]. The survey allowed quantitative data to be collected and analysed quantitatively using descriptive and inferential statistics.

Stratified random sampling was used which approach entailed the researcher stratifying the population to ensure the population within the stratum was homogenous in alignment to the characteristics on the basis of which it was stratified [27]. The sample included selecting only professionally registered construction project managers (PrCPMs) with the SACPCMP, from the most active provinces identified in the South African Cities Network report in terms of public service delivery and policy implementation in the country. These included, KwaZulu-Natal, Gauteng, and the Western Cape.

The SACPCMP database was utilised to identify the sample frame. The data base listed 1,185 PrCPMs across all three provinces, namely KwaZulu-Natal (232), Gauteng (675) and Western Cape (278). The research parameters set out to identify professionally registered PrCPMs working on public sector construction projects, specifically where the FIDPM has been released for implementation within the selected geographical regions. The data was analysed for the assessment of internal consistency and standard deviation using IBM SPSS Version 27.

6 Findings and Discussion

Out of 1185 emails circulated, 410 completed questionnaires were received indicating a response rate of 34.59%. A population size of 1000 requires a sample of 278 [28]. However, it is suggested a response rate of between 30 and 50% is appropriate [27]. It is therefore noted that the response rate received was deemed appropriate for the study.

Table 1 illustrates the years of experience of respondents working on public sector construction projects. Forty two percent of respondents have more than 30 years of experience with 38% having 20–30 years of experience. Only one respondent had between 5 and 10 years of experience indicating that the questionnaire was answered by a predominately experienced sample population.

Eighty nine percent of respondents indicated that they always work on public sector construction projects while four percent noted that they seldom work on public sector construction projects. A total of 374 respondents were employed in the private sector as either consultant or contractor, indicating that they work with or on public sector projects. While eight percent of respondents (n = 36) were employed in the public sector under a government department. Twenty eight percent of respondents were involved in infrastructure projects in Gauteng, KwaZulu-Natal and Cape Town while 19% were involved in projects in Gauteng (Johannesburg), KwaZulu-Natal (Msunduzi, EThekwini) and Cape Town. Only nine (2%) of respondents were involved on infrastructure projects in Gauteng (Johannesburg, Ekurhuleni and Tshwane). Indicating respondents were from all the target municipalities within the selected provinces.

1. Policy Development and Implementation

Policy development and implementation explored PrCPMs experience with government on public sector projects. The responses for this question are illustrated in Table 2 and were not ranked as it grouped three key statements which required respondents to rate their level of agreement on a scale where 1 (strongly disagree) to 5 (strongly agree).

Items PDI1 and PDI2 depicted mean values of 1.26 and 1.24. These two items questioned respondents experience on whether current policy development was addressing the concerns of service delivery including governments understanding

Table 1 Years of experience in public sector construction projects

Experience	Frequency	Percent (%)
05–10 Years	1	0.3
10–20 Years	79	19.3
20–30 Years	156	38.0
Above 30 Years	174	42.4
Total	410	100.0

Table 2 Policy development and implementation

Item No.	Description		1	2	3	4	5	Total	Mean	Std. Dev.
			Strongly disagree	Disagree	Neutral	Agree	Strongly Agree			
PDI1	Current policy development in SA and policy implementation addresses and deals with their targeted areas of concern regarding improving public service delivery	N	372	1	37	0	0	410	1.26	0.645
		%	90.7%	0.2%	9.0%	0.0%	0.0%	100.0%		
PDI2	Government is good at policy development, understands the policy directions and are successful at implementing policy	N	361	48	1	0	0	410	1.24	0.563
		%	88.0%	11.7%	0.2%	0.0%	0.0%	100.0%		
PDI3	The infrastructure delivery model requires a new trajectory in tackling triple challenges of poverty, unemployment and slow economic growth	N	0	0	0	15	395	410	4.83	0.511
		%	0.0%	0.0%	0.0%	3.7%	96.3%	100.0%		

and success in implementing policy. It was noted that 90.7% of respondents strongly disagreed that they had any positive experiences.

The intent of this item PD13 (m = 4.48) was to test the level of agreement respondents would have towards whether infrastructure delivery required a new trajectory in addressing service delivery challenges. The response rate indicated, 96.3% of respondents strongly agreed that the delivery model does require new trajectory suggesting that current policy is not effective and that the FIDPM may require additional support to improve implementation.

The responses concurred with the literature emphasising, current policy does not address service delivery, government are not successful in policy implementation and infrastructure delivery does require new trajectory [2, 29].

2. Effective Policy Implementation

This question investigated how often respondent's encountered poor quality public service standards in alignment to effective policy implementation on public sector projects. Table 3 ranks respondent's frequency of their encounters with government on a scale of 1 (never) to 5 (always).

Respondents encountered item EPI7 (62.2%) and EPI2 (65.6%) often when working on public sector projects. This indicates public service employees display little understanding of service standards, they do not make service standards readily available to clients and are not able to link service standards to policy reform initiatives, suggesting that government need to train, develop, and build capacity towards ensuring skilled public servants achieve service delivery targets through robust policy implementation.

Table 3 further depicts that the remaining respondents often to always encountered poor quality public service in alignment to effective policy implementation as the mean values were relatively close to the median value (5), supporting the literature that government public service standards are poor and policy implementation is not effective [30].

3. Supply Chain Management Public Procurement Challenges

The third question tested whether SCM, being implemented as a strategic tool, still faced public procurement challenges, and required respondents to rate their level of agreement on a scale of 1 (strongly disagree) to 5 (strongly agree) which were presented in Table 4.

Items SCMPPC4 and SCMPPC5 both ranked the highest with a mean value of 4.86. Emphasising 90.2% of respondents felt public servants were not held accountable while fraud and corruption still plagued government. Furthermore, SCM was not monitored and evaluated based on their performance to meet procurement targets suggesting government are not effectively managing public procurement, directly impacting service delivery negatively [18].

4. *IDMS & FIDPM*

The next question explored the PrCPMs knowledge, awareness, and application of IDMS and FIDPM.

Table 3 Effective policy implementation

Item No.	Description		1	2	3	4	5	Total	Mean	Std. Dev.	Rank
			Never	Seldom	Sometimes	Often	Always				
EPI7	Little understanding within some departments that considerable effort is required	N	0	0	0	255	155	410	4.38	0.485	1
		%	0.0%	0.0%	0.0%	62.2%	37.8%	100.0%			
EPI2	Service standards not readily available to internal or external clients	N	0	0	0	269	141	410	4.34	0.476	2
		%	0.0%	0.0%	0.0%	65.6%	34.4%	100.0%			
EPI8	Not linking the concept of service standards to Batho Pele principles	N	0	0	14	241	155	410	4.34	0.543	2
		%	0.0%	0.0%	3.4%	58.8%	37.8%	100.0%			
EPI3	Too much emphasis was placed on the responsibilities of government departments and very little emphasis was placed on the clients	N	0	0	14	255	141	410	4.31	0.532	3
		%	0.0%	0.0%	3.4%	62.2%	34.4%	100.0%			
EPI4	Service standards were sometimes unachievable, thus leading to clients having exaggerated expectations	N	0	0	14	254	142	410	4.31	0.533	3
		%	0.0%	0.0%	3.4%	62.0%	34.6%	100.0%			
EPI5	Members of staff were often uninformed of service standards	N	0	0	61	171	178	410	4.29	0.709	4
		%	0.0%	0.0%	14.9%	41.7%	43.4%	100.0%			
EPI6	Certain departments had rigid and uniform service standards across the whole department	N	0	0	37	218	155	410	4.29	0.622	4
		%	0.0%	0.0%	9.0%	53.2%	37.8%	100.0%			
EPI1	Departments not having service standards	N	0	0	0	324	86	410	4.21	0.408	5
		%	0.0%	0.0%	0.0%	79.0%	21.0%	100.0%			

Table 4 SCM public procurement challenges

Item No.	Description		1	2	3	4	5	Total	Mean	Std. Dev.	Rank
			Strongly disagree	Disagree	Neutral	Agree	Strongly agree				
SCMPPC4	Accountability, fraud and corruption	N	0	0	19	21	370	410	4.86	0.465	1
		%	0.0%	0.0%	4.6%	5.1%	90.2%	100.0%			
SCMPPC5	Inadequate monitoring and evaluation of SCM	N	0	0	15	26	369	410	4.86	0.438	1
		%	0.0%	0.0%	3.7%	6.3%	90.0%	100.0%			
SCMPPC7	Ineffectiveness of BEE policy	N	0	0	23	19	368	410	4.84	0.496	2
		%	0.0%	0.0%	5.6%	4.6%	89.8%	100.0%			
SCMPPC6	Too much decentralisation of the procurement system	N	0	0	29	34	347	410	4.78	0.562	3
		%	0.0%	0.0%	7.1%	8.3%	84.6%	100.0%			
SCMPPC3	Inadequate planning and linking of demand to budget	N	0	0	0	65	331	410	4.77	0.494	4
		%	0.0%	0.0%	0.0%	15.9%	80.7%	100.0%			
SCMPPC2	Non-compliance with SCM policy and regulations	N	0	1	18	75	316	410	4.72	0.551	5
		%	0.0%	0.2%	4.4%	18.3%	77.1%	100.0%			
SCMPPC1	Lack of proper knowledge, skills and capacity	N	0	0	22	77	311	410	4.7	0.562	6
		%	0.0%	0.0%	5.4%	18.8%	75.9%	100.0%			

Table 5 IDMS and FIDPM

Item No.	Description		1	2	3	4	5	Total	Mean	Std. Dev.
			Very poor	Poor	Average	Good	Excellent			
ISF1	National Treasury's IDMS as a model for best practice delivery of infrastructure management within the public sector	N	0	0	14	23	373	410	4.79	0.698
		%	0.0%	0.0%	3.4%	5.6%	91.0%	100.0%		
ISF2	National Treasury Instruction No. 3 of 2019/2020 which repealed the SIPDM effective from 2019 and gave legal effect to the FIDPM	N	0	0	14	23	373	410	4.79	0.698
		%	0.0%	0.0%	3.4%	5.6%	91.0%	100.0%		

Table 5 highlighted that 91% of respondents had an excellent understanding of items ISF1 and ISF2 which shared an equal mean value of 4.79. this indicates that PrCPMs have a good knowledge base, awareness, and application of the IDMS and FIDPM further suggesting, experienced project managers through their key role on projects are in a key position to assist government improve service delivery, through effective implementation of policy objectives [3, 24].

5. *Government Ability to Apply the FIDPM (Principles and Benefits)*

The final question highlighted PrCPMs experiences on public sector projects with governments ability to apply infrastructure delivery management tools offered by the IDMS through the implementation of the FIDPM. Respondents rated government's ability on a scale of 1 (far below standards) to 5 (far above standards). The mean values were ranked from highest to lowest separately for IDMS principles, presented in Table 6 and benefits presented in Table 7, to allow for a more structured discussion of the data responses.

6. *Governments Ability to apply FIDPM Principles*

Item GAAFP5 *"continuous improvement and scalability (structured to assist all)"* was rated by 55.6% of respondents as below standards and far below standards by a further 42.9% as it ranked the highest with a mean value of 1.78. While 49.3% of respondents rated item GAAFP4 *"evidence-based decision making"* and 75.6% of respondents rated GAAFP6 *"continuous capacity and capability building"* as below standards (m = 1.59).

This indicated that respondents found governments ability below standard to apply the IDMS principles through the FIDPM. Highlighting, government does not display continuous improvement and measures for evaluation through scalability while decision making, continuous capacity building was non-evident. Further reinforcing the discussion under Effective Policy Implementation, that government needs to develop and build capacity towards ensuring service delivery goals are achieved through the implementation of the FIDPM.

Governments Ability to apply FIDPM Benefits

Table 7 illustrated 303 (73.9%) respondents rated item GAAFB11 *"senior management will have a tool to hold delivery managers accountable for performance"* below standards with the highest with mean value of 1.85.

While 58.5% of respondents rated item GAAFB6 "*improved management of the procurement of supplies, services and engineering and construction works*" below standards (m = 1.66). Interestingly, only 3.7% of respondents felt item GAAFB6 met the standards. This indicated respondents found governments ability below standards when applying the IDMS benefits through the FIDPM. Highlighting, government are not accountable for senior management and delivery management as they do not have tools in place to measure performance of accountability while item GAAFB6 further emphasised the need for improved management in government.

Table 6 Government ability to apply the FIDPM (Principles)

Item No.	Description		1	2	3	4	5	Total	Mean	Std. Dev.	Rank
			Far below standards	Below standards	Meet standards	Above standards	Far above standards				
GAAFP5	Continuous improvement	N	176	228	6	0	0	410	1.78	0.54	1
		%	42.9%	55.6%	1.5%	0.0%	0.0%	100.0%			
GAAFP2	Planning alignment	N	127	252	25	0	0	410	1.74	0.563	2
		%	32.4%	61.5%	6.1%	0.0%	0.0%	100.0%			
GAAFP1	Policy consistency	N	127	275	6	2	0	410	1.72	0.539	3
		%	31.0%	97.1%	1.5%	0.5%	0.0%	100.0%			
GAAFP3	Clarity of responsibility & accountability	N	184	197	29	0	0	410	1.62	0.614	4
		%	44.9%	48.0%	7.1%	0.0%	0.0%	100.0%			
GAAFP4	Evidence-based decision making	N	189	202	19	0	0	410	1.59	0.58	5
		%	46.1%	49.3%	4.6%	0.0%	0.0%	100.0%			
GAAFP6	Continuous capacity building	N	90	310	10	0	0	410	1.59	0.522	5
		%	22.0%	75.6%	2.4%	0.0%	0.0%	100.0%			

Table 7 Government ability to apply the FIDPM (Benefits)

Item No.	Description		1	2	3	4	5	Total	Mean	Std. Dev.	Rank
			Far below standards	Below standards	Meet standards	Above standards	Far above standards				
GAAFB11	Senior management will have a tool to hold delivery managers accountable	N	85	303	22	0	0	410	1.85	0.488	1
		%	20.7%	73.9%	5.4%	0.0%	0.0%	100.0%			
GAAFB3	Consistent, reliable reporting, on the full life cycle costs of infrastructure	N	88	302	20	0	0	410	1.83	0.486	2
		%	21.5%	73.7%	4.9%	0.0%	0.0%	100.0%			
GAAFB4	Development of programme plans, using simple templates	N	100	282	28	0	0	410	1.82	0.531	3
		%	24.4%	68.8%	6.8%	0.0%	0.0%	100.0%			
GAAFB1	Understanding of delivery and procurement management processes	N	90	310	10	0	0	410	1.8	0.454	4
		%	22.0%	75.6%	2.4%	0.0%	0.0%	100.0%			
GAAFB9	Greater level of certainty achieved on how projects	N	107	282	21	0	0	410	1.79	0.519	5
		%	26.1%	68.8%	5.1%	0.0%	0.0%	100.0%			

(continued)

Table 7 (continued)

Item No.	Description		1	2	3	4	5	Total	Mean	Std. Dev.	Rank
			Far below standards	Below standards	Meet standards	Above standards	Far above standards				
GAAFB12	The quality and value for money of service delivery will improve	N	125	264	21	0	0	410	1.75	0.541	6
		%	30.5%	64.4%	5.1%	0.0%	0.0%	100.0%			
GAAFB10	A structured environment for inexperienced delivery managers to thrive	N	100	309	1	0	0	410	1.76	0.434	7
		%	24.4%	75.4%	0.2%	0.0%	0.0%	100.0%			
GAAFB5	Alignment of budgets to infrastructure programmes	N	109	294	7	0	0	410	1.75	0.471	8
		%	26.6%	71.7%	1.7%	0.0%	0.0%	100.0%			

(continued)

Table 7 (continued)

Item No.	Description		1	2	3	4	5	Total	Mean	Std. Dev.	Rank
			Far below standards	Below standards	Meet standards	Above standards	Far above standards				
GAAFB8	Greater level of uniformity of implementation, across different organs of state	N	108	299	3	0	0	410	1.74	0.453	9
		%	26.3%	72.9%	0.7%	0.0%	0.0%	100.0%			
GAAFB7	Improved oversight of the implementation of projects	N	156	225	29	0	0	410	1.69	0.597	10
		%	38.0%	54.9%	7.1%	0.0%	0.0%	100.0%			
GAAFB2	Improved control of infrastructure delivery	N	145	146	19	0	0	410	1.69	0.553	10
		%	35.4%	60.0%	4.6%	0.0%	0.0%	100.0%			
GAAFB6	Improved management of the procurement of supplies, services	N	155	240	15	0	0	410	1.66	0.547	11
		%	37.8%	58.5%	3.7%	0.0%	0.0%	100.0%			

This supports the discussion under SCM Public Procurement Challenges that government is not effectively managing public procurement, directly impacting the effective implementation of the FIDPM towards improved service delivery [3]. The literature highlighted IMDS principles and benefits to assist in successful and robust policy implementation [20]. However, the results of the data sets indicated, government is not effectively implementing the principles and benefits the IDMS offers through the implementation of the FIDPM.

7 Conclusions and Recommendations

The South African government has developed key policies and frameworks for addressing service delivery and its challenges since its democracy. Policy development and implementation are not addressing targeted areas of concern with public service delivery as government struggles with successful policy implementation. Government's public service standards were not in alignment with effective policy implementation. In addition, SCM public procurement challenges were still present. PrCPMs were aware of the IDMS and FIDPM and understood its application suggesting that they are in an opportune position to assist government improve policy implementation. Finally, government's lacks the ability to apply IDMS principles and benefits effectively through the implementation of the FIDPM towards improving service delivery.

This study indicated, through the state of current policy implementation, government is not ready. They require assistance in effective policy implementation to achieve public sector expenditure and infrastructure development. Project management should therefore be the focal point for policy makers in that they should think through the end-to-end process to translate a particular policy into delivery plans and delivery plans into desired outcomes. The involvement of construction project managers throughout the process can strengthen the deliverability of the outcomes and commitment to the policy objectives [29]. Emphasising that construction projects managers are in a key position to aid government with the implementation of the FIDPM. Further studies are to be conducted on the role of the construction project manager on public sector projects and whether they have the tools and techniques to assist government improve public service delivery.

References

1. Toor SR, Ogunlana S, Ofori G (2012) Leadership and its development in construction in developing countries. Spon Press, Abingdon, UK, pp 242–268
2. Khumalo M, Choga I, Munapo E (2017) Challenges associated with infrastructure delivery. Public Municipal Fin J 6(2):35–45
3. FIDPM (2019) Framework for infrastructure delivery and procurement management. National Treasury, October 2019

4. Janse van Rensburg JO (2014) Internal audit capability: a public sector case study. Unpublished M. Com. Dissertation. University of Pretoria, Pretoria
5. Akinboade OA, Mokwena MP, Kinfack EC (2013) Understanding citizens' participation in service delivery protests in South Africa's Sedibeng district municipality. Int J Soc Econ 40(5):458–478. https://doi.org/10.1108/03068291311315331
6. Department of Public Service and Administration (DPSA) (2018) Annual Report 2017/ 2018, Batho Pele House, Pretoria
7. Schwella E (2001) Public sector policy in the New South Africa: a critical review. Public Performance Manage Rev 24(4)
8. The South African Constitution (Act No. 108 of 1996) Constitution of the Republic of South Africa, 10 December 1996. Available at: https://www.refworld.org/docid/3ae6b5de4.html. 24 June 2019
9. White Paper on Transforming Public Service Delivery (1999) Department of Public Service and Administration, September 1997, The Republic of South Africa
10. Public Service Act (1994) Proclamation 103 published in GG 15791 of 03rd June 1994
11. Arrowsmith S (2010) Public procurement: basic concepts and the coverage of procurement rules. In: Public procurement relations: an introduction. EU Asia InterUniversity Network
12. Eyaa S, Oluka PN (2011) Explaining noncompliance in public procurement in Uganda. Int J Bus Soc Sci 2(11)
13. Bolton P (2006) Government procurement as a policy tool in South Africa. J Public Procurement 6(3):193–217
14. National Treasury (2016) Supply chain management review update 2016. National Treasury, Republic of South Africa
15. Davey J, Gatenby A (2017) The government procurement law review. The Law Reviews, Edition 5, May 2017
16. National Treasury (2003) Policy strategy to guide uniformity in procurement reforms processes in government. Republic of South Africa
17. Hanks J, Davies H, Perera O (2008) Sustainable public procurement in South Africa. Online Available: www.iisd.org. Accessed 30 May 2019
18. Ambe IM, Badenhorst-weiss JA (2012) Procurement challenges in the South African Public sector. J Transp Supply Chain Manage 242–261:2012
19. Infrastructure Delivery Management System (IDMS) (2014) Provincial infrastructure delivery management system framework contract, infrastructure framework for the health and education sectors, 2–4 June 2014 Revision
20. Infrastructure Delivery Management System (IDMS) (2019) Pilot roll out in selected municipalities through stakeholder engagement and training, training module A0: Executive Overview, Participants Module, Municipal Infrastructure Support Agent, March 2019
21. National Treasury (2019) National treasury instruction No. 03 of 2019/2020, framework for infrastructure delivery and procurement management. Republic of South Africa, 2019
22. Watermeyer R (2018) Client guide for improving infrastructure project outcomes, 1st ed. Published by the School of Construction Economics and Management, University of Witwaterstrand, Johannesburg, April 2018
23. SACPCMP (Act No. 48 of 2000) South African Council for the project and construction management professions, Guideline Scope of Services and Recommended Guideline Tariff of Fees, Board Notice 168 of 2019, September 2019
24. Knipe A, Van der Walt G, Van Niekerk D, Burger D, Nell K (2016) Project management for success. Heinemann Publishers (Pty) Ltd, South Africa
25. Pollitt C, Bouckaert G (2000) Public management reform. Oxford University Press, Oxford
26. Srivannaboon S, Milosevic DZ (2006) A two-way influence between business strategy and project management. Int J Project Manage 24(6):493–505
27. Saunders M, Lewis P, Thornhill A (2019) Research for business students, 8th ed. Pearson Education Limited
28. Collis J, Hussey R (2014) Business research: a practical guide for undergraduate & post graduate students, 4th edn. Palgrave Macmillan

29. Emuze and Smallwood (2013) Management concepts and project performance: perceptions from the South African public sector environment. J South Afr Inst Civil Eng 55
30. Public Service Commission (PSC) (2005) Evaluation of service standards in the public service, Pretoria
31. Cleland DI, Ireland LR (2002) Project management. Library of Congress Cataloguing in Publishing Data, New York

A Comparative Analysis Between the Tender Estimate and the Actual Completion Cost of High Voltage Fluid Filled Cable Decommissioning

Geraldine Smith, David J. Edwards, and John J. Posillico

Abstract This research investigates the relationship between a tender-stage cost estimate and the actual cost incurred to complete decommissioning of aging high voltage power cables connected to the transmission network. Such work is essential to preserve health and safety (H&S) of on-going power generation activities. The research adopts a mixed postpositivist and interpretivist epistemological lens to analyse longitudinal cost data accrued via a case study of an underground cable decommissioning project. Quantitative cost data and qualitative feedback from project stakeholders was analysed using summary statistical analysis and content analysis. The data from the comparative analysis then informed the basis of an interview with industry professionals and practitioners with a view to identifying the key drivers of change between tender estimate and actual cost. The interview content was analysed and key themes were extracted. Findings from the case study suggest that the niche nature of fluid-filled cable decommissioning, combined with the diminishing number of experienced practitioners in the field, contributed to the variances seen within the cost data. The interviews conducted indicate that the three key drivers for cost variance between tender and actual cost are: lack of decision around methodology at tender stage, changes in method of delivery and changes to programme and planned durations. The data suggests that many of the variances could have been foreseen and considered within risk allowances or contingencies. This research is the first published work in this area. The content provides rare insight into the electricity transmission sector and specifically the decommissioning of high voltage fluid filled cables and the cost variances that occur between the estimated and actual costs incurred.

Keywords Actual cost · Cable decommissioning · Cost uncertainty · Fluid filled cable · Tender cost · Tender estimate · Transmission voltage

G. Smith · D. J. Edwards (✉) · J. J. Posillico
Department of the Built Environment, Birmingham City University, Birmingham, UK
e-mail: drdavidedwards@aol.com

T. C. Haupt et al. (eds.), *Construction in 5D: Deconstruction, Digitalization, Disruption, Disaster, Development*, Lecture Notes in Civil Engineering 245,
https://doi.org/10.1007/978-3-030-97748-1_36

1 Introduction

The electricity transmission network of the United Kingdom is owned and maintained by National Grid Plc [34]. This was the first grid-based electricity system in the world, after its formation in the 1930s. Gale et al. [19] states that within the UK electricity transmission system, voltages that form the National Grid are 275 and 400 kV. Lower voltage networks are owned and operated by Distribution Network Operators (DNOs) separately to National Grid. During the post-Second World War economic and industrial period, supply for electricity in the UK could no longer meet demand, and so work began to install a 275 kV 'supergrid' comprising of both overhead power lines, and underground cables [46]. Underground cables are generally only used where it is not feasible to install overhead lines [37]. Overhead lines are preferable to underground cables as a method for electricity transmission due to the lower cost to supply and install tower steelwork and conductor wire [29]. The most common areas where underground cables are used are in built-up areas and town/city centres where there is no space to allow adequate clearance around the live electricity-carrying conductor, and planning permission or wayleaves will not be granted, as explored by Ituen [26]. More recently, work has begun to remove overhead lines in National Parks and Areas of Outstanding Natural Beauty (AONBs) and replace them with modern underground cables as a visual benefit; this work is being carried out in spite of the high costs involved [7]. It is therefore accepted that there is a higher cost involved, and greater uncertainty when installing below-ground infrastructure, which was identified during a study by Navrud et al. [38]. Most cable systems installed during the construction of the supergrid (circa 1950s and 1960s) used oil-cooled, paper-insulated cables [43]. These cable systems were originally installed with an intended design life of 60 years [9]. Ghafurian et al. [21] confirms that from the early 1990s onwards, many of these assets began to age past this point and were increasingly subject to electrical faults and oil leakages; such faults can lead to loss of electricity supply, create environmental contamination and/or pose a health and safety (H&S) risk. Loss of supply is a major risk to the National Grid, resulting in heavy fines from the Office of Gas and Electricity Markets (OFGEM) and reputational damage [3]. Therefore, the National Grid sought to procure specialist contractors to carry out the decommissioning of these assets, usually combined with the installation of new cable circuits [6]. The first oil cables were decommissioned in the early 1990s [56] and work continues to the present day. Oberlander [41] claims that within the transmission industry that there can be a large discrepancy between initial cost estimates for decommissioning work and the final cost expended. At tender stage, costs must be estimated based on subjective expectations of labour, plant and material requirements [22], this is commonly referred to as a 'bottom up' or 'first principles' approach. While it is possible to measure certain elements of the cost associated with cable decommissioning, it is difficult to reach a high degree of certainty while the cable remains in the ground [47]. Consequently, there is inevitably a level of unknown quantum to account for at the pre-contract stage [54]. For example, at tender stage it is possible to measure how much oil is in the system by using as-built records and draughtsman drawings, but it is difficult to predict how much oil will drain during the system purge and at what rate (cf. [39]). This unknown element affects the ability of the estimator to build up the price accurately. This subsequently causes increases in programme days, material costs and oil disposal volumes [5]. Traditionally, oil cable decommissioning would be included as a Provisional Sum within estimates, however as the New Engineering Contract (NEC)3 and NEC4 have been adopted, the use of Provisional Sums has been used less frequently and is not generally acceptable to energy sector clients [49].

Given the aforementioned contextual background, this research aims to conduct a comparative analysis between the 'bottom up' tender estimate and actual cost of completed work with a view to improve methods for calculating the costs of decommissioning high voltage fluid filled cables—and in doing, ensure H&S costs are adequately covered. Concomitant objectives are to: identify the degree of variance between both methods of estimate calculation and determine the reasons for any observable differences; accrue information and guidance on how such cost variance events could be mitigated in future cost estimates prepared; and in achieving the foregoing, ensure that adversarial relations due to cost increases are proactively avoided in future contracts undertaken by providing clients with more reliable estimates.

2 Cost Accuracy of Cable Decommissioning Projects

While there have been a small number of studies which investigate the commercial and financial aspects of electricity transmission, there is very little exploration of the challenges faced during the decommissioning of oil filled cables from a risk and cost perspective [27] Too low of a figure can leave the Contractor exposed to loss of profit (and lead to cutting corners on issues such as H&S) and negatively affect the relationship with the Client. Too high a figure can affect the competitiveness of the tender and prevent the Contractor from successfully winning the work, and even affect their ability to bid for this type of work in the future [28]. There is currently more than 1,200km of high-voltage fluid filled cable in service in the UK, with no single widely practiced methodology in place for their decommissioning [15]. Instead, methods are determined on a project-by-project basis as these circuits are taken out of service or replaced. Various methods exist, with research continuing to determine the most effective materials and practices [20]. The two key factors which must be considered to ascertain a successful method for decommissioning are the: (1) amount of oil which can be removed from the circuit (currently between 80 and 90% of oil can be removed [36], and 2 risk of environmental damage and/or H&S risks caused by the activity. Removal of oil from high-voltage cable circuits and damage (or material deterioration) to the cables represents a significant threat to the environment; particularly watercourses or aquifers where workers must follow strict H&S protocols during the works. Several high-profile incidents underscore the risks posed, for example, in 2019 the Environmental Protection Agency (Ireland) investigation into the leakage of 40,000 litres per year of hazardous oil spanning over two decades which went unreported [25, 33]. Redundant cables are often disconnected and left connected to drain pits for future maintenance. However, due to the risk of oil leakage, these cables must either be purged of oil or removed in their entirety [8]. There is consensus among other studies (cf. [36]) that removal of oil from cables is key to reducing the incidence of environmental contamination and the associated costs of remediation and maintenance—that said, the environmental risk arising from cable decommissioning represents a 'known unknown' when preparing tender costs [24].

3 Choice of Decommissioning Methodology

It is difficult to challenge and improve on methodology for oil purging, due to the electricity transmission system's reliance on live circuits remaining operational at all times—which in itself posed a significant H&S risk [55]. This means there are limited opportunities for testing new techniques, as circuit outages (necessary for trial-runs) are only used in essential circumstances and must be booked up to five years in advance. Due to the heavy fines levied by OFGEM for loss of service, or by the Environment Agency (EA) for environmental issues, transmission operators take a risk-averse approach to new techniques which further impacts the opportunity for innovation [10]. Various methods exist for the removal of oil from high-voltage cables, with no single method being generally accepted as superior to the others (cf. [36]). Two methods investigated by Caimi et al. [8] are the in-situ polymerisation of the inner channel, followed by use of water to displace the fluid in the impregnated papers. This was compared to the method of using only water and cycles of pressurisation/depressurisation to force the oil from the papers. Figure 1 shows a typical

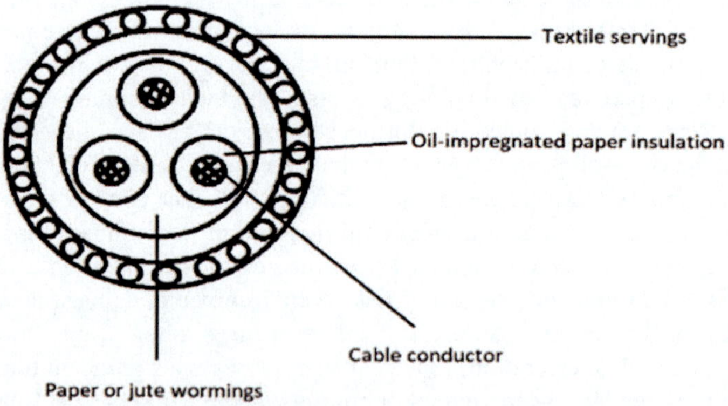

Fig. 1 Typical cross section

cross section of a three-conductor oil impregnated power cable. The conclusion of Le Poidevin et al. [36] concedes that an effective and easily applicable method for in-situ oil removal is still missing from the industry. Nitrogen (a benign gas) is used to pressurise the cables, which are then flushed with detergent (it identifies detergent as superior to water in its ability to soak the papers and displace the oil). White spot Nitrogen is to be used rather than plain compressed air due to the risk of explosion which can occur due to the 'diesel effect' caused by airborne oil droplets [31].

The H&S and logistical difficulty in arranging preventative maintenance work on these cable circuits is discussed by Langdon [34] who explains the limitations in opportunity for research and development relating to techniques for decommissioning. However, there has been relatively little change in the composition of electricity cables since the 1900s. Therefore, it could be argued that there has been ample opportunity to refine methodology and the requirement for decommissioning could be foreseen and planned for at the point of manufacture. The limitations and rarity of studies conducted to identify optimal methodology for cable decommissioning lends credence to the idea that cable decommissioning can be considered a niche activity and will therefore present difficulties to a tender estimator. The key factor contributing to this difficulty is lack of foresight of the methodology which will be used once the project reaches site, along with the relevant materials and equipment remaining unconfirmed at tender stage.

3.1 Anticipating Decommissioning Cost/Time

To estimate the cost of an activity, it is important to understand the time the activity will take to complete [2], and the resources which will be required [44]. The volume of oil which is expected to be purged from the circuit correlates directly with the amount of time the overall activity will take. Le Poidevin (2010) states that: "the greater the volume of oil to be removed, the greater the time required to carry out the process." While the study by Caimi et al. [8] further concludes that higher accuracy of oil removal can be estimated by repeated cycles of purging. Another area which impacts the time (and therefore the cost of the decommissioning), is the number of repeat cycles of purging which will be required. Repeat cycles are required due to the oil impregnated papers retaining oil, which will take time to seep out once the initial oil purge has been completed. The rate at which the oil will seep, and the volume which can be purged at any given time, is the area which cannot be accurately predicted and is dependent on temperature and gradients. The process must be repeated until the purge produces no more oil [20]. In direct conflict with the work of Caimi et al. [8], a study by EA Technology states that the concentration of oil removed is easily determined by allowing volumetric measurement of the settled oil/water mixture which is a result of the purge. However, this can only be done post-activity and not at tender stage and so this would have to be measured and recorded over time in order to build up a library of expected volumes [16]. While both the Caimi et al. [8] and Le Poidevin et al. (2010) studies identify the need for the circuits to be subject to a period of 'standing time' between flushes, neither study identifies an optimal time period and thus, there is another reason for difficulty in applying timescales to estimates.

4 Limitations of Tender Cost Estimates

There are many studies and pieces of research in the electricity transmission field, however they are heavily focused on methodology and innovation as the industry moves toward net zero carbon [45]. While some of this research is geared towards reducing cost to domestic electricity consumers, there is a gap in research to investigate the cost of electricity transmission projects themselves.

4.1 Approach to Risk in Tenders

Risk in tenders (both financial and H&S) can be split into the three areas viz. known knowns, know unknowns and unknown unknowns. The greatest threat to the alignment between an estimate and an actual cost is the approach to managing risks and unknowns. The specific approach to risk is guided by the form of contract used when the project is procured by the Client [32]. The suite of contracts most commonly used by clients in the transmission industry is the New Engineering Contract (NEC). This type of contract has a fundamental aim to encourage a collaborative approach to risk allocation and risk management with a view to avoiding disputes [35]. At the tender stage, risks may be identified by either the Client or the Tenderer—usually both [12]. A risk register prepared will contain references to specific risks which are likely to be encountered based on the scope of the project. Therefore, on a cable decommissioning project, the risk register is likely to contain an item, or multiple items, identifying potential unknowns that may occur when the cables are purged.

4.2 Drivers for Variance

Tumalla and Burchett (1999) states that: "extra high voltage (EHV) transmission estimates may be accurate to a degree of only 70–80%" and that one of the key drivers for this is that project plans are subject to much change. Variances from tender estimates naturally only occur when a project becomes live and works on site begin. Therefore, it is important that the tender team have a thorough and realistic understanding of potential changes that could occur. This may be done by having strong input from operational staff and project engineers during the tender and by gathering data from historical projects to understand how these activities work in practice. Table 1 illustrates some of the key drivers for variance.

Table 1 Key drivers for variance

Driver of variance	Description	Citation
Programme creep	Insufficient allowance has been made in the tender programme for the activities required in order to decommission the cable	[1, 11]
Scope creep	Additional scope over and above what has been allowed for is required. This could be at the request of the Client, or due to tender error	[30,48]
Physical conditions	Unforeseen physical conditions are encountered on site, for example oil leaks not previously identified	[23]
Asset condition	The asset (cable) which is being decommissioned is in worse condition than anticipated which required a change to methodology	[53]

4.3 Lack of Shared Knowledge and Experience

There is a need for knowledge-based systems to improve the quality of costs associated with EHV transmission projects [4]. However, due to the scarcity of real data, many estimates fall back on the experience of field experts. Reily [47] makes the statement that: "the public is sceptical of our ability, as a profession, to accurately estimate the final costs of large, complex public projects" and uses various case studies to support the use of 'ranges of possible cost' as opposed to fixed estimates. However, providing ranges of costs would not be compatible with the method of contracting used by clients in the UK transmission sector, who often seek price certainty due to the publicly funded nature of OFGEM regulated work. The consensus among studies is that there is a difficulty in obtaining hard data that is specific and relevant to complex works.

5 Methodology

The epistemological lens through which this research is focused, is a mixed philosophical approach [14, 40, 51] to conduct abductive reasoning [18] via a case study strategy [42]. The project case study consists of fluid filled cable circuit replacement which is currently installed in an urban area in the North-West of England. The scheme involves the decommissioning of the existing 1960s fluid filled cables and subsequent replacement with modern cross-linked polyethylene (XLPE) 275 kV cables. The overall value of the project is circa £11m (UK Sterling), and the project duration runs from November 2019 to May 2021—hence, a live project is utilized. The decommissioning of cables occurs during January 2021 and is anticipated to last for two weeks. This project is being completed by a large UK-based civil engineering tier one contractor with a turnover of £3bn and circa 7000 staff. The decommissioning aspect of the project will be broken down into the following seven activities for the purposes of the analysis: (1) white-spot nitrogen purging of the circuit—initial purge;

(2) white-spot nitrogen purging of the circuit—repeat visits; (3) removal and disposal (scrappage) of 275 kV cable joints; (4) removal and disposal (scrappage) of 275 kV cable terminations; (5) installation of fluid bridges through hydraulic sections; (6) installation of oil-collection drain pits; and (7) disposal of cable fluid.

Data was collated over two phases in order to identify and elucidate key drivers for cost variance relating to cable decommissioning projects, with a view to improving estimating techniques and following a strict ethical process (cf. [17]).

5.1 Phase One—Cost Data

In phase one, elemental cost analysis was conducted in order to provide a basis from which to identify the variances in the case study project. Historical secondary data was obtained from internal 'quantitative' unit costs for labour and plant hire, evidence of subcontract tender quotations (for activities which have been subcontracted) and material order estimates. This data will not be received in the same format as the pre-contract historical estimate data, so work will be done to normalise the two data sets. The variances identified offer a clearer understanding of the specific areas where change occurs, which will act as a driver for the interview questions in phase 2.

5.2 Phase Two—Semi-structured Interview

In phase two, as well as post cost data analysis, a semi-structured interview was used as the data collection instrument to gather qualitative feedback from members of the site team, cable system designers, key subcontractors, and the client to identify the reasons for cost variances occurring—refer to Table 2 for interview questions and Table 3 for the demographic profile of respondents. The overall sample from which participants were selected comprised the team working to deliver the project; where each participant had at least 10 years' relevant industry experience and were selected using purposive sampling.

6 Analysis

The cost comparison shows a difference of just under 18.5% between the tender cost of £76,564.72 and the actual cost to complete of £92,188.09. This is a high-level figure which indicates an increase in the overall cost of the decommissioning works, however on an activity-by-activity basis, there were increases and decreases against the tendered costs (please refer to Appendix 1 for detailed comparative analysis and Table 4 for high level breakdown).

Table 2 Demographic profile of respondents

Question No.	Question	Prompt
Q1	Have you ever previously been involved with a project involving the purging of oil cable?	Type of project? Previous experience?
Q2	Do you think the correct level of plant and labour resource required to carry out the decommissioning could have been foreseen at tender?	Additional requirement? Unknown factors?
Q3	What do you see as the elements of the cable purge that could have been predicted accurately?	Labour? Plant/Equipment?
Q4	How many subcontractors are you aware of with the capability to purge oil cable?	Limited numbers? Reasons for this?
Q5	What is your understanding of the reasons behind the change in methodology between tender and delivery?	Who drove the change? Was it necessary?
Q6	Do you think this method could have been dictated at tender stage?	Blockers to this?
Q7	Is there a disconnect between the team who dictate the scope at tender, and the team who deliver the work?	Skill sets? Experience?

The initial white-spot Nitrogen purge shows a large increase against the tender price; however, this is due to the breakdown of actual costs being given at a high-level by the Subcontractor. It was determined that the actual cost in this section includes the cost for installation of the fluid bridges which also accounts for the large reduction in that section. There was a change in scope from the Client prior to the decommissioning work taking place, which involved some of the cable joints being left in-situ, therefore the cost relating to removal and disposal of the joints shows a decrease. In practice, the method of delivering the work was changed by the Contractor (from delivery using directly employed resource, plant, and materials, to using a Subcontractor). This placed more of the Scope of work with the specialist Subcontractor than originally accounted for at tender. This change had two implications; firstly, there was less detailed cost information available for analysis and secondly, there was an additional requirement to provide Supervision to the Subcontractor's work, which carried additional costs. The result of this is an overall increase in the cost of the activity which is reflected in the analysis. There was a saving on material costs which was a result of certain elements from the excavated joint bays being reclaimed during the decommissioning process and reused elsewhere. This was an innovation suggested by Participant I. The option to reuse the materials was not anticipated at tender stage and would represent a 'known unknown' when compiling a tender price,

Table 3 Demographic profile of respondents

Participant reference	Job description	Years of industry experience	Years of experience in the current role	Qualifications/Accreditations
A	Operations Director	20	11	B.Sc., M.Sc., IOD
B	Senior Project Manager	15	15	HNC, IOSH
C	Site Manager	10	10	ILM, SMSTS
D	Project Quantity Surveyor	15	10	B.Sc., RICS
E	Senior Engineering Manager	30	22	HND, MIET
F	Principal Engineer	15	13	B.Sc., M.Sc.
G	Cable Purging Subcontractor Representative	30	20	HNC, IOSH
H	Client Representative–Technical Assurance	10	21	B.Sc., MIET
I	Lead Jointer	20	15	HNC, IOSH

Table 4 Cost variances

Activity	Tender price	Actual cost	Variance (%)
White-spot Nitrogen purge—initial	£8.064.96	£36,904.48	357.59
White-spot Nitrogen purge—revisits	£3,456.41	£7,985.52	128.43
Removal and disposal of joints	£16,550.04	£9,640.00	−41.75
Removal and disposal of terminations	£2,800.78	£6,109.50	118.14
Installation of fluid bridges	£16,788.44	£3,800.00	−77.37
Installation of oil collection pits	£16,703.74	£19,042.27	14.00
Disposal of cable fluid	£12,200.35	£8,796.32	−27.90

as the condition of the materials for re-use cannot be determined until after the items have been recovered from the ground during delivery.

7 Discussion

As expected, the findings of the analysis show that there were variances across all measured areas of the comparison. Changes in timeline, duration, delivery methodology and scope all had an impact on the actual cost of the oil decommissioning. There were no major anomalies or unplanned events that would constitute aforementioned 'unknown unknowns'. This would suggest that the variances could have been accounted for and contingencies or allowances made at tender stage if there was an experienced practitioner feeding information to the Estimator and Planner. Interviews with participants helped to provide additional clarification in this discussion.

There was an issue on the project surrounding discrepancies between the planned decommissioning works at tender stage and the client requirement on site. Initially during tender, the Main Contractor was asked to provide a price for fully comprehensive decommissioning of the cables. Subsequently, this scope was reduced upon the Client asking for Best and Final Offers (BAFO) and therefore the cost reduced in line with the reduced requirement of the Client. However, it was decided that the time duration in the submitted programme of works would remain at the previous allowance (taking into account full decommissioning), in order for the more extensive scope of works to be carried out should the Client wish to increase the scope again once the project was proceeding. In practice, due to changing of team members over time (both the Client and the Main Contractor), the time duration remained in the programme throughout the project, rather than being removed once it was confirmed that the reduced scope exercise was to be carried out. Consequently, the Client was of the impression that the activities shown in the programme had an associated cost within the tendered price. Once the reduced scope was carried out on site, the Client believed they were entitled to a reduction in cost relating to the erroneous scope items, however no such cost formed part of the tendered price. This led to a dispute between the Client and the Main Contractor and necessitated the need for the tender documents to be re-examined and interpreted by the team. This is a subjective exercise and led to professional tension between the two parties. The matter was eventually settled in favour of the Main Contractor without need for formal intervention. Discrepancies were also experienced. Even at a stage as late as the subcontract order being signed between the Main Contractor and the specialist Subcontractor, a hand-written caveat was added by the commercial representative from the Subcontractor. This demonstrates how a lack of predictability is recognised and has no formal or pre-empted solution that was defined and agreed before the two parties entered into a contract. It also further demonstrates the unknown element of cable decommissioning, which was demonstrated to be poorly understood by both commercial and planning functions during the delivery of the project.

8 Applications of the Research

The practical implications of this work are threefold; (1) a significant variance between tender cost and actual cost to deliver is confirmed—which may impact upon H&S arrangements; (2) the gap between experienced operational staff and tendering or commercial staff is identified—again impacting upon H&S management; and (3) a basis is set for the compilation of data for future use; allowing implementation of best practice in tendering. This research sets a precedent for discussions surrounding appropriate levels of input at tender stage by experienced professionals when tendering for projects containing specialist activities that pose a significant environmental and H&S risk. It demonstrates that there can still be a significant variance between tender cost and actual cost even when no major anomalous events occur. The information can be shared with tender teams who are inexperienced in fluid-filled cable decommissioning to give a guide to potential areas needing to be considered in contingency and risk allowances. This is the first work of its kind, investigating the commercial element of fluid-filled cable decommissioning and how finance (or a shortage of it) may impact upon H&S on site. Future work is however required to: expand the research to cover a wider number of case studies across the UK and internationally—so a more incisive inference can be drawn; and better tie in how costs deviations and associated contractual disputes may infringe upon the prevailing safety culture on site.

9 Conclusions

It can be concluded that the key drivers for cost variance are lack of decision around methodology at tender stage, changes in method of delivery (direct labour or subcontract) and changes to programme and duration. To improve cost certainty and lower H&S risk around cable decommissioning, there must be an improvement in the level of knowledge and experience across different functions, such as Estimating, Planning, Design and Procurement. Currently this knowledge and experience tends to sit only within the functions responsible for the on-site delivery of this type of work (whether this is operational staff for a Main Contractor, or staff of a specialist Subcontractor). The electricity transmission industry is insular and consists of a very experienced but ageing workforce who tend to have spent their careers moving between a small number of specialist Main Contractors. As such, there is an over- dependence on a small number of experienced individuals to define the scope and methodology for the work. This severely limits the ability for the cascade of knowledge to other functions or less experienced colleagues. Cable decommissioning is generally carried out during a power system outage and therefore the activity naturally falls into the final stages of a cabling project (it can only commence once a replacement circuit has been installed, commissioned, and placed into service). This may further explain the results of the data analysis, as it can be expected that by the end of a project there will have been staff turnover, changes in the scope of the project and much of the detail behind the originally tendered allowances will have been lost or superseded. Advanced digital solutions (such as those contained within the Industry 4.0 concept (cf. [13]; Sepazgosar et al. 2020; [14]) may provide a potential solutions to some of the problems reported upon—but ultimately, investments in people and knowledge sharing must come first. Without improvements in knowledge-sharing, this lack of understanding will continue to cause issues and ambiguity will remain in predicting costs and durations for this type of work. By improving practitioner involvement (Jointers and specialist Subcontractors) in the production of tender costs and the design of the decommissioning solution, it is possible to improve the techniques used for cost estimating and enable a smoother transition from tender to project delivery.

Appendix A – Cost Data Comparison – Decommissioning of High-Voltage Fluid-Filled Cable

Item	Activity	Tender Price (£)	Actual Cost (£)	Variance (£)	Variance (%)	Narrative
	Total	76,564.72	92,188.09			
1	White-spot nitrogen purging of the circuit - initial purge	8,064.96	36,904,48	28,839.52	357.59	
1.1	Labour	-	-	-	0	
1.2	Plant	-	-	-	0	
1.3	Material	-	-	-	0	
1.4	Subcontract	8,064.96	36,904,48	28,839.52	357.59	Includes bundled costs for installation of fluid cables.
2	White-spot nitrogen purging of the circuit - repeat visits	3,456.41	7,895.52	4.439.11	128.43	
2.1	Labour	-	-	-	0.00	
2.2	Plant	-	-	-	0.00	
2.3	Material	-	-	-	0.00	
2.4	Subcontract	3,456.41	7,895.52	4.439.11	128.43	Two additional purges against one anticipated.
3	Removal and disposal (scrappage) of 275kV cable joints	16,550.04	9,640.00	6,910.04	-41.75	
3.1	Labour	-	3,040.00	3,040.00	0.00	Four jointer days to supervise
3.2	Plant	-	-	-	0.00	
3.3	Material	-	-	-	0.00	
3.4	Subcontract	16,550.04	6,600.00	9,950.04	-60.12	Less joints disposed of than anticipated.
4	Removal and disposal (scrappage) of 275kV cable terminals	2,800.78	6,109.50	3,308.72	118.14	
4.1	Labour	-	1,520.00	1,520.00	0.00	Two jointer days to supervise.
4.2	Plant	-	-	-	0.00	
4.3	Material	-	-	-	0.00	
4.4	Subcontract	2,800.78	4,589.50	1,788.72	63.87	Longer than anticipated duration and temporary works.
5	Installation of fluid bridge	16,788.44	3,800.00	12,988.44	-77.37	
5.1	Labour	11,751.91	3,800.00	7,951.91	-67.66	Five jointer days to supervise.
5.2	Plant	3,357.69	-	3,357.69	-100.00	Cost not possible to separate from main purge.
5.3	Material	1,678.84	-	1,678.84	-100.00	Cost not possible to separate from main purge.
5.4	Subcontract	-	-	-	0.00	
6	Installation of oil-collection pits	16,703.74	19,042.27	2,338.52	14.00	
6.1	Labour	8,351.87	9,521.13	1,169.26	14.00	Minor increase.
6.2	Plant	5,011.12	6,013.35	1,002.22	20.00	Minor increase.
6.3	Material	3,340.75	3,507.79	167.04	5.00	Minor increase.
6.4	Subcontract	-	-	-	0.00	
7	Disposal of cable fluid	12,200.35	8,796.32	3,404.03	-27.90	
7.1	Labour	-	-	-	0	
7.2	Plant	-	-	-	0	
7.3	Material	-	-	-	0	
7.4	Subcontract	12,200.35	8,796.32	3,404.03	-27.90	Les oil recovered than anticipated.

References

1. Ansah R, Sorooshian S, Mustafa SB (2018) The 4Ps: a framework for evaluating construction project delays. J Eng Appl Sci 13(5):1222–1227. https://doi.org/10.36478/jeasci.218.1222.1227
2. Ballesteros-Pérez P, Sanz-Ablanedo E, Cerezo-Narváez A, Lucko G, Pastor-Fernández A, Otero-Mateo M, Contreras-Samper J (2020) Forecasting accuracy of in-progress activity duration and cost estimates. J Constr Eng Manage 146(9). https://doi.org/10.1061/(ASCE)CO.1943-7862.0001900
3. Bialek JW (2005) Lights out? [power generation market force effects] Power Eng 19(4):16–19
4. Bothia G, Henao E, Quintero L (2019) cost estimation of high voltage substation equipment, using artificial intelligence algorithms. In: FISE-IEEE/CIGRE Conference—living the energy transition (FISE/CIGRE). https://doi.org/10.1109/FISECIGRE48012.2019.8984963
5. Bradbury D (1992) Decommissioning of civil nuclear facilities: a world review. Energy Policy 20(8):755–760. https://doi.org/10.1016/0301-4215(92)90036-2
6. Buhari M, Levi V, Kapetanaki A (2018) Cable replacement considering optimal wind integration and network reconfiguration. IEEE Trans Smart Grid 9(6):5752–5763. https://doi.org/10.1109/TSG.2017.2696340
7. Button KS (2018) The environmental history of the national grid the process of electrification: infrastructure and influence (Doctoral thesis). https://doi.org/10.17863/CAM.31575
8. Caimi S, Colombo C, Ferrari R, Storti G, Morbidelli M (2019) Recovery of mineral oil from underground electrical cables. Int J Environ Res Public Health 16(13):2357. https://doi.org/10.3390/ijerph16132357
9. Chmura L, Jin H, Cichecki P, Smit JJ, Gulski E, Vries FD (2012) Use of dissipation factor for life consumption assessment and future life modeling of oil-filled high-voltage power cables. IEEE Electr Insul Maga 28(1):27–37. https://doi.org/10.1109/MEI.2012.6130529
10. Crognier G, Tournebise P, Ruiz M, Panciatici P (2020) Grid operation-based outage maintenance planning. Electr Power Syst Res 190. https://doi.org/10.1016/j.epsr.2020.106682
11. Dhanwate S, Jadhav P, Mane P, Rajput P (2020) Estimation and scheduling of residential building using primavera P6. Int J Adv Sci Technol 29(8)
12. Dziadosz A, Tomczyk A, Kapliński O (2015) Financial risk estimation in construction contracts. Proc Eng 122:120–128. https://doi.org/10.1016/j.proeng.2015.10.015
13. Edwards DJ, Pärn EA, Love PED, El-Gohary H (2017) Machinery, manumission and economic machinations. J Bus Res 70:391–394. https://doi.org/10.1016/j.jbusres.2016.08.012
14. Edwards DJ, Akhtar J, Rillie I, Chileshe N, Lai J, Roberts C, Ejohwomu O (2021) Systematic analysis of driverless technologies. J Eng Des Technol. https://doi.org/10.1108/JEDT-02-2021-0101
15. Energy Networks Association (2015) Environmental decommissioning—fluid-filled cables systems 1–3. Available via: https://www.energynetworks.org/industry-hub/resource-library/environmental-decommissioning-fluid-filled-cables-systems.pdf
16. Farr J, Faber I (2018) Cost estimating techniques. Engineering economics of life cycle cost analysis. CRC Press, pp 243–262. https://doi.org/10.1201/9780429466304-9
17. Fisher L, Edwards DJ, Pärn EA, Aigbavboa CO (2018) Building design for people with dementia: a case study of a UK care home. Facilities 36(7/8):349–368. https://doi.org/10.1108/F-06-2017-0062
18. Gabbay D, Kruse R (2000) Abductive reasoning and learning. Springer Netherlands, Dordrecht. ISBN: 978-90-481-5560-6
19. Gale PF, Stokoe J, Crossley PA (1997) Practical experience with travelling wave fault locators on Scottish Power's 275 & 400 kV transmission system. In: 6th international conference on developments in power systems protection, pp 192–196. https://doi.org/10.1049/cp:19970061
20. Gerdt R, Crowley J, Chato J (1978) Electrohydrodynamic pumping of cable oil. J Electrostat 5:477–488. https://doi.org/10.1016/0304-3886(78)90039-6

21. Ghafurian R, Dominguez J, Santini A, Sobel C (1999) New advances in mitigating environmental impact of pipe-type cables. IEEE Trans Power Delivery 14(2):314–318. https://doi.org/10.1109/61.754068
22. Greenhalgh B (2013) Introduction to estimating for construction. Routledge, London. https://doi.org/10.4324/9780203080061
23. Haiderali A (2020) Mitigation of ancient coal mining hazards to overhead line equipment structures. Proc Inst Civil Eng- Transp 173(4):218–231. https://doi.org/10.1680/jtran.18.00143
24. Hoseini E, Bosch-Rekveldt M, Hertogh M (2020) Cost contingency and cost evolvement of construction projects in the preconstruction phase. J Const Eng Manage 146(6). https://doi.org/10.1061/(ASCE)CO.1943-7862.0001842
25. Hosier I, Koilraj J, Vaughan A (2016) Effect of aging on the physical, chemical and dielectric properties of dodecylbenzene cable oil. In: IEEE international conference on dielectrics (ICD).https://doi.org/10.1109/ICD.2016.7547740
26. Ituen I, Sohn G (2010) The way forward – advances in maintaining right-of-way of transmission lines. GEOMATICA 64(4):451–462. https://doi.org/10.5623/geomat-2010-0047
27. Kaiser M (2019) Decommissioning forecasting and operating cost estimation, 1st ed. Gulf Professional Publishing. ISBN: 9780128181133
28. Kavishe N, Chileshe N (2020) Factors influencing bid/no bid decision of foreign contractors in Tanzania. In: 36th annual association of researchers in construction management, p 763
29. Kim J-H, Kim J-Y, Cho J-T, Song I-K, Kweon B-M, Chung I-Y, Choi J-H (2014) Comparison between underground cable and overhead line for a low-voltage direct current distribution network serving communication repeater. Energies 7:1656–1672. https://doi.org/10.3390/en7031656
30. Komal B, Janjua U, Anwar F, Madni T, Cheema M, Malik M, Shahid A (2020) The impact of scope creep on project success: an empirical investigation. IEEE Access 8:125755–125775. https://doi.org/10.1109/ACCESS.2020.3007098
31. Košík M, Likavčan L, Bílik J, Martinkovič M (2014) Diesel effect problem solving during injection moulding. Res Papers Faculty Mater Sci Technol Slovak Univ Technol 22(341):97–102. https://doi.org/10.2478/rput-2014-0014
32. Kotb P, Abo Al Anwar D, Baraka E (2020) Major barriers to risk allocation in construction contracts. SSRN Electron J. https://doi.org/10.2139/ssrn.3518988
33. Labyani D (2019) Massive oil leaks from ESB cables in Dublin investigated, 5th June, Irish Times. Available via: https://www.irishtimes.com/news/ireland/irish-news/massive-oil-leaks-from-esb-cables-in-dublin-investigated-1.3915462
34. Langdon WB (1995) Scheduling planned maintenance of the national grid. In: Fogarty TC (eds) Evolutionary computing. AISB EC. Lecture notes in computer science, vol 993. Springer, Berlin, Heidelberg. https://doi.org/10.1007/3-540-60469-3_31
35. Laryea S (2016) Risk apportionment in target cost contracts. Proc Inst Civil Eng- Manage Procurement Law 169(6):248–257. https://doi.org/10.1680/jmapl.15.00046
36. Le Poidevin K, Cummings B, Dutton T (2007) Advances in the decommissioning of end-of-life oil filled cables, CIRED, Paper0357. Available via: https://www.semanticscholar.org/paper/ADVANCES-IN-THE-DECOMMISSIONING-OF-END-OF-LIFE-OIL-Poidevin-Cummings/6358302178563eabf5dbb043b9643155186a3391
37. Mueller C, Keil S, Bauer C (2019) Underground cables vs. overhead lines: Quasi- experimental evidence for the effects on public risk expectations, attitudes, and protest behavior. Energy Policy 125:456–466. https://doi.org/10.1016/j.enpol.2018.10.053
38. Navrud S, Ready RC, Magnussen K, Bergland O (2008) Valuing the social benefits of avoiding landscape degradation from overhead power transmission lines: do underground cables pass the benefit–cost test? Landsc Res 33(3):281–296. https://doi.org/10.1080/01426390802045921
39. Neimanis R, Saha TK, Eriksson R (2000) Determination of moisture content in mass impregnated cable insulation using low frequency dielectric spectroscopy. In: 2000 power engineering society summer meeting (Cat. No.00CH37134), Seattle, WA, 1, pp 463–468. https://doi.org/10.1109/PESS.2000.867630

40. Newman C, Edwards DJ, Martek I, Lai J, Thwala WD (2020) Industry 4.0 deployment in the construction industry: a bibliometric literature review and UK-based case study. Smart Sustain Built Environ. https://doi.org/10.1108/SASBE-02-2020-0016
41. Oberlander GD (2000) Predicting accuracy of early cost estimates based on estimate quality. J Constr Eng Manage 127(3). https://doi.org/10.1061/(ASCE)0733-9364(2001)127:3(173)
42. Oludolapo IO, Edwards DJ, Chileshe N (2020) Estimating on-site emissions during ready mixed concrete (RMC) delivery: a methodology. Case Stud Constr Mater. https://doi.org/10.1016/j.cscm.2020.e00439
43. Orton H (2013) History of underground power cables. IEEE Electr Insul Maga 29(4):52–57. https://doi.org/10.1109/MEI.2013.6545260
44. Purs G, Samal N (2019) Theoretical foundations of estimated construction costs Ways to improve construction efficiency 0(42):.41–54. https://doi.org/10.32347/2707-501x.2019.42.41-54
45. Raugei M, Kamran M, Hutchinson A (2020) A prospective net energy and environmental life-cycle assessment of the UK electricity grid. Energies 13(9):2207. https://doi.org/10.3390/en13092207
46. Reading AG (1973) Overhead lines for the supergrid. Electron Power 19(15):352–356. https://doi.org/10.1049/ep.1973.0438
47. Reily JJ (2005) Underground space use: analysis of the past and lessons for the future. In: Erdem and Solak (eds) Taylor & Francis Group, London. Available via: http://citeseerx.ist.psu.edu/viewdoc/download?doi=10.1.1.475.5024&rep=rep1&type=pdf
48. Safapour E, Kermanshachi S (2019) Identifying manageable scope creep indicators and selecting best practice strategies for construction projects. In: Proceedings, annual conference of the Canadian society for civil engineering. Available via: https://csce.ca/elf/apps/CONFERENCEVIEWER/conferences/2019/pdfs/PaperPDFversion_293_0421025221.pdf
49. Seneviratne K (2016) Use of provisional sums in the UAE construction industry: an empirical study. In: The 5th works construction symposium 2016: greening environment, eco innovations & entrepreneurship, pp 29–31
50. Sepasgozar S, Shi A, Yang L, Shirowzhan S, Shafigh P, Edwards DJ (2020) Additive manufacturing applications for Industry 4.0: A systematic critical review. Buildings 10(12):231. https://doi.org/10.3390/buildings10120231
51. Smith J, Edwards DJ, Martek I, Chileshe N, Roberts C, Hayhow S (2021) The antecedents of construction project change: an analysis of design and build procurement application. J Eng Des Technol. https://doi.org/10.1108/JEDT-12-2020-0507
52. Tummala V, Burchett J (1999) Applying a risk management process (RMP) to manage cost risk for an EHV transmission line project. Int J Project Manage 17(4):223–235. https://doi.org/10.1016/S0263-7863(98)00038-6
53. Wei L, Du H, Mahesar Q, Al Ammari K, Magee D, Clarke B, Dimitrova V, Gunn D, Entwisle D, Reeves H, Cohn A (2020) A decision support system for urban infrastructure inter-asset management employing domain ontologies and qualitative uncertainty-based reasoning. Expert Syst Appl 158. https://doi.org/10.1016/j.eswa.2020.113461
54. Xiong B, Xia B (2014) Examining the impacts of early cost drivers on contingencies. In: Construction research Congress 2014: construction in a global network, pp 1518–1527. https://doi.org/10.1061/9780784413517.0155
55. Zhang F, Dong Z, Luo F, Ranzi G, Xu Y (2019) Resilient energy management for residential communities under grid outages. In: 9th international conference on power and energy systems (ICPES). https://doi.org/10.1109/ICPES47639.2019.9105455
56. Zhou C, Yi H, Dong X (2017) Review of recent research towards power cable life cycle management, HighVoltage 2(3):179–187. https://doi.org/10.1049/hve.2017.0037

Overcoming the Challenges Candidate Quantity Surveyors Face on the Route to Professional Registration

Nishani Harinarain, Dawood Joosab, and Hamza Vawda

Abstract Candidate quantity surveyors face many challenges on their road to professional registration. A candidate quantity surveyor is one who has completed a tertiary qualification in a quantity surveying programme from an accredited institution but does not yet have the required work experience to work on their own. They have to work under the supervision of a Professional Quantity Surveyor until they have acquired enough experience of between 3 and 5 years to register as a professional quantity surveyor. The aim of this paper was to identify the challenges the candidates in South Africa face during these 3–5 years and possible solutions to these challenges to help candidates be better prepared in the future. This was a quantitative study in which 52 random participants were emailed a link to the Google form questionnaire. Forty valid responses were received. The findings of the questionnaires revealed that the main challenges candidates faced were the transitioning from university into the world of work, filling out the logbooks, lack of mentorship and inadequate technological and software skills. The solutions provided to overcome these challenges were to increase mentorship to guide the candidates, encourage candidates to learn new technologies and software and that they should be willing to adapt to changes in the working environment. It was also found that involving the candidate from inception to close out gave them a better understanding of their job and the industry.

Keywords Candidate quantity surveyor · Professional registration · Routes to registration

1 Introduction

The construction industry is an industry that envelopes different professionals which includes but not limited to, architects, contractors, quantity surveyors, property

N. Harinarain (✉) · D. Joosab · H. Vawda
The School of Engineering, Construction Studies Discipline, The University of KwaZulu-Natal, Howard College, Durban, South Africa
e-mail: harinarain@ukzn.ac.za

T. C. Haupt et al. (eds.), *Construction in 5D: Deconstruction, Digitalization, Disruption, Disaster, Development*, Lecture Notes in Civil Engineering 245,
https://doi.org/10.1007/978-3-030-97748-1_37

surveyors, land surveyors, town planners, project managers, civil engineers, mechanical engineers, structural engineers and electrical engineers [1]. Their professional knowledge and experience are required to complete projects successfully and they have important responsibilities and duties in their respective roles [2].

Quantity surveyors offer a definite service in the built environment [3] by providing financial probity in the conception, planning and execution of development projects. They are the construction industry's financial consultants. They prepare contract documents and are able to advise on cost and contractual provisions through their training and experience [4]. Quantity surveyors are 'experts' in forecasting project costs and advise clients on financial implications of alternative courses of action in the planning and execution of the construction project [5].

The quantity surveying profession is experiencing many challenges such as advances in technology and client demands and is expected to adapt to these changes. This paper looks at the challenges faced by candidate quantity surveyors towards professional registration. It must be acknowledged that while challenges experienced by candidates are not confined to quantity surveyors, they are experienced in other professional qualifications such as architecture, engineering and project management, this research is limited to candidate quantity surveyor's in South Africa.

It is important that candidate QS registration eventually converts into professional registration because professional registration ensures that a certain professional standard is maintained. Professional registration is also important because it ensures that the work is of a higher and more consistent quality [6] and demonstrates one's knowledge, understanding and competence [7] thereby providing the clients with greater confidence in their ability.

Registration also shows the candidate's commitment to professional standards and their commitment to enhancing their competence while boosting one's professional credibility and prestige with current clients, and when pursuing new business opportunities or bidding on projects.

According to the South African Council for the Quantity Surveying Profession (SACQSP) there are approximately 2151 candidate quantity surveyors (QS), with around 600 applications submitted every year. Less than half of these 600 new applicants will convert into candidate QS registration [8]. One of the possible reasons for this could be the challenges that candidates experience, therefore, this study is important as it presents candidates with an idea of the challenges they may face and the possible solutions to overcome these problems.

2 The Quantity Surveyor

The quantity surveyor (QS) is one of the construction industry's team of professional advisors. Quantity surveyors work closely with other project team members including architects, contractors, engineers, suppliers, project owners, etc. The quantity surveyors' key roles are to carry out financial management, cost and contract administration of a project at all stages, from inception to completion [9]. A quantity surveyor is defined as, "*individuals who form part of the professional team who are mostly involved with the valuation and measuring of construction work being performed under a traditional construction contract where architect drawings are handed over to them to advise on possible costs, itemise the work to be carried out, prepare tender documents, negotiate construction contracts, value work as it proceeds and to prepare final accounts*" [10]. All these services are aimed at completing projects at the estimated time within its budget in order to avoid project termination and to increase the industry's contribution to the gross national product [11].

The quantity surveying profession has experienced a variety of changes over the past few decades. Quantity surveying practices around the world have undergone a gradual but steady and successful development from a simple measurer and valuer of construction works to more specialised and detailed services [12]. The quantity surveying profession has become a dynamic multidisciplinary sector with enhanced and improved skills. Quantity surveyors have gone beyond the construction industry's boundaries to pursue new opportunities [13] and have therefore opened up new career possibilities. Therefore, quantity surveyors should be prepared to address both new and current challenges.

2.1 The QS Profession in South Africa

In terms of the Quantity Surveying Professions Act 2000 (Act No. 49 of 2000) if a person performs the functions specified in identifying work for a quantity surveyor, either in training or in practice, they are required to be registered with the South African Council for the Quantity Surveying Profession (SACQSP). Persons are only permitted to undertake work according to their category of registration. This means that there are two types of quantity surveyors (QS): candidate quantity surveyor and professional quantity surveyor [8].

2.2 Professional Quantity Surveyor (PrQS)

Any individual who is registered as such in terms of the Quantity Surveying Profession Act, 2000, is regarded as a professional quantity surveyor. The PrQS can choose to work independently and provide professional services, expertise and provide consultation on construction procurement, costs and contracting [8]. A PrQS may operate in a partnership or as part of a large company, as a private sector consultant for a developer or a big organisation (such as a retailer or manufacturer) [5]. On the contracting side a PrQS could work for a local or regional general contractor, major national or international contractor and for a specialist or sub-contractor. There are a total of 2271 registered PrQS's in South Africa [8].

Registered quantity surveyors enjoy due recognition and financial reward in terms of their specialised skills and technical competence in the provision and application of primary or fundamental quantity surveying expertise. Their jobs as PrQS's will include:

- Planning for construction projects: feasibility studies, life cycle cost analyses;
- Project cost management, procurement and administration: cost management, procurement, contract administration, resource analysis;
- Specialised management services; and
- Support competencies.

2.3 Candidate Quantity Surveyor

A candidate quantity surveyor is one who has either completed a Bachelor of Science or achieved a Diploma and a Bachelor of Technology in a quantity surveying programme that has been accredited by the SACQSP. The candidate quantity surveyor does not have the required work experience to work on their own and therefore has to work under the supervision of a professional quantity surveyor until the candidate has enough experience to register as a professional quantity surveyor [14]. This will have to be done by the candidate for a period of 3–5 years, depending on their level of education. The candidate will also need to complete their assessment of professional competence (APC). Candidate QS's are constrained in terms of the Code of Professional Conduct on what they may do and their rights to practice independently are limited. Therefore, it is critical that candidates complete their professional registration as soon as possible [8].

In 2012, legislation was passed where the time limit for being a candidate QS was set at 10-years [8] which means that the candidate needs to complete their professional registration within this time period, taking into consideration of all the requirements. If the candidate fails to complete their professional registration within the 10 years, their candidacy will fall away and they will be unable to become a PrQS. It is therefore important that candidates complete their formal professional registration as soon as possible.

2.3.1 Routes to Registration

The route to professional registration for a candidate quantity surveyor is dependent on their level of tertiary education. If the candidate obtained an honours degree in quantity surveying, they will only be required to have 3–4 years of working experience before doing the assessment of professional competence (APC) interview. If the candidate only obtained an advanced diploma in property development or construction studies, they will be required to complete 18 Profession Skills Modules (PSM), together with having 4–5 years of working experience before doing the APC interview. For both types of candidates, they need to complete their daily diary and log book for the required number of years before doing their APC interview [8].

APC is the assessment of practical training and experience needed to become a PrQS. Both practical training and experience, together with academic qualifications, leads to PrQS registration [8]. The main purpose of the APC is to ensure the registered person is competent enough to practice as a PrQS.

The duration of the work-place training will be for a minimum of 3-years up to a maximum of 5-years. However, this make take longer depending on the workplace exposure that the candidate may be exposed to. The period is calculated using an 8-h day for a 240 day year (1920 h). The work-place training is competency based. The training will require the candidate to demonstrate the necessary skills and abilities needed to perform specific tasks or functions and is based on attitudes and behaviours as well as skills and knowledge [8].

Graduates are equipped to apply primary quantity surveying skills as employees, under the supervision of a registered professional quantity surveyor and are eligible for employment in numerous spheres of the economy, which include private sector consultancies and professional firms and the public sector employment. They are empowered to build their personal career paths through undertaking further learning programmes with a view to articulation towards obtaining additional qualifications associated with the built environment [15].

The amount of activities and responsibilities the candidate undertakes gradually increases as they gain more experience. The candidate would first assist with work, doing specific tasks under the supervision of a PrQS. The candidate must then be able to make contributions both individually and as part of a team. By the end of the training period, the candidate is required to perform at the level of competencies required for registration.

The candidate and supervisor must assess the level of competencies displayed by the candidate at regular intervals in order for the candidate to progress to the next level of training. These assessments may influence planning for subsequent tasks or phases. All assessments must be recorded in the QS Diary. This process will continue until the candidate is working at the level required for registration. As the candidates training moves along, the candidate's competency must develop towards that required for registration [8].

2.3.2 Challenges Faced by the Candidate QS

As the profession keeps growing, it calls for the development knowledge and skills of candidates so that they can keep up with the challenges faced by them and the construction industry. Candidates lack knowledge on various construction technologies and thereby having no understanding and experience of on-site conditions. In addition, less precedence is given to management-oriented competencies of the candidate.

The following are some of the challenges that a candidate QS faces:

- Lack of core skills and knowledge in construction;
- Inability to perform basic quantity surveyor functions such as estimating, evaluation of variations and finalising of accounts [16];
- Cost factors such as fees charges for taking the APC is high [17];
- Unable to interpret construction documents [18];
- The incapability to re-engineer and transform processes to take full benefit of technological advances [16];
- Not equipped or able to raising their services to a more proficient and higher value- adding level [19];
- Lack of mentorship and assistance [17];
- Absence of technology knowledge and inability to change and familiarise themselves with various software and applications such as WinQS, DimX and CostX [20];
- Time challenges such as lack off time to fill in the logbook [17];
- Lack off work experience in the industry [18];
- Completing skill modules while working [8];
- Qualification factor such as does not have a diploma or degree which is recognised and accredited by the board [17].

A competent quantity surveyor must be able to apply a variety of skills, knowledge and understanding in a variety of contexts and organisations. However, pressing challenges confronting today's quantity surveying profession is to enhance the relevance and level of awareness of services provided by the profession in the built environment and to expand business opportunities for continuous growth [21].

2.3.3 Possible Solutions to the Challenges Faced by Candidate Quantity Surveyors

It is vital to improve the attributes off graduate quantity surveyors and to better prepare them for the workplace. Candidates need to continuously evolve in order to minimise the effects of the challenges they may face and some of the possible solution are: utilising mentoring programmes will improve their productivity [22]; investment in information technology (IT) and infrastructure and infrastructure; develop core competencies and undertake continuous professional development (CPD) activities, research and training [8].

3 Research Method

This was a quantitative study in which 52 participants in Durban were emailed a link to the Google form questionnaire. The participants were limited to only candidate QS's and PrQS's who completed their APC in the last 5 years to ensure that the researchers obtained relevant information. The questions focused on the challenge's candidates

Table 1 Cronbach's Alpha scores

Section	Number of Items	Cronbach's Alpha
Professional competencies	16	0.961
Challenges	23	0.882
Solutions	15	0.855

faced and possible solutions of how they can be overcome. Forty valid responses were received. The data collected was analysed using SPSS v27 and Microsoft Excel. The Relative Agreement Index (RAI) was used to rank the items in each section. The formula for this tool is as follows:

- RAI = ΣW / AN
- W = the weight of each response from each participant i.e. 1 is 'Strongly Disagree' and 5 is 'Strongly Agree'
- A = highest weight, with 'Strongly Agree' being 5
- N = Number of the sample.

The final statistic included was the rank of the questions obtained through the calculation of the RAI.

The Cronbach Alpha test can be used on items in each section to check the reliability of the research. A Cronbach Alpha score of 0.6 and above will result in an acceptable reliability test [23].

With all the scores exceeding a score of 0.7 [23], the data can be considered to be reliable, acceptable and consistent for the purpose of this study.

4 Discussion of Findings

4.1 Demographic Data Analysis

From the 40 participants that answered the survey, 35% were between the ages 26–28, with 30% being greater than 31 years old. Eighty eight percent of the participants were male. Forty three percent of the participants were registered candidate quantity surveyors, 40% were registered professional quantity surveyors and 17% were not registered. Thirty three percent of the participants were registered as candidates for 5 or more years, 21% for 4 and 3 years respectively and 24% for 2 years. An Honours degree in quantity surveying was the most common tertiary qualification with 40% of the participants having an honours qualification, 35% a B. Tech, 17% a B.Sc. and 8% and M.Sc. degree.

4.2 Professional Competencies

Prior to understanding the challenges that candidate quantity surveyors face, it was important to determine the level of knowledge of candidates, relating to competencies required for the APC interview. Professional quantity surveyors will enjoy due recognition and financial reward for providing their expertise for tasks. It is important that candidates are strong in these tasks as it will better prepare them for the future and make them better professional quantity surveyors [15].

Fifty five percent of the candidate quantity surveyors had good knowledge when it came to preparing detailed estimates of construction costs as indicated in Table 2. This was ranked as the most important competency for the APC interview. The second important competency was measuring and recording site information for final account purposes. Sixty percent of the candidates had good knowledge of preparing preliminary estimates of construction costs (ranked third).

Only 27% of the candidates had adequate knowledge when it came to advising on the different forms of construction contracts available for projects (JBCC, FIDIC,

Table 2 Professional competencies

	Mean	Std. Dev	RAI	Rank
Preparing detailed estimates of construction cost	4.000	1.208	0.633	1
Measuring and recording site information for final account purposes	3.868	1.070	0.613	2
Preparing preliminary estimates of construction cost	3.842	1.128	0.608	3
Preparing valuation for final payment certificates and reconciliation	3.789	1.119	0.600	4
Advising on procurement policy for the project	3.579	1.998	0.567	5
Advising on escalation formulae and their implementation	3.579	1.368	0.567	6
Advising on selection of tenderer's	3.553	1.108	0.563	7
Preparing contract documents	3.500	1.371	0.554	8
Feasibility studies	3.421	1.154	0.542	9
Value engineering	3.395	1.079	0.538	10
Reviewing working drawings for compliance with the approved budget	3.711	1.160	0.538	11
Preparing fee accounts based on appropriate fee scale and conditions of appointment	3.541	1.346	0.538	12
Adjudication and resolving financial claims by the contractor, subcontractor of supplier	3.342	1.097	0.538	13
Advising on the different forms of construction contracts available for projects (JBCC, FIDIC, NEC, GCC)	3.053	1.374	0.538	14
Preparing schedules of predicted cash flow	3.378	1.320	0.521	15
Preparing of health and safety requirements for building projects	3.079	1.343	0.488	16

NEC, GCC). Forty two percent of the candidates felt that they were competent in preparing schedules of predicted cash flow. Only 23% of the candidates had good knowledge in preparing of health and safety requirements for building projects and this was ranked last on the list of competencies. Candidates will require more training, practice and teaching when it comes to the three tasks above. This should be done by the quantity surveying programmes in order to better prepare the candidates for the future.

4.3 *Challenges Encountered by Candidates*

There are a variety of challenges faced by candidates. The main challenge (ranked 1st) was the transition from university into the work environment. Candidates also lack technological knowledge and the inability to adjust and become familiar with different software and applications such as WinQS, DimX and CostX. Candidates find adapting to new working environments and to self-adjust into the industry a big challenge [18].

Forty two percent of the participants stated that they were not introduced to the working environment, type of work they had to do and amount of workload they had while working during their candidacy. Around 58% of the participants agreed that registration is easier for those quantity surveyors who work in a consultant firm than those who work contracting firms.

The other challenges encountered by the participants towards professional registration were insufficient time to fill in the log books and the work diary every day, completing PSM modules while working, the cost of gaining CPD points, lack of work experience and the time it takes to acquire professional registration. Participants mentioned the registration process is too stringent, long and complicated. Participants with no mentors ($n = 13$) experienced more challenges than those with mentors.

It was found that the candidates experienced challenges at each construction phase. The participants indicated they find it challenging to prepare schedules of predicted cash flows, advise on different forms of contracts, adjudicate and resolve financial claims from the contractor, dealing with incompetent contractors, variation orders, having to complete non-measured work, final account preparation and agreement (Table 3).

4.4 *Solutions to the Challenges*

Possible solutions to the challenges that could assist the candidates on their road to professional registration are indicated in Table 4.

One of the main solutions found was mentorship. It was shown that participants with mentors experienced less challenges and had more support to overcome

Table 3 Challenges encountered by candidates

	Mean	Std. Dev	RAI	Rank
The transition from university into the work environment	4.235	0.923	0.600	1
Lack technological knowledge	3.789	1.094	0.600	2
Did not receive introduction to the work environment, type of work and workload when you started working	3.711	1.160	0.588	3
Registration is easier for those quantity surveyors who work in a consultant firm than those who work contracting firms	4.118	1.122	0.583	4
You were not comfortable to ask questions in the event of any uncertainty	4.344	0.937	0.579	5
You and your mentor did not have a strong professional relationship	4.344	1.004	0.579	6
Your mentor provided support, feedback and guidance during employment	4.250	0.950	0.567	7
Your mentor did not provide clear instructions on what is expected from you	3.676	1.203	0.567	8
I have to be registered as a candidate before being registered as a PrQS	3.579	1.287	0.567	9
Admin costs and membership fees charged by the Council is high	3.556	1.027	0.533	10
You could not ask your mentor to be part of different activities	3.969	1.092	0.529	11
Renewal fees for candidate registration and professional registration is costly	3.514	1.147	0.513	12
Your mentor did not provide you with adequate information on the APC process	3.813	1.061	0.508	13
Did not have time to fill in the logbook and work diary	3.412	1.500	0.483	14
Your mentor did not assist you with the preparations of the reports you needed for the APC process	3.594	1.266	0.479	15
The fees charged for taking the APC test is too much	3.382	1.256	0.479	16
I do not have interest in the career of quantity surveyors	1.486	0.837	0.383	17
Did not have time to take the APC test	2.727	1.153	0.375	18
Did not obtain the minimum number of years of work experience required	2.559	1.637	0.363	19
I lack of the information about the registration as a professional	2.216	1.134	0.342	20
I lack of awareness about the importance of improving and furthering my qualification	2.189	1.151	0.338	21
I lack encouragement from mentors, employers, colleagues and other working professionals	1.919	0.894	0.296	22
I do not have a diploma or degree which is recognised and accredited by the professional bodies	1.829	1.403	0.267	23

Table 4 Possible solutions

	Mean	Std. Dev	RII	Rank
Having a mentor helped me during the candidate phase	4.050	0.749	0.675	1
I have tried to maintain and develop the core competencies needed	4.025	1.165	0.671	2
I have tried to learn and utilise software such as Win QS, CostX and Dimx	3.975	1.074	0.663	3
I have improved my skills and gained more experience with IT	4.361	0.798	0.654	4
I interacted with my mentor and co-workers on a personal level	3.897	0.598	0.633	5
I was willing to adapt to changes in work patterns	3.750	0.899	0.625	6
I was flexible and willing to alter my standard work practices	3.600	1.172	0.600	7
Practical and real-life activities while studying increased my skills and therefore I required less training while working	3.658	1.122	0.579	8
I have enhanced my professionalism and skills in order to maintain a competitive advantage	3.450	1.280	0.575	9
The firm I work at integrated new IT software's and provided training on how to use them	3.513	1.048	0.571	10
I have done my CPD sincerely in order to further my skills and knowledge and not to just pass it	3.385	1.067	0.550	11
I have diversified my skills and specialised in other areas	3.026	0.986	0.492	12
I encouraged the multi-procurement method to allow my clients the easier management of time and cost	2.850	1.312	0.475	13
I tried to learn how to use CAD systems	2.897	1.314	0.471	14
I have done some sort of work shadowing, problem-based learning, simulated exercises or practical work experience	2.579	1.348	0.408	15

these challenges than those who did not have mentors. Firms utilising mentoring programmes will improve their productivity because mentoring can help develop the future leaders of the construction industry [22]. Mentoring can help develop a candidate to be better at the job, help overcome challenges, achieving career satisfaction as well as being fundamental in developing a skilled workforce in a firm. The mentor validates the candidate's skills, knowledge and contributions to the firm.

Future candidates, while studying, should enter and gain experience in the work field to overcome the challenges faced when transitioning from university to the work environment. Most (75%) of the participants indicated that they have tried to maintain and develop the core competencies needed by quantity surveyors. This is important for any profession as one cannot practice without maintaining the core competencies required. Firms need to make sure that their quantity surveyors have sufficient knowledge in the core competencies and expertise of the profession and need to continue to develop these skills [19].

Approximately three quarter of the participants learned how to utilise software programs such as WinQS, CostX and DimX. It is important to learn and utilise the new technologies available as technology is the future and makes the quantity surveyors job easier and enhances the professionality of the company as a whole and enables a QS to offer better services to clients. Information technology improved the skills of 78% of the participants while also giving them more experience.

According to the findings the involvement of candidates from inception to close out gave them a better understanding of their job and the industry. Participants also tried to maintain and develop the core competencies required by quantity surveyors as well as being flexible and willing to adapt to changes in their work patterns. Firms need to make sure that their quantity surveyors have sufficient knowledge in the core competencies [19].

Only 40% of the participants tried to learn CAD systems. Construction professionals need to develop and expand their skills in CAD in order to survive in the future [19]. It was also observed that only 30% of the participants did any work shadowing, problem-based learning, simulated exercises or practical work experience to gain further competencies. Work shadowing and practical experience is important as it is the best way to gain valuable skills, such as discipline, perseverance, teamwork, leadership skills, communication skills and problem solving.

5 Conclusions

There was a wide range of skills and knowledge that the candidates possessed such as competency in preparing detailed estimates of construction cost, measuring and recording site information for final account purposes, preparing preliminary estimates and preparing valuations for final payment certificates and reconciliation. However, candidates still experience challenges on the route to professional registration. The transition from university into the work environment, the lack of technological knowledge and unfamiliarity of specialised software are just some of the challenges experienced by the candidates. Mentoring, maintaining and developing professional expertise in core competencies, learn and utilise CAD systems, diversification of skills and services, CPD programmes and developing graduate's attributes are just some of the possible solutions to overcome these challenges.

The quantity surveying profession in South Africa is experiencing change with strong emphasis on improving education, training and research. Therefore, this study is important as it gives present and future candidates an idea of the challenges they may face and the possible solutions to overcome these problems. Candidate quantity surveyors should not only be limited to conventional cost control boundaries, they should push into new environments creating new opportunities and develop new expertise.

References

1. Owolabi OSB, Olatunji AS (2014) The roles of construction professionals in the Nigeria's construction industry. IOSR J Hum Soc Sci (IOSR - JHSS) 19(11):05–10
2. Chidiebere EE, Abraham AI, Ramat S (2017) Appraisal of the perception of quantity surveying profession by non-allied construction professionals in Nigeria. Int J Adv Eng Manage Sci 3(2)
3. Ashworth A, Hogg K, Higgs C (2013). Willis's practice and procedure for the quantity surveyor, 13th ed. Wiley
4. Patten B (2003) Professional negligence in construction. Routledge
5. Towey D (2017) Construction quantity surveying: a practical guide for the contractor's QS. Wiley
6. Half R (2014). 4 reasons a professional certification benefits the entire company. Available online at: https://www.roberthalf.com/blog/management-tips/4-reasons-a-professional-certification-benefits-the-entire-company
7. University of Edinburgh (2021) Professional registration and accreditations. Available online at: https://www.ed.ac.uk/technicians/professional-registration
8. South African Council for the Quantity Surveying Profession (SACQSP) (2020) Available online at: https://www.sacqsp.org.za/. 15 Feb 2021
9. Seah E (2009) Sustainable construction and the impact on the Quantity surveyor. In 13th Pacific Association of Quantity Surveyors Congress, pp 36–48
10. Olatunde J (2006) New opportunities for quantity surveyors in Nigeria business environment. In: 21st century quantity surveying, agenda for the future, NIQS Biennial. 1–3 November
11. Udo MA, Abiola AH (2015) An assessment of the role of quantity surveying profession in the development of Nigeria. Knowl Rev 33(1):1–6
12. Mac-Barango DO (2017) The strategic imperatives for sustainable development of the 21st century quantity surveyors. World J Innov Modern Technol 1:54–67
13. Zainudeen N (2006) Career paths in quantity surveying. In: Rameezdeen R, Senavirathne I (eds) Customizing the quantity surveyor to face challenge in year 2020, pp 43–58
14. Association of South African Quantity surveyors (ASAQS) (2020) What is a QS? Available online at: https://www.asaqs.co.za/page/what_is_QS. 15 Feb 2020
15. South African Qualification Authority (SAQA) (2007) Government Gazette
16. Chong BL (2014) The services required by the Malaysian construction industry from quantity surveyors and their implications to quantity surveying graduates (Doctoral dissertation, UTAR)
17. Ching T (2016) The factors that causes delay in quantity surveyor professional registration. Universiti Teknologi Malaysia
18. Aliu J, Aigbavboa C (2016) Equipping 21st century construction graduates: a review of key skills in fostering infrastructural development
19. Smith P (2004) Trends in the Australian quantity surveying profession: 1995–2003. Int Roundup 19(1):1–14
20. Shen Q, Li H, Shen L, Derek D, Jacky C (2003) Benchmarking the use of information technology by the QS profession. Benchmarking 10(6):581–596
21. Oke AE, Ogunsemi DR (2012) Training of Nigerian quantity surveyors for value management practice. Int J Constr Project Manage 4(2)
22. Nkomo M, Thwala W (2013). Assessment of mentoring of construction workers in the South African construction industry: a literature review. In: International conference on education, law and humanities
23. Tavakol M, Dennick R (2011) Making sense of Cronbach's alpha. Int J Med Educ 2(1):53–55